# Metal and Metal Oxide Nanoparticles

This book presents synthesis, characterization, as well as biosensing and optoelectronic applications of metal and metal oxide nanoparticles. It examines surface functionalization and modification techniques for enhancing the performance and functionality of metals and metal oxides. The book also provides an overview of the biocompatibility of surface-functionalized metal and metal oxide nanomaterials and explores the factors influencing their biocompatibility. The initial chapters introduce various synthesis methods used for the production of metals and metal oxide nanoparticles, as well as the techniques used for their characterization. The subsequent chapters explore recent advances in metals and metal oxide-based biosensors, their biomedical applications, the detection of gases, and their role in optoelectronic devices. Last, the chapters cover the basic mechanisms, types of biosensors and their applications, advancements, and advantages and disadvantages. This book is intended for researchers, scientists, academicians, students, and professionals working in the fields of nanotechnology, biosensors, materials science, biomedical engineering, and optoelectronics.

**Key Features**

- Explores synthesis, characterization, biosensing, and optoelectronic applications of metal and metal oxide nanoparticles
- Presents various techniques used for the synthesis and characterization of metals and metal oxide nanoparticles
- Discusses surface modification techniques for enhancing performance of metals and metal oxide nanoparticles
- Offers insights into the biocompatibility of surface-functionalized nanoparticles and examines the factors influencing their compatibility with biological systems
- Highlights the latest advancements in metals and metal oxide-based biosensors and their applications in biomedicine and gas detection
- Examines the role of metal and metal oxide nanoparticles in optoelectronic devices

# Metal and Metal Oxide Nanoparticles

## Biological Sensing and Optoelectronic Applications

Edited by
Radhakrishnan EK, Riju K Thomas, and Divya NG

CRC Press
Taylor & Francis Group
Boca Raton London New York

CRC Press is an imprint of the
Taylor & Francis Group, an **informa** business

Designed cover image: Shutterstock

First edition published 2026
by CRC Press
2385 NW Executive Center Drive, Suite 320, Boca Raton FL 33431

and by CRC Press
4 Park Square, Milton Park, Abingdon, Oxon, OX14 4RN

*CRC Press is an imprint of Taylor & Francis Group, LLC*



ISBN: 978-1-032-64329-8 (hbk)
ISBN: 978-1-032-66062-2 (pbk)
ISBN: 978-1-032-66063-9 (ebk)

DOI: 10.1201/9781032660639

Typeset in Times
by Apex CoVantage, LLC

# Contents

# Preface

Nanotechnology has already had a revolutionary impact in diverse sectors due to the innovative research ranging from its synthesis to application. Among the various nanomaterials used for diverse applications, metal and metal oxide nanoparticles are of special interest due to their remarkable properties. The availability of a range of methods for synthesis, including green synthesis, and the remarkable material properties of metal nanoparticles have already made them highly sought-after nanomaterials. *Metal and Metal Oxide Nanomaterials: Biological Sensing and Optoelectronic Applications* has been designed to provide detailed insight into the applications of metal and metal oxide nanoparticles in diverse sectors so that it will be a primary reference for those who are working on metal nanoparticles. With the recent advances in the knowledge economy, there is always a challenge to connect the applications of advanced scientific research to real-world applications. This book forms an ideal model in which advanced research has been effectively correlated with applications that have commercial potential. The book presents the wider scope of metal nanotechnology, as its core theme is cost-effective and rapid sensors and diagnostic method development. The chapter contributors significantly communicate these objectives through the impressive presentation of technological advances and their applications in diverse areas ranging from sensor development to cancer diagnosis and optoelectronic application to bioimaging. The use of metal nanoparticles for medical diagnosis has already formed the basis of rapid diagnosis of various clinical conditions. At the same time, the antimicrobial nature of metal nanoparticles has been effective even against antibiotic-resistant pathogens. This book presents both the use of microorganisms for the fabrication of nanoparticles and their antimicrobial nature and the applications of the formed nanoparticles. The chapters of the book have been well organized, starting from synthesis to applications, so that continuity from the basics to advanced applications is ensured for readers.

# About the Editors

**Dr. Radhakrishnan EK** is currently working as Associate Professor at School of Biosciences of Mahatma Gandhi University, Kottayam, as Hon. Director of Business Innovation and Incubation Centre (BIIC), and as Joint Director of the Inter-University Centre for Organic Farming and Sustainable Agriculture (IUCOFSA). In addition, he is also the Hon. Director of Dr. N. Radhakrishnan Foundation for Medical Innovation (NRICMI) and Hon. Director of Industry Academia Interface Centre (IAIC). He completed his doctoral degree at the Rajiv Gandhi Centre for Biotechnology, Thiruvananthapuram, Kerala, India, in 2008 and post-doctoral studies at the University of Tokyo, Japan, in 2010. During his 13 years of research, he has published more than 150 journal articles, 55 book chapters, and 3 review papers and edited 10 books. Recently, he has been listed in the top 2% of scientists according to the Stanford University, USA, and Elsevier, the Netherlands database published in September 2022. He has also been selected for Kairali Gaveshana Puraskaram by the Kerala Higher Education Council, Govt. of Kerala, in 2022. So far, 14 PhDs have been completed under his guidance. He has credit for four filed patents and two technology transfera. His team, in collaboration with ABTEC, Pvt, Ltd, Kottayam, successfully launched an agriproduct in the market under the brand name Nanopower in 2022. His research areas include plant–microbe interactions, microbial natural products, microbial synthesis of metal nanoparticles, antimicrobial resistance gene prediction, and development of polymer-based nanocomposites with antimicrobial effects for food packaging and medical applications.

**Dr. Riju K Thomas** is an assistant professor at Bharata Mata College, Thrikakkara, Kerala, India. He received his M.Phil. (2013) from Kerala University from the Department of Optoelectronics and Ph.D. (2019) in Physics from Mahatma Gandhi University, Kerala, India. His research interests are in the interaction of drug-functionalized nanoparticles with biomolecules and their therapeutic applications. He also has expertise in biomolecular sensing applications based on functionalized metal nanoparticles.

**Dr. Divya NG** is an assistant professor, Bharata Mata College, Thrikakkara, Kerala, India. She received her M.Phil. (2015) and Ph.D. (2020) in Physics from Cochin University of Science and Technology (CUSAT), Kerala, India. Her research expertise and interests include nanocomposite devices, including sensors/detectors, especially electrochemical sensors and photovoltaic cells based on metal oxide semiconductor nanostructures.

# 1 Metal Nanoparticles
## *Synthesis, Characteristics, and Applications*

*Shibil S. and Aparna C. M.*
School of Pure and Applied Physics, Mahatma Gandhi University, Kottayam, Kerala

### 1.1 INTRODUCTION

In the extensive realm of scientific exploration, the emergence of nanoscience and technology has ignited a revolution that surpasses the boundaries of conventional scientific disciplines. At the core of this revolution lies the investigation of the nanoscale, a domain in which the properties of matter assume distinct and often unexpected attributes. Nanoscience, the study of phenomena at the scale of a billionth of a meter, delves into the fundamental principles that govern the behaviour of materials and systems. At this scale, quantum effects become pronounced, and the properties of matter demonstrate a remarkable deviation from classical physics. The manipulation and regulation of individual atoms and molecules empower scientists to fabricate materials with customized properties, paving the way for unparalleled advancements in various domains. Nanotechnology has emerged as a field of great transformative potential, enabling the manipulation and engineering of materials at the nanoscale. This offers unparalleled opportunities for innovation across various scientific disciplines. Within this vast array of applications, particular attention has been given to metal nanoparticles due to their unique properties. These nanoparticles serve as fundamental building blocks in the realms of nanoscience and nanotechnology. At the nanoscale, materials exhibit distinct physicochemical characteristics that differ from their bulk counterparts. This discrepancy arises from the effects of quantum size and the increased surface area-to-volume ratios. Metal nanoparticles showcase exceptional electrical, optical, and catalytic properties that significantly diverge from their macroscale equivalents. These distinctive properties originate from the confinement of electrons within the nanoparticle, resulting in heightened reactivity and novel functionalities. The precise control of synthesis and assembly techniques of metal nanoparticles opens avenues for tailoring their properties to meet specific application requirements in electronics, medicine, catalysis, and environmental remediation.

### 1.2 HISTORY OF NANOPARTICLES

Nanoparticles, frequently lauded as a significant achievement of contemporary scientific inquiry, possess historical origins that reach far back into antiquity. For thousands of years, our ancestors have been harnessing the power of nanotechnology. Nevertheless, the precise moment when they initially seized the benefits of nanoparticles across various domains remains uncertain.[1] Nanoparticles are widely regarded as the fundamental units of nanotechnology and are defined as particles with at least one dimension. Particles within these size ranges have been utilized by various industries and humankind for an indefinite period. However, there has been a recent resurgence in their usage due to the ability to manufacture and control such materials. Nanoscale materials are employed in a diverse range of fields, including electronics, magnetism and optoelectronics, biomedicine, pharmaceuticals, cosmetics, energy, environment, catalysis, and materials. As a result of the tremendous potential offered by this technology, there has been a global increase in investment in research and development in the domain of nanotechnology.[2]

Even before the fourth century AD, the utilization of nanoparticles in the realm of painting was evident, considering their optical attributes. A prominent instance that is widely recognized is the Lycurgus Cup (Figure 1.1).

This cup is currently positioned within the esteemed British Museum, which is situated in the city of London. This remarkable cup serves as a prime example from a historical standpoint. It is fashioned out of an extraordinary variant of glass, commonly referred to as dichroic glass, which possesses the unique ability to undergo a chromatic transformation when subjected to the influence of light. Specifically, when light is incident upon the cup at an angle of 90°, the previously opaque green hue of the cup transmutes into resplendent translucence. This cup is composed of a minuscule amount of gold and silver. A prominent manifestation of early nanotechnology was the achievement of a deep crimson tint. Throughout the medieval period, it found usage in the creation of stained-glass windows. However, the skilled craftsmen of that era were unaware of the fact that they were employing nanotechnology. Jin and his colleagues identified a specific procedure for generating an exquisite chromatic effect. They explained the correlation between particles and the respective colours that were linked to specific particles.[4] They succinctly summarized the interrelationship between the colour of the stained glass and the shape and dimensions of the nanoparticles. Faraday was the first to synthesize gold nanoparticles. These gold particles have been preserved at the Royal Institution, London. In the Middle Ages and earlier, the soluble gold

DOI: 10.1201/9781032660639-1

**FIGURE 1.1** Schematic representation of the Lycurgus Cup.

**Source: Reprinted from Ref.[3] Copyright © 2015 Optical Society of America.**

was employed due to its remarkable curative properties in the treatment of various ailments, including dysentery, epilepsy, and heart diseases. In addition, it proved advantageous in the detection of diseases such as syphilis. Daniel and Astruc Ages have provided a comprehensive overview of the historical progression of nanoparticle implementation from ancient civilizations through the Middle Ages.[5] Today, the production of nanomaterials can greatly enhance the properties of large-scale materials, particularly in terms of their strength, conductivity, durability, and lightweight nature. Furthermore, these nanomaterials possess advantageous characteristics, such as self-repairing capabilities, self-cleansing properties, resistance to freezing, and antibacterial attributes. Additionally, they can serve as reinforcing agents in construction or as sensing components to ensure safety. During the initial ten-year period of the twentieth century, Richard Adolf Zsigmondy utilized an ultramicroscope with the darkfield technique to effectively observe and gauge the dimensions of gold and other nanoparticles. Primarily, he coined the term "nanometre" to describe these measurements. As a testament to his distinctive contributions, he was awarded the Nobel Prize in Chemistry in 1925. In 1974, Nario Taniguchi introduced the concept of nanotechnology, wherein he elucidated the properties of materials with dimensions smaller than a millimetre.[6]

## 1.3 NANOMATERIALS

Nanomaterials are substances that possess at least one dimension (length, width, or height) on the nanoscale, typically characterized by dimensions smaller than 100 nanometres. At this scale, materials can exhibit distinctive and often improved attributes in comparison to their bulk counterparts. The prefix "nano" denotes a billionth of a meter. Various types of nanomaterials can be classified, encompassing nanoparticles, nanotubes, nanowires, nanosheets, and more. They can be composed of diverse constituents, such as metals, semiconductors, polymers, or composites. The characteristics of nanomaterials can significantly deviate from those exhibited by macroscale materials because of quantum effects and increased surface area. Nanomaterials have applications in a broad array of fields, including electronics, medicine, energy, materials science, and environmental science.[7]

Nanomaterials (NMs) are categorized into different groups based on their morphology, size, and form.

Some of the most important classes are discussed here.

### 1.3.1 Dimension-Based Classification

The property of dimensionality holds significant importance, as it enables the differentiation of various structures of nanomaterials. Nanomaterials always possess at least one dimension within the range of 1–100 nm.[8]

- Zero dimensional (0D)
  All three dimensions of nanomaterials are confined within the range of the nanoscale. This classification encompasses nanoparticles (NPs), quantum dots, nanospheres, and nanoclusters.
- One dimensional (1D)
  Two dimensions are in the vicinity of the nanoscale, but the last one is not. This classification includes nanorods (NRs), nanotubes (NTs), and nanowires.
- Two dimensional (2D)
  One of the dimensions will be in the nanoscale, while the other two dimensions will lie outside

this scale. Illustrative examples of this include nanofilms and nanolayers.

- Three dimensional (3D)
  Bulk nanomaterials with dimensions larger than the nanoscale (1–100 nm) are referred to as three-dimensional nanomaterials. One-dimensional, two-dimensional, and zero-dimensional nanomaterials serve as the fundamental components of 3D nanomaterials. Within this category, there are various types of nanomaterials, such as nanocrystals (NCs), core-shell structures, bundles of nanowires and nanotubes, and multi-layered nanomaterials.[9]

### 1.3.2 Composition Based

#### 1.3.2.1 Carbon-Based Material

These nanomaterials consist of carbon and exhibit a diverse range of structural features. Fullerenes, carbon nanotubes (CNTs), graphene (GR), carbon nanofibers, and carbon black are a few examples of various carbon-based nanomaterials.

- Fullerene
  Fullerene, specifically denoted as C60, consists of carbon atoms that are interconnected through sp2 hybridization, resulting in a spherical configuration. These carbonaceous structures are constructed from nanomaterial and possess a unique characteristic of forming a hollow enclosure. Their appeal results from their exceptional electrical conductivity, robustness, arrangement, electron affinity, and adaptability. Fullerenes are a composition of pentagonal and hexagonal carbon constituents. It is feasible for fullerenes to exist as either single-strata or multilayered configurations.[10]
- Carbon nanotubes
  Carbon nanotubes, formerly referred to as buckytubes, are elongated cylindrical structures composed of carbon atoms. These structures possess diameters on the nanometre scale and lengths ranging from micrometres to millimetres. CNTs are a type of nanofoil that is derived from graphene, a material consisting of a single layer of carbon atoms arranged in a honeycomb lattice pattern. Depending on their folding pattern, CNTs can be classified into two categories: single-walled nanotubes (SWNTs) and multi-walled nanotubes (MWNTs). These structures exhibit a remarkable level of durability, surpassing that of steel, and have the capacity to enhance the overall structural integrity.[11]
- Graphene
  Graphene is a nanostructure that represents an allotrope of carbon where a single layer of atoms is meticulously arranged in a hexagonal lattice. The name of this structure is derived from "graphite" and the suffix-ene, which symbolizes the presence of multiple double bonds in the carbon allotrope of graphite. In the graphene sheet, each atom establishes connections with its three nearest neighbours through $\sigma$-bonds and a delocalized $\pi$-bond, contributing to the formation of a valence band that spans the entire sheet. This bonding pattern is analogous to the one observed in carbon nanotubes, polycyclic aromatic hydrocarbons, and to a partial extent fullerenes and glassy carbon.[12]

#### 1.3.2.2 Semiconductor Nanoparticles

Metal-based nanoparticle semiconductors are semiconductor materials that incorporate metal nanoparticles. These materials have garnered interest in a multitude of fields, including electronics, optics, and catalysis, due to their distinctive properties that arise from the amalgamation of metal and semiconductor characteristics. Metals, known for their excellent electrical conductivity, are combined with semiconductors, which exhibit intermediate conductivity. The inclusion of metal nanoparticles within a semiconductor matrix has the potential to alter the overall electrical properties of the material. The synthesis of metal-based nanoparticle semiconductors can be achieved through a variety of chemical methods, such as sol-gel processes, chemical reduction, or deposition techniques.

#### 1.3.2.3 Ceramic Nanoparticles

Metallic nanoparticles can be integrated into ceramics to amplify their attributes. This amalgamation frequently yields what is referred to as nanoceramics or nanostructured ceramics. The infusion of metallic nanoparticles can bestow distinctive features upon the ceramic substance, including enhanced mechanical robustness, heightened electrical conduction, catalytic reactivity, and optical characteristics. Titanium dioxide ($TiO_2$), zinc oxide (ZnO), and other nanoparticles composed of metal oxides possess the capability to be introduced into ceramics to augment their mechanical and photocatalytic characteristics. To enhance the electrical conductivity of ceramics, silver (Ag) or copper (Cu) nanoparticles are frequently integrated. This aspect becomes of the utmost significance in the realm of electronic devices and sensors.

#### 1.3.2.4 Polymer Nanoparticles

Polymer metal nanoparticles are materials that combine polymer matrices with metal nanoparticles. These materials frequently exhibit distinct characteristics and possess a broad array of applications because of the synergistic effects of both metal nanoparticles and polymers. In-situ polymerization entails the direct synthesis of metal nanoparticles within a polymer matrix during the polymerization process.

Post-synthesis incorporation involves the integration of pre-synthesized metal nanoparticles into a polymer matrix through various techniques such as blending, solution mixing, or melt processing. The utilization of gold and silver nanoparticles is widespread due to their distinctive optical, electronic, and catalytic properties. Additionally, metal oxide nanoparticles like titanium dioxide or zinc oxide are often coupled with polymers to enhance their properties.

#### 1.3.2.5 Lipid Nanoparticles

Lipid-based nanoparticles have arisen as multifaceted conveyors with substantial potential for pharmaceutical administration and diverse biomedicinal utilities. Consisting of lipids, which are inherently existing or artificial molecules that exhibit amphiphilic properties, these nanoparticles provide an unparalleled amalgamation of biocompatibility, durability, and the capacity to enclose a wide array of therapeutic substances.

## 1.4 METAL NANOPARTICLES

Metal nanoparticles are a type of nanomaterial consisting solely of one element. They can exist as either individual atoms or groups of multiple atoms. Elements such as Au, Ag, Pt, Cu, Pd, Re, Zn, Ru, Co, Cd, Al, Ni, and Fe are among the most frequently synthesized metal nanoparticles. Metal nanoparticles can be produced through various methods, such as the bio-assisted method, hydrothermal method, and microwave-assisted method. These methods yield colloidal fluids or solid nanoparticles. Metal nanoparticles possess notable characteristics, including localized surface plasmon resonance (LSPR), high reactivity, and a wide range of electromagnetic spectrum absorption. Due to their enhanced optical, optoelectrical, catalytic, and antimicrobial/cancer/viral properties, metal nanoparticles are highly intriguing materials with numerous practical applications.[13] Metal nanoparticles display intriguing linear optical properties that bear resemblance to molecular systems in terms of their effects, although they arise from distinct physical processes. The absorption of plasmon resonance, which imparts metal nanoparticles with their distinctive colour, is greatly influenced by the shape of the particles as well as the interaction between them. The electrochemical template synthesis technique has offered a convenient means for investigating the optical implications of interactions between pairs of metal nanoparticles.[14] Through the utilization of uncomplicated electrodeposition methods, it becomes feasible to regulate the distance between the members of the pair, thereby inducing systematic shifts in the maximum plasmon resonance. By manipulating the relative sizes of the pair members, it becomes feasible to amplify the second-order nonlinear optical activity of minute nanoparticle systems. Metal nanoparticles (MNPs) are produced exclusively through the use of metallic precursors and can be generated via chemical, physical, or biological means. Metal nanoparticles have gained significant interest because of their size-dependent physicochemical feature in comparison to their counterparts.[15] They find utility in various domains, including cosmetics, pharmaceuticals, biomedical apparatus, and environmental remediation.[16]

### 1.4.1 Gold Nanoparticles

Gold (Au) nanoparticles are extensively utilized in various fields due to their distinctive optical and physical characteristics, such as surface plasmon oscillations, which are employed for labelling, imaging, and sensing purposes. Notably, significant progress has been achieved in the biomedical domain, particularly in terms of enhanced biocompatibility for disease diagnosis and therapeutics.[17] It is feasible to fabricate Au-NPs and associate them with a plethora of functionalizing agents, including polymers, surfactants, ligands, dendrimers, drugs, DNA, RNA, proteins, peptides, and oligonucleotides. This comprehensive analysis focuses on the utilization of gold nanoparticles and surface functionalization through a diverse array of molecules. Expanding and enhancing the applications of gold nanoparticles allow for more efficient targeting of drugs in photothermal therapy, resulting in reduced cytotoxicity in various cancer types, gene therapy, and numerous other diseases. Collectively, Au-NPs exhibit great potential as a vehicle for drug delivery and therapeutic interventions. Colloidal gold and the resulting nanoparticles that emerge from colloidal deposition have been a subject of study in the scientific community for quite some time.[18] Early researchers acknowledged their significance and identified several of their properties. The presence of Au nanoparticles with different sizes and shapes in suspension leads to distinct colours in the Au suspensions due to the impact they have on light scattering at visible wavelengths. This characteristic was initially exploited in the manufacturing of stained glass. In contemporary times, there is a growing trend in the production of Au nanoplates and nanoparticles from gold chloride solutions through the utilization of reductive reagents, often of biological origin.[19] These reagents facilitate the formation of colloidal suspensions and subsequent depositions of Au, as well as Au/Ag nanoparticles. The various shapes of Au nanoplates that can be generated through these processes encompass triangles, hexagons, octahedra, squares, boxes, and spheres, each exhibiting unique properties in comparison to their bulk counterparts. The characteristics of having a large surface area and being reactive are crucial factors in their utilization in the field of medicine for the purpose of delivering drugs. Additionally, these properties play a vital role in various manufacturing processes, particularly catalysis. Likewise, these attributes hold immense importance when contemplating the mechanisms involved in the transportation and deposition of Au within both hydrothermal and supergene environments at low temperatures.[20]

### 1.4.2 Silver Nanoparticles

Silver (Ag) possesses an abundance of contemporary industrial applications and is widely regarded as a repository of wealth. Nevertheless, the narrative surrounding this renowned precious metal originates from its utilization by ancient civilizations. Silver boasts numerous attributes that rendered it highly valuable to early societies. Its capacity for malleability, ductility, resilience, conductivity, antibacterial properties, and scarcity all contributed to its desirability. Furthermore, it served as a precious commodity in the realms of currency, adornments, jewellery, electrical contacts, and photography, among other domains. Although bulk silver is renowned for its resplendent surfaces and hues, a profound disparity in colour manifests when the metal undergoes dimensional reduction. Numerous methodologies for fabricating silver NPs have been documented, including laser ablation, gamma irradiation, electron irradiation, chemical reduction, photochemical techniques, microwave processing, and biological synthesis approaches.[21] An understanding of the techniques for creating silver nanomaterials is crucial due to the broad range of applications and perspectives in which they can be utilized. A significant challenge in the synthesis of silver nanoparticles lies in controlling their physical properties, including achieving a uniform distribution of particle sizes, consistent shape and morphology, appropriate coating or stabilizing agents, specific chemical composition or type, and desired crystal structure. These methods can be organized and classified based on common approaches, while also recognizing their differences, such as the reactants and reaction conditions employed.

### 1.4.3 Aluminium Nanoparticles

Aluminium (Al) nanoparticles are minute aluminium particles that possess dimensions on the nanometre scale. These nanoparticles generally exhibit a spherical morphology, with their size varying within the range of a few to tens of nanometres. Due to their expansive surface area, aluminium nanoparticles display a high degree of reactivity. As a result, they can readily engage in chemical reactions and hold promising potential for utilization in catalysis.[22] In the realm of chemical reactions, aluminium nanoparticles find application as catalysts, facilitating the reduction of organic compounds and the degradation of pollutants. Moreover, they are integrated into polymers, coatings, and composites to augment their mechanical and thermal properties. Additionally, aluminium nanoparticles are employed as fuel additives to enhance combustion efficiency and mitigate emissions. Within the electronics industry, aluminium nanoparticles are harnessed for the advancement of conductive inks, coatings, and nanocomposites. Furthermore, ongoing research endeavours are being conducted to explore the prospective utilization of aluminium nanoparticles in drug delivery and medical imaging. While the utilization of Aluminium nanoparticles spans a broad range of applications, their safety remains a source of concern.[23] The handling of aluminium nanoparticles necessitates the implementation of precautionary measures to avert inhalation or contact with the skin, as they may potentially pose health hazards. The synthesis of aluminium nanoparticles can be achieved through a variety of methods, such as the chemical reduction of aluminium salts, ball milling, and laser ablation. The selection of the synthesis technique can influence the dimensions, configuration, and surface characteristics of the nanoparticles. Although their distinct attributes render them highly advantageous for an extensive array of applications, their utilization must be meticulously administered to ensure both safety and environmental accountability.[24]

### 1.4.4 Iron Nanoparticles

As previously mentioned, iron (Fe) nanoparticles are exceedingly minute, typically at the nanoscale. This diminutive size gives them a substantial surface area-to-volume ratio, which is one of the contributing factors to their distinctive attributes. Iron nanoparticles can manifest properties such as intensified magnetic behaviour, heightened reactivity, and catalytic activity. These characteristics render them invaluable in diverse applications, encompassing the realms of materials science, medicine, and environmental remediation.[25] The synthesis of iron nanoparticles can be accomplished through a variety of methods, including chemical reduction, thermal decomposition, and green synthesis techniques. The selection of the method employed can exert an influence on the size, shape, and properties of the nanoparticles.[26] Iron nanoparticles are harnessed in the advancement of targeted drug delivery systems, as contrast agents for magnetic resonance imaging (MRI), and in the treatment of cancer via hyperthermia. Iron nanoparticles are also employed in the eradication of impurities from water and soil through processes such as adsorption and chemical reduction. Iron nanoparticles possess a substantial surface area and distinctive reactivity, which enables them to function as catalysts in diverse chemical reactions. These nanoparticles find utility in an array of applications, including but not limited to data storage and sensors, as they are incorporated into magnetic materials. When employed in biomedical and environmental contexts, concerns arise regarding the safety of iron nanoparticles. Therefore, it becomes crucial to understand their potential toxicity and exercise necessary precautions during their handling.[27] Due to their inherent nature, iron nanoparticles exhibit a propensity for high reactivity and rapid oxidation, particularly when exposed to air and moisture. To counteract this, surface coatings or stabilization techniques are often employed to forestall oxidation and preserve their unique properties. Fascinating and versatile, iron nanoparticles represent a captivating nanomaterial with a multitude of potential applications.[28]

### 1.4.5 Zinc Nanoparticles

Zinc (Zn) nanoparticles manifest distinct characteristics in comparison to bulk zinc because of their dimensions and expansive surface area. These characteristics encompass heightened catalytic activity, enhanced electrical conductivity, and modified optical properties. Zinc nanoparticles are employed as catalysts in diverse chemical reactions such as hydrogenation and oxidation processes. They have been extensively scrutinized for their antimicrobial attributes and potential utility in coatings, textiles, and medical applications.[29] Zinc nanoparticles can be utilized in sensors for the detection of gases, biomolecules, and other analytes. They may find utility in the advancement of electronic devices and conductive inks for printed electronics.[30] While zinc is a vital micronutrient for human beings, the potential toxicity of zinc nanoparticles can be a cause for concern. Prolonged exposure to elevated concentrations of zinc nanoparticles can result in health-related complications, necessitating a meticulous evaluation of their potential toxicity across a wide range of applications.[31] Zinc nanoparticles can be fabricated via different methodologies, encompassing chemical reduction, sol-gel processes, and green synthesis employing plant extracts or microorganisms. Each technique possesses its own merits and demerits in terms of regulating dimensions, morphology, and stability. The environmental ramifications of zinc nanoparticles are a matter of concern, particularly when utilized in consumer goods and industrial operations.

### 1.4.6 Titanium Nanoparticles

Titanium (Ti) nanoparticles are particles of nanoscale dimension that are composed of titanium, a chemical element renowned for its remarkable strength, low density, and resistance to corrosion. Upon reduction to the nanoscale, titanium manifests distinct properties and shows utility in diverse fields. Presented here are several pivotal facets of titanium nanoparticles. Typically, titanium nanoparticles possess a size range spanning from 1 to 100 nanometres. These nanoparticles can adopt various configurations and structures, encompassing spherical, rod-like, or irregular forms. Owing to their diminutive size, titanium nanoparticles have the capacity to exhibit heightened reactivity, thereby conferring utility in catalytic applications. Based on their size and shape, titanium nanoparticles can manifest distinctive optical properties, inclusive of plasmonic effects, which find applications in the fields of sensors and imaging. Investigations are currently being undertaken to explore the potential applications of nano-sized titanium dioxide in the realms of drug delivery, cancer therapy, and imaging. This particular variant of titanium nanoparticle finds utilization in sunscreens and cosmetics owing to its remarkable UV-blocking characteristics.[32] Titanium nanoparticles can be incorporated into various materials to augment their mechanical, electrical, or thermal properties. Although titanium dioxide nanoparticles find extensive use in consumer products, apprehensions exist regarding their potential impact on health and the environment. Ongoing research aims to gain a better understanding of the safety aspects of nanoparticles and establish guidelines for their employment.[33] Titanium nanoparticles can be synthesized through a range of techniques, including chemical reduction, sol-gel processes, and physical methods such as laser ablation or ball milling.

### 1.4.7 Copper Nanoparticles

There exists a keen interest in the synthesis of metallic copper (Cu) and copper (I) oxide nanoparticles (NPs) owing to their catalytic activity, sensing capabilities, and surface properties. Copper NPs find utility as antibacterial and antifungal agents in various applications such as coatings, plastics, textiles, nanowires, and nanofibers. Furthermore, these CuNPs possess advantageous features, including low production cost, as well as optical and magnetic properties. Consequently, Cu nanowires are employed in the field of nanoelectronics and find applications in magnetic devices, nanosensors, electron emitters, and other electronic domains.[34] Additionally, CuNPs serve as nanoprobes in the realms of medicine and bioanalytical fields. Nevertheless, the synthesis of CuNPs presents challenges due to their propensity for oxidation and the increased stability of Cu oxides. Therefore, the synthesis process necessitates meticulous preparation under an inert atmosphere to ensure the protection of CuNPs from oxidation, often accomplished by utilizing a capping agent to minimize this undesired phenomenon. However, the use of stabilizers or capping agents has the potential to mitigate the issue of oxidation, although it cannot completely eliminate this problem. Moreover, various inert media, including nitrogen and argon, have been employed to overcome the drawback of oxidation during synthesis. Several techniques have been explored in this regard, such as pulsed laser ablation, vacuum vapor deposition, microemulsion techniques, electrochemical methods, microwave-assisted processes, and hydrothermal processes. Additionally, the achievement of biological or biosynthesis of CuNPs has been made possible through the utilization of ascorbic acid as both a protecting and reducing agent. This approach not only ensures an environmentally friendly procedure but also proves economically viable and non-toxic. Furthermore, the combination of microwave-assisted synthesis and green synthesis has shown promise, as it enhances the yield of crystalline CuNPs and improves their stability for a period of approximately four months.[35]

## 1.5 SYNTHESIS OF NANOMATERIALS WITH VARIOUS APPROACHES AND METHODS

Methods for the synthesis of nanomaterials are categorized into two main approaches: top-down and bottom-up methodologies.

### 1.5.1 Top-Down Methods

The top-down approach, also known as the descending approach, entails the fragmentation of the bulk material into nanosized structures or particles. The techniques employed in top-down synthesis are an extension of those utilized to produce micron-sized particles. These top-down approaches are inherently simpler and rely on either the removal or division of bulk material or the miniaturization of bulk fabrication processes to achieve the desired structure with appropriate properties. However, the top-down approach's main drawback lies in the surface structure's imperfections. For instance, nanowires created through lithography exhibit roughness and may contain numerous impurities and structural defects on their surface. Various techniques, including high-energy wet ball milling, electron beam lithography, atomic force manipulation, gas-phase condensation, and aerosol spray, exemplify this approach.[36]

#### 1.5.1.1 Thermal Decomposition Method

It is a thermochemical process characterized by the production of chemical decomposition through the application of heat. The application of heat results in the rupture of the chemical bond within the compound. The temperature at which this chemical decomposition occurs is referred to as the decomposition temperature, which is specific to each element. The formation of nanoparticles is a direct consequence of the decomposition of the metacompound at a particular temperature.[37]

#### 1.5.1.2 Mechanical Method/Ball-Milling Method

The production of nanoparticles from a large quantity of material can be achieved using a cost-effective approach. The simplest mechanical technique, known as ball milling, is employed for this purpose.[38] Through the process of attrition, ball milling can generate nanoparticles. This method involves the transfer of kinetic energy from the grinding medium to the material being reduced. By utilizing the industrial-scale processes of consolidation and compaction, materials with improved properties can be obtained. These processes essentially "reconstruct" the nanoparticles into a cohesive form. Moreover, this method can also be employed to create alloys composed of various metals.[39]

#### 1.5.1.3 Lithographic Methods

Lithographic techniques comprise top-down approaches that have the capacity to generate micron-sized features for the most part, although they are energy intensive and necessitate expensive equipment. Lithography has been employed for several decades in the production of printed circuits and computers. Nanoimprint lithography represents a distinct variant of lithography, differing from the conventional form. It closely resembles template synthesis, whereby a template material is initially generated, followed by the stamping of a pliable polymeric material to produce a pattern. The manufacture of the stamped material utilizes a top-down method. Nanosphere lithography employs latex spheres to create a templated matrix.[40] There exist various categories of lithography methods, such as photo-lithography, electron beam lithography, soft lithography, focused ion lithography, nano-imprint lithography, and dip-pin lithography. Photo-lithography encompasses contact and proximity printing as well as projection printing.[41]

#### 1.5.1.4 Laser Ablation

Laser ablation synthesis in solution represents a straightforward methodology to produce nanoparticles utilizing various solvents. Employing a laser beam to irradiate different metal entities immersed in the solution results in the condensation of plasma, thereby generating the desired nanoparticles. This top-down approach distinguishes itself from traditional chemical techniques by facilitating the reduction of metals into nanoparticles. Moreover, the utilization of laser ablation techniques enables the synthesis of stable nanoparticles without the need for any stabilizing agents or chemicals.[42]

### 1.5.2 Bottom-Up Methods

The "bottom-up" approach, an alternative method, possesses the potential to generate less waste, thus making it more cost effective. The bottom-up approach involves the gradual buildup of a substance from its smallest units, such as atoms, molecules, or clusters. Although still in the developmental stage or just at the initial stages of commercial production, numerous techniques are being utilized for manufacturing nanoparticles. Some well-known bottom-up techniques employed for synthesizing luminescent nanoparticles include the organometallic chemical route, revere-micelle route, sol-gel synthesis, colloidal precipitation, hydrothermal synthesis, template-assisted sol-gel, and electrodeposition. The bottom-up method includes chemical vapor deposition (CVD), sol-gel, spinning, pyrolysis, and biological synthesis.[43]

#### 1.5.2.1 Chemical Vapor Deposition Method

The deposition of a gaseous reactant as a thin film on a substrate is referred to as the chemical vapor deposition method. The act of depositing the thin film occurs within a reaction chamber. The interaction between the combined gas and the heated substrate results in a chemical reaction. Subsequently, a thin film of product is formed on the substrate's surface because of this reaction. This resultant thin film is then retrieved and utilized. The advantages of the CVD method encompass the production of hard, strong, uniform, and highly pure nanoparticles. Conversely, the CVD method entails certain drawbacks, such as the necessity for specialized equipment and the generation of highly toxic gaseous by-products.[44]

#### 1.5.2.2 Sol-Gel Method

The sol-gel method, a combination of the terms "sol" and "gel", involves the formation of a colloid called a sol, which consists of solid particles suspended in a continuous liquid. On the other hand, gel refers to a solid macromolecule that is dissolved in a solvent. This method, known for its simplicity, is widely preferred as a bottom-up approach for synthesizing nanoparticles. It entails the utilization of a suitable chemical solution as a precursor. The most used precursors in the sol-gel method are metal oxides and chlorides.[45] Through various techniques such as stirring, sonication, and shaking, the precursor is dispersed in a host liquid. The resulting solution comprises a solid phase and a liquid phase, which can be separated using methods like filtration, sedimentation, and centrifugation, ultimately allowing for the recovery of the nanoparticles.[46]

#### 1.5.2.3 Spinning

In this method, nanoparticles are produced through the process of spinning. The synthesis of these nanoparticles is achieved by employing a spinning disc reactor (SDR), which comprises a rotating disc capable of regulating physical parameters like temperature. To prevent any chemical reactions and eliminate the presence of oxygen, the reactor is filled with nitrogen or inert gases. The chamber or reactor is then infused with a liquid, such as water and precursor compounds.[47]

#### 1.5.2.4 Pyrolysis

The method predominantly employed in industry to produce nanoparticles is known as pyrolysis. In this method, the precursor is subjected to combustion with a flame. The precursor can exist in either a liquid or vapor state. Subsequently, the precursor is introduced into a high-pressure furnace, where the nanoparticles are subsequently obtained.[48]

#### 1.5.2.5 Biological Synthesis

Biological synthesis entails the fabrication of nanoparticles through the utilization of plant extract and microorganisms, including bacteria and fungi. Phyto nanotechnology has introduced a fresh domain to produce nanoparticles, characterized by its environmentally friendly nature, simplicity, and cost-effectiveness. The advantages of phyto nanotechnology encompass scalability, bio-compatibility, and the utilization of water as a universal solvent for the synthesis of nanoparticles, serving as reducing agents. Phyto nanotechnology employs plants as a means for the fabrication of nanoparticles.[49] Distinct portions of the plant, such as the root, fruit, stem, seed, and leaf, are utilized for the synthesis of nanoparticles. The precise mechanism underlying the synthesis of nanoparticles using plants remains to be fully elucidated. It has been elucidated that the synthesis of various types of nanoparticles is attributed to organic acids, proteins, vitamins, and secondary metabolites, including alkaloids, flavonoids, terpenoids, polysaccharides, and heterocyclic compounds.

Microorganisms are regarded as nanofactories with extensive potential as a cost-effective tool, environmentally friendly in nature, capable of avoiding the use of toxic and harsh chemicals, as well as reducing energy demand during the process of synthesizing nanoparticles. The presence of various reductase enzymes within these microorganisms allows them to accumulate and detoxify heavy metals.[50] These reductase enzymes play a crucial role in the reduction of metal salts, leading to the formation of nanoparticles. In recent years, microorganisms such as yeast, fungi, and bacteria have been extensively studied in this regard. Proteins, reducing cofactors, metal-resistant genes, enzymes, and organic materials all play significant roles as reducing and capping agents in the synthesis of nanoparticles.

## 1.6 CHARACTERISTICS OF METAL NANOPARTICLES

### 1.6.1 Size-Dependent Characteristics

Metal nanoparticles display distinctive characteristics that are dependent on their size and differ from their bulk counterparts as a consequence of the quantum confinement phenomenon and an augmented ratio of surface area to volume. As the size of metal nanoparticles diminishes, there is a modification in their electronic characteristics. The energy levels of electrons become discretized, thus resulting in an elevation of the bandgap. Consequently, this leads to modifications in the optical, electronic, and catalytic attributes.[51] As the size of metal nanoparticles decreases, profound changes occur in their electronic properties. The energy levels of electrons undergo quantization, leading to an increase in the bandgap. As a result, there are alterations in the optical, electronic, and catalytic characteristics. Owing to their diminutive size and extensive surface area, metal nanoparticles exhibit an escalated catalytic efficiency. They possess the capacity to function as highly effective catalysts for a diverse array of reactions, encompassing hydrogenation, oxidation, and reduction. Additionally, they find widespread utilization in various industries such as environmental remediation and energy generation. The alteration in the hue of metal nanoparticles can be attributed to their dimensions and configuration. This phenomenon, acknowledged as the Mie scattering effect, is accountable for the wide array of colours perceived in metal nanoparticle solutions. To illustrate, gold nanoparticles can exhibit a red, blue, or purple appearance contingent upon their magnitude. Metal nanoparticles of smaller size are inclined to be more responsive and susceptible to clumping or oxidation. As a result, stabilizing agents or coatings are frequently employed to uphold their stability and hinder aggregation. The size attributes of metallic nanoparticles render them exceedingly adaptable substances, finding utility in domains such as nanotechnology, materials science, catalysis, biomedicine, and environmental science. Scholars persist in investigating and capitalizing on these exceptional characteristics for a broad spectrum of applications.[52]

### 1.6.2 High Surface Area-to-Volume Ratio

Metal nanoparticles exhibit unique properties characterized by a significant surface area-to-volume ratio, which stems from their small size and a notable surface-to-volume correlation. This exceptional attribute exerts a profound influence on both their physical and chemical properties, rendering them invaluable in diverse applications. Here are some of the primary traits associated with the high surface area-to-volume ratio of metal nanoparticles. The enlarged ratio of surface area to volume creates a greater number of active sites for chemical reactions.[53] This augmented reactivity imparts an advantage for catalytic processes and various chemical reactions, as a greater quantity of atoms or molecules can engage with the surface of the nanoparticle. Metal nanoparticles find extensive utility as catalysts, courtesy of their elevated surface area. This enables efficient catalysis as a larger number of reactants can be adsorbed onto the surface of the nanoparticle, thereby leading to enhanced catalytic activity. The substantial surface area of metal nanoparticles can engender unique optical properties, exemplified by surface plasmon resonance (SPR). The large surface area of metal nanoparticles can be effectively utilized in drug delivery systems due to its capability of accommodating a substantial quantity of therapeutic agents on the surface of the nanoparticles. This phenomenon has the potential to enhance drug solubility, bioavailability, and the ability to specifically target delivery to particular cells or tissues. Moreover, the ample surface area allows for efficient surface functionalization, enabling the attachment of various molecules, ligands, or coatings to the nanoparticle surface. This attribute is of great value, as it allows for the customization of nanoparticle properties and the improvement of their stability in diverse environments. Metal nanoparticles' high surface area-to-volume ratio is a fundamental property that underpins many of their unique features and applications. Because of their extraordinary reactivity and surface-related effects, these nanoparticles are widely used in domains such as catalysis, nanoelectronics, optics, sensing, and drug delivery.[54]

### 1.6.3 Plasmonic Properties

Metal nanoparticles display a diverse range of intriguing plasmonic characteristics owing to the interaction between light and the unbound electrons within the metal. These plasmonic properties arise from the collective oscillations of these unbound electrons, referred to as surface plasmon resonances. The most notable plasmonic characteristic of metal nanoparticles is the SPR, which emerges when the frequency of incident light aligns with the inherent frequency of the collective oscillations of unbound electrons at the metal's surface.[55] This SPR results in pronounced absorption and scattering of light at specific wavelengths, rendering metal nanoparticles visually vibrant and highly effective at light absorption. The plasmonic properties of metal nanoparticles are heavily reliant on their dimensions and configuration. The resonance wavelength of SPR can be adjusted by modifying the size or shape of the particles. Smaller nanoparticles demonstrate blue-shifted resonances, whereas larger ones exhibit red-shifted resonances. This adjustability holds significant value for applications such as sensing and imaging. Plasmonic properties can be influenced by temperature, and this characteristic is utilized in photothermal applications. The illumination of metal nanoparticles by laser radiation can induce thermal changes in their SPR and consequently modify their optical properties. Metal nanoparticles possessing plasmonic properties are extensively employed in biomedical applications, including cancer therapy (photothermal therapy), drug delivery, and bioimaging (as contrast agents and labels). They are also utilized in plasmonic sensors for the detection of diverse analytes, encompassing biomolecules and gases.[56]

### 1.6.4 Thermal and Electrical Conductivity

Metal nanoparticles possess distinctive thermal and electrical conductivity attributes, owing to their unique electronic structure. Generally, metal nanoparticles exhibit higher thermal conductivity in comparison to bulk materials composed of the same metal. This disparity arises due to the increased phonon mean free path, which denotes the distance a phonon can traverse before scattering in nanoparticles. Consequently, as the particle size diminishes, the number of grain boundaries and defects increases, thereby facilitating a more efficient transfer of heat.[57] The thermal conductivity of nanoparticles can also be influenced by quantum confinement effects. It is observed that the thermal conductivity of metal nanoparticles is contingent upon their size, with smaller nanoparticles typically exhibiting greater thermal conductivity. However, it is noteworthy that this behaviour may not be uniform across all materials, as the exact relationship may vary depending on the specific metal and the surrounding environment. Furthermore, metal nanoparticles often manifest enhanced electrical conductivity relative to bulk materials composed of the same metal.[58] This enhancement is attributed to the elevated surface-to-volume ratio exhibited by nanoparticles, which facilitates a more efficient transport of charge. Moreover, the electrical conductivity of metal nanoparticles can be impacted by quantum size effects. Specifically, when the size of a metal nanoparticle becomes comparable to the wavelength of an electron, discrete energy levels emerge, resulting in the quantization of the electronic structure. Consequently, changes in electrical conductivity occur as a function of particle size. In addition, metal nanoparticles can exhibit surface plasmon resonance, which denotes the collective oscillation of electrons on the surface of the nanoparticle. This resonance engenders unique electrical properties, rendering nanoparticles valuable in applications such as sensing and catalysis. Furthermore, the electrical conductivity of metal nanoparticles can be influenced by the type and properties of the surface ligands or capping agents. Organic ligands can either augment or impede charge transport, contingent upon their electronic properties. In summation, the thermal and electrical conductivity characteristics of metal nanoparticles are profoundly reliant on the specific metal, nanoparticle size, and the surrounding environment or surface coatings. These distinctive properties endow metal nanoparticles with considerable value in a multitude of applications, encompassing electronics, catalysts, and materials science.[59]

## 1.7 APPLICATIONS OF METAL NANOPARTICLES

Metal nanoparticles have a wide range of applications across various fields due to their unique properties and versatile characteristics, and here we are going to discuss some of their important applications in medicine, biosensing, water purification, wound dressing, and electronics.

### 1.7.1 Applications in Medicine

Noble metals and their compounds have been used as therapeutic agents in medicine since ancient times to cure a variety of infections. Significant advances in nanobiotechnology have been made in recent years, resulting in the development of various nanomaterials with a wide range of uses.[60] Among these nanomaterials, noble metal nanoparticles have demonstrated promising potential for biomedical purposes. Owing to their small size, nanoparticles can readily interact with biomolecules both on the surface and within cells, resulting in enhanced signal detection and targeted specificity for diagnostics.[61] Here, we aim to concentrate on the biomedical applications of noble metal nanoparticles, specifically silver, gold, and platinum, in the diagnosis and treatment of dreaded diseases such as cancer, human immunodeficiency virus (HIV), tuberculosis (TB), and Parkinson's disease. Furthermore, we critically examine the role of silver nanoparticles as novel antimicrobials, gold nanoparticles (AuNPs) as efficient drug carriers, and the uses of platinum nanoparticles in bone allografts and dentistry, among others.

Gold metal nanoparticles find utility as contrast agents in a diverse range of imaging techniques, including but not limited to magnetic resonance imaging and computed tomography (CT) scans.[62] The medical practice associated with the use of gold is commonly referred to as "chrysotherapy". Upon delving into the annals of history, one discovers that ancient civilizations, such as those of Egypt, India, and China, employed gold for the treatment of ailments such as smallpox, skin ulcers, syphilis, and measles. Furthermore, gold is utilized in the field of medicine for various purposes such as the creation of medical devices like pacemakers and gold-plated stents for the management of heart disease.[63] Metal nanoparticles can be engineered to enclose drugs and transport them to specific sites within the body. This specific drug delivery method serves to minimize any potential side effects and enhance the effectiveness of the drugs in question. In photothermal therapy, these nanoparticles are targeted towards cancer cells and, upon exposure to near-infrared light, generate heat, which selectively destroys the cancer cells while minimizing any harm to healthy tissue. Metal nanoparticles are also employed in the creation of biosensors used for detecting biomolecules and disease markers. By observing any changes in the optical, electrical, or magnetic properties of these nanoparticles in the presence of specific biomolecules, it becomes possible to detect diseases early on.[64] Silver nanoparticles are notable for their strong antimicrobial properties, making them ideal for use in wound dressings and coatings to prevent infections and expedite the healing process. By inhibiting the growth of bacteria and fungi, silver nanoparticles can effectively reduce the risk of complications. An emerging field known as theragnostics combines therapy and diagnostics, with metal nanoparticles playing a crucial role. Nanoparticles can be designed to not only deliver therapeutic agents but also serve as imaging contrast agents, allowing for real-time monitoring of the treatment's efficacy. Furthermore, metal nanoparticles can serve as carriers for antigens in vaccine development. Their unique properties enable improved vaccine stability, controlled release of antigens, and enhanced immune responses. In the realm of gene delivery systems, metal nanoparticles can be employed to facilitate the introduction of genetic material into cells for therapeutic purposes. This is particularly relevant in gene therapy, where precise delivery is essential for the treatment of genetic disorders. The application of metallic nanoparticles in the field of medicine is continuously developing, presenting novel resolutions for enhanced diagnostic techniques, focused treatments, and individualized medical approaches.[64]

### 1.7.2 Applications in Catalysis

Metal nanoparticles, possessing a significantly greater surface area per unit volume or weight of metal compared to bulk metal, have emerged as promising materials for catalysis, given that industrial catalysts predominantly operate on the metal surface.[65] Typically, industrial catalysts consist of metals deposited on inorganic supports, which are often fabricated via heat treatment of metal ions on the support at elevated temperatures, occasionally in the presence of hydrogen.[66] These catalysts exhibit intricate structures, such as mixtures of metal particles with diverse sizes and shapes. Catalysts play a crucial role not only in the realm of industrial chemical processes but also in the occurrence of chemical reactions (specifically enzymatic reactions) within the human body. Industries commonly rely on heterogeneous catalysts supported by inorganic substances at elevated temperatures, occasionally utilizing homogeneous catalysts at lower temperatures.[67]

Metal nanoparticles, namely platinum (Pt), palladium (Pd), and gold (Au), are extensively utilized as catalysts in hydrogenation reactions. They facilitate the reduction of diverse organic compounds, encompassing alkenes and nitro compounds. Metal nanoparticles possess the capability to serve as catalysts for oxidation reactions. For instance, the oxidation of alcohols to aldehydes or ketones is accomplished through the utilization of silver (Ag) and palladium (Pd) nanoparticles. Metal nanoparticles play a pivotal role in numerous carbon-carbon coupling reactions, including Suzuki, Heck, and Sonogashira couplings.[68] Specifically, palladium nanoparticles are frequently employed as catalysts in these cross-coupling reactions. The synthesis of metal nanoparticles often employs green chemistry approaches, wherein they serve as both catalysts and reaction participants.[69] This approach minimizes the utilization of hazardous reagents and diminishes environmental impact. Metal nanoparticles, including platinum and gold, find utility as electrocatalysts in fuel cells and electrolyzers. These materials enhance the efficiency of electrochemical reactions, such as the oxygen reduction reaction (ORR) and hydrogen evolution reaction (HER). The distinctive properties inherent to metal nanoparticles, such as their elevated surface area, adjustable size and shape, and surface reactivity, render them versatile catalysts with applications spanning a multitude of industries and research fields.[70]

### 1.7.3 Applications in Biosensing

In the era of nanotechnology, noble metal nanoparticles have played a crucial role in the advancement of novel biosensors and/or the augmentation of existing biosensing techniques to meet the demand for more specific and highly sensitive biomolecular diagnostics.[71] The distinctive physicochemical characteristics of such metals at the nanoscale have resulted in the creation of a wide range of biosensors, encompassing: (i) nanobiosensors for point-of-care disease diagnosis; (ii) nanoprobes for in vivo sensing/imaging, cell tracking, and monitoring disease pathogenesis or therapy monitoring; and (iii) other nanotechnology-based tools that contribute to scientific research on fundamental biology.[72] In essence, NPs are generally considered one of the most prevalent nanotechnology-based strategies for developing biosensors, due to their simplicity, physiochemical malleability, and large surface areas. The vast majority of the biosensing applications rely on nanoparticles which have been formulated on gold NPs.[73] This preference is primarily attributed to the distinctive optical characteristics as well as their convenient modification with various biomarkers in an aqueous medium. Metal nanoparticles, specifically gold and silver nanoparticles, display localized surface plasmon resonance, resulting in alterations in their optical characteristics upon interaction with target biomolecules. This occurrence is employed in surface plasmon resonance biosensors to achieve label-free analyte detection. The presence of metal nanoparticles can elicit a modification in colour due to their localized surface plasmon resonance. This particular attribute is harnessed in colorimetric biosensors, where the alteration in colour signifies the existence or concentration of the targeted analyte. This technique is frequently employed for the purpose of easy and swift detection. Metal nanoparticles are capable of augmenting the fluorescence of neighbouring fluorophores. This capability is utilized in fluorescence-based biosensors to enhance sensitivity.[73] Quantum dots serve as an example of metal nanoparticles exploited for the purpose of fluorescence enhancement. Metal nanoparticles, such as gold and platinum nanoparticles, are commonly utilized as electrode modifiers in electrochemical biosensors. They enhance the rate of electron transfer and expand the surface area of the electrode, thus leading to improved sensitivity and lower limits of detection. Magnetic metal nanoparticles, such as iron oxide nanoparticles, are implemented in biosensors for the purpose of magnetic separation and detection. Metal nanoparticles can also serve as a support matrix for immobilizing enzymes in biosensors. This enhances the stability and catalytic activity of the enzymes, thus resulting in improved sensor performance. The distinctive characteristics of metal nanoparticles render them suitable for the development of point-of-care biosensors, facilitating rapid and on-site detection of diseases, pathogens, or specific biomarkers.[74] Metal nanoparticles, particularly gold and silver nanoparticles, demonstrate robust surface-enhanced Raman scattering (SERS) effects. In SERS biosensors, the presence of target molecules elicits variations in the Raman signals, thereby enabling highly sensitive analyte detection. The multifunctionality of metal nanoparticles renders them indispensable instruments in the progression of biosensors, thereby enhancing the field of medical diagnostics, environmental surveillance, and bioanalytical inquiry.[75]

### 1.7.4 Applications in Water Purification

Metal nanoparticles have gained significant attention in the field of water purification due to their unique properties. The elimination of heavy metal ions from water has been an issue for decades without ever being solved. Conversely, in recent times, nanotechnology has been integrated with various innovative methodologies for the extraction of heavy metals from water systems to enhance removal efficiency.[76] Several techniques for the purification of heavy metals in water are obtainable, including chemical precipitation and coagulation, flocculation, electrochemical procedures, photocatalytic degradation, membrane filtration, ion exchange, bioremediation, and adsorption.[77] The primary focus of the adsorption technique lies in the interaction between nanoparticles and adsorbents, which is reliant on their physicochemical characteristics. The term "physicochemical" encompasses aspects such as particle dimensions, surface area, surface charge, aggregation, structure, surface coating, and so forth. Magnetic nanoparticles offer benefits owing to their considerable surface area, size, and shape-dependent catalytic properties and can be used to remove magnetic waste from aqueous solutions. Metal nanoparticles such as silver (Ag), gold (Au), and copper (Cu) nanoparticles possess exceptional catalytic characteristics.[78] They possess the capability to serve as catalysts for the degradation of organic contaminants found in water, including dyes, pesticides, and pharmaceutical remnants. Notably, silver nanoparticles are renowned for their potent antibacterial attributes. Consequently, they can be incorporated into water treatment systems to impede the proliferation of bacteria and other microorganisms. This is particularly advantageous for disinfection purposes in both drinking water treatment and wastewater treatment. Metal nanoparticles are capable of effectively adsorbing heavy metals from water, thereby acting as adsorbents.[79] For instance, iron nanoparticles have been employed to eliminate heavy metals such as lead, cadmium, and arsenic through adsorption mechanisms. This serves to diminish the concentration of toxic metals in water. Furthermore, metal nanoparticles can be applied as coatings on membranes employed in water filtration systems. Through this coating, the effectiveness of the membrane is heightened, as it prevents fouling, reduces biofilm formation, and enhances the overall performance of the membrane in eliminating contaminants. Commonly used in photocatalytic water treatment processes, titanium dioxide ($TiO_2$) nanoparticles can generate reactive oxygen species (ROS) upon exposure to UV or sunlight.[80] These ROS can subsequently oxidize and degrade organic contaminants, pathogens, and even select heavy metals. By incorporating metal nanoparticles into various nanocomposite materials, such as carbon nanotubes or graphene oxide, hybrid materials can be created that exhibit augmented adsorption and catalytic qualities. These composites can be customized to cater to specific water treatment applications. Furthermore, metal nanoparticles can be utilized in sensors designed for the detection of contaminants in water. These sensors offer rapid and sensitive detection of pollutants, thereby enabling real-time monitoring of water quality. Silver nanoparticles, owing to their antimicrobial properties, can be harnessed to eradicate pathogens such as bacteria and viruses from water. This is particularly valuable in guaranteeing the safety of drinking water.[81]

### 1.7.5 Applications in Food Packing

The emergence of nanotechnology has brought about a significant transformation in various fields of science and technology, with a particular impact on the food packaging industry. Consequently, certain particles at the nanoscale have the potential to be incorporated into materials that come into contact with food, thereby prolonging the preservation of food products.[82] Numerous inorganic and metal particles have already been employed in the production of active food packaging materials, resulting in an extended shelf life for food items. The utilization of nanocomposites containing these particles in packaging offers various benefits, including a reduction in the need for preservatives and an increased capacity to inhibit the growth of microorganisms.[83] Silver nanoparticles possess well-documented antimicrobial attributes. They can be incorporated into materials used for food packaging in order to impede the proliferation of bacteria and other microorganisms, thus prolonging the duration of time in which the packaged food remains consumable.[84] The inclusion of metal nanoparticles within polymer films augments their resistance against oxygen, moisture, and other gases, thus assisting in the prevention of both the oxidation and spoilage of food products. Gold nanoparticles can be harnessed within sensors that are integrated into packaging materials to effectively monitor the freshness of food.[85] Modifications in the colour or conductivity of these nanoparticles can indicate fluctuations in temperature, humidity, or the emission of specific gases by the food during its spoilage. Metal nanoparticles have the potential to form an integral part of packaging systems that are intelligent in nature and are capable of altering their colour or emitting signals as a response to particular environmental circumstances.[86] A conspicuous change in colour could, for instance, serve as an indication of whether the packaged food has been subjected to excessively high temperatures or if the packaging itself has been compromised. Certain metal oxide nanoparticles, such as zinc oxide or titanium dioxide, possess advantageous antioxidant characteristics. These nanoparticles may be integrated into the composition of packaging materials to effectively inhibit the process of fat and oil oxidation in food, thereby preserving the overall quality of the product.[87] Metal nanoparticles can also be utilized to enhance the mechanical attributes of packaging materials, consequently bolstering their resilience and longevity. For instance, copper nanoparticles can be employed in modified atmosphere packaging (MAP) systems to assimilate oxygen and subsequently discharge antimicrobial agents, thereby engendering a precisely regulated environment that protracts the shelf life of perishable food items. Nanoparticles composed of specific metals, namely zinc and copper, can be assimilated into edible coatings that are employed for fruits and vegetables. These coatings exhibit the ability to reduce microbial contamination and sustain the freshness of the produce.[88]

### 1.7.6 Applications in Wounds and Dressing

The increasingly prevalent problem of microbial resistance has sparked an extensive investigation into metal nanoparticles as potential agents for combating bacteria due to their well-documented efficacy in hindering and eradicating these microorganisms.[89] When compared to eukaryotes, prokaryotes are recognized as being more vulnerable to the detrimental impacts of silver. The larger size of eukaryotic cells, as well as the redundancy in structure and function within these cells, necessitates higher concentrations of silver to exterminate eukaryotes.[45] To promote the healing of wounds, metal nanoparticles have been subjected to testing either on their own or in conjunction with other substances as a strategy to enhance their antibacterial capabilities. Metal nanoparticles, such as silver nanoparticles, possess potent antimicrobial characteristics.[90] They can effectively impede the proliferation of a wide range of microorganisms, encompassing bacteria, fungi, and viruses. This attribute renders them highly valuable in the prevention and treatment of wound infections. Gold nanoparticles, akin to metal nanoparticles, have demonstrated the capability to expedite the wound-healing process.[91] They have the potential to augment cellular proliferation, migration, and tissue regeneration, culminating in expedited and more efficient healing. Certain metal nanoparticles, such as gold and silver nanoparticles, exhibit anti-inflammatory properties. These substances can aid in the mitigation of inflammation in the vicinity of the wound, a critical factor in fostering proper healing and averting the development of chronic wounds.[92] Metal nanoparticles can serve as carriers for the delivery of drugs to the wound site. They can encapsulate therapeutic agents and release them in a controlled manner, ensuring a sustained therapeutic impact and diminishing the necessity for frequent administration. The incorporation of metal nanoparticles into wound dressings can improve their mechanical attributes. This enhancement can promote the flexibility, tensile strength, and overall durability of the dressing, thereby optimizing its ability to conform to the shape of the wound and provide superior protection.[93]

### 1.7.7 Applications in Electronics

Metal nanoparticles, such as silver nanoparticles, are utilized in the production of conductive inks utilized in the field of printed electronics. This application proves to be particularly advantageous in the realm of flexible electronics, as well as in the development of Radio Frequency Identification (RFID) tags, these are small electronic devices that use radio waves to transmit data for identifying and tracking object), and other scenarios where traditional wiring may prove to be unfeasible.[94] In the scope of nanoelectronics, metal nanoparticles are employed to fabricate minuscule interconnects that connect different components on a chip. This technique is instrumental in reducing the size of electronic devices and enhancing their overall performance.[95] Metal nanoparticles serve as catalysts in the manufacturing of electronic components. Their presence facilitates the growth of nanowires and other nanostructures, which are critical in the development of electronic devices on a nanoscale. Metal nanoparticles find application in the creation of NEMS devices, which function on a scale smaller than microelectromechanical systems (MEMS).[96] These devices have potential uses in sensors, actuators, and other electronic components. Metal nanoparticles exhibit localized surface plasmon resonance effects, which can be effectively utilized for a myriad of applications. In the field of electronics, plasmonic nanoparticles are employed in sensors and imaging devices and also to enhance the performance of solar cells and light-emitting diodes (LEDs). Certain metal nanoparticles, such as gold nanoparticles, have been investigated for their potential in non-volatile memory devices. The distinctive electronic properties of nanoparticles present advantageous prospects in the development of high-density and fast-access memory storage.[97] Metal nanoparticles can be employed in the creation of remarkably sensitive sensors that can detect various substances. Metal nanoparticles are currently being researched for their potential use in improving the performance of batteries and supercapacitors.[98] They can augment conductivity and offer expansive surface areas that consequently lead to improved energy storage and faster charge/discharge rates. Metal nanoparticles can be integrated into thermal interface materials to effectively improve heat dissipation in electronic devices. This is of the utmost importance in maintaining optimal operating temperatures and preventing overheating. Metal nanoparticles play a significant role in enhancing the performance of optoelectronic devices such as photodetectors and solar cells. They can enhance light absorption and conversion efficiency in these devices.[99]

## 1.8 CONCLUSION

In conclusion, this chapter offers a comprehensive overview of the dynamic and rapidly evolving realm of metal nanoparticle research. By examining different methods of synthesis, including chemical, physical, and biological approaches, readers acquire knowledge about the diverse techniques employed to fabricate nanoparticles with tailored properties. The exploration of the characteristics of metal nanoparticles delves into their distinct size-dependent properties, surface plasmon resonance, and exceptional catalytic activities. The chapter underscores the importance of understanding these characteristics within the context of designing and optimizing nanoparticles for specific applications. Moreover, the chapter sheds light on the broad range of applications of metal nanoparticles in various fields, such as medicine, electronics, catalysis, and environmental science. The immense potential of these nanoparticles in drug delivery, imaging, sensing, and pollution remediation highlights their significance in addressing contemporary challenges. As researchers continue to push the boundaries of nanotechnology, this chapter serves as a valuable resource for scientists, engineers, and students who seek a profound understanding of metal nanoparticle synthesis, their fascinating characteristics, and the numerous ways in which they can be utilized for innovative applications. The interdisciplinary nature of this field encourages collaboration and ignites imagination, paving the way for future breakthroughs and advancements in the captivating realm of metal nanoparticles.

## Keywords

Metal Nanoparticles
Polymer Nanoparticles
Synthesis Methods of Metal Nanoparticles
Applications of Metal Nanoparticles

## REFERENCES

1. Jeevanandam, Jaison, Ahmed Barhoum, Alian Dufresne, and Michael Danquah. "Review on nanoparticles and nanostructured materials: history, sources, toxicity, and regulations". *Beilstein J. Nanotechnol.* 9 (2018): 1050–1074.
2. Haroon, H. B., A. C. Hunter, Z. S. Farhangrazi, and S. M. Moghimi. "A brief history of long-circulating nanoparticles". *Adv. Drug Deliv. Rev.* (2022): 114396.
3. Barchiesi, Dominique. "Lycurgus cup: inverse problem using photographs for characterization of matter". *JOSA A* 32.8 (2015): 1544–1555.
4. Ebrahiminezhad, Alireza, Mohammad Javad Raee, Zahra Manafi, Abdolreza Sotoodeh Jahromi, and Younes Ghasemi. "Ancient and novel forms of silver in medicine and biomedicine". *JAMSAT* 2.1 (2016): 122–128.
5. Schaming, Delphine, and Hynd Remita. "Nanotechnology: from the ancient time to nowadays". *Found. Chem.* 17 (2015): 187–205.

6. Baig, Nadeem, Irshad Kammakakam, and Wail Falath. "Nanomaterials: a review of synthesis methods, properties, recent progress, and challenges". *Mater. Adv.* 2.6 (2021): 1821–1871.
7. Zainab D. Abd Ali, Ahmed A. Ahmed, and Osama Abdul Azeez Dakhil. "A review on nanomaterials for environmental remediation". *Energy Environ. Sci.* 5.8 (2012): 8075–8109.
8. Ariga, Katsuhiko, Mao Li, Gary J. Richards, and Jonathan P. Hill. "Nano architectonics: a conceptual paradigm for design and synthesis of dimension-controlled functional nanomaterials". *J. Nanosci. Nanotechnol.* 11.1 (2011): 1–13.
9. Paras, Kushal Yadav, Prasanth Kumar, Dharmasanam Ravi Teja, Sudipto Chakraborty, Monojit Chakraborty, Soumya Sanjeeb Mohapatra, Abanti Sahoo, Mitch M. C. Chou, Chi-Te Liang, and Da-Ren Hang. "A review on low-dimensional nanomaterials: nanofabrication, characterization and applications". *Nanomaterials* 13.1 (2022): 160.
10. Husen, Azamal, and Khwaja Salahuddin Siddiqi. "Carbon and fullerene nanomaterials in plant system". *J. Nanosci. Nanotechnol* 12.1 (2014): 1–10.
11. Arora, Bharti, and Pankaj Attri. "Carbon nanotubes (CNTs): a potential nanomaterial for water purification". *J. Compos. Sci.* 4.3 (2020): 135.
12. Liao, Chengzhu, Yuchao Li, and Sie Chin Tjong. "Graphene nanomaterials: synthesis, biocompatibility, and cytotoxicity". *Int. J. Mol. Sci.* 19.11 (2018): 3564.
13. Chen, Ye, et al. "Two-dimensional metal nanomaterials: synthesis, properties, and applications". *Chem. Rev.* 118.13 (2018): 6409–6455.
14. Fu Bo, Jingxuan Sun, Yuan Cheng, Hao Ouyang, Giuseppe Compagnini, Peng Yin, Songrui Wei, Shaojuan Li, and Dabing Li. "Recent progress on metal-based nanomaterials: fabrications, optical properties, and applications in ultrafast photonics". *Adv. Funct. Mater.* 31.49 (2021): 2107363.
15. Chen, Ye, Zhanxi Fan, Zhicheng Zhang, Wenxin Niu, Cuiling Li, Nailiang Yang, Bo Chen, and Hua Zhang. "Two-dimensional metal nanomaterials: synthesis, properties, and applications". *Chem. Rev.* 118.13 (2018): 6409–6455.
16. Wang, Jiaqing, and Hongwei Gu. "Novel metal nanomaterials and their catalytic applications". *Molecules* 20.9 (2015): 17070–17092.
17. Sardar, Rajesh, Alison M. Funston, Paul Mulvaney, and Royce W. Murray. "Gold nanoparticles: past, present, and future". *Langmuir* 25.24 (2009): 13840–13851.
18. Madkour, Loutfy H. "Applications of gold nanoparticles in medicine and therapy". *Pharm. Pharmacol. Int. J* 6.3 (2018): 157–174.
19. Huang, Xiaohua, Prasant K. Jain, Ivan H. El-Sayed, and Mostafa A El-Sayed. "Gold nanoparticles: interesting optical properties and recent applications in cancer diagnostics and therapy". *Nanomedicine* (2007): 681–693.
20. Hough, R. M., R. R. P. Noble, and M. Reich. "Natural gold nanoparticles". *Ore Geol. Rev.* 42.1 (2011): 55–61.
21. Abou El-Nour, Kholoud M. M., Ala a Eftaiha, Abdulrahman Al-Warthan, and Reda A. A. Ammar. "Synthesis and applications of silver nanoparticles". *Arab. J. Chem.* 3.3 (2010): 135–140.
22. Ghorbani, Hamid Reza. "A review of methods for synthesis of Al nanoparticles". *Orient. J. Chem.* 30.4 (2014): 1941–1949.
23. Li, Wanpeng, Jian Mao, and Jie Feng. "Aluminium grain refinement by Ti (C, N) nanoparticles additions: principles, advantages and drawbacks". *Metallurgical Res. Technol.* 116.2 (2019): 212.
24. Ghorbani, Hamid Reza. "A review of methods for synthesis of Al nanoparticles". *Orient. J. Chem.* 30.4 (2014): 1941–1949.
25. Huber, Dale L. "Synthesis, properties, and applications of iron nanoparticles". *Small* 1.5 (2005): 482–501.
26. Aragaw, Tadele Assefa, Fekadu Mazengiaw Bogale, and Belete Asefa Aragaw. "Iron-based nanoparticles in wastewater treatment: a review on synthesis methods, applications, and removal mechanisms". *J. Saudi Chem. Soc.* 25.8 (2021): 101280.
27. Ali, Attarad, Hira Zafar, Muhammad Zia, Ihsan Ul Haq, Abdul Rehman Phull, and Joham Sarfaraz Ali. "Synthesis, characterization, applications, and challenges of iron oxide nanoparticles". *Nanotechnol. Sci. Appl.* (2016): 49–67.
28. Campos, Eunice Aparecida, Denise Vilela Barcza Stockler Pinto, Jose Ir Elizabeth da Costa Mattos, and Rita de Cassia Lazzarini Dutra. "Synthesis, characterization, and applications of iron oxide nanoparticles-a short review". *JATM* 7 (2015): 267–276.
29. Uikey, Prateek, and Kirti Vishwakarma. "Review of zinc oxide (ZnO) nanoparticles applications and properties". *IJETCSE* 21.2 (2016): 239–242.
30. Majee, Subimal, Mikael C. F. Karlsson, Pawel Jerzy, Anurak Sawatdee, Mohammad Yusuf Mulla, Naveed Ul Hassan Alvi, Peter Dyreklev, Valerio Veni, and David Nilsson. "Low-temperature chemical sintering of inkjet-printed Zn nanoparticles for highly conductive flexible electronic components". *NPJ Flex. Electron.* 5.1 (2021): 14.
31. Król, A. P. Pomastowski, K. Rafinska, V. Railean-Pliugaru, and B. Buszewski. "Zinc oxide nanoparticles: synthesis, antiseptic activity, and toxicity mechanism". *Adv. Colloid Interface Sci.* 249 (2017): 37–52.
32. Wu, Fan, Zheng Zhou, and Andrea L. Hicks. "Life cycle impact of titanium dioxide nanoparticle synthesis through physical, chemical, and biological routes". *Environ. Sci. Technol.* 53.8 (2019): 4078–4087.
33. Elizabeth, Elmy, et al. "ZnO nanoparticle incorporated nanostructured metallic titanium for increased mesenchymal stem cell response and antibacterial activity". *Nanotechnology* 25.11 (2014): 115101.
34. Din, Muhammad Imran, and Rida Rehan. "Synthesis, characterization, and applications of copper nanoparticles". *Anal. Lett.* 50.1 (2017): 50–62.
35. Parveen, Farhana, Basavaraja Sannakki, and Mohanrao V. Mandke. "Copper nanoparticles: synthesis methods and its light-harvesting performance". *Sol. Energy Mater. Sol. Cells Solar Cells* 144 (2016): 371–382.
36. Jamkhande, Prasad Govindrao, Namrata W. Ghule, Abdul HaqueBamer, and Mohan G. Kalaskar. "Metal nanoparticles synthesis: an overview on methods of preparation, advantages and disadvantages, and applications". *Drug. Delv. Technol.* 53 (2019): 101174.
37. Salavati-Niasari, Masoud, and Fatemeh Davar. "Synthesis of copper and copper (I) oxide nanoparticles by thermal decomposition of a new precursor". *Mater. Lett.* 63.3–4 (2009): 441–443.
38. De Carvalho, J. F. S. N. de Medeiros, M. A. Morales, A. L. Dantas, and A. S. Carrico. "Synthesis of magnetite nanoparticles by high energy ball milling". *Appl. Surf. Sci.* 275 (2013): 84–87.
39. Xing, Tan, Jaka Sunarso, Wenrong Yang, Yongbai Yin, Alexey M. Glushenkov, Lu Hua Li, Patrick C. Howlett, and Ting Chen. "Ball milling: a green mechanochemical approach for the synthesis of nitrogen-doped carbon nanoparticles". *Nanoscale* 5.17 (2013): 7970–7976.

40. Zhang, Xiaoyu, Alyson Whitney, and Zhao Jing. "Advances in contemporary nanosphere lithographic techniques". *J. Nanosci. Nanotechnol.* 6.7 (2006): 1920–1934.
41. Zhang, Xiaoyu, Erin M. Hicks, and Richard P. Van Duyne. "Advances in contemporary nanosphere lithographic techniques". *J. Nanosci. Nanotechnol.* 6.7 (2006): 1920–1934.
42. Sadrolhosseini, Amir Reza, Mohd Adzir Mahdi, Farideh Alizadeh, and Surya Abdul Rashid. "Laser ablation technique for synthesis of metal nanoparticle in liquid". *Laser Technol. Appl.* (2019): 63–83.
43. de Oliveira, Paulo F. M. "Challenges and opportunities in the bottom-up mechanochemical synthesis of noble metal nanoparticles". *J. Mater. Chem A* 8.32 (2020): 16114–16141.
44. Shah, Khurshed A., and Bilal A. Tali. "Synthesis of carbon nanotubes by catalytic chemical vapor deposition: a review on carbon sources, catalysts, and substrates". *Mater. Sci. Semicond. Process.* 41 (2016): 67–82.
45. Vijayakumar, Veena, Sushanta K. Samal, Smita Mohanty, and Sanjay K. Nayak. "Recent advancements in biopolymer and metal nanoparticle-based materials in diabetic wound healing management". *Int. J. Biol. Macromol.* 122 (2019): 137–148.
46. Parashar, Mritunjaya, Vivek Kumar Shukla, and Ranbir Singh. "Metal oxides nanoparticles via sol-gel method: a review on synthesis, characterization, and applications". *J. Mater. Sci. Mater. Electron.* 31 (2020): 3729–3749.
47. Mohammadi, Somaieh, Adam Harvey, and Kamelia V. K. Boodhoo. "Synthesis of TiO2 nanoparticles in a spinning disc reactor". *J. Chem. Eng.* 258 (2014): 171–184.
48. Majerič, Peter, and Rebeka Rudolf. "Advances in ultrasonic spray pyrolysis processing of noble metal nanoparticles". *Materials* 13.16 (2020): 3485.
49. Thakkar, Kaushik N., Snehit S. Mhatre, and Rasesh Y. Parikh. "Biological synthesis of metallic nanoparticles". *Nanomed. Nanotechnol. Biol. Med.* 6.2 (2010): 257–262.
50. Gericke, M., and A. Pinches. "Biological synthesis of metal nanoparticles". *Hydrometallurgy* 83.1–4 (2006): 132–140.
51. Grassian, Vicki H. "When size really matters: size-dependent properties and surface chemistry of metal and metal oxide nanoparticles in gas and liquid phase environments". *J. Phys. Chem. C* 112.47 (2008): 18303–18313.
52. Rao, C. N. R., U. Kulkarni, P. John Thomas, and Peter P. Edwards. "Size-dependent chemistry: properties of nanocrystals". *Chem. Eur. J.* 8.1 (2002): 28–35.
53. Grassian, Vicki H. "When size really matters: size-dependent properties and surface chemistry of metal and metal oxide nanoparticles in gas and liquid phase environments". *J. Phys. Chem. C* 112.47 (2008): 18303–18313.
54. Horie, Masanori, Katsuhide Fujita, Harushisa Kato, Shigehisa Endoh, and Keiko Nishio. "Association of the physical and chemical properties and the cytotoxicity of metal oxide nanoparticles: metal ion release, adsorption ability, and specific surface area". *Metallomics* 4.4 (2012): 350–360.
55. Wang, Lu, Morteza Hasanzadeh Kafshgari, and Michel Meunier. "Optical properties and applications of plasmonic-metal nanoparticles". *Adv. Funct. Mater.* 30.51 (2020): 2005400.
56. Zhang, Jin Z., and Cecilia Noguez. "Plasmonic optical properties and applications of metal nanostructures". *Plasmonics* 3 (2008): 127–150.
57. Wessels, Jurina M. Heinz-Georg Nothofer, William E. Ford, Florian Von Wrochem, Frank Scholz, Tobias Vossmeyer, Andrea Schroedter, Horst Weller, and Akio Yasuda. "Optical and electrical properties of three-dimensional interlinked gold nanoparticle assemblies". *JACS* 126.10 (2004): 3349–3356.
58. Asadikia Amir, Seyed Ali Agha Mirjalily, Navid Nasirizadeh, and Hadi Kargarsharifabad. "Characterization of thermal and electrical properties of hybrid nanofluids prepared with multi-walled carbon nanotubes and Fe2O3 nanoparticles". *ICHMT* 117 (2020): 104603.
59. Coetzee, Divan, et al. "Influence of nanoparticles on thermal and electrical conductivity of composites". *Polymers* 12.4 (2020): 742.
60. Klębowski, Bartosz, Joanna Depciuch, Magdalena Prlinska-Wojtan, and Jarek Baranet al. "Applications of noble metal-based nanoparticles in medicine". *Int. J. Mol. Sci.* 19.12 (2018): 4031.
61. Shnoudeh, Abeer Jabra, Isla, Hamad, Ruwaida W. Abdo, Lana Qadumii, Abdulmutallab Yousaef Jaber, Haiba Salim Surchi, and Sahd Z. Alkelany. "Synthesis, characterization, and applications of metal nanoparticles". In *Biomaterials and Bionanotechnology*. Academic Press, 2019. 527–612.
62. Pareek, Vikram, Arpit Bhargava, Rinki Gupta, Navin Jain, and Jitendra Panwar. "Synthesis and applications of noble metal nanoparticles: a review". *ASEM* 9.7 (2017): 527–544.
63. Conde, João, Gonçalo Doria, and Pedro Baptista. "Noble metal nanoparticles applications in cancer". *Drug. Deliv. Technol.* 2012 (2012).
64. Pareek, Gupta Rinki, and Bhargava Navin. "Synthesis and applications of noble metal nanoparticles: a review". *ASEM* 9.7 (2017): 527–544.
65. Campelo, J. M., D. Luna, R. Luque, J. M. Marinas & A. A. Romero. Sustainable preparation of supported metal nanoparticles and their applications in catalysis. *ChemSusChem* (2009): 2(1), 18–45.
66. Campelo, Rafael Luque, and Antonio A. Romero. "Sustainable preparation of supported metal nanoparticles and their applications in catalysis". *ChemSuchem*.
67. Astruc, Didier. "Transition-metal nanoparticles in catalysis: from historical background to the state-of-the-art". *Nanopart. Catal.* 1 (2008): 1–48.
68. Xia, Younan, Hong Yang, and Charles T. Campbell. "Nanoparticles for catalysis". *Acc. Chem. Res.* 46.8 (2013): 1671–1672.
69. Chandran, S. P., Chaudhary, M., Pasricha, R., Ahmad, A. & Sastry, M. "Synthesis of gold nanotriangles and silver nanoparticles using Aloe vera plant extract". *Biotechnol. Prog.*, 22 (2006), 577–583.
70. Ishida, T. and Haruta, M. Gold Catalysts: Towards Sustainable Chemistry. *Angew. Chem.* 119 (2007): 7288–7290; *Angew. Chem. Int. Ed.* 46 (2007): 7154–7156.
71. Doria, Gonçalo, Joao Conde, Bruno Veigas, Leticia Giestas, Carina Almeida, Maria Aassuncao, Joao Rosa, and Pedro V. Baptista. "Noble metal nanoparticles for biosensing applications". *Sensors* 12.2 (2012): 1657–1687.
72. Wang, Jing. "Electrochemical biosensing based on noble metal nanoparticles". *Microchim. Acta* 177 (2012): 245–270.
73. Malekzad, Hedieh, Parham shaadi Zangabad, Hamed Mirshekari, Mahdi Karimi, and Michael R. Hamblin. "Noble metal nanoparticles in biosensors: recent studies and applications". *Nanotechnol. Rev.* 6.3 (2017): 301–329.
74. Kaushal, Sandeep, Sitantu Sekhar Nanda, Shashadhar Samal, and Dong Kee Yi. "Strategies for the development of metallic-nanoparticle-based label-free biosensors and their biomedical applications". *ChemBioChem* 21.5 (2020): 576–600.
75. Holzinger, Michael, Alan Le Goff, and Serge Cosnier. "Nanomaterials for biosensing applications: a review". *Front. Chem.* 2 (2014): 63.
76. Pradeep, T. "Noble metal nanoparticles for water purification: a critical review". *Thin Solid Films* 517.24 (2009): 6441–6478.

77. Sayan, B., Saha Indrani, Mukhopadhyay Aniruddha, and C Dhurbajyothi. "Role of nanotechnology in water treatment and purification: potential applications and implications". *Int. J. Chem. Sci. Technol.* 3.3 (2013): 59.
78. Savage, Nora, and Mamadou S. Diallo. "Nanomaterials and water purification: opportunities and challenges". *J. Nanopart. Res.* 7 (2005): 331–342.
79. Sreeprasad, Theruvakkattil Sreenivasan, and Shihabudheen M. K. P. Lisha. "Reduced graphene oxide–metal/metal oxide composites: facile synthesis and application in water purification". *J. Hazardous Mater.* 186.1 (2011): 921–931.
80. Theron, Jacques, Joseph Adrian Walker, and Thomas Eugene Cloete. "Nanotechnology and water treatment: applications and emerging opportunities". *Crit. Rev. Microbiol.* 34.1 (2008): 43–69.
81. Kar, Soumitra, M. Subramanian, A. K. Ghosh, R. C. Bindal, S. Prabhakar, and J. Nuwad. "Potential of nanoparticles for water purification: a case study on anti-biofouling behavior of metal-based polymeric nanocomposite membrane". *Desalinat. Water Treat.* 27 (2011).
82. Kumar, Antul. "Metal-based nanoparticles, sensors, and their multifaceted application in food packaging". *J. Nanobiotechnol.* 19.1 (2021): 256.
83. Hoseinnejad, Mahmoud, Seid Mahdi Jafari, and Iman Katouzian. "Inorganic and metal nanoparticles and their antimicrobial activity in food packaging applications". *Crit. Rev. Microbiol.* 44.2 (2018): 161–181.
84. Adeyemi, Jerry O., and Olaniyi A. Fawole. "Metal-based nanoparticles in food packaging and coating technologies: a review". *Biomolecules* 13.7 (2023): 1092.
85. Jagadish, Krishnegowda, Yallappa Shiralgi, Bananakere N. Chandrashekar, and Bhadrapura L. Dhananjaya. "Ecofriendly synthesis of metal/metal oxide nanoparticles and their application in food packaging and food preservation". In *Impact of Nanoscience in the Food Industry.* Academic Press, 2018. 197–216.
86. Jafarzadeh, Shima, and Seid Mahdi Jafari. "Impact of metal nanoparticles on the mechanical, barrier, optical and thermal properties of biodegradable food packaging materials". *Crit. Rev. Food Sci. Nutr.* 61.16 (2021): 2640–2658.
87. Garcia, Coralia V., Gye Hwa Shin, and Jun Tae Kim. "Metal oxide-based nanocomposites in food packaging: applications, migration, and regulations". *Trends Food Sci. Technol.* 82 (2018): 21–31.
88. Costa, Cristina, A. Conte, M. Alessandro, and D. Nobile. "Use of metal nanoparticles for active packaging applications". In *Antimicrobial Food Packaging.* Academic Press, 2016. 399–406.
89. Shurygina, Irina A., and Michael G. Shurygin. "Nanoparticles in wound healing and regeneration". *Metal Nanopart. Pharma* (2017): 21–37.
90. Kwiatkowska, Angelika, Monika Drabik, Agata Lipko, Anna Grzeczkowicz, Radoslaw Stachowiak, Anna Marszalik, and Ludomira H. Granica "Composite membrane dressings system with metallic nanoparticles as an antibacterial factor in wound healing". *Membranes* 12.2 (2022): 215.
91. Krishnan, Priya Dharshini, Dominik Banas, Ramya Devi Durai, Danili Kabanov, Bozena Hosnedlova, Marta Kepinska, Carlos Fernandez, Awaiz Farid, Jiri Sochor, Vedha Hari B. Narayana, and Rene Kizek. "Silver nanomaterials for wound dressing applications". *Pharmaceutics* 12.9 (2020): 821.
92. Mendes, Carolini. "The use of metallic nanoparticles in wound healing: new perspectives". *Int. J. Mol. Sci.* 23 (2022): 15376.
93. Kumar, Sathish Sundar Dhilip. "Recent advances on silver nanoparticle and biopolymer-based biomaterials for wound healing applications". *Int. J. Biol. Macromol.* 115 (2018): 165–175.
94. Tan, Hong Wei. "Metallic nanoparticle inks for 3D printing". *ASEM* 5.5 (2019): 1800831.
95. Ko, Seung H., Inkyu Park, Heng Pan, Costas P. Grigoropoulos, and Albert P. Pisano. "Direct nanoimprinting of metal nanoparticles for nanoscale electronics fabrication". *Nano Lett.* 7.7 (2007): 1869–1877.
96. Balantrapu, Krishna, and Dan V. Goia. "Silver nanoparticles for printable electronics and biological applications". *JMR* 24.9 (2009): 2828–2836.
97. Männl, U., B. Magunje, M. Harting, D. T. Britton, and S. Jones. "Nanoparticle composites for printed electronics". *Nanotechnology* 25.9 (2014): 094004.
98. Xerox Research Centre of Canada. *Silver Nanoparticle & Silver Ink.* https://xrcc.external.xerox.com/electronic-materials.html (accessed January 2019).
99. Hsien-Hsueh L., Kan-Sen C., and H. Kuo-Cheng. "Inkjet printing of nano-sized silver colloids. Nanotechnology", 16 (2005): 2436–2441.

## LIST OF ABBREVIATIONS

| | |
|---|---|
| **0D** | Zero Dimension |
| **1D** | One Dimension |
| **2D** | Two Dimension |
| **3D** | Three Dimension |
| **Ag** | Silver |
| **Al** | Aluminium |
| **Au** | Gold |
| **CNTs** | Carbon Nanotubes |
| **CT** | Computed Tomography |
| **Cu** | Copper |
| **CVD** | Chemical Vapour Deposition |
| **Fe** | Iron |
| **GR** | Graphene |
| **HER** | Hydrogen Evolution Reaction |
| **HIV** | Human Immunodeficiency Virus |
| **LED** | Light-Emitting Diode |
| **LSPR** | Localized Surface Plasmon Resonance |
| **MAP** | Modified Atmosphere Packaging |
| **MEMS** | Micro Electro Chemical Systems |
| **MRI** | Magnetic Resonance Imaging |
| **MWNTs** | Multiple-Walled Nanotubes |
| **NCs** | Nanocrystals |
| **Ni** | Nickel |
| **NMs** | Nanomaterials |
| **NPs** | Nanoparticles |
| **NRs** | Nanorods |
| **NTs** | Nanotubes |
| **ORR** | Oxygen Reduction Reaction |
| **Pl** | Palladium |
| **Pt** | Platinum |
| **ROS** | Reactive Oxygen Species |
| **SDR** | Spinning Disc Reactor |
| **SERS** | Surface-Enhanced Raman Scattering |
| **SPR** | Surface Plasmon Resonance |
| **SWNTs** | Single-Walled Nanotubes |
| **TB** | Tuberculosis |
| **Ti** | Titanium |
| **$TiO_2$** | Titanium Dioxide |
| **Zn** | Zinc |
| **ZnO** | Zinc Oxide |

# 2 Chemical and Physical Methods for Synthesis of Metal and Metal Oxide Nanoparticles

*Aparna V. M., Aparna Raj, and C. Sudarsanakumar*
School of Pure and Applied Physics, Mahatma Gandhi University, Kottayam, Kerala, India

## 2.1 INTRODUCTION

Nanoparticles have emerged as a valuable tool in material research over the years due to their broad spectrum of applications. The ease with which one can control their properties by changing their size, shape, and functionalization makes them more desirable. Physics Nobel Laureate Richard P. Feynman gave the first educational talk about nanotechnology in his famous lecture "There's Plenty of Room at the Bottom" at December 1959 meeting of the American Physical Society. This was the beginning of manipulating materials on the atomic scale to make them suitable for applications of interest.[1,2] This small scale has opened tremendous possibilities and led to revolutionary discoveries in various fields, from electronics and medicine to material science and environmental sustainability. Nanomaterials have grown into an exciting group of materials with various beneficial applications and are in high demand worldwide. The length of a nanometer can be defined as the length of 10 hydrogen atoms aligned, which is equal to $10^{-9}$ m. Materials with one of their dimensions in the 1–100 nm range are often called nanomaterials.[3] Depending on the type of substances, nanoparticles have far-reaching applications like drug delivery, biosensors, bioimaging, molecular tagging, food technology, textile manufacturing, anti-microbial coating, quantum computers, quantum lasers, energy, and the environment. However, the usage of nanoparticles dates back to ancient times. The ancient Egyptians used Pb nanoparticles in their hair-dyeing formulas. The Lycurgus cup used by ancient Romans shows color variation due to the presence of Ag and Au nanoparticles, and a colorant named Purple of Cassius used in the 17$^{th}$ century was made up of gold nanoparticles and tin dioxide.[1] They are unique since their size can affect the physico-chemical properties of their parent material. They have properties such as high surface-to-volume ratio, surface charge with density, high surface area, quantum confinement effect, and high reactivity.

The size, morphology, and chemical composition of nanoparticles allow for their division into different classes. Depending on the shape, nanoparticles can be 0D, 1D, 2D, or 3D. Carbon-based, ceramic, semiconductor, polymeric metal, and metal oxide nanoparticles are some of the well-known divisions among them. Among these, metal and metal oxide nanoparticles are worth special attention due to their unique optical, electronic, biological, and catalytic properties. The optoelectronic characteristics of these nanoparticles are mainly due to confined surface plasmon resonance.[4] Metals like aluminum, cobalt, cadmium, copper, lead, zinc, silver, and gold are commonly used to synthesize nanoparticles. Among these, gold has been used in cancer treatment; silver is a key component in the field of electronics; palladium and platinum are used as catalysts in various chemical reactions; and iron and zinc are used in the field of water purification, making them an irreplaceable tool for environmental cleanup. In the case of metal oxides, $TiO_2$ is the central part of the development of solar cells and lithium batteries due to its photoelectric properties and storage capacity.[5] ZnO is used in water purification, and iron oxide nanoparticles play a crucial role in drug delivery when coated with suitable polymers. Apart from these, metal and metal oxide nanoparticles are also used for gas sensors, monitoring air quality, and food packaging.[4,6]

Nanoparticles are synthesized in a controlled atmosphere to acquire successful and reproducible functionalities for specific applications. Top-down and bottom-up are two synthesis approaches for nanoparticles. In the top-down method, a large component is broken down into smaller ones, while in the bottom-up process, the basic principle is molecular condensation. Physical methods are energy consuming; they require high vacuum and temperature in addition to expensive types of equipment.[7] The conventional wet chemical methods involve developing nanoparticles in a liquid medium with reducing agents and a metal salt. The reaction mixture is also provided stabilizing agents to prevent the agglomeration of nanoparticles. Wet chemical methods usually increase particle reactivity and toxicity due to the absorption of these chemicals into the surface of nanoparticles and produce toxic wastes that might harm human health and the environment.[8] Although

DOI: 10.1201/9781032660639-2

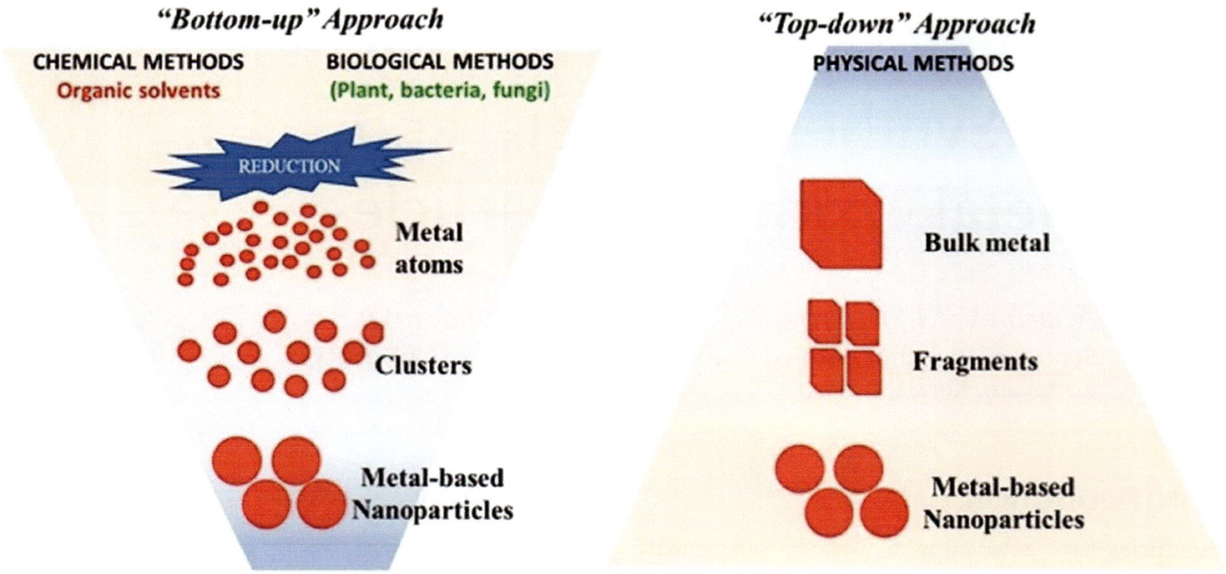


**FIGURE 2.1** Schematic diagram for top-down and bottom-up approaches to nanoparticle synthesis.

**Source: Reprinted from Ref.[10] Copyright © the author(s) 2022.**

metal nanoparticles have been made via physical and chemical methods, recent developments in this area show the importance of biological agents in the synthesis of nanoparticles. It is a cost-effective method for producing metal nanoparticles and a hazard-free process. The possibilities in the field of metal and metal oxide nanoparticles are endless. Due to their distinct features, they have applications in the areas of electronics, medicine, and environmental cleanup. In the future, we may expect many groundbreaking discoveries and triumphs in the field of nanotechnology.[9]

## 2.2 TOP-DOWN AND BOTTOM-UP APPROACHES

Depending on the size of the starting material used for the preparation, nanoparticle synthesis is mainly categorized into two methods:

- Top-down approach
- Bottom-up approach

In top-down methods, bulk materials are used for the synthesis of nanomaterials using different mechanical and lithographical techniques such as etching, cutting, ball milling, photolithography, and electron beam lithography.[5] Figure 2.2 represents different synthesis methods for the preparation of nanoparticles. The top-down method belongs to the physical method of nanoparticle synthesis, where mechanical pressure, high-energy radiation, thermal energy, or electrical energy is applied to reduce the size of the bulk material. This method produces high-purity nanoparticles with excellent physical properties on a large scale. However, top-down methods such as ball milling demand high energy input and extended milling durations.[3] Figure 2.1 shows a schematic diagram of top-down and bottom-up approaches for nanoparticle synthesis.[10]

In the bottom-up approach, nanoparticles are synthesized by joining small molecules, atoms, or particles. In this method, nanostructured building blocks are initially formed and then assembled to produce desired nanoparticles. This method is mainly used to obtain nanostructures with fewer defects, more homogenous chemical composition, and the ability to control the particle size. Nucleation is the main working principle behind this approach.[1] Table 2.1 lists the advantages and disadvantages of the top-down and bottom-up approaches.

## 2.3 PHYSICAL METHODS

Physical methods involve transformation the bulk material into the nanoscale via their interaction with photons (laser ablation), heat (inert gas condensation), ions (sputtering), or even mechanical milling. The commonly used physical methods for nanoparticle synthesis are mechanical and physical vapor deposition methods. Mechanical methods involve mechanically grinding the bulk material to produce nanomaterials. Physical vapor deposition involves heating the material in a tube furnace and cooling the vapor to produce nanoparticles. This method can be considered green synthesis since it has no by-products harmful to the environment. The advantages of physical methods over other synthesis methods are the absence of solvent contamination and the uniformity of the obtained nanoparticles. However,

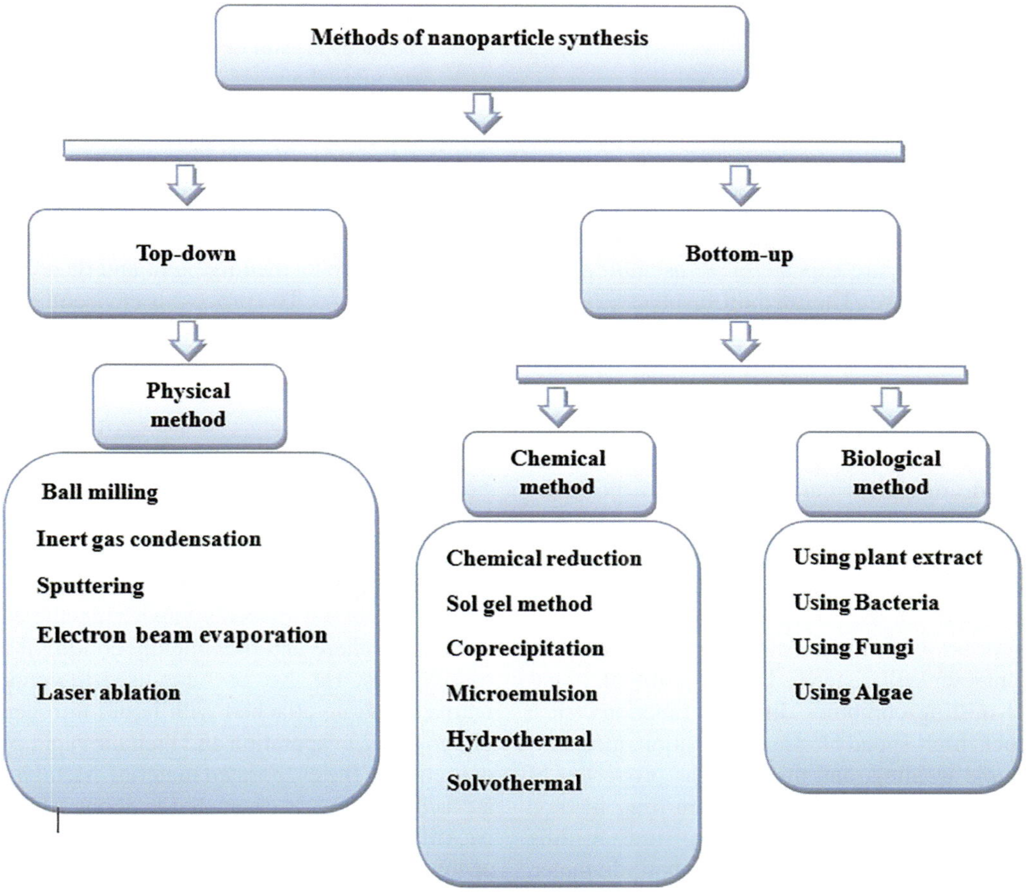


**FIGURE 2.2** Schematic diagram of synthesis methods to produce nanoparticles.

**TABLE 2.1**
**Advantages and Disadvantages of Top-Down and Bottom-Up Approaches**

| Synthesis Method | Advantages | Disadvantages |
|---|---|---|
| Top-down | • Large-scale production<br>• Chemical purification is not required<br>• No solvent contamination<br>• Uniform distribution<br>• Controlled surface area<br>• Control over particle size | • Broad size distribution<br>• Non-uniform particle size<br>• Presence of impurities and defects<br>• Less economical<br>• Loss of properties of bulk material<br>• Limited to certain materials |
| Bottom-up | • Control over size and composition<br>• Narrow size distribution<br>• Cheaper technique<br>• Structural diversity<br>• Less energy intensive<br>• Less damage to nanoparticles | • Large-scale production is difficult<br>• Chemical purification is required<br>• Requires a deeper understanding of chemical reaction<br>• Batch-to-batch variation<br>• Limited control over shape |

they also have disadvantages, like high energy consumption and the large size of furnace tubes.

### 2.3.1 Ball Milling

Ball milling/mechanical milling is yet another cost-effective method to produce nanomaterials. John Benjamin developed the synthesis of oxide dispersion-strengthened alloys in 1970, and they are capable of withstanding high temperatures and pressure. The method involves grinding or milling materials in a ball mill to reduce their size to the nanoscale. In this method, a ball mill is employed, which consists of a rotating drum filled with milling balls made of steel or tungsten.[11] The material to be processed along with the milling balls is placed inside the drum, and the rotation of the drum causes the balls to grind the material, leading to particle size reduction and the formation of nanoparticles. Here, the kinetic energy of the moving ball is transferred on to the material, which results in the breaking of its chemical bond and the formation of small particles.[12] During this process, collision between the tiny rigid balls in a sealed container will generate high pressure and temperature. Ball milling combines different functions, such as thermal shock, high-speed plastic deformation, mechanical grinding, cold welding, and mixing. The properties of the nanoparticles synthesized by the ball milling method depend on parameters such as the milling time, milling media, ball-to-powder ratio (usually 2:1), ball diameter, and type of milling.[13] Generally, the milling atmosphere and milling time determine the energy transfer between the balls and material to be ground and thus affect the physical and morphological properties of the resulting material. Sometimes certain chemicals are added into the milling container during the processes to increase the nanomaterials' quality and functionalize the nanoparticles.[14]

High-energy ball milling is a type of ball milling where the material placed in the container undergoes high-energy collision with the balls. In this method, a magnet is placed close to the container to apply a strong magnetic force on the magnetic milling balls. Therefore, the impact energy is much higher than conventional ball milling energy.[5] High-energy ball milling requires a longer milling time to get the material's desired structural and chemical properties. This method produces a fine, uniform dispersion of nanoparticles, which conventional synthesis methods cannot make.[15,16]

#### Advantages

- Cost effective
- Mass production
- Simple operation

#### Disadvantages

- High energy required
- Long milling time
- Contamination due to ball and milling atmosphere
- Produces crystal defects
- Irregular shape

### 2.3.2 Physical Vapor Deposition

Physical vapor deposition (PVD) is a bottom-up method of producing materials at the nanometer scale. The material is heated above its melting point and then deposited on the substrate surface. The process is environmentally friendly.[7] The most-used PVD methods for nanoparticle synthesis are

1. Inert gas condensation
2. Sputtering
3. Laser ablation
4. Electron beam evaporation

#### 2.3.2.1 Inert Gas Condensation

Inert gas condensation is a process used to synthesize nano-sized material. It involves vaporizing the material in an inert atmosphere and then rapidly cooling it using an inert gas. Usually, He, Xe, or Ar is used to condense material onto a substrate attached with liquid nitrogen.[17] It consists of two steps: evaporation and then the rapid controlled condensation of the evaporated material. Metal is evaporated in a crucible, which is evacuated to about 10–7 Torr and then backfilled with an inert gas to lower pressure. Evaporation of the metal can be achieved either by thermal evaporation or electron beam evaporation, depending on the metal.[18] Sputtering, plasma heating, laser vaporization, and electric arc discharge are often used for vaporizing metals in this technique. In the case of metals with high melting points (W, Ta, C, etc.), electron beam evaporation is used. This method can be used to synthesize nanoparticles from any materials that can be vaporized. When the evaporated metal atoms achieve supersaturation, homogeneous nucleation takes place, resulting in the formation of tiny particles. This particle experiences collision due to random motion and grows into a larger particle. Usually, helium is used in inert gas condensation techniques due to its high thermal conductivity. The particles formed are then collected on a substrate, which is cooled using liquid nitrogen. Liquid nitrogen is used to enhance the deposition efficiency.[19]

The produced nanoparticles' size, morphology, and crystallinity depend on the source temperature, forced gas flow, inert gas type, and inert gas pressure. The size of the particle formed can vary depending on the time the nuclei spend on the growth region. The pressure exerted by the inert gas also plays a vital role in determining the size and morphology. An increase in the pressure will increase the number of collisions, which may lead to agglomeration of the nanoparticles formed; pressure must be kept to a minimum to obtain nanoparticles with small size.[21] This method mainly synthesizes metal nanoparticles like manganese (Mn), nickel nanoparticles (Ni), iron and iron oxide nanoparticles, and size-controlled gold and palladium nanoparticles. Inert gas condensation produces high-purity

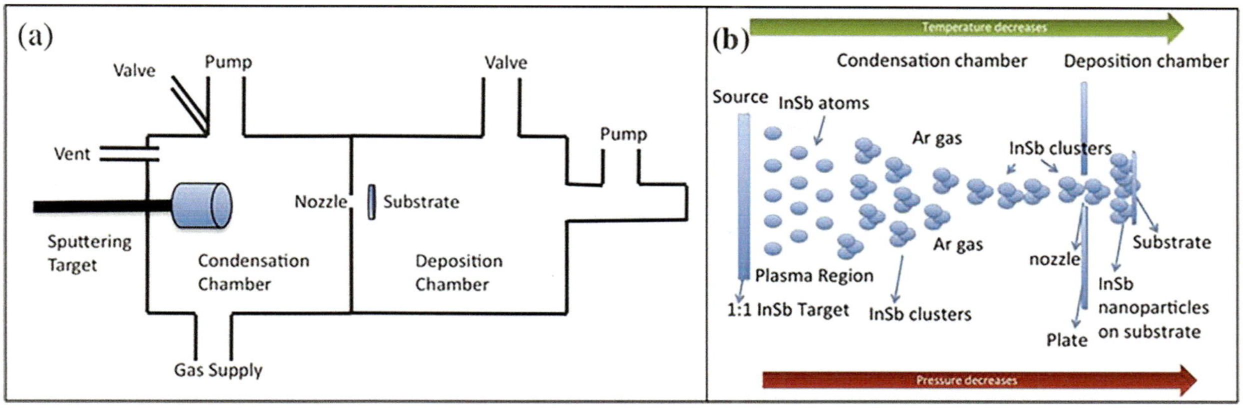


**FIGURE 2.3** (a) Schematic of the inert gas condensation instrument. (b) Diagram explaining the formation of nanoparticles during the inert gas condensation process.

**Source: Reprinted from Ref.[20] Copyright © 2015, Pandya and Kordesch.**

metal and metal oxide nanoparticles with high thermal stability.[22,23] Figure 2.3 shows the typical instrumentation for inert gas condensation and the formation of nanoparticles during the condensation process.[20]

***Advantages:*** It is possible to produce any material that can be vaporized, it is possible to control the size distribution by changing the processing parameters, no batch-to-batch variation, and less contamination.

***Disadvantages:*** Probability of agglomeration is high, high cost, maintaining a vacuum is not always easy, not suitable for all materials, and the amount of powder produced per batch is limited.

#### 2.3.2.2 Sputtering

Sputtering is a vacuum-based PVD method that produces nanomaterials by depositing materials with high melting points. It is a physical process where the interaction of matter with ions is utilized to synthesize nanomaterial. This transformation is based on the momentum transfer between ions and the target. It pulls out atoms or small clusters of the target material and directs them to a substrate, where they nucleate and grow.[24]

Sputtering is conducted in an evacuated vacuum chamber where a large negative potential (generally −100 to −1000 V) is applied to the target. An inert gas is introduced to the chamber at a typical pressure range of 70–7000 Pa. The commonly used inert gas is argon, which is naturally abundant and relatively safe. The usage of argon also increases the sputter yield. Once a voltage of sufficient magnitude is applied, positive gas ions and electrons are generated. These positive ions are accelerated to the negatively biased cathode or target material. Momentum exchange occurs between the positive ions and the atoms in the target material, resulting in the release of atoms from the target, thus creating a vapor of the target material at a low pressure. This vapor will condense atom by atom on the surface of the substrate, which can be heated to control the nucleation.[26] The main advantage of the sputtering process is that a negative potential can be applied to the substrate to control the kinetic energy of the incident ions; this can be achieved by either a direct current or a radio frequency. Thus, nucleation and growth kinetics can be controlled by changing the applied potential and substrate temperature. The sputter yield – defined as the average number of target atoms ejected per incident ion – depends on several factors: the kinetic energy of the incoming ion, the masses of both the ion and the target atoms, the surface binding energy of the target atoms, and the ion's angle of incidence.[27] Discharge voltage (V), discharge current (i), distance between the target and the substrate (d), and working pressure (P) are the main parameters that control the sputtering process. These variables can be adjusted to control the rate of sputtering, the velocity of the ejected atoms, and any potential energy losses. This method is frequently applied in analytical, etching, and thin-film deposition processes. Sputtering can be done in various ways, with the three primary methods being DC-diode, RF-diode, and magnetron sputtering.[24,28]

DC sputtering is the simplest form where a DC potential is applied between the substrate and the target. Here, plasma can be generated when sufficient electric field is applied. Despite the simplicity, DC sputtering has issues like the generation of sparks within the plasma and difficulty in maintaining glow discharge; hence, it is not suitable for insulating targets. To avoid this, RF configuration is used. In this case, the applied potential is alternating.[27] This allows operation at a lower pressure and can also be used for insulating materials. In the case of magnetron sputtering, a magnet is placed behind the target. This will increase the deposition rate and prevent the target from overheating. The operating pressure can be further reduced by using a DC, AC, or RF electric field. Under the influence of a magnetic field, the electrons follow a helical path; this will increase the ionizing collision near the target and thus lead to a higher deposition rate. Figure 2.4 illustrates the basic components of magnetron sputtering instrumentation.[25]

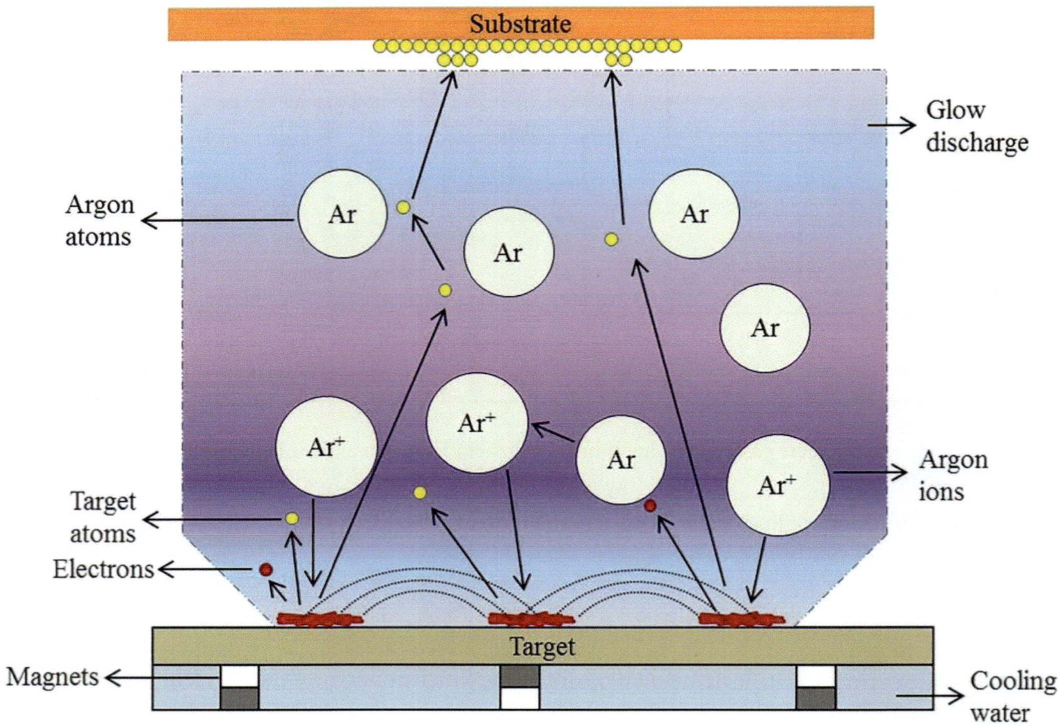


**FIGURE 2.4** Schematic diagram illustrating the basic components of a magnetron sputtering system.

**Source: Reprinted from Ref.[25] Copyright © 2014 by author(s) and Scientific Research Publishing Inc.**

Reactive sputtering is the name given to sputtering that occurs in a reactive gas ($N_2$, $O_2$) environment instead of an inert gas environment. The reactive gases initiate chemical reactions with the target vapors to form oxide or nitride films. Metal nitrides such as ScN, TiN, and AlN are sputtered using this method; in the case of boron nitride, argon and nitrogen are combined. Intense plasma particles interact with nitrogen molecules to trigger their dissociation and ionization, which results in the formation of reactive species. These reactive species can readily bond to every surface in the chamber and alter the chemistry of the metal layer as it forms.[28] In ion beam sputtering, an external ion source vaporizes the target. The main advantage of this technique is the ability to change the energy and flow of ions separately. Moreover, the ions that strike the target are made up of neutral atoms or molecules; hence, this technique applies to insulating and conducting materials.[29]

***Advantages:*** Composition of the material remains the same, large area deposition, thin film of alloys/composites of materials with a high melting point, and no damage due to X-rays generated from electron beam evaporation.

***Disadvantages:*** Complicated setup, requires vacuum, contamination due to plasma and lower vacuum level, and the film morphology may be rougher and damaged due to bombardment and cluster formation.

#### 2.3.2.3 Laser Ablation

Laser ablation is a versatile technique for the synthesis of nanoparticles. The term "laser ablation" is used to differentiate the nonequilibrium vapor/plasma conditions formed at the surface by an intense laser pulse from "laser evaporation", which is the heating and evaporation of material in thermodynamic equilibrium.[30] During the laser ablation process, a laser beam (continuous wave or pulsed laser) is used as an energy source to ablate material from a target placed in a liquid, vacuum, or gaseous environment. This ablated material is then condensed to form nanoparticles. Nd: YAG (neodymium-doped yttrium aluminum garnet) laser at 106 μm output and its harmonic Ti: Sapphire (Titanium-doped sapphire) laser and copper vapor lasers are mainly used for laser ablation technique. The experimental setup is simpler than many PVD methods, yet the laser ablation process is rather complicated.[31] The experimental design comprises a pulsed laser, a vacuum chamber with a spinning target holder, and a substrate holder with a heater. It may be done in a vacuum at various substrate temperatures. A tiny lens is used to concentrate laser pulses on the surface of a solid object during film deposition. This will produce a laser spot with high irradiance and a local electric field. The local electric field is strong enough to promote atomic bond breaking and dielectric breakdown of the target materials, hence initiating vaporization of the target materials. This vapor will absorb the remaining portion of the laser pulse, and the electric field of the pulse will accelerate electrons and ions inside the vapor, causing fast heating and the formation of laser plasma.[32,33]

Pulsed laser ablation in the liquid phase has gained popularity for generating monodisperse colloidal nanoparticles, as it provides a chemical-free environment that forms clean nanoparticles. Here, the plasma plume formed is further confined by the high pressure of surrounding liquid. Nanoparticles formed by pulsed laser ablation into liquid have catalytic applications.[34] Since the catalytic efficiency of metal nanoparticles depends on the surface features, a surface without capping agents is highly recommended. Nanoparticles synthesized by this method possess higher catalytic activity than their chemically synthesized counterparts.[30] Metal nanoparticles, carbon nanomaterials, oxide composites, and ceramics can be produced through this technique.[35] Figure 2.5 shows a schematic diagram of nanoparticle synthesis using the laser ablation technique.[36]

#### 2.3.2.4 Electron Beam Evaporation

Electron vapor deposition is a vacuum-based physical vapor deposition method for synthesizing metal and metal oxide nanoparticles. Both thin films and nanoparticles can be prepared using this method.[9] Here, a metal target is hit with high-energy electrons to generate vaporized atoms or molecules that are then condensed on a substrate to form nanoparticles. The experimental setup consists of a metal target, a vacuum chamber, an electron gun, and a substrate. The target, the metal to be deposited on the substrate, is placed in the vacuum chamber. The chamber is maintained at a low pressure using a high vacuum. The electron gun produces a stream of high-energy electrons. These electrons hit the target, causing the evaporation of the metal target. The vaporized metal atoms travel across the vacuum chamber and deposit on the substrate.[37] To synthesize metal oxide nanoparticles, an oxygen-containing gas such as water vapor is introduced in the vacuum chamber during the deposition. The metal atoms will react with oxygen to form metal oxides. It is possible to modify the properties of nanoparticles to fit a particular need by altering reaction conditions and parameters such as substrate temperature, pressure, deposition time, and choice of metal target. Compared to thermal evaporation, this method is helpful for the deposition of materials with high melting points and has a high deposition rate.[5,38]

## 2.4 CHEMICAL METHODS

The chemical method involves reducing a metal salt or precursor into nanoparticles using a reducing agent. This method usually involves organic and inorganic substances for reducing and stabilizing nanoparticles. Simple and inexpensive instrumentation and no need for high vacuum, temperature, or pressure are the main advantages of this method.[3] The biological method is also considered a subset of the chemical method where the reducing agent used is naturally occurring biological agents instead of chemicals.[39] In addition, capping of nanoparticles with foreign molecules is possible, which may help to reduce the toxicity. Some of the most used chemical methods are

- Solgel method
- Chemical reduction
- Hydrothermal method
- Microemulsion technique
- Coprecipitation
- Biological method

### 2.4.1 Sol-Gel Method

The sol-gel method is a synthesis strategy for high-purity metal oxide nanoparticles and mixed oxide composites. This method provides precise control over the size and surface features of the nanoparticles.[40] In this method, a network is formed by colloidal suspension and gelatine. Metal alkoxides and alkoxysilanes are used as the precursors for colloids. Tetraethoxysilane is commonly used to form silica gel. Nanoparticle synthesis using the sol-gel method is a five-step process that includes hydrolysis, condensation, aging, drying, and thermal decomposition.[41] Hydrolysis is when the metal precursor dissolves in water or alcohol. Precursors for various metals like aluminum and titanium are insoluble in water; in these cases, oxygen, which is necessary for the formation of metal oxides, is provided by organic solvents (e.g. alcohol). Water-mediated synthesis is generally termed the aqueous sol-gel method, and if the

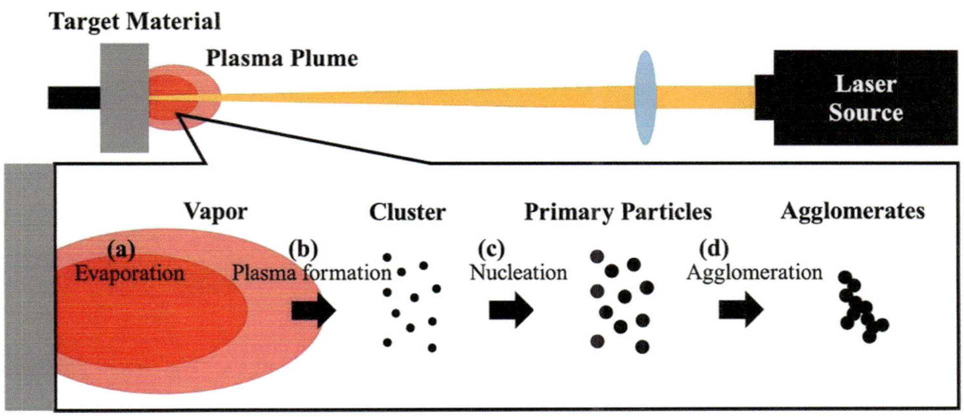


FIGURE 2.5 Schematic diagram of nanoparticle synthesis using laser ablation.

Source: Reprinted from Ref.[36] *Copyright © 2017 The authors*

medium is organic solvents, it is called the nonaqueous sol-gel method.[42] In addition to water and alcohol, an acid or a base can also help hydrolysis of metal precursors. The chemical reaction for the hydrolysis is given as:

**M-OR + $H_2O \rightarrow$ MOH + ROH (hydrolysis)**

**where M is metal, and R is an alkyl group ($C_nH_{2n+1}$)**

The second step removes water molecules to form metal oxide linkages. The process of elimination of water/alcohol molecules is called condensation. The general chemical formula for the condensation is as follows:

**M-OH + XO-M $\rightarrow$ M-O-M + XOH (condensation)**

**where M is Metal and X is Hydrogen or alkyl group**

Condensation results in the formation of gel, a porous liquid-like structure. The size and cross-linking of the colloidal particle depend mainly on the precursor and pH of the solution. A catalyst is sometimes added to increase the reaction rate at a particular pH. The aging process again changes the gel's structure and properties, resulting in the gel's thickening and reduced porosity. In the drying process, water or alcohol molecules are uncoupled from the gel, which again alters its structure.[44–46] Different drying processes include atmospheric/thermal, supercritical, and freeze-drying. The type of drying process has a unique impact on the gel structure. Heating the gel at a high temperature may decrease the pores in the gel and cause densification. In the case of freeze drying, this densification is much lower than in thermal drying. In the case of supercritical drying, the original structure of the gel remains almost unchanged. Finally, calcination or thermal treatment is employed to remove the residual water content from the sample. Calcination temperature is an essential parameter in determining the pore size and density of the material.[40] Figure 2.6 illustrates different stages in the sol-gel synthesis process of nanoparticles.[43] Sol-gel is one of the easy and coherent methods to synthesize metal oxide nanoparticles. This method mainly synthesizes $TiO_2$, ZnO, $ZnO_2$, $SnO_2$, $WO_3$ nanoparticles, and magnetic nanoparticles such as $Fe_3O_4$ nanoparticles for different applications.[40,47]

### 2.4.2 Chemical Reduction

One of the most common synthesis methods for metal nanoparticles is chemical reduction, where we use organic and inorganic reducing agents to reduce the metal salt into zero-valent metal ions, which then combine to form metal nanoparticles.[48] Reducing agents like sodium borohydride, sodium hydroxide, ethylene glycol, and sodium citrates convert metal ions into nanoparticles in a suitable medium. As these nanoparticles can bind together, stabilizing agents are used to avoid the agglomeration of dispersive nanoparticles during preparation. The formation mechanism of colloidal nanoparticles from metal ions consists of two steps: nucleation and growth. Nucleation requires high activation energy, but the activation energy required for the growth mechanism is minimal.[49] Particle size and shape can vary depending on nucleation and growth rate. We can change the size and shape of the nanoparticle by changing the reaction parameters such as temperature and concentration of reducing agent as well as metal salt, pH of the reaction medium, and reaction time; generally, particle size has a linear relation with reaction time and temperature. An increase in these parameters will increase the size of nanoparticles due to the production of more monomers and an increase in reaction rate, respectively.[50]

Among the metal nanoparticles synthesized via chemical reduction, silver nanoparticles are worth attention due to their applications in the biomedical field, especially their antimicrobial and anti-viral properties. The antibacterial activity of silver nanoparticles has been confirmed in Gram-negative and Gram-positive bacteria.[48] Anti-fungal

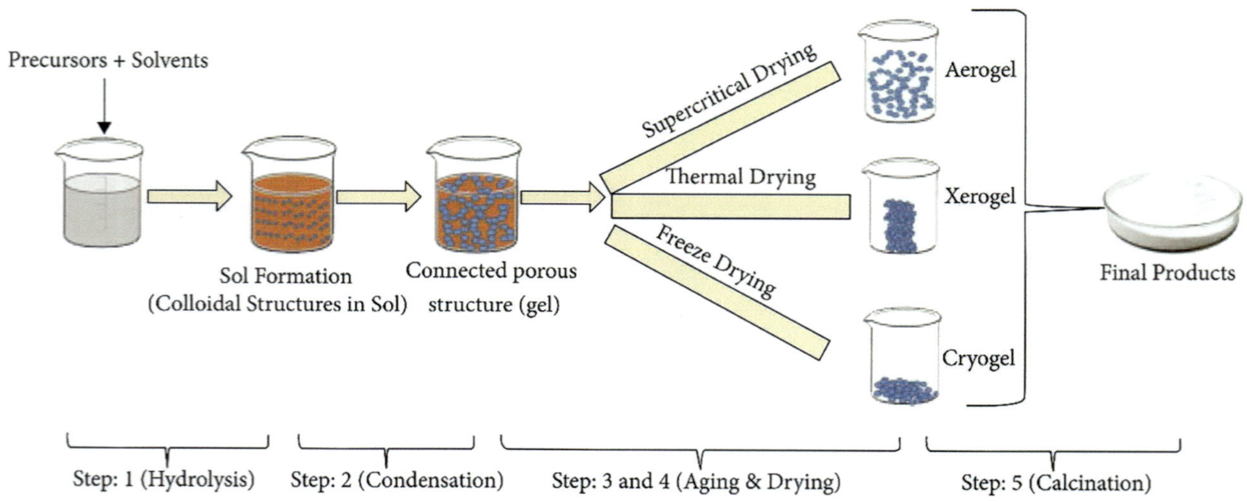


FIGURE 2.6 Schematic of different stages of sol-gel process.

Source: Reprinted from Ref.[43] *Copyright © 2021 Dmitry Bokov* et al.

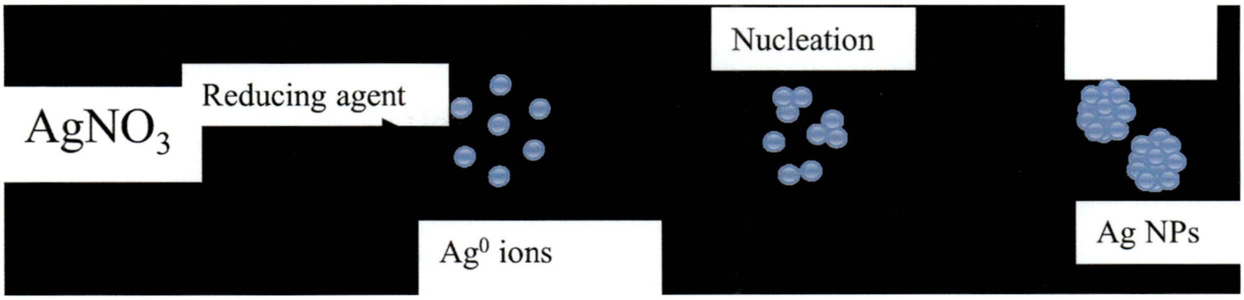


**FIGURE 2.7** Formation of Ag nanoparticles via chemical reduction.

and wound-healing properties of silver nanoparticles are also reported. They have applications in sunscreen lotion, burn treatment, wound dressing, textiles, dental materials, biosensor material, bone implants, and medical device coatings. Sodium borohydride ($NaBH_4$), hydrazine ($N_2H_4$), and formaldehyde can be used as reducing agents.[51] Figure 2.7 represents the basic mechanism involved in the synthesis of silver nanoparticles using chemical reduction.

Chemical reduction comes with difficulties, such as the need for large amounts of solvents and pricey reagents. The reduction process may require relatively high temperatures, and the presence of other components makes the separation of nanoparticles difficult. The presence of hazardous reducing and capping agents also increases the risk. Since these are primarily made for medical use, we prefer a green method to reduce and stabilize nanoparticles by substituting environmentally safe materials for hazardous reducing agents.[52–54]

### 2.4.3 Hydrothermal and Solvothermal Methods

The hydrothermal method is one of the most-used techniques for nanoparticle preparation. The materials with high vapor pressure around the melting point are expected to form nanoparticles only through this method.[55] Here, the nanoparticles are prepared by a heterogeneous reaction in an aqueous medium in a sealed vessel. High pressure and temperature near the critical temperature will be maintained in the vessel to complete the process.[56] In most cases, this temperature varies between 150 and 250°C. Hydrothermal synthesis is carried out in a closed container called an autoclave, a stainless-steel chamber lined with Teflon, platinum, gold, or silver to protect the autoclave from highly corrosive solvents maintained at high pressure and temperature. In hydrothermal synthesis, the materials are dissolved in water and are transferred to an autoclave. A specific temperature is maintained in the autoclave to increase the solvents' internal pressure and initiate nucleation and growth. Figure 2.8 shows a diagram of typical hydrothermal equipment.[57] This method is widely used to develop high-purity crystalline nanomaterials with desired phases, such as $CO_3O_4$, NiO, $ZrO_2$, $\alpha$-$Fe_2O_3$, and $\gamma$-$Fe_2O_3$.[58] It is broadly used in growing single crystals like quartz and many kinds of metal oxides, metal, and semiconductor powders. This method is also helpful in obtaining particles with controlled grain size, morphology, surface properties, and polymorphism by optimizing pressure, temperature, solution composition, additives, and aging time.[59]

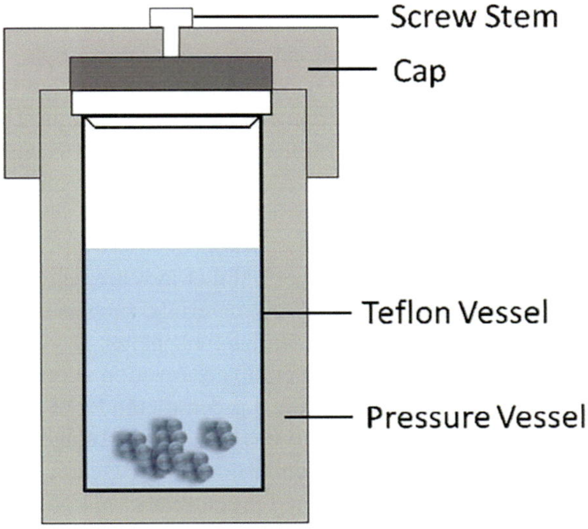


**FIGURE 2.8** Schematic diagram of hydrothermal equipment.

**Source: Reprinted from Ref.**[57] ***Copyright © 2019 by author(s) and Scientific Research Publishing Inc.***

In the case of solvothermal method, a nonaqueous medium is used. This synthesis procedure often uses organic solvents like methanol, ethanol, and polyol as the reactive medium. The reaction occurs in a pressure vessel where the solvent is heated above its boiling point. Compared to hydrothermal method, solvothermal synthesis often produces metal nanoparticles with a small size distribution at a low temperature. Microwave-assisted hydrothermal/solvothermal is an easy one-pot synthesis method that can increase the reaction rate, lead to the formation of new phases, and eliminate metastable phases. This method is used to generate various types of functional materials, such as metal oxides, metal-oxide composites, and inorganic biomaterials.[60] In addition, this method can reduce the reaction time from days to minutes and screen the experimental condition to increase yield and material properties. The nucleation and growth mechanism of nanoparticles synthesized

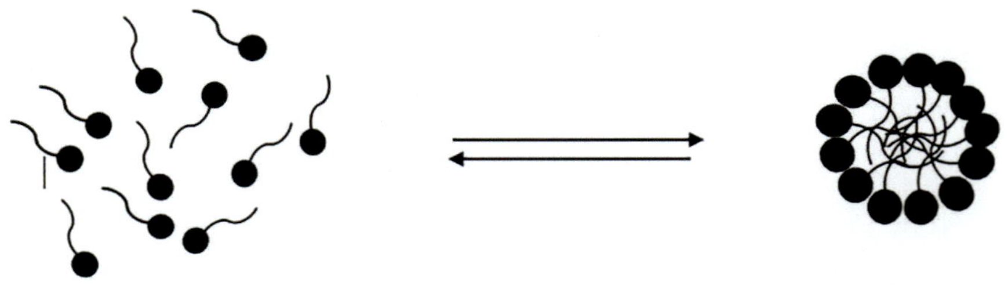


**FIGURE 2.9** Formation of reverse micelles in aqueous solution.

by microwave-assisted hydrothermal synthesis often differs from that of the conventional method. For example, the growth mechanism of $VO_2$ nanoparticles synthesized via microwave-assisted hydrothermal reaction is unique from that of the usual hydrothermal synthesis route.[61] It is widely applied to synthesize silver nanoparticles and nanostructures of Pt, Pd, Ag, and Au using polyethylene glycol or methanol as a reducing agent.

Hydrothermal and solvothermal methods are convenient for generating various nanosized materials, such as nanowires, nanorods, nanosheets, and nanospheres. It is environmentally sound since the complete reaction occurs in a closed container. Besides being a green method, the hydrothermal/solvothermal method has several advantages. It is energy efficient, used components can be recycled, it produces high-purity material, and precipitants are not always required. But at the same time, it's challenging to attain reproducibility, and the process is difficult to control.[62]

### 2.4.4 Microemulsion Techniques

An emulsion combines two immiscible liquids, mainly water and oil, in the presence of an appropriate surfactant. When the mixture of these three is mechanically agitated, a dispersion of nano-sized domains of one liquid in the other is formed.[63] The stability of the emulsion is determined by the surfactant molecules, which also decide the continuous phase. If the surfactant is hydrophilic, oil droplets will be dispersed in a continuous water phase, and if the surfactant is lipophilic, water droplets will be distributed in the oil phase. They are milky or turbid in appearance due to droplets of size ranging from 0.1 to 1 micron. If a surfactant can balance the hydrophilic and lipophilic properties at some concentration, the emulsion is called a microemulsion.[64] Microemulsions are thermodynamically stable and optically isotropic, have low interfacial tension, and have a sizeable interfacial area. Microemulsions are classified as oil in direct/water, reverse/water in oil, and bicontinuous based on the component proportion and hydrophilic-hydrophobic balance (HLB) value of the surfactant. Water in oil emulsion is also called reverse micelles; they are capable of solubilizing both hydrophilic and hydrophobic substances. The formation of reverse micelles from the spontaneous self-assembly of surfactants is shown in Figure 2.9. Microemulsions are formed only under certain conditions, and the HLB of the surfactant molecule plays an important role in selecting the components that form microemulsions. Temperature, pressure, concentration of reagents, and salt addition can greatly influence microemulsion stability.[65]

Microemulsions containing metal salt and reducing agents are mixed for nanoparticle synthesis, leading to the collision of water droplets. These collisions result from the van der Waals, osmotic, and elastic force between the micelles, inducing the mixing of reactants. The reaction between the solubilizers reduces the metal salt, facilitating the initial nucleation. Both nucleation and particle growth happen inside the micelles. The developed nanoparticle's size and morphology depend on the nanodroplets' size and shape, the type of surfactants, and the stabilizing molecules attached to the surfactants. In another method called single microemulsion, the metal salt is solubilized in the reverse micelle, and the reducing agent is added directly to the microemulsion.[66] The formation of nanoparticles in the microemulsion is diagrammatically represented in Figure 2.10.[67] Since the inorganic metal salts are mostly water soluble, the reverse micelle process is the most suitable method for the nanoparticles of such materials. Many surfactants, such as cetyltrimethylammonium bromide (CTAB), bis(2-ethylhexyl)sulfosuccinate (AOT), sodium dodecyl benzene sulfonate (SDBS), Triton X-100, sorbitan monooleate Span80, nonylphenol ether, and lauryl sodium sulfate (SDS), have been used to form reverse micelles. However, the most-used surfactant is AOT. $Fe_2O_3$ has been synthesized using the reverse micelle method by mixing water and a stoke solution of AOT in n-heptane.[63] The size of the nanoparticle is certainly affected by the size of emulsion drops, which in turn depends on the viscosity of the mixture. Thus, viscosity, temperature, and stirring rate can influence the prepared nanoparticles' size and polydispersity.[68]

Apart from the advantages like control of size, shape, and composition, the technique is worth mentioning due to its simple instrumentation and soft reaction conditions. Also, the presence of surfactant molecules protects the nanoparticle from possible agglomeration. The microemulsion

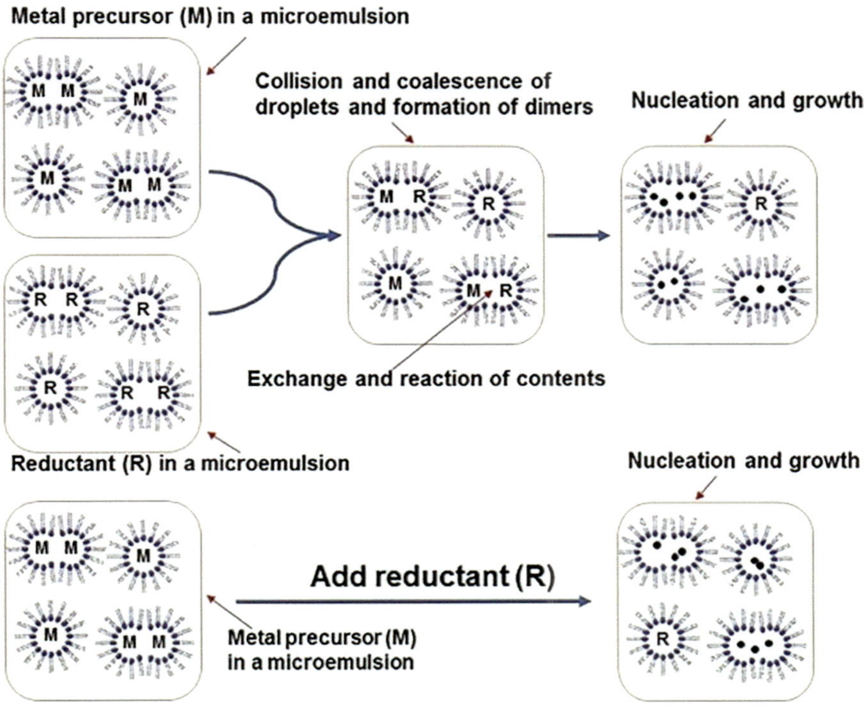


**FIGURE 2.10** Schematic diagram of nanoparticle synthesis using microemulsion.

**Source: Reprinted from Ref.**[67] ***Copyright © 2019 by author(s) and Scientific Research Publishing Inc.***

method also has disadvantages; synthesis of nanoparticles via this method requires a large quantity of surfactants, and factors like temperature and pH can also influence the formation of stable nanoparticles.[69,70]

### 2.4.5 Coprecipitation

Coprecipitation is a commonly used and effective method for nanoparticle synthesis. It has been used to prepare metal and metal oxide nanoparticles using inorganic salts as the starting material. This method is popular because of the inexpensive chemicals and simple experimental procedure. The absence of byproducts and moderate reaction temperature make it even more favourable.[71–73] The base solution is added slowly to the precursor solution upon stirring; this will result in the formation of metal hydroxide nanoparticles. Once the nanoparticles are formed, they can be separated from the solution using either centrifugation or filtration. Wash them multiple times using distilled water to remove unreacted precursors. Finally, dry the washed nanoparticles at a suitable temperature using an oven or air drying. The result will be metal hydroxide or metal oxide. The nature of the nanoparticle will depend on the reaction conditions and the starting materials used.[74,75]

The pH, temperature, stirring time, and concentration of the precursor and base solution can influence the size and morphology of the nanoparticles. Commonly used bases for precipitation metal salts are NaOH, $NH_4OH$, and KOH. Metal oxides like $Fe_2O_3$, NiO, and $CeO_2$ can be synthesized using this method.[76,77] It is also possible to functionalize nanoparticles with bioactive molecules with this method. Coprecipitation is a one-pot synthesis method that can be used to prepare nanoparticles without using stabilizing agents. It is an appropriate method for synthesizing metal oxide nanopowders, which are highly sensitive in low-temperature sintering. Coprecipitation method has high economic value due to its ability to scale up and its eco-friendly nature. However, the complicated growth mechanism leads to a broad size distribution.[78–80] A schematic diagram of the synthesis of $Fe_3O_4$ using coprecipitation is given in Figure 2.11.

## 2.5 BIOLOGICAL METHODS

The biological technique, also known as green synthesis, is an environmentally friendly and sustainable approach to producing metal and metal oxide nanoparticles. This method starts from the bottom and utilizes biological agents instead of chemical-reducing agents.[6] The resulting nanoparticles' size, shape, and dimensions are influenced by factors such as the biological agent; concentration; the reducing agent; and parameters such as temperature, pH, salt concentration, and exposure time. The capping of biological molecules enhances the biocompatibility of the nanoparticles and mitigates the likelihood of agglomeration.[4,81] Figure 2.12 represents different biological reducing agents commonly used to prepare nanoparticles. The biological method enables the recycling of metal salt found

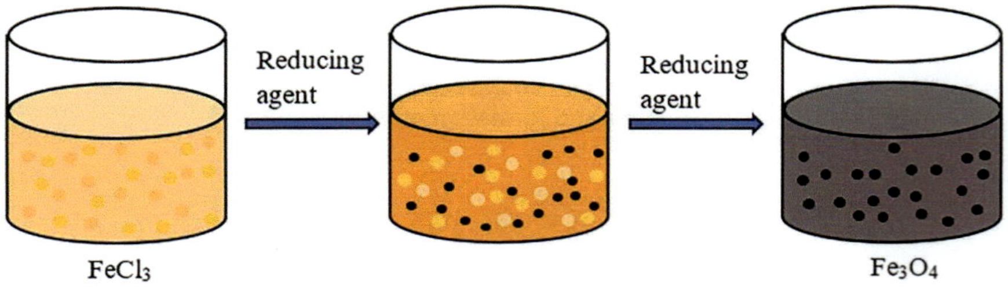


FIGURE 2.11 Schematic diagram of synthesis of $Fe_3O_4$ using coprecipitation.

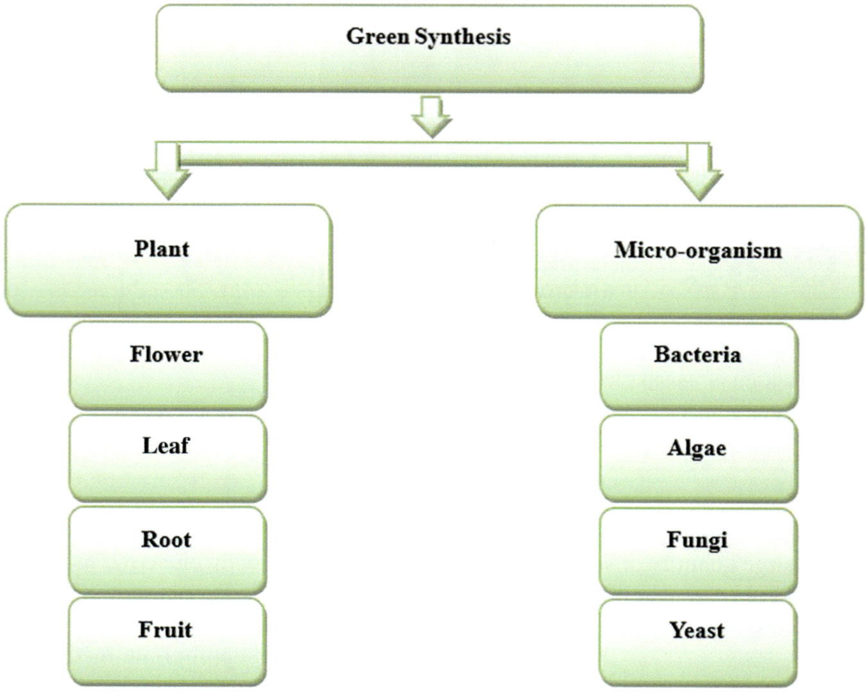


FIGURE 2.12 Schematic representation of various natural reducing agents.

in water bodies and facilitates the easy separation of synthesized nanoparticles from the reaction medium through centrifugation. The bio-assisted techniques can be categorized into two distinct groups.

- Synthesis using micro-organisms
- Synthesis using plants

Phytosynthesis, which uses plant components as biological factories to produce metallic nanoparticles, is a more economical strategy for nanoparticle production. This approach eliminates the need for complicated, multistep operations like microbial isolation, culturing, and maintenance. This is also a rapid and inexpensive method that can be readily scaled up for large-scale manufacturing of nanoparticles.[81] Usually, plant parts are dried, cut into pieces, and mixed with distilled water at room temperature or elevated temperature to obtain a plant extract. In some cases, solvents like ethanol and methanol are also used. Phytochemicals present in these extracts serve the role of a natural reductant and a stabilizer. The extract can be further purified by filtration and centrifugation.[82] Plant extract and metal salt in different ratios are used to prepare nanoparticles. Temperature, pH, volume, concentration of metal salt, and phytochemicals present in the extracts are crucial in determining the synthesized metal nanoparticles' size morphology and biological activity. The phytochemicals responsible for the synthesis of nanoparticles are terpenoids, flavonoids, saponins, and alkaloids.[83] Figure 2.13 shows a schematic diagram of nanoparticle synthesis using plant extract as a reducing agent.[84]

Microorganisms like bacteria, fungi, algae, and yeast can collect and detoxify heavy metals due to various reductase enzymes inside them. These enzymes can convert metal salts to metal nanoparticles with a small size distribution. For example, the nitrate reductase enzyme found in *B. lichenifomis* reduced silver nitrate to form silver nanoparticles.[85,86] Extracellular synthesis is more popular

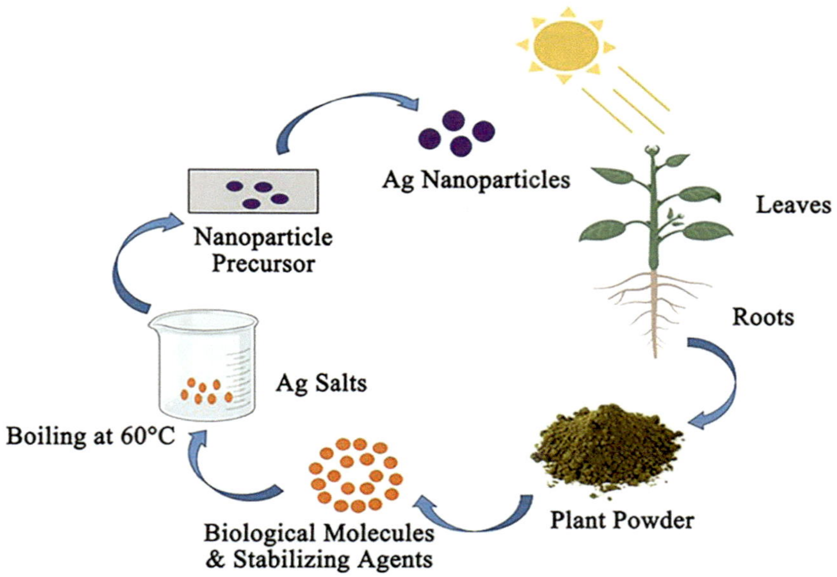


FIGURE 2.13 Schematic diagram for nanoparticle synthesis using plant extract as reducing agent.



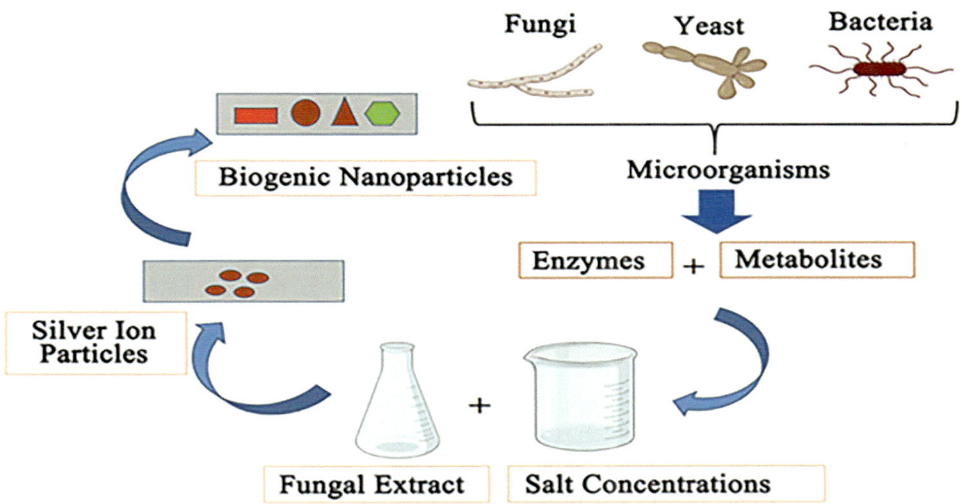


FIGURE 2.14 Schematic diagram of nanoparticle synthesis using microorganisms.



because extraction of nanoparticles produced is much easier than intracellular synthesis, where we need to separate the nanoparticles from biomass by sonication and several stages of centrifugation and washing. The procedure for microorganism-mediated nanoparticle synthesis is shown in Figure 2.14.[84]

The reduction of metal and metal oxide nanoparticles in algae is mediated by amino acids, carboxyl function groups, and enzymes under ambient circumstances.[87] Nearly every algal type, including blue-green algae, green algae, diatoms, brown algae, and red algae, has been observed to produce nanoparticles. As they are more exposed to salts in saltwater, marine microalgae have a more remarkable ability to convert metal salts to nanoparticles. Several fungi have been reported to yield intracellular and extracellular silver, gold, platinum, and cadmium nanoparticles.[88] Intracellular synthesis includes the input of metal precursors to fungal culture, which are then absorbed in the biomass. Extracellular synthesis requires the incorporation of a metal precursor to an aqueous fungal filtrate having just the biomolecules, leading to the formation of nanoparticles in the medium. Fungi also release more proteins than bacteria, which immediately leads to a better yield of nanoparticles.[89,90] Some examples of biosynthesized nanoparticles with reducing agents and their potential application in the field of medicine are given in Table 2.2.

**TABLE 2.2**
**Nanoparticles Synthesized through Biological Methods**

| NPs | Organism | Size (nm) | Applications | Reference |
|---|---|---|---|---|
| Bacteria | | | | |
| Ag | *Pseudomonas* sp. | 20–70 | Antibacterial | 91 |
| | *Bacillus thuringiensis* | 43.5–142.9 | Larvicidal action | 92 |
| Au | *Mycobacterium* sp. | 5–55 | Anticancer | 93 |
| ZnO | *Staphylococcus aureus* | 10–50 | Antimicrobial | 94 |
| Fungi | | | | |
| Au | *Cladosporium Cladosporioides* | 60 | Antibacterial and antioxidant | 95 |
| Ag | *Candida glabrata* | 2–15 | Antibacterial | 96 |
| | *Trichoderma longibrachiatum* | 10 | Antimicrobial | 97 |
| Algae | | | | |
| Au | *Cystoseira baccata* | 8.4 | Anticancer | 98 |
| ZnO | *Sargassum muticum* | 30–57 | Anticancer | 99 |
| Ag | *Portieria hornemannii* | 60–70 | Antibacterial | 100 |
| Plant | | | | |
| Ag | *Morus alba* L. | 80–150 | Antibacterial | 101 |
| | *Dolichos lablab* | 4–16 | Antimicrobial and anticancer | 102 |
| Au | *Camellia sinensis* | 10 | Antibacterial | 103 |
| $TiO_2$ | *Trigonella foenum graecum* | 20–90 | Antimicrobial | 104 |
| ZnO | *Aloe socotrina* | 15–50 | Drug delivery | 105 |

## 2.6 CONCLUSION

The current chapter discusses different types of chemical and physical techniques used to synthesize metal and metal oxide nanoparticles and their various applications in nanotechnology. The realm of metal and metal oxide nanoparticles provides endless potential and opportunities for developing biomedical applications, industrial applications, and sensing applications. In the field of nanotechnology, exploring synthesis methods is one of the significant topics. Over the past few decades, there has been a shift towards renewable and eco-friendly synthesis methods, making nontoxic and eco-friendly nanoparticles with fewer byproducts. So the green chemistry approach in synthesizing nanoparticles makes the production eco-friendly and more suitable for various applications. Soon, combining the properties of multicomponent nanoparticles will make a revolutionary change in industries, mainly in textiles, food, energy, and medicine.

**Keywords:**

Nanoparticle synthesis
Chemical methods
Physical methods
Physical vapor deposition
Hydrothermal method
Biological method

## REFERENCES

1. Heiligtag, F. J. & Niederberger, M. The fascinating world of nanoparticle research. *Mater Today* 16, 262–271 (2013).
2. Ijaz, I., Gilani, E., Nazir, A. & Bukhari, A. Detail review on chemical, physical and green synthesis, classification, characterizations and applications of nanoparticles. *Green Chem Lett Rev* 13, 223–245 (2020).
3. Dhand, C., Dwivedi, N., Loh, X. J., Ying, A. N. J., Verma, N. K., Beuerman, R. W., Lakshminarayanan, R. & Ramakrishna, S. Methods and strategies for the synthesis of diverse nanoparticles and their applications: a comprehensive overview. *RSC Adv* 5, 105003–105037 (2015).
4. Iravani, S., Korbekandi, H., Mirmohammadi, S. V. & Zolfaghari, B. Synthesis of silver nanoparticles: chemical, physical and biological methods. *Res Pharm Sci* 9 (2014).
5. Jamkhande, P. G., Ghule, N. W., Bamer, A. H. & Kalaskar, M. G. Metal nanoparticles synthesis: an overview on methods of preparation, advantages and disadvantages, and applications. *J Drug Deliv Sci Technol* 53, 101174 (2019).
6. Jeevanandam, J., Chan, Y. S. & Danquah, M. K. Biosynthesis of metal and metal oxide nanoparticles. *ChemBioEng Rev* 3, 55–67 (2016).
7. Swihart, M. T. Vapor-phase synthesis of nanoparticles. *Curr Opin Colloid Interface Sci* 8, 127–133 (2003).
8. Dizaj, S. M., Lotfipour, F., Barzegar-Jalali, M., Zarrintan, M. H. & Adibkia, K. Antimicrobial activity of the metals and metal oxide nanoparticles. *Mater Sci Eng C* 44, 278–284 (2014).
9. Baig, N., Kammakakam, I. & Falath, W. Nanomaterials: a review of synthesis methods, properties, recent progress, and challenges. *Mater Adv* 2, 1821–1871 (2021).
10. Amiri, M. R., Alavi, M., Taran, M. & Kahrizi, D. Antibacterial, antifungal, antiviral, and photocatalytic

activities of $TiO_2$ nanoparticles, nanocomposites, and bionanocomposites: recent advances and challenges. *J Public Health Res* 11, 227990362211041 (2022).
11. de Carvalho, J. F., de Medeiros, S. N., Morales, M. A., Dantas, A. L. & Carriço, A. S. Synthesis of magnetite nanoparticles by high energy ball milling. *Appl Surf Sci* 275, 84–87 (2013).
12. Xing, T., Sunarso, J., Yang, W., Yin, Y., Glushenkov, A. M., Li, L. H., Howlett, P. C. & Chen, Y. Ball milling: a green mechanochemical approach for synthesis of nitrogen doped carbon nanoparticles. *Nanoscale* 5, 7970 (2013).
13. Carneiro, J. O., Azevedo, S., Fernandes, F., Freitas, E., Pereira, M., Tavares, C. J., Lanceros-Méndez, S. & Teixeira, V. Synthesis of iron-doped TiO2 nanoparticles by ball-milling process: the influence of process parameters on the structural, optical, magnetic, and photocatalytic properties. *J Mater Sci* 49, 7476–7488 (2014).
14. Lam, C., Zhang, Y. F., Tang, Y. H., Lee, C. S., Bello, I. & Lee, S. T. Large-scale synthesis of ultrafine Si nanoparticles by ball milling. *J Cryst Growth* 220, 466–470 (2000).
15. Salah, N., Habib, S. S., Khan, Z. H., Memic, A., Azam, A., Alarfaj, E., Zahed, N. & Al-Hamedi, S. High-energy ball milling technique for ZnO nanoparticles as antibacterial material. *Int J Nanomed* 863 (2011).
16. de Lima Barizão, A. C., de Oliveira, J. P., Gonçalves, R. F. & Cassini, S. T. Nanomagnetic approach applied to microalgae biomass harvesting: advances, gaps, and perspectives. *Environ Sci Pollut Res* 28, 44795–44811 (2021).
17. Gordon, W. O., Morris, J. R. & Tissue, B. M. Control of morphology in inert-gas condensation of metal oxide nanoparticles. *J Mater Sci* 44, 4286–4295 (2009).
18. Hemben, A., Chianella, I. & Leighton, G. J. T. Surface engineered iron oxide nanoparticles generated by inert gas condensation for biomedical applications. *Bioengineering* 8, 38 (2021).
19. Ayesh, A. I. Production of metal-oxide nanoclusters using inert-gas condensation technique. *Thin Solid Films* 636, 207–213 (2017).
20. Pandya, S. G. & Kordesch, M. E. Characterization of InSb nanoparticles synthesized using inert gas condensation. *Nanoscale Res Lett* 10, 258 (2015).
21. Pérez-Tijerina, E., Gracia Pinilla, M., Mejía-Rosales, S., Ortiz-Méndez, U., Torres, A. & José-Yacamán, M. Highly size-controlled synthesis of Au/Pd nanoparticles by inert-gas condensation. *Faraday Discuss* 138, 353–362 (2008).
22. Ward, M. B., Brydson, R. & Cochrane, R. F. Mn nanoparticles produced by inert gas condensation. *J Phys Conf Ser* 26, 296–299 (2006).
23. Gracia-Pinilla, M., Martínez, E., Vidaurri, G. S. & Pérez-Tijerina, E. Deposition of size-selected Cu nanoparticles by inert gas condensation. *Nanoscale Res Lett* 5, 180 (2010).
24. Grammatikopoulos, P., Steinhauer, S., Vernieres, J., Singh, V. & Sowwan, M. Nanoparticle design by gas-phase synthesis. *Adv Phys X* 1, 81–100 (2016).
25. Maurya, D., Sardarinejad, A. & Alameh, K. Recent developments in R. F. Magnetron sputtered thin films for pH sensing applications – an overview. *Coatings* 4, 756–771 (2014).
26. Wender, H., Migowski, P., Feil, A. F., Teixeira, S. R. & Dupont, J. Sputtering deposition of nanoparticles onto liquid substrates: recent advances and future trends. *Coord Chem Rev* 257, 2468–2483 (2013).
27. Verma, M., Kumar, V. & Katoch, A. Synthesis of ZrO2 nanoparticles using reactive magnetron sputtering and their structural, morphological and thermal studies. *Mater Chem Phys* 212, 268–273 (2018).
28. Verma, M., Singh, K. P. & Kumar, A. Reactive magnetron sputtering based synthesis of WO3 nanoparticles and their use for the photocatalytic degradation of dyes. *Solid State Sci* 99, 105847 (2020).
29. Magudapathy, P., Srivastava, S. K., Gangopadhyay, P., Amirthapandian, S., Saravanan, K., Das, A. & Panigrahi, B. K. Alloying of metal nanoparticles by ion-beam induced sputtering. *Chem Phys Lett* 667, 38–44 (2017).
30. Hahn, A. Influences on nanoparticle production during pulsed laser ablation. *J Laser Micro Nanoeng* 3, 73–77 (2008).
31. Inogamov, N. A., Petrov, Yu. V., Khokhlov, V. A. & Zhakhovskii, V. V. Laser ablation: physical concepts and applications (review). *High Temp* 58, 632–646 (2020).
32. Ghorbani, H. A review of methods for synthesis of Al nanoparticles. *Orient J Chem* 30, 1941–1949 (2014).
33. Zhang, J., Chaker, M. & Ma, D. Pulsed laser ablation based synthesis of colloidal metal nanoparticles for catalytic applications. *J Colloid Interface Sci* 489, 138–149 (2017).
34. Kupracz, P., Coy, E., Grochowska, K., Karczewski, J., Rysz, J. & Siuzdak, K. The pulsed laser ablation synthesis of colloidal iron oxide nanoparticles for the enhancement of TiO2 nanotubes photo-activity. *Appl Surf Sci* 530, 147097 (2020).
35. Rivera-Chaverra, M. J., Restrepo-Parra, E., Acosta-Medina, C. D., Mello, A. & Ospina, R. Synthesis of oxide iron nanoparticles using laser ablation for possible hyperthermia applications. *Nanomaterials* 10, 2099 (2020).
36. Kim, M., Osone, S., Kim, T., Higashi, H. & Seto, T. Synthesis of nanoparticles by laser ablation: a review. *KONA Powder Part J* 34, 80–90 (2017).
37. Bernard, B., Quet, A., Bianchi, L., Joulia, A., Malié, A., Schick, V. & Rémy, B. Thermal insulation properties of YSZ coatings: suspension plasma spraying (SPS) versus electron beam physical vapor deposition (EB-PVD) and atmospheric plasma spraying (APS). *Surf Coat Technol* 318, 122–128 (2017).
38. Durmanov, N. N., Guliev, R. R., Eremenko, A. A., Boginskaya, I. A., Ryzhikov, I. A., Trifonova, E. A., Putlyaev, E. V., Mukhin A. N., Kalnov S. L., Balandina M. V., Tkachuk A. P., Gushchin V. V., Sarychev, A. K., Lagarkov, A. N., Rodionov, I. A., Gabidullin, A. R. & Kurochkin, I. N. Non-labeled selective virus detection with novel SERS-active porous silver nanofilms fabricated by electron beam physical vapor deposition. *Sens Actuat B Chem* 257, 37–47 (2018).
39. Vijaya Kumar, P., Jafar Ahamed, A. & Karthikeyan, M. Synthesis and characterization of NiO nanoparticles by chemical as well as green routes and their comparisons with respect to cytotoxic effect and toxicity studies in microbial and MCF-7 cancer cell models. *SN Appl Sci* 1, 1083 (2019).
40. Chen, D.-H. & He, X.-R. Synthesis of nickel ferrite nanoparticles by sol-gel method. *Mater Res Bull* 36, 1369–1377 (2001).
41. Parashar, M., Shukla, V. K. & Singh, R. Metal oxides nanoparticles via sol–gel method: a review on synthesis, characterization and applications. *J Mater Sci Mater Electron* 31, 3729–3749 (2020).
42. Hasnidawani, J. N., Azlina, H. N., Norita, H., Bonnia, N. N., Ratim, S. & Ali, E. S. Synthesis of ZnO NANOSTRuctures using sol-gel method. *Procedia Chem* 19, 211–216 (2016).
43. Bokov, D., Jalil, A. T., Chupradit, S., Suksatan, W., Ansari, M. J., Shewael, I. H., Valiev, G. H. & Kianfar, E. Nanomaterial by sol-gel method: synthesis and application. *Adv Mater Sci Eng* 2021, 1–21 (2021).

44. García-Barrasa, J., López-de-Luzuriaga, J. & Monge, M. Silver nanoparticles: synthesis through chemical methods in solution and biomedical applications. *Open Chem* 9, 7–19 (2011).
45. Habte, L., Shiferaw, N., Mulatu, D., Thenepalli, T., Chilakala, R. & Ahn, J. W. Synthesis of nano-calcium oxide from waste eggshell by sol-gel method. *Sustainability* 11, 3196 (2019).
46. Dubey, R. S. Temperature-dependent phase transformation of TiO2 nanoparticles synthesized by sol-gel method. *Mater Lett* 215, 312–317 (2018).
47. Nwanna, E. C., Imoisili, P. E. & Jen, T.-C. Fabrication and synthesis of SnOX thin films: a review. *Int J Adv Manuf Technol* 111, 2809–2831 (2020).
48. Yaqoob, A. A., Umar, K. & Ibrahim, M. N. M. Silver nanoparticles: various methods of synthesis, size affecting factors and their potential applications–a review. *Appl Nanosci* 10, 1369–1378 (2020).
49. Zahoor, M., Zahooe, M., Nazir, N., Naz, M. I. S., Zekker, I., Burlakovs, J., Uddin, S., Kamran, A. W., Kallistova, A., Pimenov, N. & Khan, F. A. A review on silver nanoparticles: classification, various methods of synthesis, and their potential roles in biomedical applications and water treatment. *Water* 13, 2216 (2021).
50. Gudikandula, K. & Charya Maringanti, S. Synthesis of silver nanoparticles by chemical and biological methods and their antimicrobial properties. *J Exp Nanosci* 11, 714–721 (2016).
51. Quintero-Quiroz, C., Acevedo, N., Zapata-Giraldo, J., Botero, L. E., Quintero, J., Zárate-Triviño, D., Saldarriaga, J. & Pérez, V. Z. Optimization of silver nanoparticle synthesis by chemical reduction and evaluation of its antimicrobial and toxic activity. *Biomater Res* 23, 27 (2019).
52. Thomas, R. K., Sukumaran, S. & Sudarsanakumar, C. An insight into the comparative binding affinities of chlorogenic acid functionalized gold and silver nanoparticles with ctDNA along with its cytotoxicity analysis. *J Mol Liq* 287, 110911 (2019).
53. Raj, A., Thomas, R. K., Vidya, L., Aparna, V. M., Neelima, S. & Sudarsanakumar, C. Exploring the cytotoxicity on human lung cancer cells and DNA binding stratagem of camptothecin functionalised silver nanoparticles through multi-spectroscopic, and calorimetric approach. *Sci Rep* 13, 9045 (2023).
54. Raj, A., Thomas, R. K., Vidya, L., Aparna, V. M., Neelima, S. & Sudarsanakumar, C. A minor groove binder with significant cytotoxicity on human lung cancer cells: the potential of hesperetin functionalised silver nanoparticles. *J Fluoresc* 13, 9045 (2023).
55. Hosokawa, S. Synthesis of metal oxides with improved performance using a solvothermal method. *J Ceramic Soc Japan* 124, 870–874 (2016).
56. Liang, M.-T., Wang, S.-H., Chang, Y.-L., Hsiang, H.-I., Huang, H.-J., Tsai, M.-H., Juan, W.-C. & Lu, S.-F. Iron oxide synthesis using a continuous hydrothermal and solvothermal system. *Ceram Int* 36, 1131–1135 (2010).
57. Yang, G. & Park, S.-J. Conventional and microwave hydrothermal synthesis and application of functional materials: a review. *Materials* 12, 1177 (2019).
58. Titirici, M.-M., Antonietti, M. & Thomas, A. A generalized synthesis of metal oxide hollow spheres using a hydrothermal approach. *Chem Mater* 18, 3808–3812 (2006).
59. Ndlwana, L., Raleie, N., Dimpe, K. M., Ogutu, H. F., Oseghe, E. O., Motsa, M. M., Msagati, T. A. M. & Mamba, B. B. Sustainable hydrothermal and solvothermal synthesis of advanced carbon materials in multidimensional applications: a review. *Materials* 14, 5094 (2021).
60. Mishra, D., Arora, R., Lahiri, S., Amritphale, S. S. & Chandra, N. Synthesis and characterization of iron oxide nanoparticles by solvothermal method. *Protect Metals Phys Chem Surf* 50, 628–631 (2014).
61. Okada, M., Kuno, M. & Yamada, Y. Fast synthesis of monoclinic VO2 nanoparticles by microwave-assisted hydrothermal method and application to high-performance thermochromic film. *Solar Energy Mater Solar Cells* 255, 112311 (2023).
62. Köçkar, H., Karaagac, O. & Özel, F. Effects of biocompatible surfactants on structural and corresponding magnetic properties of iron oxide nanoparticles coated by hydrothermal process. *J Magn Magn Mater* 474, 332–336 (2019).
63. Richard, B., Lemyre, J.-L. & Ritcey, A. M. Nanoparticle size control in microemulsion synthesis. *Langmuir* 33, 4748–4757 (2017).
64. Tojo, C., de Dios, M. & Barroso, F. Surfactant effects on microemulsion-based nanoparticle synthesis. *Materials* 4, 55–72 (2010).
65. Solanki, J. N. & Murthy, Z. V. P. Controlled size silver nanoparticles synthesis with water-in-oil microemulsion method: a topical review. *Ind Eng Chem Res* 50, 12311–12323 (2011).
66. Stubenrauch, C., Wielpütz, T., Sottmann, T., Roychowdhury, C. & DiSalvo, F. J. Microemulsions as templates for the synthesis of metallic nanoparticles. *Colloids Surf A Physicochem Eng Asp* 317, 328–338 (2008).
67. Soleimani Zohr Shiri, M., Henderson, W. & Mucalo, M. R. A review of the lesser-studied microemulsion-based synthesis methodologies used for preparing nanoparticle systems of the noble metals, Os, Re, Ir and Rh. *Materials* 12, 1896 (2019).
68. Chen, S., Xu, H., He, X., Su, Y., Liu, Q. & Zhang, B. One-pot synthesis of polydopamine/Ag microspheres through microemulsion environment and its methylene blue removal application. *J Polym Res* 29, 34 (2022).
69. Asab, G., Zereffa, E. A. & Abdo Seghne, T. Synthesis of silica-coated $Fe_3O_4$ Nanoparticles by microemulsion method: characterization and evaluation of antimicrobial activity. *Int J Biomater* 2020, 1–11 (2020).
70. Haq, S., Ahmad, P., Khandaker, M. U., Faruque, M. R. I., Rehman, W., Waseem, M. & Din, S. U. Antibacterial, antioxidant and physicochemical investigations of tin dioxide nanoparticles synthesized via microemulsion method. *Mater Res Express* 8, 035013 (2021).
71. Abdul-Ameer, Z. N. Novel co-precipitation method for synthesis of nanostructured nickel oxide in accordance to PH: structural and optical properties as an active optical filter. *Ibn AL-Haitham J Pure Appl Sci* 32, 1–6 (2019).
72. Tazikeh, S., Akbari, A., Talebi, A. & Talebi, E. Synthesis and characterization of tin oxide nanoparticles via the co-precipitation method. *Mater Sci Poland* 32, 98–101 (2014).
73. Yazid, N. A. & Joon, Y. C. Co-precipitation synthesis of magnetic nanoparticles for efficient removal of heavy metal from synthetic wastewater. In *AIP Conference Proceedings* (2019). https://doi.org/10.1063/1.5117079.
74. Ahn, T., Kim, J. H., Yang, H.-M., Lee, J. W. & Kim, J.-D. Formation pathways of magnetite nanoparticles by coprecipitation method. *J Phys Chem C* 116, 6069–6076 (2012).
75. Besenhard, M. O., LaGrow, A. P., Hodzic, A., Kriechbaum, M., Panariello, L., Bais, G., Loizou, K., Damilos, S., Cruz, M. M., Thanh, N T. K. & Gavriilidis, A. Co-precipitation synthesis of stable iron oxide nanoparticles with NaOH:

new insights and continuous production via flow chemistry. *Chem Eng J* 399, 125740 (2020).
76. Tofighi, G., Lichtenberg, H., Gaur, A., Wang, W., Wild, S., Delgado, K. H., Pitter, S., Dittmeyer, R., Grunwaldt, J. D. & Doronkin, D. E. Continuous synthesis of Cu/ZnO/$Al_2$ $O_3$ nanoparticles in a co-precipitation reaction using a silicon based microfluidic reactor. *React Chem Eng* 7, 730–740 (2022).
77. Hui, B. H. & Salimi, M. N. Production of iron oxide nanoparticles by co-precipitation method with optimization studies of processing temperature, pH and stirring rate. *IOP Conf Ser Mater Sci Eng* 743, 012036 (2020).
78. Rahman, M. A., Radhakrishnan, R. & Gopalakrishnan, R. Structural, optical, magnetic and antibacterial properties of Nd doped NiO nanoparticles prepared by co-precipitation method. *J Alloys Compd* 742, 421–429 (2018).
79. Li, L., Mak, K. Y., Leung, C. W., Chan, K. Y., Chan, W. K., Zhong, W. & Pong, P. W. T. Effect of synthesis conditions on the properties of citric-acid coated iron oxide nanoparticles. *Microelectron Eng* 110, 329–334 (2013).
80. Mohammadi, H., Nekobahr, E., Akhtari, J., Saeedi, M., Akbari, J. & Fathi, F. Synthesis and characterization of magnetite nanoparticles by co-precipitation method coated with biocompatible compounds and evaluation of in-vitro cytotoxicity. *Toxicol Rep* 8, 331–336 (2021).
81. Akhtar, M. S., Panwar, J. & Yun, Y.-S. Biogenic synthesis of metallic nanoparticles by plant extracts. *ACS Sustain Chem Eng* 1, 591–602 (2013).
82. Iravani, S. Green synthesis of metal nanoparticles using plants. *Green Chem* 13, 2638 (2011).
83. Mittal, A. K., Chisti, Y. & Banerjee, U. C. Synthesis of metallic nanoparticles using plant extracts. *Biotechnol Adv* 31, 346–356 (2013).
84. Khan, N., Ali, S., Latif, S. & Mehmood, A. Biological synthesis of nanoparticles and their applications in sustainable agriculture production. *Nat Sci* 14, 226–234 (2022).
85. Iravani, S. Bacteria in nanoparticle synthesis: current status and future prospects. *Int Sch Res Notices* 2014, 1–18 (2014).
86. Srivastava, S., Usmani, Z., Atanasov, A. G., Singh, V. K., Singh, N. P., Abdel-Azeem, A. M., Prasad, R., Gupta, G., Sharma, M. & Bhargava, A. Biological nanofactories: using living forms for metal nanoparticle synthesis. *Mini Rev Med Chem* 21, 245–265 (2021).
87. Yadi, M., Mostafavi, E., Saleh, B., Davaran, S., Aliyeva, I., Khalilov, R., Nikzamir, M., Nikzamir, N., Akbarzadeh, A., Panahi, Y. & Milani, M. Current developments in green synthesis of metallic nanoparticles using plant extracts: a review. *Artif Cells Nanomed Biotechnol* 46, 336–343 (2018).
88. Singh, P., Kim, Y.-J., Zhang, D. & Yang, D.-C. Biological synthesis of nanoparticles from plants and microorganisms. *Trends Biotechnol* 34, 588–599 (2016).
89. Thakkar, K. N., Mhatre, S. S. & Parikh, R. Y. Biological synthesis of metallic nanoparticles. *Nanomedicine* 6, 257–262 (2010).
90. Parthasaradi, V., Kavitha, M., Sridevi, A. & Rubia, J. J. Novel rare-earth Eu and La co-doped ZnO nanoparticles synthesized via co-precipitation method: optical, electrical, and magnetic properties. *Journal of Materials Science: Materials in Electronics* 33, 25805–25819 (2022).
91. Banu, A. N. & Balasubramanian, C. Optimization and synthesis of silver nanoparticles using Isaria fumosorosea against human vector mosquitoes. *Parasitol Res* 113, 3843–3851 (2014).
92. John, M. S., Nagoth, J. A., Ramasamy, K. P., Mancini, A., Giuli, G., Natalello, A., Ballarini, P., Miceli, C. & Puccuarelli, S. Synthesis of bioactive silver nanoparticles by a Pseudomonas strain associated with the antarctic psychrophilic protozoon Euplotes focardii. *Mar Drugs* 18, 38 (2020).
93. Camas, M., Sazak Camas, A. & Kyeremeh, K. Extracellular synthesis and characterization of gold nanoparticles using Mycobacterium sp. BRS2A-AR2 isolated from the aerial roots of the ghanaian mangrove plant, Rhizophora racemosa. *Indian J Microbiol* 58, 214–221 (2018).
94. Rauf, Mohd. A., Owais, M., Rajpoot, R., Ahmad, F., Khan, N. & Zubair, S. Biomimetically synthesized ZnO nanoparticles attain potent antibacterial activity against less susceptible S. aureus skin infection in experimental animals. *RSC Adv* 7, 36361–36373 (2017).
95. Joshi, C. G., Danagoudar, A., Poyya, J., Kudva, A. K. & BL, D. Biogenic synthesis of gold nanoparticles by marine endophytic fungus-Cladosporium cladosporioides isolated from seaweed and evaluation of their antioxidant and antimicrobial properties. *Process Biochem* 63, 137–144 (2017).
96. Jalal, M., Ansari, M. A., Alzohairy, M. A., Ali, S. G., Khan, H. M., Almatroudi, A. & Raees, K. Biosynthesis of silver nanoparticles from oropharyngeal Candida glabrata isolates and their antimicrobial activity against clinical strains of bacteria and fungi. *Nanomaterials* 8, 586 (2018).
97. Elamawi, R. M., Al-Harbi, R. E. & Hendi, A. A. Biosynthesis and characterization of silver nanoparticles using Trichoderma longibrachiatum and their effect on phytopathogenic fungi. *Egypt J Biol Pest Control* 28, 28 (2018).
98. González-Ballesteros, N., Prado-López, S., Rodríguez-González, J. B., Lastra, M. & Rodríguez-Argüelles, M. C. Green synthesis of gold nanoparticles using brown algae Cystoseira baccata: its activity in colon cancer cells. *Colloids Surf B Biointerfaces* 153, 190–198 (2017).
99. Sanaeimehr, Z., Javadi, I. & Namvar, F. Antiangiogenic and antiapoptotic effects of green-synthesized zinc oxide nanoparticles using Sargassum muticum algae extraction. *Cancer Nanotechnol* 9, 3 (2018).
100. Fatima, R., Priya, M., Indurthi, L., Radhakrishnan, V. & Sudhakaran, R. Biosynthesis of silver nanoparticles using red algae Portieria hornemannii and its antibacterial activity against fish pathogens. *Microb Pathog* 138, 103780 (2020).
101. Razavi, R., Molaei, R., Moradi, M., Tajik, H., Ezati, P. & Yordshahi, A. S. Biosynthesis of metallic nanoparticles using mulberry fruit (Morus alba L.) extract for the preparation of antimicrobial nanocellulose film. *Appl Nanosci* 10, 465–476 (2020).
102. Kahsay, M. H., RamaDevi, D., Kumar, Y. P., Mohan, B. S., Tadesse, A., Battu, G. & Basavaiah, K. Synthesis of silver nanoparticles using aqueous extract of Dolichos lablab for reduction of 4-Nitrophenol, antimicrobial and anticancer activities. *OpenNano* 3, 28–37 (2018).
103. Onitsuka, S., Hamada, T. & Okamura, H. Preparation of antimicrobial gold and silver nanoparticles from tea leaf extracts. *Colloids Surf B Biointerfaces* 173, 242–248 (2019).
104. Subhapriya, S. & Gomathipriya, P. Green synthesis of titanium dioxide (TiO2) nanoparticles by Trigonella foenum-graecum extract and its antimicrobial properties. *Microb Pathog* 116, 215–220 (2018).
105. Fahimmunisha, B. A., Ishwarya, R., AlSalhi, M. S., Devanesan, S., Govindarajan, M. & Vaseeharan, B. Green fabrication, characterization and antibacterial potential of zinc oxide nanoparticles using Aloe socotrina leaf extract: a novel drug delivery approach. *J Drug Deliv Sci Technol* 55, 101465 (2020).

## LIST OF ABBREVIATIONS

| | |
|---|---|
| **0D** | Zero Dimensional |
| **1D** | One Dimensional |
| **2D** | Two Dimensional |
| **3D** | Three Dimensional |
| **AC** | Alternating Current |
| **AOT** | Bis(2-ethylhexyl)sulfosuccinate |
| **CTAB** | Cetyltrimethylammonium Bromide |
| **DC** | Direct Current |
| **HLB** | Hydrophilic-Hydrophobic Balance |
| **PVD** | Physical Vapor Deposition |
| **RF** | Radio Frequency |
| **SDBS** | Sodium Dodecyl Benzene Sulfonate |
| **SDS** | Lauryl Sodium Sulfate |

# 3 Applications of Green Synthesized Metal and Metal Oxide Nanoparticles

*Varsha V., Akshay G., Chithrathara M., Jerrit Spary T., and Anil Kumar T. R.*
Inter University Centre for Evolutionary and Integrative Biology, University of Kerala, Kariavattom, Kerala, India

## 3.1 ANTIMICROBIAL APPLICATIONS

One of the most-studied properties of metal nanoparticles (MNPs) is their antimicrobial activity [1], especially due to their potential use against multi-drug-resistant microorganisms (MDRMs) [2]. In addition to this, their relatively easy functionalization, which further facilitates their applications and manipulation (e.g., through immobilization), also contributes to the interest in studying and trying to develop functionalized MNPs with enhanced properties. Biosynthesis of MNPs, a green alternative to chemical methods, offers an eco-friendly approach with lower environmental impact [3]. Extracellular biosynthesis allows efficient one-pot reactions at room temperature, enabling simultaneous functionalization with organic molecules to enhance antimicrobial activity and stability [1]. Despite these advantages, understanding the mechanisms of biosynthesis and MNP antimicrobial activity remains incomplete.

Nano-metals and oxides like silver, copper, and zinc exhibit broad-spectrum antibacterial activity, gaining attention for biomedical applications. However, challenges persist in developing green synthesis, understanding antibacterial mechanisms, enhancing biosafety, and expanding applications [4]. Metal oxide nanoparticles, such as nano-silver, nano-copper, and nano-zinc, have garnered attention for their broad-spectrum antibacterial activity in biomedical applications. Despite their current use in medical antibacterial and cancer treatment, challenges persist in developing green preparation methods, understanding antibacterial mechanisms, improving biosafety, and expanding application fields. Metal oxide nanoparticles, like $Fe_3O_4$, $TiO_2$, CuO, and ZnO, exhibit notable antibacterial efficiency, with mechanisms including reactive oxygen species formation, metal-ion release, particle internalization, and direct bacterial cell wall/membrane destruction, showcasing their potential for bacterial infection control. Research on polymetallic oxides shows promise for synergistic effects, combining strong antibacterial efficiency with reduced cytotoxicity [5].

Metal oxide nanoparticles can exhibit either bacteriostatic or bactericidal effects. Bacteriostatic effects halt bacterial growth without causing cell death, allowing resumption upon removal from nanoparticle solutions, while bactericidal effects result in the elimination of bacterial colonies upon re-plating onto nanoparticle-free agar. Depending on the experimental conditions, nanoparticle concentration, and bacterial strain, a particular type of metal oxide nanoparticle may have a bacteriostatic or bactericidal effect, as shown for ZnO [6] or $TiO_2$ [7]. NPs derived from plants that show promising antibacterial activities have high potential to be developed into future antibacterials, mainly due to their low toxic effects [8]. NPs have been previously reported to inhibit Gram-positive bacteria such as *Staphylococcus* spp. [9, 10], *Streptococcus* spp. [11, 12], and *Bacillus* spp. [13, 14] and Gram-negative bacteria such as *Escherichia* spp. [15, 16], *Pseudomonas* spp. [17, 18], *Salmonella* spp. [17, 19] *Shigella* spp. [15, 20], *Proteus* spp. [21, 22], and *Vibrio* spp. [20, 23]. More promisingly, NPs have also been shown to inhibit antibiotic-resistant bacteria such as methicillin-resistant *S. aureus* (MRSA) [24, 25] and drug-resistant *E. coli* [26].

Reactive oxygen species (ROS) generated on metal oxide nanoparticle surfaces, including ions, small molecules (e.g., $H_2O_2$), and free radicals (e.g., OH, $^1O_2$), induce oxidative stress, causing damage to bacterial cells, including DNA disruption, cell cycle interruption, and death (example: $Ag_2O$ nanoparticles) [27]; also, CuO nanoparticles induce ROS, particularly superoxide anions, on bacterial cell surfaces or upon internalization, leading to bactericidal effects in both Gram-positive (*S. aureus*) and Gram-negative (*E. coli*) bacteria [28]. Metal oxide nanoparticles' physicochemical features, like size and crystal structure, directly impact antibacterial effects. Synthesis and treatment methods influence characteristics, influencing antibacterial efficacy, for instance, smaller nanoparticles (<20 nm), can penetrate bacterial cells effectively [27]. Smaller metal oxide nanoparticles are generally more efficient antibacterial agents, but enhanced aggregation can compromise efficacy. Surface defects, like point defects, contribute to antibacterial efficiency. For instance, nano-ZnO dissolution in water forms surface defects, causing uneven texture with rough edges and corners, damaging *E. coli* membranes. Crystallographic orientation and surface plane type also influence ZnO

DOI: 10.1201/9781032660639-3

nanowires' antibacterial efficiency, with randomly oriented nanowires more effective against *E. coli* [27]. Metal oxide nanoparticles' physicochemical characteristics, such as size and crystal structure, significantly impact antibacterial effects. Smaller nanoparticles (<20 nm) are generally more efficient antibacterial agents, though aggregation may compromise efficacy. Surface defects, like point defects, contribute to antibacterial efficiency, as seen with nano-ZnO causing membrane damage to *E. coli*. Crystallographic orientation and surface plane type also influence antibacterial efficiency, with randomly oriented ZnO nanowires showing superior efficacy against *E. coli* [27]. Surface charge influences membrane damage and internalization. Positively charged nano-ZnO interacts with negatively charged C. *jejuni*, causing membrane disruption and leakage. Even negatively charged particles can damage membranes through non-electrostatic interactions. Metal oxide nanoparticles may bind specific membrane moieties, inducing bacterial death. Notably, nanoparticles remain bound to damaged or dead bacteria, altering effective concentration. Variations in physicochemical characteristics result in diverse antibacterial mechanisms. Combining two or more metals in oxide nanoparticles can efficiently eliminate various bacterial strains, including highly resistant ones [29].

ZnO nanoparticles synthesized with *Ziziphorateniri* and *Asteraceae* plant extracts demonstrated strong antibacterial properties against *E. coli* and *Staphylococcus aureus*, with minimum inhibitory concentrations (MICs) of 0.125 mg/ml (*E. coli*) and 0.062 mg/ml (*S. aureus*), while minimum bactericidal concentrations (MBCs) were 0.500 mg/ml (*E. coli*) and 0.250 mg/ml (*S. aureus*) [30]. In nano-fluids, ZnO exhibited bacteriostatic behavior against *E. coli*, showing heightened efficacy at higher nanoparticle concentrations and smaller sizes, with polyethylene glycol and polyvinylpyrrolidone dispersants minimally impacting antibacterial activity but enhancing stability. Electron microscopy reveals membrane damage, suggesting a potential antibacterial mechanism, while electrochemical analyses indicate direct interactions between ZnO nanoparticles and bacterial membranes at higher concentrations [31]. Bacterial suspensions exposed to ZnO-coated glass surfaces and plated on sub-MIC antibiotic concentrations exhibit increased susceptibility. Compared to uncoated controls, *E. coli* shows a 43% reduction in viable counts with chloramphenicol and a 34% reduction with ampicillin. *S. aureus* demonstrates a moderate increase in sensitivity, particularly with chloramphenicol [28].

Polyaniline/Cu(0.05)Zn(0.95)O (PANI/CZO) nanocomposites exhibited outstanding antibacterial efficacy against *Staphylococcus aureus, Escherichia coli,* and *Candida albicans*, as indicated by significant inhibition zones and low minimum inhibitory and minimal bactericidal concentrations [32]. *Leuconostoc* sp. strain C2–derived nanoparticles (C2-AuNPs and C2-ZnONPs) exhibited potent antibacterial effects, inhibiting biofilm formation and virulence in *Pseudomonas aeruginosa* and *Staphylococcus aureus*. They suppressed bacterial motility, reduced virulence factor production, and disrupted mature biofilms. Green synthesis demonstrated dose-dependent inhibition of virulence factors, with C2-AuNPs displaying superior antibiofilm and antivirulence properties [33]. A multifunctional nanocomposite, formed by incorporating pectin into zinc oxide nanoparticles using *Psidium guajava* leaf extract, demonstrated heightened antibacterial efficacy. Additionally, it exhibited enhanced mechanical strength in biofilms and excellent water vapor blocking capacity, showcasing its potential for versatile applications [34]. Zinc oxide nanoparticles (ZnO NPs) green synthesized using *Scoparia dulcis* leaf extract displayed potent antibacterial and antifungal activities against *E. coli, Staphylococcus aureus, Candida albicans,* and *Aspergillus niger*. The simplicity, cost-effectiveness, and eco-friendliness of plant extract-assisted synthesis highlight the potential of ZnO NPs in biomedical and agricultural sectors. With their high antioxidant activity, these NPs offer promise for multifunctional biomedical applications [35]. Chitosan-stabilized copper oxide nanoparticles, synthesized from *Ficus carica* leaf extract, demonstrated enhanced antibacterial efficacy. The addition of chitosan significantly boosted antimicrobial activity against various Gram-positive and Gram-negative bacteria, suggesting potential applications in biomedicine [36]. Additionally metallic nanoparticles synthesized by *Aspergillus kambarensis*, encompassing silver, gold, copper, and zinc oxide, exhibited potent antifungal, antibacterial, and anticancer activities. Inhibition zones ranged from 13.1 to 44.2 mm, with IC50 values of 62.01–77.03 µg/ml against HepG-2 cancer cells [37]. Another study introduced casein/carboxymethyl cellulose nanocomposite with enhanced antibacterial activity was synthesized using an aqueous bark extract from *Syzygium cumini* (Java plum). Gold nanoparticles were incorporated into the nanocomposite using the extract as a stabilizing and reducing agent. The resulting blend demonstrated strengthened antibacterial activity, making it particularly suitable for applications in food packaging [38].

Carboxymethyl cellulose and polyvinyl alcohol (CMC/PVA) films, formed by incorporating green-synthesized silver nanoparticles, using *Plectranthus amboinicus* leaf extract as a reducing and stabilizing agent, into CMC/PVA containing AgNPs, exhibited enhanced antibacterial capacity against *Salmonella typhi, Staphylococcus aureus, Bacillus spizizenii,* and *Escherichia coli*. Additionally, these films showed improved morphological and mechanical properties compared to the neat CMC/PVA film [39]. Carboxymethyl cellulose and silver nanoparticles (CMC-AgNPs) demonstrated effective antibacterial activity against both Gram-positive (*Bacillus cereus* and *Staphylococcus aureus*) and Gram-negative bacteria (*Klebsiella oxytoca* and *Escherichia coli*). Antifungal activity was observed against *Aspergillus fumigatus, Aspergillus niger,* and *Aspergillus terreus*. The nanocomposite displayed good cytotoxic activity against HepG2 hepatocellular tumor cells, inducing necrosis and fatality through increased caspase-8 and -9 activities and reduced VEGFR-2 levels [40]. Copper oxide nanoparticles (CuO NPs) have been shown to be more efficient in in-vitro studies such as cytotoxicity and antimicrobial when compared to other metal oxides. Green CuO NPs (synthesized using *Coix lacryma-jobi* leaf extract) showed substantial anticancer activity and were proved on human cervical and lung cancer cell lines with IC50 values of 31.88 and 15.61 µg/ml, respectively [41].

## 3.2 BIOMEDICAL APPLICATIONS

Nanotechnology, an interdisciplinary field, has led to the miniaturization of molecules for various applications. The synthesis of nanoparticles involves various philosophical perspectives and is primarily focused on their properties. Metal and metal oxide nanoparticles are popular due to their potential use in biomedicine. Green synthesis, an environmentally friendly approach, is being increasingly studied for its safe and biocompatible qualities. Green synthesis offers advantages like cost effectiveness and large-scale production.

### 3.2.1 Tissue Engineering

Nanotechnology has significantly improved tissue engineering by synthesizing biomaterials for intelligent medication delivery systems and tissue repair or reconstruction. Green synthesis offers improved biological effects and eliminates drawbacks of traditional synthesis techniques. Metal and metal oxide nanoparticles are being biogenerated, optimized, and used in tissue engineering applications. One of the most extensively studied uses is the integration of nanoparticles into scaffolds, which provide the framework for regeneration and bioactive chemicals [42]. NPs can also be added to porous sponges to improve their functionality. The antibacterial activity of NPs is widely recognized, with their high surface area-to-volume ratio increasing their effectiveness. Researchers have used NPs to prevent wound infections, enhance stability and antibacterial capabilities, and create electrospun polyurethane scaffolds with groundnut oil and green-generated nickel oxide nanoparticles. These scaffolds have shown better mineral deposition in bone tissues and improved physicochemical characteristics, making them a great option for bone tissue engineering [43].

### 3.2.2 Drug Delivery Systems

Green-synthesized nanoparticles have gained popularity as carriers for drug delivery due to their small size and high surface area, which improves the pharmacokinetics of therapeutic agents. NPs can overcome traditional delivery methods, such as intracellular trafficking and biodistribution, by improving the stability and solubility of molecules, encouraging membrane transport, and lengthening circulation periods [44]. NPs can be employed in combination therapies to optimize therapeutic efficacy against specific macromolecules, target specific stages of the cell cycle, or counteract drug resistance mechanisms [45]. Plant extracts, such as algae and fungi, can be used as safe, non-toxic, and non-volatile media for NP synthesis. ZnO NPs can serve as effective carriers for targeted delivery of anticancer medicines into malignant tumor cells, with their efficacy being dose dependent [46]. Biomolecules from plant, algae, or fungal extracts can cover the surface of Au green-produced NPs, allowing plants to absorb or functionalize the biosynthesized NPs. Green-synthesized AuNPs are more suitable as a drug delivery vector due to their biocompatibility with both normal and aberrant cells [47]. Imaging

Bioimaging is a non-invasive technique used to view interior structures and physiological processes in living cells or organisms in real-time. It is safe and effective, allowing tissue exploration at subcellular and multicellular levels and providing information about the sample's three-dimensional nanostructure. Nanomaterials, particularly nanoprobes, are considered ideal for this purpose due to their ease of release and precise description using nuclear magnetic resonance or gel permeation chromatography. Despite limited research on nanomaterials, scientists continue to explore new materials, such as magnetic resonance, fluorescence, and X-ray CT [48]. Nanomaterials can be functionalized to be observable across multiple imaging modalities, providing synergistic benefits. Gd-doped ZnO quantum dots can be used for magnetic resonance imaging and optical imaging, while $Fe_3O_4$-ZnO magnetic quantum dots have been used for imaging and cancer therapy [49]. Recent studies have examined the optical and biocompatibility characteristics of Co-Au nanocomposites, which have been applied to in vitro cancer cell detection and have shown continuous brightness under UV radiation without causing cytotoxicity [50].

### 3.2.4 Anti-Cancerous Activity

Cancer is a group of diseases with different signaling mechanisms, including cell proliferation, angiogenesis, and metastasis. Current anticancer therapies have adverse effects on normal cells due to excessive drug administration and radiation exposure. To address these therapeutic requirements, nano-sized molecular tools are being developed to distinguish between malignant and non-malignant cells and deliver lethal payloads [51]. Current research has focused on the discovery of novel transition metal compounds, such as metallic nanoparticles, which have been used in drug delivery for cancer treatment. Nanoparticles display specific physicochemical attributes, making them valuable by inducing cytotoxicity in the treatment of cancer cells. Biosynthesized nanoparticles are mostly biocompatible and highly suitable for biomedical applications. Biogenic metal oxide nanoparticles have shown promising results in cancer therapy by inducing cytotoxicity in cancerous cells without affecting normal cells. Black bean-synthesized copper oxide nanoparticles (CuO NPs) and *Nepeta deflersiana*-synthesized AgNPs have shown efficient anticancer activity by increasing intracellular reactive oxygen species concentration-dependently [52, 53]. MgO NPs have biomedical properties and are highly stable under harsh conditions [54]. Zinc oxide nanoparticles have been reported to possess anticancer properties, with some studies showing reduced levels of oxidative stress markers and hepatocyte integrity [55]. Spherical copper oxide nanoparticles have been synthesized using various plant extracts, and their cytotoxicity activity has been confirmed by MTT assays against various cancer cell lines [56]. Nanosilver immobilized on $TiO_2$ nanometric fibers ($Ag/TiO_2$) has shown cytotoxicity in both cancer cells and macrophages. Environmentally safe synthesizing tellurium nanoparticles has also been explored, with cytotoxicity assays showing that the nanoparticles possess anticancer properties towards melanoma cells without being toxic to healthy cells [57]. Selenium nanoparticles have shown anticancer potential and have been functionalized with hyaluronic acid (HA-SeNPs) [58, 59]. Magnesium oxide nanoparticles combined with human serum albumin show excellent photocatalytic, antibacterial, and anticancer performance [60].

### 3.2.5 Anti-Diabetic Activity

Diabetes is a metabolic disorder caused by insufficient insulin production, leading to sugar accumulation in the blood. Enzymes like α-glucosidase and α-amylase play a role in this disease, and bio-synthesized metal oxide nanoparticles can help treat it by suppressing enzyme secretion levels [61]. AgNPs were synthesized from *Lonicera japonica*, which has medicinal properties such as antiviral, anti-inflammatory, and anti-diabetic activities. Silver nanoparticles from *Pouteria sapota* leaf extracts inhibited α-amylase action, and non-enzymatic glycosylation of hemoglobin confirmed their anti-diabetic activity. Selenium nanoparticles are known to cure diabetes by reducing oxidative damage and preventing hypoglycemic activity. Chitosan-stabilized selenium nanoparticles (Cs-SeNPs) and ZnONPs synthesized through microwave-assisted methods showed efficient anti-diabetic activities [62, 63].

### 3.2.6 Anti-Inflammatory Properties

Inflammation is a defense mechanism that responds to external stimuli, such as pathogens, allergens, irritants, and damaged cells. Metal-based nanoparticles have shown excellent penetrating capacity in epithelial cells and inflammatory cells, making them effective treatments for diseases like cardiovascular diseases and gastric ulcers [64]. Green-synthesized metal and metal oxide nanoparticles have shown anti-inflammation activity, with peptide-based AuNPs hybrid libraries inhibiting toll-like receptor (TLR) signaling. European cranberry bush fruit extracts have shown effective release of IL-1α and enhanced anti-inflammatory activity [65]. Cerium oxide nanoparticles exhibit free radical scavenging activity, while polysaccharide-modified selenium nanoparticles and gum-arabic stabilized selenium nanoparticles have anti-inflammatory properties. Other metals like titanium oxide and zinc oxide nanoparticles also exhibit anti-inflammatory properties, with biofabricated zinc oxide nanoparticles showing wound-healing properties and titanium oxide nanoparticles decreasing platelets and increasing thrombin-anti-thrombin levels [66]. Researchers should focus on finding the mode of mechanism for these anti-inflammatory properties in biogenic metal and metal oxide nanoparticles.

### 3.2.7 Antioxidant Properties

Nanoparticles play a significant role in biomedical applications due to their antioxidant properties, which help fight off free radicals. Cerium oxide nanoparticles, which exist in two oxidation states, have been found to increase thiol content and initiate caspase-3 activity in brain tissues of rats [67]. They also exhibit anti-tumor, anti-inflammatory, and water solubility properties when coated with levan polysaccharide [68]. Nanoparticles can act as synthetic antioxidants due to their small size and large surface-to-volume ratio, making them suitable for various therapeutic applications. Examples include synthesizing zinc oxide nanoparticles from plant extracts like *citrus × paradisi* and *Cassia fistula*, which have shown excellent antioxidant activity through DPPH radical scavenging. Green-synthesized copper oxide nanoparticles from *Matricaria chamomilla, Mangifera indica,* and *Vitis rotundifolia* also show excellent antioxidant activity [69, 70]. Lanthanum oxide nanoparticles ($La_2O_3$ NPs) from *Moringa oleifera* also show excellent antioxidant activity, as they easily transport free electrons to free radicals of $N_2$ [70, 71].

### 3.2.8 Anticoagulant Activity

Coagulation, the process of blood turning from a liquid to a gel-like substance to form clots, can be harmful in certain diseases like allergies, injuries, and cardiovascular diseases. Researchers have synthesized biogenic silver and gold nanoparticles, cobalt oxide nanoparticles, nickel oxide nanoparticles, and titanium oxide nanoparticles to improve blood coagulation and thrombolytic activity [72, 73]. These nanoparticles have shown efficacy in maintaining red blood cell morphology and can be used to synthesize various metal and metal oxide nanoparticles with anti-coagulation properties [74, 75].

### 3.2.9 Neurogenerative Therapy

Neuronal activity is crucial in neurodegenerative diseases, leading to new treatments like those for Parkinson's disease [76, 77]. Biosynthesized nanoparticles of metal and metal oxides have potential applications in treating these diseases. Silver nanoparticles showed extraordinary neurodegenerative activity in human neuroblastoma cells [78]. Selenium treatment can reduce the risk of neurodegenerative diseases in animal models. Selenium nanoparticles exhibit antioxidant activity on neurons and the brain, reducing $H_2O_2$ and preventing oxidative damage. Early detection of selenium deficiency can help develop therapeutic approaches [79, 80]. Cerium oxide nanoparticles are excellent antioxidant agents, exhibiting neutron shielding effects and protecting against ischemic stroke. However, there is limited data on green synthesized metal and metal oxide nanoparticles for treating neurodegenerative diseases [81].

### 3.2.10 Bone-Related and Dental Therapeutics

Researchers have reported the use of nano-zirconia and silver in bone tissue regeneration, with $CeO_2$-incorporated mesoporous calcium-silicate ($CeO_2$-MCS) being a promising material for drug delivery [82]. Osteoporosis, a musculoskeletal ailment, has gained attention due to its lack of effective treatments. ZnO nanoparticles, derived from *Artemisia annua*, have shown potential in treating osteoporosis by stimulating osteoblast differentiation in MG-63 cells [83]. Since green synthesis produces nanoparticles that are biologically trustworthy, the dentistry industry uses them extensively these days. Instead of employing ceramic materials, biocompatible hydroxyapatite and titanium embedded into the alveolar bone and artificial tooth guarantee safety as a dental implant replacement. Alumina/zirconia nanocomposites were used to generate new implants that were more effective than ceramic materials [84]. In addition, it was shown that zirconia oxide nanoparticles functioned as polishing agents in dental offices while also demonstrating strong anti-biofilm activity against germs like *Enterococcus faecalis* [85]. Although more research is required, several additional metals and metal oxides produced by the biogenic method may also be able to treat tooth issues. Other potential metal and metal oxide nanoparticles, such as silver, copper, cerium, zinc, magnesium, and vanadium, have antimicrobial properties but need further research.

### 3.2.11 Cardiovascular Diseases

Cardiovascular diseases, including coronary heart diseases, atherosclerosis, angina pectoris, and myocardial infarction, can be diagnosed using biomarkers like myoglobin. AuNPs can inhibit vascular endothelial growth factor-stimulated angiogenesis and reduce vascular permeability, showing anti-angiogenic properties [86, 87]. Biogenic metals and metal oxide nanoparticles can be used to treat these diseases, and researchers should validate their utility and determine the mode of mechanism data for curing this deadly disease [72].

### 3.2.12 Wound Healing

Metal nanomaterials have shown potential wound-healing properties when combined with other wound dressing materials. Topical application of chitosan-capped AgNPs has been shown to accelerate the healing of burn wounds, decrease inflammation, and reduce the length of the repair phase [88]. AgNPs have been extensively used in wound dressings, pressure ulcer wound dressings, and clinical devices for microbial infection prevention. In vitro studies have shown a reduction in inflammatory cytokines and oxidative stress, promoting healing in human keratinocytes and dermal fibroblasts. In vivo studies have shown total healing achieved within two weeks in treated samples. However, long-term use may be hampered by toxicity, blue-grey coloration, and the emergence of silver-resistant microbes. Conjugation of nanomaterials with pathogen-specific antibodies may be applied for photothermal therapy, improving healing in infected wounds [89–91].

### 3.2.13 Antimicrobial

Drug/antibiotic resistance is a growing issue, necessitating the development of new drugs against microbes. Resistance can be attributed to gene transfer, antibiotic target adjustments, and mutational changes in biofilm formation and efflux pumps. Nanotechnology offers a platform to develop nanoparticles with potential applications as cell labelers, biomarkers, bioimaging contrast agents, antimicrobial agents, and drug delivery schemes. Nanomaterials have unique physicochemical characteristics, making them effective against multidrug-resistant bacteria. They can damage cell membranes and induce reactive oxygen species, destroying cell molecular machinery. Metallic nanoparticles, such as copper, titanium, magnesium, gold, silver, and zinc, have been proven to be antibiotic resistant at nano levels [92–94].

### 3.2.14 Biosensors

Sensors are increasingly important in various industries due to technological advancements and the need for reliable information. Nanoparticles, with their unique surface plasmonic properties, are primarily used in biosensing applications due to their sensitivity to size, shape, and dielectric medium. Surface plasma resonance spectroscopy measures minute changes in refractive index at or near the surface of nanoparticles [95, 96]. AgNPs have demonstrated remarkable glucose sensing capability, while Pt NPs and Pd NPs are enzyme-free electrochemical sensors for hydrogen peroxide, uric acid, and glucose. Metal oxide nanoparticles are also used in biosensing applications, including tumor cell molecular and serological diagnosis [97–99]. $TiO_2$ nanoparticles have also been applied to a wide range of sensing applications, such as biosensors for biological materials, gas sensors for volatile chemical and gas detection, and sensors for soluble organics in water as chemical oxygen demand sensors [95]. Penicillamine in aqueous solution was detected using CdO-based sensors [100].

### 3.2.15 Antiviral

The medical, pharmaceutical, and biotechnology industries face significant challenges due to viruses being the main cause of human disease and mortality. Despite advancements in antiviral chemotherapy drugs, preventive measures and treatment plans still fall short. Nanomaterials, such as silver nanoparticles (AgNPs), have shown strong antiviral properties against various viruses due to their unique properties and small sizes. AgNPs have a lower likelihood of developing resistance compared to traditional antiviral medications and can prevent viral entry into cells through multivalent interactions with cell membrane receptors and viral surface components. AgNPs may improve the effectiveness of current antiviral treatments and address viral infections [63, 64, 65, 101–103].

## 3.3 TARGETED DRUG DELIVERY

Nanoparticles are vesicular molecules which are at the nanoscale. When combined using colloidal systems, they can be used in medicinal form. The nanoscale is considered to be 10 to 1000 nm [104], but usually drugs at size 200 nm are not viable for therapeutic effects. Nanoparticles have several advantages over traditional drug delivery mechanisms, which is very important for curing future diseases. The distinct optical, mechanical, catalytic, and biological features of nanomaterials give rise to a wide range of possible applications. Electrochemistry, medical devices, cosmetics, and the textile industry are just a few of the industries that use metal and metal oxide nanoparticles (MONPs) extensively as a form of nanomaterial due to their high specific surface area, biocompatibility, UV light absorption, scattering, and antibacterial qualities [105].

### 3.3.1 Drug Delivery

Targeting specific tissue for a disease is always a difficult process, since the human body consists of more than 1 trillion cells, which is quite a large number of cells. Cells can have a problem anywhere, which can eventually lead to problems like cancer, or a small change in the folding of a protein moiety can lead to seriously damaged protein folding, which, on accumulation, leads to serious damage. All these diseases can be cured completely by targeting that specific tissue. The identification of nanoparticles provided insights on therapeutic interventions for curing various diseases. Thus, the advancement of nanotechnology enables us to overcome many limitations of conventional drug delivery systems such as restricted biodistribution and inefficient intracellular trafficking within the cells.

In early times, drug design and delivery mechanisms were not well established or studied well enough to produce solid particles. Alec Douglas Bangham was the first person who put forward the idea of using liposomes as a drug in the form of NPs [106]. NPs are usually used in colloidal form for maximum diffusion into the cell. There are reports on employing solid NPs as a carrier, which improves the loading and unloading of cargo in cells more efficiently than liposomes, which was the primary study that influenced the study of metal and metal oxide NPs [107]. MNPs/MONPs have the advantage of having a surface charge that forms an electrostatic attraction to a drug and efficiently loads drugs; even loading multiple drugs is possible. The presence of a magnetic field inside the MNP very advantageous over traditional delivery since they have an attraction to cells that are infected [107]. This reduces accumulation inside the healthy cells, which is an advantage. Some MNPs have a magnetic effect, which is surprisingly helpful to attract them directly to damaged cells for more targeted attacks using an external magnetic field. This will produce localized death of cells that are damaged, and other cells remain unharmed [108].

Recent works concluded that the traditional drug delivery mechanism has several disadvantages, such as poor efficiency, lack of distribution, and sometimes poor sensitivity. These can be mitigated by a controlled drug delivery system. They can reduce toxicity due to biocompatibility, drug accumulation, and other damage, which is helpful to reduce the after-effects of the drug. Later studies put forward the idea of using drug vehicles, which is also an improvement to the mechanism. Currently, using plasmid vectors, drugs are targeted to cells, which produces a synergetic effect for the NPs. These new advantages are categorized as the delivery of NPs based on (1) using agents like mRNA, siRNA, and plasmid; (2) using a co-delivery mechanism; (3) stimulus-responsive drug delivery; and 4) receptor-targeted cells [109, 110].

### 3.3.2 Future Perspectives of Drug Delivery

In the 21st century, we use drug targeting systems, but most of them are chemical in nature, which sometimes causes damage to other organs. To replace chemicals, we can use biological, environmentally friendly plant extracts [111]. NP robots and other carriers are the future of medicine, but we still need to find answers to many questions and need better solutions to other problems like aggregation. Some green synthesis pathways require a complicated process of extractions and purification of metals, which is time consuming. The unpredictability of many of the pathways still needs to be explored. If we knew the pathways well enough, we could alter them according to our needs. The potentially toxic effects of many MNPs and MONPs should be studied well, and alternative methods should be established.

The size of NPs is very significant in some cases. More efficient mechanisms need to be invented. Right now, we know that metal and metal oxide nanoparticles are very efficient in preventing cancer, but their cytotoxic effects and biocompatibility needed to be studied further. To prevent damage to normal cells, vital investigations and trials are required. Nanotechnology offers safe, reliable, and effective therapeutic approaches for addressing incurable diseases with unparalleled precision.

## 3.4 SENSORS

In the rapidly advancing technological landscape, the demand for precise information across various fields has significantly elevated the importance of sensors, with key attributes including high sensitivity, quick response, and reliable selectivity. Nanoparticles, owing to their extremely small size and large surface-to-volume ratio, have emerged as crucial components capable of substantially enhancing sensor sensitivity compared to traditional materials [95]. The application of immunoassays based on biomolecules proves instrumental in measuring pesticide residue in soil, offering high selectivity, sensitivity, reliability, and rapidity [112]. Gold nanoparticles (30 nm) employed in dipstick competitive immunoassays demonstrate a sensitivity of 27 mg $ml^{-1}$ for the detection of organochlorine pesticides such as DDT [113].

Green synthesis techniques, utilizing natural extracts, have been explored for the creation of advanced sensor materials. For instance, the aqueous extract of the *Ficus johannis* plant serves as the basis for the synthesis of cadmium telluride quantum dots (CdTe QDs), which, in turn, can be employed in the detection of the herbicide 2,4-Dn [114]. Similarly, the green synthesis of $ZrO_2$ using bacteria, fungi, algae, and plants offers versatility in applications, including the detection of dimethoate, an organophosphorus pesticide [115, 116]. Copper oxide (CuO) nanoparticles synthesized from *Fortunella japonica* fruit extract demonstrate effectiveness in sensors designed for the detection of 4-nitrophenol (4-NP). The resulting sensor, based on CuO nanoparticles modified screen-printed electrodes (SPEs), exhibits a linear range from 10 nM to 10 mM, demonstrating high reproducibility and selectivity for 4-NP. The small size of the SPE/CuONPs allows for the development of a compact device suitable for on-site detection of 4-NP [117].

Innovative approaches involving biopolymers are also evident in sensor development. Luo et al. utilized xylan for the green synthesis of silver nanoparticles (AgNPs), acting as both a stabilizing and reducing agent. These xylan-AgNP composites exhibit high sensitivity to $Hg^{2+}$ ions, surpassing WHO recommended levels, making them effective sensors for detecting harmful metal ions like $Hg^{2+}$. Another noteworthy sensor incorporates gold nanoparticles (AuNPs) in tin dioxide ($SnO_2$) and exhibits rapid response and excellent sensitivity to ethanol vapor within the 10 to 120 ppm range at room temperature. The AuNP-doped $SnO_2$ sensor demonstrates a twofold improvement in ethanol vapor sensing compared to the pure $SnO_2$ sensor [118].

Efficient reduction of graphene oxide (GO) using *Azadirachta indica* extract results in the synthesis of reduced GO (rGO)/mono and bimetallic composites, featuring silver (Ag) and silver–gold (Ag–Au) nanoparticles. Characterized by a crystalline structure, these nanostructures show promise in electrocatalytic applications, particularly in the reduction of hydrogen peroxide ($H_2O_2$). The resulting sensor demonstrates a remarkable linear response to $H_2O_2$ concentrations within a specific range, highlighting its potential as a non-enzymatic electrochemical sensor with a low detection limit [119].

In addition, innovative nanocomposites, such as Ag nanoparticle-graphene nanocomposites (AgNPs-G), synthesized through an eco-friendly one-pot approach with tannic acid, exhibit exceptional surface-enhanced Raman scattering (SERS) activity and remarkable catalytic performance in $H_2O_2$ reduction. These nanocomposites demonstrate rapid amperometric response, a wide linear range, and a low detection limit, showcasing their potential in enzymeless sensors and biosensing applications [120]. Moreover, the utilization of natural extracts, such as locust bean gum (LBG), for the creation of polysaccharide-based silver nanoparticles (AgNPs) demonstrates their potential as optical sensors. These LBG-stabilized AgNPs exhibit sensitivity to lower concentrations of $H_2O_2$, making them promising candidates for cost-effective, stable, and portable biosensors for industrial $H_2O_2$ detection [121].

Advancements in microbial detection technology involve the use of silica-based nanoparticles (60 nm) filled with a fluorescent dye and conjugated to specific antibodies for targeted detection of microorganisms [122]. Triangular silver nanoparticles, created through nanosphere lithography, serve as highly sensitive and selective nanoscale affinity biosensors, incorporating favorable aspects of surface plasmon resonance (SPR) spectroscopy. These nanosensors exhibit distinct characteristics, including moderate refractive sensitivity and a short-range sensing length scale, resulting in a mass sensitivity area ranging from 100 to 1000 $pg/mm^2$ [123].

Furthermore, metal oxide nanoparticles derived from oat biomass are employed in the creation of modified sensors, specifically CPX-PM-$MnO_2$ NPs and CTX-PT-CuO NPs. These electroactive materials enhance sensitivity, allowing accurate quantification of antibacterial drugs across a broad range of concentrations, making them suitable for routine analysis in pharmaceutical and central research laboratories [124].

Label-free biosensing and surface biofunctionalization with user-friendly, non-toxic silver nanoparticles have been demonstrated by assessing the adsorption of bovine serum albumin (BSA) in a model sensing experiment. While slightly less responsive than lithographically fabricated nanostructures, these particles offer advantages as label-free biosensors [98]. Copper oxide nanoparticles (CuO NPs), synthesized using an extract from the leaves of *Amaranthus caudatus*, exhibit significant sensitivity towards glucose sugar, making them potential candidates for glucose sensing applications [125].

Researchers emphasize the advantages of metal nanoparticles deposited on reduced graphene oxide in sensor applications due to their improved electrochemical properties [122, 126]. Carbonaceous materials, including graphite, graphene oxide, reduced graphene oxide, and multi-walled carbon nanotubes (MWCNTs), serve as supports for depositing metal nanostructures. Among these, silver nanoparticles (AgNPs) have gained attention for their excellent electrocatalytic activity in sensor applications [126]. In conclusion, the integration of green synthesis methods and the utilization of diverse nanomaterials hold immense promise in the development of advanced sensors with enhanced sensitivity, selectivity, and rapid response times. These innovative sensor technologies pave the way for applications in various fields, including environmental monitoring, industrial research, and healthcare.

## REFERENCES


1. Wang, L., C. Hu, and L. Shao, The antimicrobial activity of nanoparticles: present situation and prospects for the future. *International Journal of Nanomedicine*, 2017. 12: p. 1227–1249.
2. Rudramurthy, G.R., et al., Nanoparticles: alternatives against drug-resistant pathogenic microbes. *Molecules*, 2016. 21(7).
3. Singh, J., et al., Green' synthesis of metals and their oxide nanoparticles: applications for environmental remediation. *Journal of Nanobiotechnology*, 2018. 16(1): p. 84.
4. Zuo, J., et al., Preparation, properties and antibacterial applications of medical nano-metals and their oxides: a review. *Sheng wu gong cheng xue bao = Chinese Journal of Biotechnology*, 2023. 39(4): p. 1462–1476.
5. Stankic, S., et al., Pure and multi metal oxide nanoparticles: synthesis, antibacterial and cytotoxic properties. *Journal of Nanobiotechnology*, 2016. 14(1): p. 73.
6. Padmavathy, N. and R. Vijayaraghavan, Enhanced bioactivity of ZnO nanoparticles-an antimicrobial study. *Science and Technology of Advanced Materials*, 2008. 9(3): p. 035004.
7. Li, J., et al., Enhanced bioactivity and bacteriostasis effect of TiO2 nanofilms with favorable biomimetic architectures on titanium surface. *RSC Advances*, 2013. 3(28): p. 11214–11225.
8. Gupta, A., et al., Synergistic antimicrobial therapy using nanoparticles and antibiotics for the treatment of multidrug-resistant bacterial infection. *Nano Futures*, 2017. 1(1): p. 015004.
9. Jiménez Pérez, Z.E., et al., Ginseng-berry-mediated gold and silver nanoparticle synthesis and evaluation of their in vitro antioxidant, antimicrobial, and cytotoxicity effects on human dermal fibroblast and murine melanoma skin cell lines. *International Journal of Nanomedicine*, 2017. 12: p. 709–723.
10. Kelkawi, A.H.A., A. Abbasi Kajani, and A.K. Bordbar, Green synthesis of silver nanoparticles using Mentha pulegium and investigation of their antibacterial, antifungal and anticancer activity. *IET Nanobiotechnology*, 2017. 11(4): p. 370–376.
11. Dehghanizade, S., J. Arasteh, and A. Mirzaie, Green synthesis of silver nanoparticles using Anthemis atropatana extract: characterization and in vitro biological activities. *Artificial Cells, Nanomedicine, and Biotechnology*, 2018. 46(1): p. 160–168.
12. Manikandan, V., et al., Green synthesis of silver oxide nanoparticles and its antibacterial activity against dental pathogens. *3 Biotech*, 2017. 7(1): p. 72.
13. Chahardoli, A., et al., Green approach for synthesis of gold nanoparticles from Nigella arvensis leaf extract and evaluation of their antibacterial, antioxidant, cytotoxicity and catalytic activities. *Artificial Cells, Nanomedicine, and Biotechnology*, 2018. 46(3): p. 579–588.
14. Skandalis, N., et al., The effect of silver nanoparticles size, produced using plant extract from arbutus unedo, on their antibacterial efficacy. *Nanomaterials (Basel)*, 2017. 7(7).
15. Patra, J.K. and K.H. Baek, Antibacterial activity and synergistic antibacterial potential of biosynthesized silver nanoparticles against foodborne pathogenic bacteria along with its anticandidal and antioxidant effects. *Frontiers in Microbiology*, 2017. 8: p. 167.
16. Yuan, C.G., et al., Green synthesis of gold nanoparticles using Citrus maxima peel extract and their catalytic/antibacterial activities. *IET Nanobiotechnology*, 2017. 11(5): p. 523–530.
17. Kanjikar, A.P., A.L. Hugar, and R.L. Londonkar, Characterization of phyto-nanoparticles from Ficus krishnae for their antibacterial and anticancer activities. *Drug Development and Industrial Pharmacy*, 2018. 44(3): p. 377–384.
18. Naraginti, S. and Y. Li, Preliminary investigation of catalytic, antioxidant, anticancer and bactericidal activity of green synthesized silver and gold nanoparticles using Actinidia deliciosa. *Journal of Photochemistry and Photobiology B: Biology*, 2017. 170: p. 225–234.
19. Zia, M., et al., Green synthesis of silver nanoparticles from grape and tomato juices and evaluation of biological activities. *IET Nanobiotechnology*, 2017. 11(2): p. 193–199.
20. Banumathi, B., et al., Toxicity of herbal extracts used in ethno-veterinary medicine and green-encapsulated ZnO nanoparticles against Aedes aegypti and microbial pathogens. *Parasitology Research*, 2017. 116(6): p. 1637–1651.
21. Dinesh, M., et al., Phyllanthus emblica seed extract mediated synthesis of PdNPs against antibacterial, heamolytic and cytotoxic studies. *Journal of Photochemistry and Photobiology B: Biology*, 2017. 167: p. 64–71.
22. Rajakumar, G., et al., Evaluation of anti-cholinesterase, antibacterial and cytotoxic activities of green synthesized silver nanoparticles using from Millettia pinnata flower extract. *Microbial Pathogenesis*, 2017. 103: p. 123–128.
23. Kumar, V.A., et al., Photo-induced and phytomediated synthesis of silver nanoparticles using Derris trifoliata leaf extract and its larvicidal activity against Aedes aegypti. *Journal of Photochemistry and Photobiology B: Biology*, 2017. 171: p. 1–8.
24. Azizi, S., R. Mohamad, and M. Mahdavi Shahri, Green microwave-assisted combustion synthesis of zinc oxide nanoparticles with Citrullus colocynthis (L.) Schrad: characterization and biomedical applications. *Molecules*, 2017. 22(2).
25. Jadhav, K., et al., Green and ecofriendly synthesis of silver nanoparticles: characterization, biocompatibility studies and gel formulation for treatment of infections in burns. *Journal of Photochemistry and Photobiology B: Biology*, 2016. 155: p. 109–115.
26. Khan, F.U., et al., Visible light inactivation of E. coli, Cytotoxicity and ROS determination of biochemically capped gold nanoparticles. *Microbial Pathogenesis*, 2017. 107: p. 419–424.
27. Sirelkhatim, A., et al., Review on zinc oxide nanoparticles: antibacterial activity and toxicity mechanism. *Nanomicro Letters*, 2015. 7(3): p. 219–242.
28. Applerot, G., et al., Understanding the antibacterial mechanism of CuO nanoparticles: revealing the route of induced oxidative stress. *Small*, 2012. 8(21): p. 3326–3337.
29. Xie, Y., et al., Antibacterial activity and mechanism of action of zinc oxide nanoparticles against Campylobacter jejuni. *Applied and Environmental Microbiology*, 2011. 77(7): p. 2325–2331.
30. Imade, E.E., et al., Green synthesis of zinc oxide nanoparticles using plantain peel extracts and the evaluation of their antibacterial activity. *Scientific African*, 2022. 16: p. e01152.
31. Zhang, L., et al., Investigation into the antibacterial behaviour of suspensions of ZnO nanoparticles (ZnO nanofluids). *Journal of Nanoparticle Research*, 2007. 9(3): p. 479–489.
32. Liang, X., et al., Preparation and antibacterial activities of polyaniline/Cu0.05Zn0.95O nanocomposites. *Dalton Transactions*, 2012. 41(9): p. 2804–2811.

33. Kang, M.G. and F. Khan, Antibiofilm and antivirulence activities of gold and zinc oxide nanoparticles synthesized from kimchi-isolated Leuconostoc sp. *Strain C2*. 2022. 11(11).
34. Kommidi, D.R., et al., Novel carbapenem chalcone derivatives: synthesis, cytotoxicity and molecular docking studies. *Organic & Biomolecular Chemistry*, 2015. 13(14): p. 4344–4350.
35. Alarfaj, N., et al., Prospective of green synthesized Oleum cumini Oil/PVP/MgO bionanocomposite film for its antimicrobial, antioxidant and anticancer applications. *Journal of Polymers and the Environment*, 2020. 28.
36. Syame, S., et al., Synthesis of copper-chitosan nanocomposites and its application in treatment of local pathogenic isolates bacteria. *Oriental Journal of Chemistry*, 2017. 33: p. 2959–2969.
37. Gholami-Shabani, M. and F. Sotoodehnejadnematalahi, Physicochemical properties, anticancer and antimicrobial activities of metallic nanoparticles green synthesized by Aspergillus kambarensis. *IET Nanobiotechnology*, 2022. 16(1): p. 1–13.
38. Yadollahi, M., et al., Synthesis and characterization of antibacterial carboxymethyl cellulose/ZnO nanocomposite hydrogels. *International Journal of Biological Macromolecules*, 2015. 74: p. 136–141.
39. Nguyen, T., et al., Green synthesis of silver nanoparticles using plectranthus amboinicus leaf extract for preparation of CMC/PVA nanocomposite film. *Journal of Renewable Materials*, 2021. 9: p. 1393–1411.
40. Salem, S.S., et al., Synthesis of silver nanocomposite based on carboxymethyl cellulose: antibacterial, antifungal and anticancer activities. *Polymers*, 2022. 14(16): p. 3352.
41. Devi, K.S., et al., Green synthesis of copper oxide nanoparticles using coix lacryma jobi leaves extract and screening of its potential anticancer activities. *Journal of Pharmaceutical Research International*, 2021. 33(52A): p. 128–139.
42. Amiryaghoubi, N., et al., Hydrogel-based scaffolds for bone and cartilage tissue engineering and regeneration. *Reactive and Functional Polymers*, 2022. 177: p. 105313.
43. Li, J. and S. Zhuang, Antibacterial activity of chitosan and its derivatives and their interaction mechanism with bacteria: current state and perspectives. *European Polymer Journal*, 2020. 138: p. 109984.
44. Mitchell, M.J., M.M. Billingsley, and R.M. Haley, Engineering precision nanoparticles for drug delivery. *Nature Reviews Drug Discovery*, 2021. 20(2): p. 101–124.
45. Zhang, Di, Guoxun Wang, Xueliang Yu, Tuo Wei, Lukas Farbiak, Lindsay T. Johnson, Alan Mark Taylor et al. "Enhancing CRISPR/Cas gene editing through modulating cellular mechanical properties for cancer therapy." *Nature nanotechnology* 17, no. 7 (2022): 777–787.
46. Zeghoud, S., et al., A review on biogenic green synthesis of ZnO nanoparticles by plant biomass and their applications. *Materials Today Communications*, 2022. 33: p. 104747.
47. Khan, T., et al., Plant-based gold nanoparticles; a comprehensive review of the decade-long research on synthesis, mechanistic aspects and diverse applications. *Advances in Colloid and Interface Science*, 2019. 272: p. 102017.
48. Harish, V. and D. Tewari, Review on nanoparticles and nanostructured materials: bioimaging, biosensing, drug delivery, tissue engineering, antimicrobial, and agro-food applications. *Nanomaterials*, 2022. 12(3).
49. Kalpana, V.N. and V. Devi Rajeswari, A review on green synthesis, biomedical applications, and toxicity studies of ZnO NPs. *Bioinorganic Chemistry and Applications*, 2018. 2018: p. 3569758.
50. Radulescu, Denisa-Maria, Vasile-Adrian Surdu, Anton Ficai, Denisa Ficai, Alexandru-Mihai Grumezescu, and Ecaterina Andronescu. "Green synthesis of metal and metal oxide nanoparticles: a review of the principles and biomedical applications." *International Journal of Molecular Sciences* 24, no. 20 (2023): 15397.
51. Gmeiner, W.H. and S. Ghosh, Nanotechnology for cancer treatment. *Nanotechnology Reviews*, 2015. 3(2): p. 111–122.
52. Saravanakumar, K., et al., Biosynthesis and characterization of copper oxide nanoparticles from indigenous fungi and its effect of photothermolysis on human lung carcinoma. *Journal of Photochemistry and Photobiology B: Biology*, 2019. 190: p. 103–109.
53. Al-Sheddi, E.S. and N.N. Farshori, Anticancer potential of green synthesized silver nanoparticles using extract of Nepeta deflersiana against human cervical cancer cells (HeLA). *Bioinorganic Chemistry and Applications*, 2018. 2018: p. 9390784.
54. Pugazhendhi, A., et al., Anticancer, antimicrobial and photocatalytic activities of green synthesized magnesium oxide nanoparticles (MgONPs) using aqueous extract of Sargassum wightii. *Journal of Photochemistry and Photobiology B: Biology*, 2019. 190: p. 86–97.
55. Hannachi, E. and F.A. Khan, In vitro antimicrobial and anticancer peculiarities of ytterbium and cerium co-doped zinc oxide nanoparticles. *Biology*, 2022. 11(12).
56. Rehana, D., et al., Evaluation of antioxidant and anticancer activity of copper oxide nanoparticles synthesized using medicinally important plant extracts. *Biomedicine & Pharmacotherapy*, 2017. 89: p. 1067–1077.
57. Bonan, R.F., et al., In vitro antimicrobial and anticancer properties of TiO(2) blow-spun nanofibers containing silver nanoparticles. *Materials Science and Engineering: C*, 2019. 104: p. 109876.
58. Kumar, S., M.S. Tomar, and A. Acharya, Carboxylic group-induced synthesis and characterization of selenium nanoparticles and its anti-tumor potential on Dalton's lymphoma cells. *Colloids Surf B Biointerfaces*, 2015. 126: p. 546–552.
59. Ma, Y., et al., Targeting strategies for enhancing paclitaxel specificity in chemotherapy. *Frontiers in Cell and Developmental Biologyt*, 2021. 9: p. 626910.
60. Karthik, K., et al., Fabrication of MgO nanostructures and its efficient photocatalytic, antibacterial and anticancer performance. *Journal of Photochemistry and Photobiology B: Biology*, 2019. 190: p. 8–20.
61. Almatroudi, A., Silver nanoparticles: synthesis, characterisation and biomedical applications. *Open Life Sciences*, 2020. 15(1): p. 819–839.
62. Bayrami, A., et al., Enriched zinc oxide nanoparticles by Nasturtium officinale leaf extract: joint ultrasound-microwave-facilitated synthesis, characterization, and implementation for diabetes control and bacterial inhibition. *Ultrasonics Sonochemistry*, 2019. 58: p. 104613.
63. Prabhu, S., et al., Retracted: Evaluation of antidiabetic activity of biologically synthesized silver nanoparticles using Pouteria sapota in streptozotocin-induced diabetic rats. *Journal of Diabetes*, 2018. 10(1): p. 28–42.
64. Viscido, A., et al., Nanotechnology in the treatment of inflammatory bowel diseases. *Journal of Crohn's and Colitis*, 2014. 8(9): p. 903–918.

65. Wang, L., et al., Manipulation of macrophage polarization by peptide-coated gold nanoparticles and its protective effects on acute lung injury. *Journal of Nanobiotechnology*, 2020. 18(1): p. 38.
66. Yadav, E., et al., Ameliorative effect of biofabricated ZnO nanoparticles of Trianthema portulacastrum Linn. on dermal wounds via removal of oxidative stress and inflammation. *RSC Advances*, 2018. 8(38): p. 21621–21635.
67. Singh, Kshitij RB, Vanya Nayak, Tanushri Sarkar, and Ravindra Pratap Singh. "Cerium oxide nanoparticles: properties, biosynthesis and biomedical application." *RSC advances* 10, no. 45 (2020): 27194–27214.
68. Assis, Mariane Brandão Da Silva, Gabriela Nestal De Moraes, and Kátia Regina De Souza. "Cerium oxide nanoparticles: Chemical properties, biological effects and potential therapeutic opportunities." *Biomedical Reports* 20, no. 3 (2024): 48.
69. Singh, K.R. and V. Nayak, Cerium oxide nanoparticles: properties, biosynthesis and biomedical application. *RSC Advances*, 2020. 10(45): p. 27194–27214.
70. Gowda, U., et al., Tinospora cordifolia mediated facile green synthesis of cupric oxide nanoparticles and their photocatalytic, antioxidant and antibacterial properties. *Materials Science in Semiconductor Processing*, 2015. 33: p. 81–88.
71. Das, D., et al., Synthesis and evaluation of antioxidant and antibacterial behavior of CuO nanoparticles. *Colloids Surf B Biointerfaces*, 2013. 101: p. 430–433.
72. Zhang, X.F., et al., Silver nanoparticles: synthesis, characterization, properties, applications, and therapeutic approaches. *International Journal of Molecular Sciences*, 2016. 17(9).
73. Gurunathan, S., et al., Multidimensional effects of biologically synthesized silver nanoparticles in Helicobacter pylori, Helicobacter felis, and human lung (L132) and lung carcinoma A549 cells. *Nanoscale Research Letters*, 2015. 10(1): p. 35.
74. Chopra, H., et al., Green metallic nanoparticles: biosynthesis to applications. *Frontiers in Bioengineering and Biotechnology*, 2022. 10.
75. Akinola, P.O., et al., Multifunctional titanium dioxide nanoparticles biofabricated via phytosynthetic route using extracts of Cola nitida: antimicrobial, dye degradation, antioxidant and anticoagulant activities. *Heliyon*, 2020. 6(8): p. e04610.
76. Casson, R., et al., Definition of glaucoma: clinical and experimental concepts. *Clinical & Experimental Ophthalmology*, 2012. 40: p. 341–349.
77. Dunnett, S.B. and A. Björklund, Prospects for new restorative and neuroprotective treatments in Parkinson's disease. *Nature*, 1999. 399(6738 Suppl): p. A32–A39.
78. Abdal Dayem, A., et al., The role of reactive oxygen species (ROS) in the biological activities of metallic nanoparticles. *International Journal of Molecular Sciences*, 2017. 18(1).
79. Aaseth, J., et al., Treatment strategies in Alzheimer's disease: a review with focus on selenium supplementation. *Biometals*, 2016. 29(5): p. 827–839.
80. Huang, B., et al., Free radical scavenging efficiency of Nano-Se in vitro. *Free Radical Biology and Medicine*, 2003. 35(7): p. 805–813.
81. Kim, C.K., et al., Ceria nanoparticles that can protect against ischemic stroke. *Angewandte Chemie*, 2012. 51(44): p. 11039–11043.
82. Zhang, J. and Y. Zhu, Synthesis and characterization of CeO2-incorporated mesoporous calcium-silicate materials. *Microporous and Mesoporous Materials*, 2014. 197: p. 244–251.
83. Sabir, S., et al., Biosynthesis of ZnO nanoparticles using Bacillus subtilis: characterization and nutritive significance for promoting plant growth in Zea mays L. *Dose-Response*, 2020. 18(3): p. 1559325820958911.
84. Bartolomé, José F., Antonio H. De Aza, Antonia Martín, Jose Y. Pastor, Javier Llorca, Ramon Torrecillas, and Giovanni Bruno. "Alumina/zirconia micro/nanocomposites: a new material for biomedical applications with superior sliding wear resistance." *Journal of the American Ceramic Society* 90, no. 10 (2007): 3177–3184.
85. Pushpalatha, C., et al., Zinc oxide nanoparticles: a review on its applications in dentistry. *Frontiers in Bioengineering and Biotechnology*, 2022. 10: p. 917990.
86. Shumyantseva, V.V., et al., Electrochemical methods in biomedical studies. *Biochemistry (Moscow) Supplement Series B: Biomedical Chemistry*, 2015. 9(3): p. 228–243.
87. Mukherjee, P., et al., Antiangiogenic properties of gold nanoparticles. *Clinical Cancer Research*, 2005. 11(9): p. 3530–3534.
88. Oryan, A., et al., Topical delivery of chitosan-capped silver nanoparticles speeds up healing in burn wounds: a preclinical study. *Carbohydrate Polymers*, 2018. 200: p. 82–92.
89. Nascimento, E.G., et al., Evaluation of chitosan gel with 1% silver sulfadiazine as an alternative for burn wound treatment in rats. *ACTA Cirúrgica Brasileira*, 2009. 24(6): p. 460–465.
90. Rice, L.B., The clinical consequences of antimicrobial resistance. *Current Opinion in Microbiology*, 2009. 12(5): p. 476–481.
91. Lateef, A., The microbiology of a pharmaceutical effluent and its public health implications. *World Journal of Microbiology and Biotechnology*, 2004. 20(2): p. 167–171.
92. Lateef, A., et al., Akara Ogbomoso: microbiological examination and identification of hazards and critical control points. *Food Science and Technology International*, 2010. 16(5): p. 389–400.
93. Khan, A.U. and S. Parvez, Detection of bla(NDM-4) in Escherichia coli from hospital sewage. *Journal of Medical Microbiology*, 2014. 63(Pt 10): p. 1404–1406.
94. Qayyum, S. and A.U. Khan, Biofabrication of broad range antibacterial and antibiofilm silver nanoparticles. *IET Nanobiotechnology*, 2016. 10(5): p. 349–357.
95. Bai, J. and B. Zhou, Titanium dioxide nanomaterials for sensor applications. *Chemical Reviews*, 2014. 114(19): p. 10131–10176.
96. Arai, T., et al., An optical biosensor based on localized surface plasmon resonance of silver nanostructured films. *Journal of Optics A: Pure and Applied Optics*, 2007. 9: p. 699.
97. Haes, A.J. and R.P. Van Duyne, A nanoscale optical biosensor: sensitivity and selectivity of an approach based on the localized surface plasmon resonance spectroscopy of triangular silver nanoparticles. *Journal of the American Chemical Society*, 2002. 124(35): p. 10596–10604.
98. Sotiriou, G.A., et al., Non-toxic dry-coated nanosilver for plasmonic biosensors. *Advanced Functional Materials*, 2010. 20(24): p. 4250–4257.
99. Chen, X., et al., Advances in enzyme-free electrochemical sensors for hydrogen peroxide, glucose, and uric acid. *Microchimica Acta*, 2013. 181: p. 689–705.
100. Khalilzadeh, B., et al., Preparation of a new electrochemical sensor based on cadmium oxide nanoparticles and application for determination of penicillamine. *International Journal of Electrochemical Science*, 2011. 6: p. 4164–4175.

101. Strasser, P., et al., Lattice-strain control of the activity in dealloyed core-shell fuel cell catalysts. *Nature Chemistry*, 2010. 2(6): p. 454–460.
102. Rai, M., et al., Metal nanoparticles: the protective nanoshield against virus infection. *Critical Reviews in Microbiology*, 2016. 42(1): p. 46–56.
103. Sharma, V., et al., Green synthesis of silver nanoparticles from medicinal plants and evaluation of their antiviral potential against chikungunya virus. *Applied Microbiology and Biotechnology*, 2019. 103(2): p. 881–891.
104. Vijayaram, S., et al., Applications of green synthesized metal nanoparticles – a review. *Biological Trace Element Research*, 2024. 202(1): p. 360–386.
105. Shenashen, M., S. El-Safty, and E. Elshehy, Synthesis, morphological control, and properties of silver nanoparticles in potential applications. *Particle & Particle Systems Characterization*, 2014. 31.
106. Ge, X., et al., Advances of non-ionic surfactant vesicles (niosomes) and their application in drug delivery. *Pharmaceutics*, 2019. 11(2).
107. McBain, S.C., H.H. Yiu, and J. Dobson, Magnetic nanoparticles for gene and drug delivery. *International Journal of Nanomedicine*, 2008. 3(2): p. 169–180.
108. Amreddy, N., et al., Recent advances in nanoparticle-based cancer drug and gene delivery. *Advances in Cancer Research*, 2018. 137: p. 115–170.
109. Müller, R.H., K. Mäder, and S. Gohla, Solid lipid nanoparticles (SLN) for controlled drug delivery – a review of the state of the art. *European Journal of Pharmaceutics and Biopharmaceutics*, 2000. 50(1): p. 161–177.
110. Xu, J., A. Singh, and M.M. Amiji, Redox-responsive targeted gelatin nanoparticles for delivery of combination wt-p53 expressing plasmid DNA and gemcitabine in the treatment of pancreatic cancer. *BMC Cancer*, 2014. 14(1): p. 75.
111. Radulescu, D.M. and V.A. Surdu, Green synthesis of metal and metal oxide nanoparticles: a review of the principles and biomedical applications. *International Journal of Molecular Sciences*, 2023. 24(20).
112. Gabaldón, J.A., A. Maquieira, and R. Puchades, Current trends in immunoassay-based kits for pesticide analysis. *Critical Reviews in Food Science and Nutrition*, 1999. 39(6): p. 519–538.
113. Lisa, M., et al., Gold nanoparticles based dipstick immunoassay for the rapid detection of dichlorodiphenyltrichloroethane: an organochlorine pesticide. *Biosensors and Bioelectronics*, 2009. 25(1): p. 224–227.
114. Vinayaka, A.C., S. Basheer, and M.S. Thakur, Bioconjugation of CdTe quantum dot for the detection of 2,4-dichlorophenoxyacetic acid by competitive fluoroimmunoassay based biosensor. *Biosensors and Bioelectronics*, 2009. 24(6): p. 1615–1620.
115. Gan, N., et al., A disposable organophosphorus pesticides enzyme biosensor based on magnetic composite nanoparticles modified screen printed carbon electrode. *Sensors (Basel)*, 2010. 10(1): p. 625–638.
116. Tran, T.V., et al., Green synthesis of ZrO(2) nanoparticles and nanocomposites for biomedical and environmental applications: a review. *Environmental Chemistry Letters*, 2022. 20(2): p. 1309–1331.
117. Singh, S., et al., Electrochemical sensing and remediation of 4-nitrophenol using bio-synthesized copper oxide nanoparticles. *The Chemical Engineering Journal*, 2016. 313.
118. Tagad, C., et al., Green synthesis of polysaccharide stabilized gold nanoparticles: chemo catalytic and room temperature operable vapor sensing application. *RSC Advances*, 2014. 4: p. 24014–24019.
119. Kumar, G., et al., A facile one-pot green synthesis of reduced graphene oxide and its composites for non-enzymatic hydrogen peroxide sensor applications. *RSC Advances*, 2014. 4: p. 7944.
120. Zhang, Y., et al., One-pot green synthesis of Ag nanoparticles-graphene nanocomposites and their applications in SERS, H2O2, and glucose sensing. *RSC Advances*, 2011. 2: p. 538–545.
121. Tagad, Chandrakant K., Hyeong U. Kim, R. C. Aiyer, Pooja More, Taesung Kim, Sang Hyun Moh, Atul Kulkarni, and Sushma G. Sabharwal. "A sensitive hydrogen peroxide optical sensor based on polysaccharide stabilized silver nanoparticles." *RSC advances* 3, no. 45 (2013): 22940–22943.
122. Zhao, L., et al., Green preparation of Ag-Au bimetallic nanoparticles supported on graphene with alginate for non-enzymatic hydrogen peroxide detection. *Nanomaterials*, 2018. 8(7).
123. Ahmed, S., et al., A review on plants extract mediated synthesis of silver nanoparticles for antimicrobial applications: a green expertise. *Journal of Advanced Research*, 2016. 7(1): p. 17–28.
124. Al-Tamimi, S., Biogenic green synthesis of metal oxide nanoparticles using oat biomass for ultrasensitive modified polymeric sensors. *Green Chemistry Letters and Reviews*, 2021. 14: p. 165–178.
125. Ansari, P., et al., Green synthesis of copper oxide nanoparticles using Amaranthus caudatus leaf extract and its non-enzymatic glucose sensor application. *Applied Physics A*, 2023. 129.
126. Lorestani, F., et al., One-step hydrothermal green synthesis of silver nanoparticle-carbon nanotube reduced-graphene oxide composite and its application as hydrogen peroxide sensor. *Sensors and Actuators B: Chemical*, 2014. 208.

# 4 Harnessing Metallic Nanoparticles for Next-Generation Sensor Technologies

*Jincemon Cyriac[1, 2], Sheeja V. N.[2], Gayathri M. J.[2], Bindhu R. G.[3], Riju K. Thomas[4], Ginu V. G.[4], and K. Sreelatha[5]*

[1]Department of Physics, Christ College (Autonomous), Calicut University, Irinjalakuda, Thrissur, Kerala, India

[2]Department of Physics, Marian College (Autonomous), Kuttikkanam P.O., Peermade, Idukki, Kerala, India

[3]Department of Physics, NSS College, Nilamel, Kollam, Kerala, India

[4]Department of Physics, Bharata Mata College, Thrikkakara, Edappally, Kochi, Kerala, India

[5]Department of Physics, Sree Narayana College, Chengannur, Kerala, India

## 4.1 INTRODUCTION

Metal nanomaterials are metal-based substances manipulated at the nanoscale, typically measuring 1 to 100 nanometers. These materials showcase distinct properties from their larger counterparts due to their small size. They encompass various metals like gold, silver, iron, and platinum, including alloys and metal oxides. Their unique properties make them valuable in catalysis, electronics, medicine, and energy applications, thanks to their customizable optical, electronic, and catalytic characteristics. Researchers worldwide are fascinated by their versatility, driving innovations across industries.

Nanomaterials play a crucial role in sensing due to their distinct physical and chemical properties at the nanoscale. Their high surface area-to-volume ratio enhances interactions with analytes, boosting sensitivity. Quantum confinement effects alter their electronic and optical properties, enabling precise analyte detection. Moreover, nanomaterials can exhibit surface plasmon resonance, improving molecule detection. Their customizable nature allows for tailored sensor designs, ensuring superior selectivity. Overall, nanomaterial-based sensors offer swift, precise, and reliable detection, essential for healthcare diagnostics, environmental monitoring, and various industries, ultimately enhancing the quality of life.

## 4.2 UNIQUE PROPERTIES OF METAL NANOPARTICLES

Nanomaterials exhibit a wide range of unique properties that distinguish them from their bulk counterparts, offering unprecedented opportunities for innovation and advancement across various fields, including materials science, electronics, healthcare, and energy.

**Elastic properties**: At the nanoscale, materials often exhibit enhanced elastic properties, including increased strength, flexibility, and resilience. This is primarily attributed to the smaller grain sizes and higher density of defects in nanomaterials, which hinder the dislocation motion and improve the mechanical properties. For instance, carbon nanotubes possess exceptional tensile strength and stiffness, making them suitable for reinforcing composite materials and creating lightweight yet durable components. Guisbiers et al. derived an equation for the melting temperature of nanostructures and also determined the elastic moduli and thermal expansivity of the nanomaterials [1]. Venkatraman et al.'s experiments on aluminum films demonstrated that both film thickness and grain size independently influence the strength of the film [2, 3].

**Melting point:** Nanomaterials can display distinct melting characteristics in comparison to the bulk materials owing to their diminished dimensions and heightened surface-to-volume ratio. In certain instances, nanomaterials might exhibit lower melting thresholds than their bulk counterparts due to considerations of surface energy and diminished cohesive forces. Conversely, specific nanomaterials could demonstrate increased melting points or bolstered stability at elevated temperatures, particularly if they possess unique crystal structures or surface coatings. Both experimental and theoretical investigations have explored the melting points of metals [4–10], revealing a trend where the melting point diminishes as particle size decreases, with a pronounced decrease when particles reach the nanoscale.

**Magnetic properties**: Nanomaterials can display a wide range of magnetic properties, including diamagnetism, paramagnetism, ferromagnetism, antiferromagnetism, ferrimagnetism, and super paramagnetism [11, 12]. These properties arise from the arrangement of magnetic moments within the material and can be influenced by factors such as size, shape, composition, and surface modification.

The distinctive attributes of MNPs stem from their nanoscale magnetic properties, which diverge from those of bulk materials owing to their elevated surface-to-volume



DOI: 10.1201/9781032660639-4

ratios [13, 14]. Magnetic nanoparticles, for example, are utilized in various applications such as data storage, magnetic resonance imaging (MRI), sensing [15, 16], drug delivery [17–19], and hyperthermia [20] due to their tunable magnetic properties and biocompatibility.

**Diffusivity**: Nanomaterials often exhibit enhanced diffusivity compared to bulk materials, enabling faster transport of atoms, ions, or molecules. This enhanced diffusivity can be attributed to the shorter diffusion pathways and higher surface-to-volume ratio of nanomaterials, which promote rapid interatomic or intermolecular interactions.

Research has investigated the increase in thermal diffusivity values of solvents like ethanol, water, ethylene glycol, and cellular culture medium (DMEM/F-12) when exposed to gold nanoparticles [21, 22]. This effect arose from the electrostatic interactions between the cationic solution and the negatively charged Au nanoparticles. Such a characteristic has proved beneficial in catalysis, gas sensing, and energy storage applications, where fast diffusion kinetics are crucial for efficient chemical reactions or ion transport processes.

**Grain growth characteristics**: Nanomaterials may display distinct grain growth characteristics, including finer grain sizes, increased grain boundary density, and unique grain boundary structures [23, 24]. These characteristics can impart superior mechanical properties, such as higher hardness, strength, and ductility, as well as enhanced thermal and electrical conductivity. Control over grain growth is crucial in optimizing the performance of nanomaterials for various applications, including structural materials, coatings, and electronic devices.

**Electric properties**: Nanomaterials exhibit a wide range of unique electrical properties, including size-dependent conductivity, quantum confinement effects, and tunable electronic band structures. These characteristics arise from their crystalline arrangement and the presence of delocalized electrons, commonly referred to as an electron gas.

For example, semiconductor nanoparticles such as quantum dots demonstrate size-dependent optical and electronic properties due to quantum confinement effects, making them valuable for applications in optoelectronics, photovoltaics, and light-emitting diodes (LEDs).

Tunable chemical disorder leads to tunable alloy properties. In basic metals or diluted alloys, randomness alters local symmetry, influencing electronic, magnetic, and vibrational excitations [25], thereby allowing for fundamental control over tunability. The electrical conductivity of clustered metallic particles or nano-sized thin films relies on the filling factor of interconnected and empty spaces. At low filling factors, particles may exist individually, and conductivity on the thin film primarily involves tunnelling between localized insulating states [26, 27], resulting in high resistivity at low temperatures. However, as the filling factor increases enough for particles to cluster together and form aggregates, and if there is at least one conducting path connecting from one side of the sample to the other, the sample starts to conduct electricity.

**Optical properties:** Nanomaterials showcase intriguing optical characteristics stemming from their nanoscale dimensions, encompassing size-dependent absorption, emission, scattering, and plasmonic phenomena. This is attributed to the elevated ratio of surface atoms in particles of nano dimensions. The heightened impact of irregular surfaces, featuring a dense array of kinks, edges, and corners, results in the presence of surface states capable of trapping electrons and holes, thereby significantly altering the material's optical and electronic properties. Shifts in the band positions observed in the optical absorption spectra of the metallic nanoparticles, varying with the particle size, mirror those observed in the semiconductor clusters.

Quantum dots, for instance, display tunable optical properties based on their size, shape, and composition, making them suitable for applications such as displays, biological imaging, and light-emitting devices. Similarly, metal nanoparticles exhibit localized surface plasmon resonance, leading to enhanced light absorption and scattering properties, which find applications in sensing, imaging, and photothermal therapy. The existence of delocalized conduction electrons, capable of collective excitation into surface plasmons, significantly impacts the development and origin of specific absorption bands in the spectra [28]. On rough surfaces and gratings with suitable parameters, surface plasmons can be converted back into light. This emitted light serves as a crucial factor in assessing the roughness of metal surfaces [29].

The color emitted by the colloidal solution originates from a large proportion of surface atoms within particles of such small dimensions. As metal particles shrink and develop band gaps, transitioning from metallic to semiconductor or insulator properties, electron inter band transitions become crucial. Additionally, the existence of delocalized conduction electrons, which can collectively excite into surface plasmons, significantly impacts the development and origin of certain absorption bands in spectra [28].

## 4.3 APPLICATIONS OF METAL NANOMATERIALS IN SENSING

### 4.3.1 Surface Plasmon Resonance Sensors

Surface plasmon resonance sensors have revolutionized the field of biosensing by enabling label-free, real-time detection of biomolecular interactions with exceptional sensitivity and specificity. Harnessing the unique optical properties of surface plasmons, SPR sensors offer a versatile platform for studying molecular binding events in various biological and chemical systems. In this comprehensive review, we explore the principles, recent advancements, applications, and future prospects of SPR sensors, highlighting their significance in biomedical research, clinical diagnostics, and beyond.

### 4.3.2 Principles of SPR Sensors

Surface plasmon resonance occurs when a photon from incident light interacts with a metal surface, often made of gold. At certain angles of incidence, some of the light energy is absorbed by electrons in the metal layer, prompting their movement. These electron motions, called plasmons, propagate alongside the metal surface. Consequently, plasmon oscillations create an electric field that extends roughly 300 nm from the interface between the metal surface and the sample solution [30].

SPR sensors utilize the surface plasmon phenomenon – collective electron oscillations at the junction of a metal film and a dielectric medium – to detect variations in the refractive index near the sensor's surface. Typically, these sensors comprise a thin metal film (commonly gold or silver) deposited on a glass substrate. When polarized light of a particular wavelength hits the metal surface at the resonance angle, it stimulates surface plasmons, leading to a reduction in the intensity of reflected light. Changes in the nearby refractive index due to biomolecular binding events on the sensor surface induce alterations in the resonance angle, which can be continuously monitored to quantify the molecular interactions in real-time.

#### 4.3.2.1 Recent Advancements in SPR Sensor Technology

Recent developments in SPR sensor technology have concentrated on enhancing sensitivity, resolution, and multiplexing abilities, alongside improving integration with microfluidic systems and signal processing algorithms. Innovations include the development of novel sensor geometries, plasmonic nanostructures, and surface functionalization strategies for enhancing detection performance and specificity. Moreover, the advancements in instrumentation and data analysis techniques have enabled high-throughput screening of biomolecular interactions and the detection of complex analytes in heterogeneous samples. Integration with complementary techniques such as mass spectrometry, fluorescence spectroscopy, and surface-enhanced Raman scattering (SERS) further expands the capabilities of SPR sensors for multidimensional analysis of biological systems.

#### 4.3.2.2 Applications of SPR Sensors

SPR sensors find diverse applications across various fields, including drug discovery, biomolecular interaction analysis, clinical diagnostics, environmental monitoring, and food safety. In drug discovery and development, SPR sensors are used for screening small molecule inhibitors, characterizing protein-ligand interactions, and studying antibody-antigen binding kinetics. In clinical diagnostics, SPR sensors enable label-free detection of biomarkers in complex biological samples, offering potential applications in disease diagnosis, prognosis, and therapeutic monitoring. Surface plasmon resonance has proven highly effective in investigating biomolecular interactions through affinity binding analysis across diverse bonds. These include antibody-antigen interactions [31] ligand-receptor kinetics [32–39], enzyme-substrate reactions [40, 41], and epitope mapping [42].

Additionally, SPR sensors are employed in environmental monitoring for detecting pollutants, toxins, and pathogens in air, water, and soil, as well as in food safety for detecting allergens, toxins, and adulterants in food products.

Nanomaterials like graphene and carbon nanotubes have found application in enhancing signals in SPR biosensors. In 2010, a study introduced the concept of enhancing SPR signals using a graphene-coated gold sensing film [43]. Wu et al. modified the conventional gold film sensing substrate by applying multiple layers of graphene onto it. Graphene, an extraordinary two-dimensional material, enables swift electron transfer even at room temperature. The movement of charge from graphene's surface to the gold thin film surface enhances the sensing surface's sensitivity to changes in the surrounding environment. This heightened sensitivity stems from the potent electric field created on the gold thin film surface by graphene. Introducing ten layers of graphene increased the SPR signal by up to 25%.

### 4.3.3 Electrochemical Sensors

Electrochemical sensors rely on ion conduction for their operation and can be classified based on their electrical characteristics into potential sensors, conductivity sensors, electricity sensors, polarographic sensors, and electrolytic sensors. They are primarily utilized for analyzing gas, liquid, or solid components dissolved in liquids, as well as measuring liquid pH, conductivity, and oxidation-reduction potential (Figure 4.1).

Electrochemical sensors have become essential instruments across numerous domains, from environmental surveillance to medical testing, due to their sensitivity, specificity, and user-friendly nature. Utilizing electrochemical principles, these sensors provide swift and dependable detection of substances within intricate samples. This thorough examination delves into the basic tenets, recent progressions, wide-ranging uses, and forthcoming potential of electrochemical sensors, underscoring their pivotal role in advancing scientific comprehension and tackling societal issues.

#### 4.3.3.1 Principles of Electrochemical Sensors

Electrochemical sensors operate based on the principles of electrochemistry, utilizing the interactions between analytes and electrodes to generate measurable electrical signals. Common types of electrochemical sensors include amperometric, potentiometric [44, 45], and impedimetric sensors [46]. Amperometric sensors measure current generated by electrochemical reactions at the electrode surface, whereas potentiometric sensors detect changes in electrode potential. In contrast, impedimetric sensors detect alterations in the impedance caused by the interaction between analytes and the electrode interface. These sensors offer distinct advantages in terms of sensitivity, selectivity, and

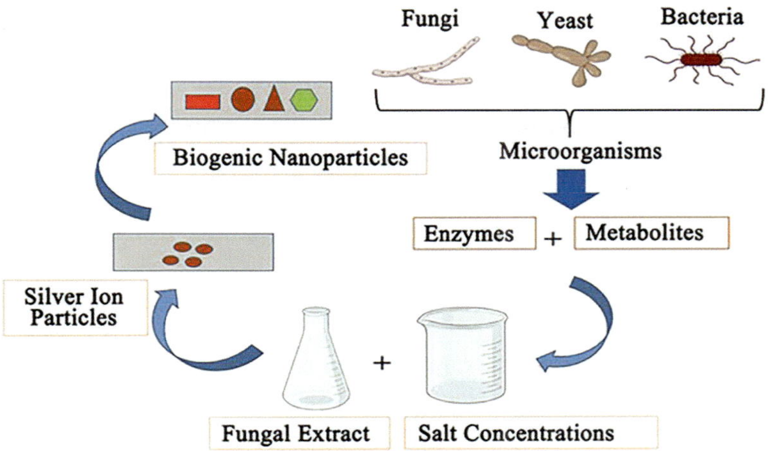


**FIGURE 4.1** Schematic representation of the key elements of a typical sensor configuration.

detection limits, depending on the specific application requirements.

#### 4.3.3.2 Recent Advancements in Electrochemical Sensor Technology

Recent advancements in electrochemical sensor technology have focused on improving sensor performance, miniaturization, and integration with emerging technologies. Innovations include the development of new electrode materials, such as nanomaterials and conductive polymers, to enhance sensitivity and stability. Furthermore, the advancements in microfabrication techniques have enabled the miniaturization of sensor devices, leading to the development of portable and wearable sensors for on-site analysis. Integration with wireless communication and data analysis platforms has facilitated real-time monitoring and remote sensing applications, opening new avenues for environmental and healthcare monitoring.

#### 4.3.3.3 Applications of Electrochemical Sensors

Electrochemical sensors are utilized across a spectrum of sectors, spanning environmental surveillance, medical diagnostics, food safety, and industrial process regulation. In the field of environmental monitoring, these sensors are vital for detecting pollutants, heavy metals, and noxious gases found in the atmosphere, water bodies, and soil.

The detection of inorganic ions, metal cations, organic chemicals, and biomolecules can be achieved through a sensitive electroanalytical technique called electrochemical stripping analysis (ESA). This method involves preconcentrating the target analyte(s) or compounds onto a suitable working electrode, followed by the removal of any remaining accumulated analyte using an electroanalytical procedure [47]. ESA has been acknowledged as an effective method for quantifying trace metal concentrations [48–52]. The process involves two stages: deposition, where a minor quantity of metal ions in solution is electrolytically placed onto a mercury electrode for preconcentration, and stripping (measurement), which involves dissolving (stripping) the deposit. The exact technique for stripping analysis differs based on the properties of the deposition and measurement phases [53]. In healthcare, these sensors play a vital role in point-of-care diagnostics, enabling the detection of biomarkers associated with diseases such as diabetes, cardiovascular diseases, and infectious diseases.

Advanced biorecognition platforms for biomolecule detection utilize methods such as cyclic voltammetry (CV), square wave voltammetry (SWV), differential pulse voltammetry (DPV), and electrochemical impedance spectroscopy (EIS) for the analysis of targets like DNA, thrombin, MMP-2, estrogen, and human chronic gonadotrophin, among others [54–62].

Additionally, electrochemical sensors are employed in food safety for detecting contaminants, allergens, and adulterants, ensuring the quality and safety of food products [63]. In industrial settings, electrochemical sensors are utilized for process monitoring, quality assurance, and safety compliance, contributing to the efficient production processes and risk mitigation.

## 4.4 CLASSIFICATION OF SENSORS IN VIEW OF THEIR APPLICATIONS IN INDUSTRIES/MANUFACTURING

**Pressure Sensors**: Pressure sensors typically consist of electro-mechanical components utilized for gauging the pressure of gases or liquids, transmitting signals to control and display devices. Their technology, design, applicability, performance, and cost can vary considerably. These sensors find extensive use across diverse everyday applications for control and monitoring purposes [64–67].

**Temperature Sensors:** A temperature sensor is an electronic apparatus designed to identify thermal conditions in its surroundings and translate this information into electronic signals for recording, monitoring, or indicating temperature fluctuations. Such sensors commonly employ resistance temperature detectors (RTDs) or thermocouples for temperature detection. Thermometers, the prevailing type of temperature sensor, gauge levels of warmth or chilliness. These sensors find widespread application across various sectors including automotive, aviation, medical, computing, culinary appliances, and everyday activities [68–70].

**Proximity Sensors:** Proximity sensors are electronic instruments utilized for detecting the existence of nearby objects without direct contact. Typically, these sensors emit an electromagnetic field and analyze alterations in the field or the returning signal. They play a crucial role in various manufacturing processes, particularly for identifying the presence of parts and machine components. For instance, they are integral in machine vibration monitoring to assess fluctuations in distance between a shaft and its supporting bearing [71–73].

**Humidity Sensors:** Humidity sensors gauge the moisture levels in the surrounding environment and translate these measurements into usable signals for input. Their applications span a broad spectrum and are commonly found in meteorology, medical, automotive, heating, ventilation, and air conditioning (HVAC), and manufacturing sectors [74–76].

**Force Sensors:** Force sensors, also known as load cells, are instruments that convert mechanical forces, such as tension and compression, into digital output signals, representing the applied force magnitude. They find extensive use across diverse applications, including consumer products, computer peripherals, musical instruments, medical equipment, automotive systems, sports equipment, and various other fields [77–79].

**Flow Sensors:** Flow sensors, whether electronic or electro-mechanical, are employed to quantify the movement of fluids like gases or liquids, furnishing output signals to controllers. Given the critical role of flow measurement in regulating numerous industrial operations, these sensors are extensively utilized in automotive and medical sectors, among others [80–82].

**Gas Sensors:** Gas sensors, whether stationary or portable, are electronic devices designed to detect the presence and characteristics of various gases, transmitting signals to the controller in the form of output signals. They serve as crucial tools for detecting harmful gases, offering an essential means of monitoring gas concentration and environmental data, particularly in contexts where safety is paramount [83, 84].

**Flaw Sensors**: Flaw sensors or detectors are utilized across a range of manufacturing processes to detect irregularities on surfaces or within different underlying materials. Typically, these sensors employ ultrasonic, acoustic, or similar methods to identify defects in materials. Regarded as vital tools in quality control during manufacturing, they play a key role in non-destructive testing and material analysis, inspecting components for flaws and ensuring product integrity [85–87].

**Color Sensors:** Color sensors belong to the category of photoelectric sensors, functioning by emitting light from a transmitter and subsequently detecting the reflected light from the object using a receiver. They play a crucial role in developing color-sensing applications across various fields such as image processing, color identification, industrial object tracking, medical diagnostic systems, health and fitness monitoring systems, and industrial process control, among others [88–90].

**Light Sensors:** A light sensor, also referred to as a photoelectric device, converts detected light energy (photons) into electrical energy (electrons). These sensors are indispensable tools across various industries. However, different types of light sensors, including photoresistors, photodiodes, phototransistors, and others, operate with slight variations and cater to unique applications [91–93].

### 4.4.1 Colorimetric Sensors

Color-based sensors have garnered significant interest across various sensing scenarios due to their precision, heightened sensitivity, affordability, user-friendly nature, quick analysis, straightforward operation, and easily observable results without the need for additional tools [94]. Colorimetric sensors have emerged as powerful tools for detecting and quantifying a wide range of analytes in diverse fields such as environmental monitoring, healthcare, food safety, and industrial process control. These sensors harness the principles of color changes induced by specific chemical reactions or interactions, offering simplicity, affordability, and portability for rapid and sensitive detection. Here, we delve into the principles, advancements, applications, and future prospects of colorimetric sensors, highlighting their significance in addressing pressing societal challenges and driving innovation in various sectors.

#### 4.4.1.1 Principles of Colorimetric Sensors

Colorimetric sensors operate based on the principle of generating visible color changes in response to the presence or concentration of a target analyte [95–99]. The amount of light is absorbed or transmitted by a colored solution in accordance with Beer's law and Lambert's law. The combined Beer's–Lambert's law is denoted as: the amount of light transmitted through a colored solution decreases exponentially with an increase in the concentration of the colored solution and an increase in the path length of a cuvette or thickness of the colored solution [100] (Figure 4.2). These changes can result from various mechanisms, including chemical reactions, molecular recognition events, or physical interactions. Typically, colorimetric sensors consist of indicator molecules or materials that undergo specific transformations upon interacting with the analyte, leading to observable changes in color. The intensity, hue, or saturation of the resulting color can be correlated with the concentration of the analyte, enabling qualitative and quantitative analysis.

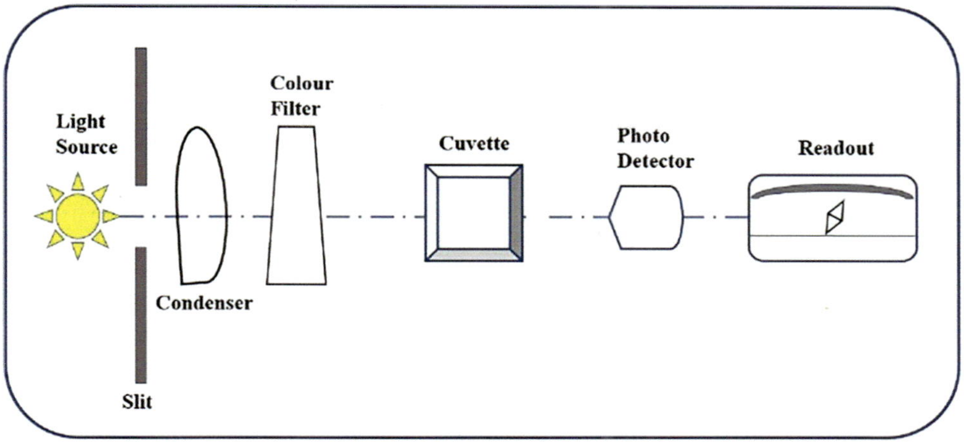


**FIGURE 4.2** Working principles of colorimetry.

#### 4.4.1.2 Advancements in Colorimetric Sensor Technology

Recent advancements in colorimetric sensor technology have significantly enhanced its sensitivity, selectivity, and versatility. Innovations include the development of novel indicator molecules, functionalized nanoparticles, and microfluidic platforms for enhanced detection performance. Moreover, integration with smartphone-based imaging systems and machine learning algorithms has enabled rapid and remote analysis, democratizing access to colorimetric sensing technology. Furthermore, efforts to engineer stimuli-responsive materials and smart sensing platforms hold promise for expanding the applicability of colorimetric sensors in dynamic and complex environments.

#### 4.4.1.3 Applications of Colorimetric Sensors

Colorimetric sensors find diverse applications across various sectors, including environmental monitoring, healthcare diagnostics, food safety, and industrial quality control. In environmental monitoring, they are utilized for detecting pollutants, heavy metals, and harmful gases in air, water, and soil. In healthcare, colorimetric sensors play a vital role in point-of-care diagnostics, enabling the detection of biomarkers associated with diseases such as diabetes, cancer, and infectious diseases. Additionally, they are employed in food safety to detect contaminants, allergens, and spoilage indicators, ensuring the quality and integrity of food products. In industrial settings, colorimetric sensors are used for process monitoring, quality assurance, and safety compliance, contributing to the efficient production processes and risk mitigation.

To enhance the compatibility of sample preparation techniques with the principles of "green" analytical chemistry, Smirnova et al. devised a swift extraction and color-based testing method for synthetic food dyes. This method relies on an aqueous two-phase system (ATPS) comprising a blend of cationic and anionic surfactants. The approach involves micro-extracting from ATPS followed by spectrophotometric or colorimetric analysis of the food dyes. Moreover, colorimetric assessments can be directly conducted within the extract using a smartphone [101]. This approach proves suitable for detecting dyes in food samples and wastewater from food processing industries.

Chai et al. [102] introduced a novel nanozyme, Tris(hydroxymethyl)aminomethane-Cu (Tris-Cu), to address the risk of phenolic compound pollution in ecosystems. This nanozyme, synthesized using a rapid method, effectively catalyzed various phenolic compounds and enabled color-based detection of 2,4-dichlorophenol (2,4-DP) within a concentration range of 10–400 μM. Additionally, Tris-Cu demonstrated efficient removal of six phenolic compounds, showcasing its potential for environmental pollutant monitoring and mitigation.

Furthermore, adenine phosphate-Cu (AP-Cu) nanozyme, possessing multienzyme mimicking attributes, was synthesized for efficient degradation of phenolic compounds and detection of hydrogen peroxide, epinephrine, and glutathione [102].

### 4.4.2 Surface-Enhanced Raman Scattering Sensors

Surface-enhanced Raman scattering sensors have become a potent analytical method for detecting molecules with high sensitivity and selectivity at extremely low concentrations. They utilize plasmonics to glean intricate chemical details of molecules or molecular assemblies adhered to or affixed onto nanostructured metallic surfaces. By harnessing the enhancement of Raman scattering signals induced by nanostructured metallic surfaces, SERS sensors offer unprecedented sensitivity, specificity, and multiplexing capabilities.

Here, we explore the principles, recent advancements, applications, and future prospects of SERS sensors, highlighting their significance in various fields of science and technology.

#### 4.4.2.1 Principles of Surface-Enhanced Raman Scattering

Surface plasmons (SPs) refer to the collective oscillation of free conducting electrons stimulated by electromagnetic radiation at the interface between a metal and a dielectric material [103]. These phenomena are sustained by thin films of noble metals or the surfaces of nanoparticles (NPs). Investigating the interaction between light and metallic nanostructures constitutes a rapidly evolving field of study known as plasmonics [104–107]. By deliberately designing plasmonic nanostructures, we gain the capability to regulate and manage visible light on the scale of nanometers [108–110]. This potential offers avenues for practical uses like incorporating and reducing the size of electronic elements, optical links, and exceedingly sensitive analytical instruments.

There are two types of SPs: propagating and non-propagating [103, 111]. Surface plasmon polaritons (SPPs) are the propagating form of surface plasmons, emerging on thin films of noble metals like gold or silver with thicknesses ranging from 10 to 200 nm. In contrast, localized surface plasmon resonances (LSPRs) represent the non-propagating type of surface plasmons, generated on the surfaces of nanoparticles typically 10 to 200 nm in diameter or formed through nanosphere lithography. The electromagnetic field associated with SPPs on noble metallic thin films travels approximately 10 to 100 μm in the lateral (x and y) directions and 200 to 300 nm in the perpendicular (z) direction along the metal-dielectric interface. The extent of propagation is contingent upon factors such as the metal type, film thickness, and surface roughness [107, 112].

SERS relies on the enhancement of Raman scattering signals when molecules are adsorbed onto nanostructured metallic surfaces, typically gold or silver nanoparticles. This enhancement arises from two main mechanisms: electromagnetic enhancement due to LSPR and chemical enhancement resulting from charge transfer interactions between the molecule and the metal surface. The combination of these effects leads to significant amplification of Raman signals, enabling detection of molecules at single-molecule levels and enhancing the vibrational fingerprinting capabilities of Raman spectroscopy.

#### 4.4.2.2 Recent Advancements in SERS Sensor Technology

Recent advancements in SERS sensor technology have focused on improving sensitivity, reproducibility, and scalability for practical applications. Strategies include the design and synthesis of novel SERS-active substrates with controlled nanostructures, such as nanospheres, nanorods, and nanostars, to optimize signal enhancement and stability. Moreover, functionalization of SERS substrates with specific capture molecules or ligands enables selective detection of target analytes, facilitating applications in biosensing, environmental monitoring, and medical diagnostics. Integration with microfluidic devices, spectroscopic imaging techniques, and machine learning algorithms further enhances the capabilities of SERS sensors for high-throughput and multiplexed analysis.

#### 4.4.2.3 Applications of SERS Sensors

SERS sensors find diverse applications across various fields, including biomedical research, environmental monitoring, food safety, and homeland security. In biomedical research, SERS sensors enable label-free detection and imaging of biomolecules, cells, and tissues with high sensitivity and spatial resolution, opening new avenues for disease diagnosis and drug discovery. Several scientists worked on DNA [113, 114], RNA [115, 116], cancer markers [117], drugs [118], pathological markers on cellular membranes and tissues [119], other biomolecules, ion concentrations, redox potential in cells [120], and even in vivo SERS measurements on mice [121]. In environmental monitoring, SERS sensors are used for detecting pollutants, toxins, and hazardous chemicals in air, water, and soil, contributing to environmental protection and public health [122, 123]. Additionally, SERS sensors are employed in food safety for detecting contaminants, adulterants, and allergens, ensuring the quality and safety of food products [124].

### 4.4.3 Gas Sensors

A gas detection system, also known as an electronic nose [125], has the capability to identify specific gases either qualitatively or quantitatively. This ability holds significant importance across a range of applications, including detecting industrial pollutant gas leaks, environmental surveillance, therapeutic diagnostics, food safety, and security measures. The effectiveness of an optimal gas sensor is assessed based on several criteria including its high sensitivity, rapid response and recovery times, consistent stability, excellent selectivity, suitability for operation at ambient temperatures, affordability, and ease of manufacturing for practical deployment [126–132].

The foundation of the basic sensing system known as an "electronic nose" lies in a gas sensor array, where each individual sensor doesn't exhibit specificity towards a particular gas but instead displays a distinct sensing response pattern. When integrated with a pattern recognition module, this setup serves as a general olfaction structure for characterizing odors [133]. These sensors, which are commonly utilized in single gas applications, are also the building blocks of electronic noses.

An electronic nose typically consists of several components: a multi-sensor array, an information-processing unit, software containing pattern-recognition algorithms, and a digital reference library database containing aroma fingerprint signatures. These systems are valuable across diverse domains where detecting a wide range of gases is necessary.

Gas sensors play a crucial role in various industries and applications, enabling the detection and monitoring of gases in diverse environments. From ensuring workplace safety to environmental monitoring and medical diagnostics, gas sensors provide invaluable insights into the presence and concentration of hazardous or beneficial gases. In this comprehensive review, we explore the principles, recent advancements, applications, and future prospects of gas sensors, highlighting their significance in addressing pressing societal challenges and driving innovation in multiple sectors [126–132].

#### 4.4.3.1 Principles of Gas Sensors

Gas sensors operate based on different principles, including chemical reactions, physical adsorption, and optical absorption, depending on the type of sensor and target gas [133]. Common types of gas sensors include electrochemical sensors, semiconductor sensors, and optical sensors. Electrochemical sensors rely on the detection of changes in electrical properties resulting from gas interactions with electrodes, while semiconductor sensors detect changes in conductivity or resistance upon exposure to gases. Optical sensors utilize the absorption or scattering of light by gases to quantify their concentration. Understanding the underlying principles of gas sensing is crucial for designing sensors with optimal sensitivity, selectivity, and reliability.

Electrochemical cells comprise distinct electrodes where chemical reactions occur involving electron exchange with charged ions. Illustrated in Figure 4.3, the anode serves as the site for oxidation (electron loss to the circuit), while the cathode facilitates reduction (electron gain from the circuit). These reactions allow for charge equilibrium via electron transfer within electrical circuits. Furthermore, the cells generate essential electrical voltage and exhibit alterations in resistance when exposed to gases, forming the basis for chemical sensors reliant on resistance changes.

Semiconductors of the n-type exhibit a transition in resistance, shifting from its highest to lowest value in the presence of gas, while p-type semiconductors experience the opposite, with resistance increasing from its lowest to highest value. Impurities play a crucial role in enhancing sensor properties by modifying reaction pathways, enhancing sensitization and selectivity, and expanding gas detection limits. Certain impurities act as catalysts and activators for these properties. Factors such as particle size and surface porosity also contribute to improving gas sensitivity. These considerations are fundamental in the initial selection of materials for chemical sensors [134].

#### 4.4.3.2 Recent Advancements in Gas Sensor Technology

Recent advancements in gas sensor technology have focused on enhancing sensitivity, selectivity, and response time, as well as reducing power consumption and miniaturizing sensor devices. Nanomaterials such as metal oxides, carbon nanotubes, and graphene have been extensively investigated for their potential to improve gas sensing performance due to their high surface-to-volume ratio and unique electronic properties. Functionalization of sensor surfaces with specific receptors or coatings enables selective detection of target gases, while advancements in microfabrication techniques have facilitated the development of miniaturized and portable sensor platforms. Furthermore, integration with wireless communication and Internet of things (IoT) technologies allows for real-time monitoring and data analysis, paving the way for smart sensing solutions in various applications [135].

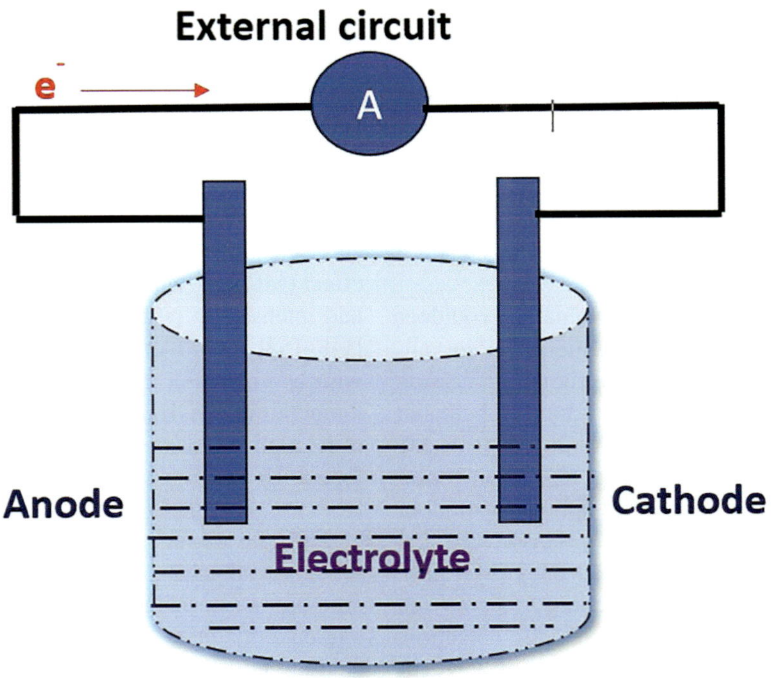


FIGURE 4.3 Components of an electrochemical cell.

#### 4.4.3.3 Applications of Gas Sensors

Gas sensors find diverse applications across multiple sectors, including industrial safety, environmental monitoring, healthcare, automotive, and consumer electronics [136]. In industrial settings, gas sensors are used for detecting toxic, flammable, or explosive gases to ensure workplace safety and compliance with regulatory standards.

Environmental monitoring applications include detection of air pollutants, greenhouse gases, and volatile organic compounds (VOCs) in urban, industrial, and indoor environments. Current studies have concentrated on creating a nanowire array composed of p-type semiconducting metal oxides such as CuO, NiO, and $Cr_2O_3$, aimed at detecting volatile organic compounds [137].

Gas sensors are crucial in healthcare for medical diagnostics, breath analysis, and anesthesia monitoring. The exhalation of human breath contains a multitude of volatile organic compounds, typically detected at very low concentrations in individuals who are in good health. While the correlation between certain diseases and specific VOCs has been recognized for over a century, modern instruments, particularly semiconductor gas sensors, provide the quantitative measurements necessary for rigorous clinical practice. Independent clinical trials have demonstrated the potential of breath analysis in detecting serious illnesses, including various cancers, diabetes, Alzheimer's and Parkinson's disease, multiple sclerosis, tuberculosis, and chronic kidney disease [138].

Portable acetone analyzers, which assess ketone levels and estimate the rate of body fat burning solely from exhaled breath, are widely employed in numerous diet and fitness regimes. These compact devices are utilized to monitor fat metabolism and ketosis levels. One instance of semiconductor materials applied in breath analysis involves nanofibers made of $SnO_2$ that are functionalized with reduced graphene (rGO). These nanofibers were tested for their ability to detect acetone and hydrogen sulfide in human breath [139].

Additionally, gas sensors are integrated into automotive exhaust systems for emissions control and air quality monitoring, as well as in consumer electronics for applications such as air purifiers, indoor air quality monitors, and smart home devices.

Mobile robots with gas detectors are valuable in outdoor environments for pollution monitoring, industrial surveillance, and landfill oversight. A novel signal processing technique enhances their sensitivity to acetone and ethanol [140]. Portable electronic noses are advantageous for IoT and wireless sensor networks due to their on-demand operation [141]. While drones offer another option for outdoor gas monitoring, they face limitations in sensor size and power [142]. However, micro- and nanodrones [143] have been employed for various tasks, including volcanic gas sampling, fugitive emission detection, early fire detection, precision agriculture, landfill monitoring, and mine blasting assistance.

### 4.4.4 Magnetic Sensors

Magnetic sensors play a pivotal role in various industries and applications, enabling the detection, measurement, and monitoring of magnetic fields. From navigation systems to industrial automation, medical devices, and consumer electronics, magnetic sensors provide critical data for a wide range of applications. In this review, we examine the principles, recent advancements, applications, and future prospects of magnetic sensors, highlighting their significance in modern technology and industry.

#### 4.4.4.1 Principles of Magnetic Sensors

Magnetic sensors operate based on the principle of detecting changes in magnetic fields using various sensing mechanisms. When semiconducting magnetic oxides or magnetic materials come into contact with gas molecules, either through reduction or oxidation processes, the absorption of ions/molecules on the surface causes a shift in the vacancy and defect concentration within the materials. This alteration significantly influences the saturation magnetization (Ms), remanent magnetization (Mr), and coercivity (Hc) of the materials [144]. Adsorbed ions or molecules on the surface induce alterations in the electronic structure, thereby modifying the overall magnetic behavior of the material. This change occurs due to the rise in uncompensated surface spins.

Common types of magnetic sensors include Hall effect sensors, Kerr effect-based magneto resistive sensors, fluxgate sensors, and superconducting quantum interference device (SQUID) sensors.

Hall effect sensors utilize the Hall effect, which describes the generation of a voltage perpendicular to the flow of current in a conductor when subjected to a magnetic field. The Hall effect is classified into the ordinary Hall effect (OHE) and extraordinary Hall effect (EHE), depending on its origin in magnetic materials. Due to their advantageous properties, both OHE and EHE concepts are employed for the fabrication of magnetic gas sensors [145–147].

Magneto resistive sensors, on the other hand, rely on changes in electrical resistance induced by magnetic fields. Gas sensing applications utilize the magneto optic Kerr effect (MOKE), which involves changes in the polarization and intensity of reflected light from a magnetized surface [148–152]. This method employs a MOKE magnetometer with a very small magnetic field (around 50 Oe) at room temperature. MOKE offers significant advantages because of its high surface sensitivity, and the signal remains unaffected by paramagnetic and diamagnetic contributions from the substrates.

Fluxgate sensors measure changes in magnetic flux. The fundamental principle of a functional fluxgate system is the ratio of the external magnetic field ($B_{ext}$) as measured to the reference magnetic field ($B_{ref}$) [153]. Converting magnetic field strength into electrical signals can be accomplished directly to measure $B_{ext}$. However, the fluxgate magnetic sensor takes an indirect approach, employing a reference magnetic field, $B_{ref}$, for comparison with the measured magnetic field, $B_{ext}$. This comparison is facilitated by containers filled with core material.

Superconducting quantum interference devices are considered highly sensitive magnetic flux detectors. The field resolution of SQUID devices is 10–17 T, and the device can be operated at low temperatures such as cryogenic temperatures with quantum-limited sensitivity. The SQUID device is the most fundamentally used application of flux quantization and the Josephson phenomena.

SQUID devices are mainly used to identify very small amounts of magnetic field, current, voltage, inductance, and magnetic susceptibility. These devices show excellent sensitivity in the magnetic flux detection. They have been used to detect weak magnetic fields. These devices are used in the form of magnetometers and gradiometers.

In practice, a SQUID serves as a converter of magnetic flux into voltage with extremely low magnetic flux noise. By measuring the voltage across the SQUID and knowing the area of the ring, the magnetic field can be determined. Larger ring areas result in improved sensitivity to magnetic fields. However, increasing the area of the superconducting ring is limited by the magnetic flux noise, which is directly related to the ring inductance. Therefore, special configurations are employed to enhance the magnetic field sensitivity of a SQUID device when used as a magnetometer. One commonly used configuration involves a superconducting flux transformer, comprising a pickup coil (often square or circular) connected in series with a multiturn coil that is magnetically coupled to the SQUID ring [154, 155].

**Various uses of SQUID devices:**

- **Magnetoencephalography (MEG):** Weak magnetic fields are produced in the brain by the presence of the electrical currents occurring naturally in the brain. SQUID magnetometers are used to measure this magnetic field and thus perform brain activity mapping; this technique is known as MEG [156].
- **Magnetogastrography:** In this technique, SQUID devices are used to record the magnetic fields produced due to electrical activity from the stomach [157].
- **Magnetic marker monitoring (MMM)**: SQUID sensors are used to detect the dosage form of an orally applied drug through the intestinal tract, containing a small amount of magnetite ($Fe_3O_4$), magnetized by a high-energy magnetic field [158].
- **Magnetic property measurement systems (MPMS):** Measurement of magnetic properties of systems is the fundamental commercial use of SQUID sensors [159].
- **Superparamagnetic relaxometry (SPMR), NMR, and EPR:** In these techniques, a sample is placed in the center of the SQUID detection coils, and by applying an external field or RF excitation to the sample, corresponding signals are measured [160, 161].
- **Scanning SQUID microscope:** By moving a SQUID sensor across an examination area, weak magnetic fields are measured [162].
- **Magnetic anomaly detector (MAD):** In this technique, SQUID magnetometers are used to detect infinitesimal deviations in the Earth's magnetic field. The technique is most commonly used by military forces to detect submarines [163].

Understanding these principles is essential for optimizing sensor performance and selecting the most suitable sensor type for specific applications [164].

#### 4.4.4.2 Recent Advancements in Magnetic Sensor Technology

Recent progress in magnetic sensor technology has concentrated on enhancing sensitivity, resolution, and dependability while also decreasing power consumption and shrinking sensor devices. These advancements involve the exploration of new materials and fabrication methods to enhance sensor performance, such as incorporating giant magnetoresistance (GMR) and tunnel magnetoresistance (TMR) materials into magneto resistive sensors. Furthermore, advancements in microelectromechanical systems (MEMS) technology have enabled the miniaturization of magnetic sensors, making them suitable for integration into compact and portable devices. Integration with signal processing algorithms and wireless communication technologies has facilitated the development of smart and connected sensor systems for real-time monitoring and data analysis.

Magnetic sensors offer several advantages:

1. They do not require electrical contacts to detect gas, reducing the risk of explosion in environments with hydrogen-powered vehicles or reactive chemicals/pollutants.
2. Magnetic response is notably faster compared to chemi resistive sensors.
3. The sensors can operate at room temperature and can be tuned to very low or high temperatures by selecting magnetic materials with different Curie temperatures (Tc) [144, 165–167].

#### 4.4.4.3 Applications of Magnetic Sensors

Magnetic sensors have a wide range of applications across numerous industries and sectors, including automotive, aerospace, healthcare, robotics, and consumer electronics. Various magnetism effects, such as the Hall effect, Kerr effect, magnetization, spin change effects, ferromagnetic resonance (FMR) effect, magneto-plasmonic effect, and magnetostatic spin-wave (MSW) effect, are utilized for magnetic gas sensing applications [168–172]. Powder, thin films, nanomaterials with magnetic (ferro- and anti-ferromagnets), diluted magnetic semiconducting (DMS) properties, and Pd alloys with transition metals (Co, Ni, Fe, Mn, and Cu) have been employed for magnetic gas sensing applications using different magnetic effects.

In the automotive industry, magnetic sensors are used for vehicle navigation, speed sensing, position detection, and anti-lock braking systems (ABSs). In aerospace applications, magnetic sensors enable attitude and heading reference systems (AHRSs), magnetometer-based navigation, and magnetic anomaly detection. In healthcare, magnetic sensors are utilized in magnetic resonance imaging (MRI) machines, medical implants, and diagnostic devices. Additionally, magnetic sensors are integrated into consumer electronics such as smartphones, tablets, and gaming consoles for compass navigation, gesture recognition, and augmented reality applications.

## 4.5 FUTURE PROSPECTS AND CHALLENGES

Despite significant progress, several challenges remain in the development and deployment of sensor technology. These include optimizing sensor performance for robustness, improving sensor selectivity, stability, reproducibility, and scalability, as well as addressing issues related to non-specific binding, surface fouling, and sample matrix effects. Furthermore, efforts to standardize assay protocols, validate sensor performance, and ensure regulatory compliance are essential for widespread adoption and commercialization. Looking ahead, interdisciplinary collaborations, advancements in materials science, and integration with emerging technologies such as artificial intelligence, the Internet of Things, and lab-on-a-chip platforms hold promise for overcoming these challenges and unlocking new opportunities for sensors to explore fresh possibilities across diverse applications.

## REFERENCES

1. Guisbiers, G., Mejía-Rosales, S., & Deepak, F.L. (2012). Nanomaterial properties: size and shape dependencies. *Journal of Nanomaterials*, 2012. https://doi.org/10.1155/2012/180976.
2. Venkatraman, R., Chen, S., & Bravman, J.C. (1991). The effect of laser reflow on the variation of stress with thermal cycling in aluminum thin films. *J. Vat. Sci. Technol. A*, 9(1991), 2536. https://doi.org/10.1116/1.577269
3. Venkatraman, R., & Bravman, J.C. (1992). Separation of film thickness and grain boundary strengthening effects in Al thin films on Si. *Journal of Materials Research*, 7(8), 2040–2048. https://doi.org/10.1557/JMR.1992.2040
4. Buffat, P., & Borel, J.P. (1976). Size effect on the melting temperature of gold particles. *Phys. Rev. A*, 13(1976), 2287–2298. https://doi.org/10.1103/PhysRevA.13.2287
5. Lai, S.L., Carlsson, J.R.A., & Allen, L.H. (1998). Melting point depression of Al clusters generated during the early stages of film growth: nanocalorimetry measurements. *Appl. Phys. Lett.*, 72(1998), 1098–1100. https://doi.org/10.1063/1.120946
6. Peters, K.F., Cohen, J.B., & Chung, Y.W. (1998). Melting of Pb nanocrystals. *Phys. Rev. B*, 57(1998), 13430–13438. https://doi.org/10.1103/PhysRevB.57.13430
7. Allen, G.L., Bayles, R.A., Gile, W.W., & Jesser, W.A. (1986). Small particle melting of pure metals. *Thin Solid Films*, 144(1986), 297–308. https://doi.org/10.1016/0040-6090(86)90422-0
8. Skripov, V.P., Koverda, V.P., & Skokov, V.N. (1981). Size effect on melting of small particles. *Phys. Status Solidi (A)*, 66(1981), 109–118. https://doi.org/10.1002/pssa.2210660111
9. Kim, E.H., & Lee, B.J. (2009). Size dependency of melting point of crystalline nano particles and nano wires: a thermodynamic modelling. *Met. Mater. Int.*, 15(2009), 531–537. https://doi.org/10.1007/s12540-009-0531-8
10. Luo, W., & Hu, W. (2013). Gibbs free energy, surface stress and melting point of nanoparticle. *Phys. B*, 425(2013), 90–94. https://doi.org/10.1016/j.physb.2013.05.025
11. Gignoux, D. (2005). *Phenomenology of Magnetism at the Macroscopic Scale*. Springer. https://doi.org/10.1007/978-0-387-23062-7_3
12. Jeong, U., Teng, X., Wang, Y., Yang, H., & Xia, Y. (2007). Superparamagnetic colloids: controlled synthesis and niche applications. *Adv. Mater.*, 19, 33–60. https://doi.org/10.1002/adma.200600674
13. Colombo, M., Carregal-Romero, S., Casula, M.F., Gutierez, L., Morales, M.P., Bohm, I.B., . . . Parak, W.J. (2012). Biological applications of magnetic nanoparticles. *Chem. Soc. Rev.*, 41, 4306–4334. https://doi.org/10.1039/c2cs15337h
14. Lu, A.H., Salabas, E.L., & Schuth, F. (2007). Magnetic nanoparticles: synthesis, protection, functionalization, and application. *Angew. Chem. Int. Ed.*, 46, 1222–1244. https://doi.org/10.1002/anie.200602866
15. Frey, N.A., Peng, S., Cheng, K., & Sun, S. (2009). Magnetic nanoparticles: synthesis, functionalization, and applications in bioimaging and magnetic energy storage. *Chem. Soc. Rev.*, 38, 2532–2542. https://doi.org/10.1039%2Fb815548h
16. Koh, I., & Josephson, L. (2009). Magnetic nanoparticle sensors. *Sensors,* 9, 8130–8145. https://doi.org/10.3390/s91008130
17. Arruebo, M., Fernandez-Pacheco, R., Ibarra, M., & Santamaria, J. (2007). Magnetic nanoparticles for drug delivery. *Nanotoday,* 2, 22–32. https://doi.org/10.1016/S1748-0132(07)70084-1
18. Kami, D., Takeda, S., Itakura, Y., Gojo, S., Watanabe, M., & Toyoda, M. (2011). Application of magnetic nanoparticles to gene delivery. *Int. J. Mol. Sci.*, 12, 3705–3722. https://doi.org/10.3390%2Fijms12063705
19. Dobson, J. (2006). Magnetic nanoparticles for drug delivery. *Drug Dev. Res.*, 67, 55–60. https://doi.org/10.1002/ddr.20067
20. Gazeau, F., Levy, M., & Wilhelm, C. (2008). Optimizing magnetic nanoparticle design for nano thermotherapy. *Nanomedicine,* 3,831–844. https://doi.org/10.2217/17435889.3.6.831
21. Perez, J.L.J., Fuentes, R.G., Ramirez, J.F.S., & Orea, A.C. (2008). Study of gold nanoparticles effect on thermal diffusivity of nanofluids based on various solvents by using thermal lens spectroscopy. *Eur. Phys. J. Spec. Top.*, 153, 159–161. https://doi.org/10.1140/epjst/e2008-00417-5
22. Perez, J.L.J., Fuentes, R.G., Alvarad, E.M., Gallegos, E.R., Orea, A.C., Cordova, J.T., & Alvarez, J.G.M. (2008). Enhancement of the thermal transport in a culture medium with Au nanoparticles. *Appl. Surf. Sci.*, 255, 701–702. https://doi.org/10.1016/j.apsusc.2008.07.012

23. Yamakov, V., Wolf, D., Phillpot, S.R., Mukherjee, A.K., & Gleiter, H. (2002). Dislocation processes in the deformation of nanocrystalline aluminium by molecular-dynamics simulation. *Nat. Mater.*, 1, 45–49. https://doi.org/10.1038/nmat700
24. Hall, E.O. (1954). Variation of hardness of metals with grain size. *Nature,* 173, 948–949. https://ui.adsabs.harvard.edu/link_gateway/1954Natur.173.948H/https://doi.org/10.1038/173948b0
25. Wu, Z., Troparevsky, M.C., Gao, Y.F., Morris, J.R., Stocks, G.M., & Bei, H. (2017). Phase stability, physical properties and strengthening mechanisms of concentrated solid solution alloys. *Curr. Opin. Solid State Mater. Sci.*, 21(5), 267–284. https://doi.org/10.1016/j.cossms.2017.07.001
26. Chen, L.J., Tyan, J.H., & Lue, J.T. (1994). The non-metallic conductivity of Nb3Ge films sputtered on silicon substrates. *Phys. Chem. Solids,* 55, 871. https://doi.org/10.1016/0022–3697(94)90011–6
27. Reiss, G., Vancea, J., & Hoffmann, H. (1986). Grain-boundary resistance in polycrystalline metals. *Phys. Rev. Lett.,* 56, 2100. https://doi.org/10.1103/PhysRevLett.56.2100
28. Roduner, E. (2006). *Nanoscopic Materials: Size Dependent Phenomena.* RSC Publishing. https://doi.org/10.1039/9781839168970
29. Raether, H. (1988). *Surface Plasmons on Smooth and Rough Surfaces and on Gratings.* New York, NY: Springer. https://ui.adsabs.harvard.edu/link_gateway/1988STMP..111.....R/https://doi.org/10.1007/BFb0048317
30. Homola, J., Yee, S.S., & Gauglitz, G. (1999). Surface plasmon resonance sensors: review. *Sens. Actuat. B Chem.*, 54, 3–15. https://doi.org/10.1016/S0925–005(98)00321–9
31. Zeder-Lutz, G., Zuber, E., Witz, J., & van Regenmortel, M.H. (1997). Thermodynamic analysis of antigen-antibody binding using biosensor measurements at different temperatures. *Anal. Biochem.*, 246, 123–132. https://doi.org/10.1006/abio.1996.9999
32. Evans, S.V., & MacKenzie, R. (1999). Characterization of protein-glycolipid recognition at the membrane bilayer. *J. Mol. Recog.*, 12, 155–168. https://doi.org/10.1002/(sici)1099–1352(199905/06)12:3%3C155::aid-jmr456%3E3.0.co;2-s
33. Cooper, M.A., Hansson, A., Löfås, S., & Williams, D.H. (2000). A vesicle capture sensor chip for kinetic analysis of interactions with membrane-bound receptors. *Anal. Biochem.*, 277, 196–205. https://doi.org/10.1006/abio.1999.4389
34. Cooper, M.A., & Williams, D.H. (1999). Kinetic analysis of antibody-antigen interactions at a supported lipid monolayer. *Anal. Biochem.*, 276, 36–47. https://doi.org/10.1006/abio.1999.4333.
35. Cooper, M.A., Try, A.C., Carroll, J., Ellar, D.J., & Williams, D.H. (1998). Surface plasmon resonance analysis at a supported lipid monolayer. *Biochim. Biophys. Acta Biomembr.*, 1373, 101–111 https://doi.org/10.1016/S0005–2736(98)00091–1
36. Saenko, E., Sarafanov, A., Greco, N., Shima, M., Loster, K., Schwinn, H., & Josic, D. (1999). Use of surface plasmon resonance for studies of protein-protein and protein-phospholipid membrane interactions. Application to the binding of factor VIII to von Willebrand factor and to phosphatidylserine-containing membranes. *J. Chromatogr. A*, 852, 59–71. https://doi.org/10.1016/S0021–9673(99)00491–4
37. Baird, C.L., Courtenay, E.S., & Myszka, D.G. (2002). Surface plasmon resonance characterization of drug/liposome interactions. *Anal. Biochem.*, 310, 93–99. https://doi.org/10.1016/S0003–2697(02)00278–6
38. Hubbard, J.B., Silin, V., & Plant, A.L. (1998). Self assembly driven by hydrophobic interactions at alkanethiol monolayers: mechanisms of formation of hybrid bilayer membranes. *Biophys. Chem.*, 75, 163–176. https://doi.org/10.1016/S0301–4622(98)00199–9
39. Pattnaik, P. (2005). Surface plasmon resonance: applications in understanding receptor-ligand interaction. *Appl. Biochem. Biotechnol.*, 126, 79–92. https://doi.org/10.1385/ABAB:126:2:079 https://doi.org/10.1385/abab:126:2:079.
40. Fong, C.-C., Lai, W.-P., Leung, Y.-C., Lo, S.C.-L., Wong, M.-S., & Yang, M. (2002). Study of substrate-enzyme interaction between immobilized pyridoxamine and recombinant porcine pyridoxal kinase using surface plasmon resonance biosensor. *Biochim. Biophys. Acta Gener. Subj.*, 1596, 95–107. https://doi.org/10.1016/S0167–4838(02)00208-X
41. Wegner, G.J., Wark, A.W., Lee, H.J., Codner, E., Saeki, T., Fang, S., & Corn, R.M. (2004). Real-time surface plasmon resonance imaging measurements for the multiplexed determination of protein adsorption/desorption kinetics and surface enzymatic reactions on peptide microarrays. *Anal. Chem.*, 76(19), 5677–5684. https://doi.org/10.1021/ac0494275
42. Sibille, P., & Strosberg, A.D. (1997). A FIV epitope defined by a phage peptide library screened with a monoclonal anti-FIV antibody. *Immunol. Lett.*, 59(3), 133–137. https://doi.org/10.1016/s0165–2478(97)00111–9
43. Wu, L., Chu, H.S., Koh, W.S., & Li, E.P. (2010). Highly sensitive graphene biosensors based on surface plasmon resonance. *Opt. Express*, 18(14), 14395–14400. https://doi.org/10.1364/OE.18.014395
44. Shetti, N.P., Nayak, D.S., Reddy, K.R., & Aminabhavi, T.M. (2019). Graphene–clay-based hybrid nanostructures for electrochemical sensors and biosensors. In T.M. Aminabhavi (Ed.), *Graphene-Based Electrochemical Sensors for Biomolecules* (pp. 235–274). Amsterdam, The Netherlands: Elsevier. https://doi.org/10.1016/B978-0-12-815394-9.00010-8
45. Neiva, E.G.C., Bergamini, M.F., Oliveira, M.M., Marcolino, L.H., & Zarbin, A.J.G. (2014). PVP-capped nickel nanoparticles: synthesis, characterization and utilization as a glycerol electrosensor. *Sens. Actuat. B Chem.*, 196, 574–581. https://doi.org/10.1016/j.snb.2014.02.041
46. Thévenot, D.R., Toth, K., Durst, R.A., & Wilson, G.S. (2001). Electrochemical biosensors: recommended definitions and classification. *Biosens. Bioelectron.*, 16(1–2), 121–131. https://doi.org/10.1081/AL-100103209
47. Arino, C., Banks, C.E., Bobrowski, A., Crapnell, R.D., Economou, A., Królicka, A., Pérez-Ràfols, C., Soulis, D., & Wang, J. (2022). Electrochemical stripping analysis. *Nat. Rev. Methods Prim.*, 2, 63. https://doi.org/10.1038/s43586-022-00143-5
48. Rahman, M., Kumar, P., Park, D.-S., & Shim, Y.-B. (2008). Electrochemical sensors based on organic conjugated polymers. *Sensors*, 8(1), 118–141. https://doi.org/10.3390/s8010118
49. Feng, Q.-M., Zhang, Q., Shi, C.-G., & Xu, J.-J. (October 2013). Using nanostructured conductive carbon tape modified with bismuth as the disposable working electrode for stripping analysis in paper-based analytical devices. *Talanta,* 115, 235–240. https://doi.org/10.1016/j.talanta.2013.04.071
50. Liu, Q., Liu, L., Perdicakis, M., & Walcarius, A. (2021). Electrochemical stripping analysis from micro-counter electrode. *Electrochim. Acta*, 393, 139095. https://doi.org/10.1016/j.electacta.2021.139095

51. Mohamad Nor, N., Arivalakan, S., Zakaria, N.D., Nilamani, N., Lockman, Z., & Abdul Razak, K. (2022). Self-assembled iron oxide nanoparticle-modified APTES-ITO electrode for simultaneous stripping analysis of Cd(II) and Pb(II) ions. *ACS Omega*, 7, 3823–3833. https://doi.org/10.1021/acsomega.1c07158
52. Jovanovski, V., Xhanari, K., & Finšgar, M. (2022). Editorial: Recent advances of metal-film electrodes for trace electrochemical analysis. *Front. Chem.*, 10, 973672. https://doi.org/10.3389/fchem.2022.973672
53. Moro, G., De Wael, K., & Moretto, L.M. (2019). Challenges in the electrochemical (bio)sensing of nonelectroactive food and environmental contaminants. *Curr. Opin. Electrochem.*, 16, 57–65. https://doi.org/10.1016/j.coelec.2019.04.019
54. Cinti, S., Proietti, E., Casotto, F., Moscone, D., & Arduini, F. (2018). Paper-based strips for the electrochemical detection of single and double stranded DNA. *Anal. Chem.*, 90, 13680–13686. https://doi.org/10.1021/acs.analchem.8b04052
55. Dutta, S., Dutta Chowdhury, A., Biswas, S., Park, E.Y., Agnihotri, N., De, A., & De, S. (2018). Development of an effective electrochemical platform for highly sensitive DNA detection using MoS2 – polyaniline nanocomposites. *Biochem. Eng. J.*, 140, 130–139. https://doi.org/10.1016/j.bej.2018.09.016
56. Cheng, L., Xu, C., Cui, H., Liao, F., Hong, N., Ma, G., Xiong, J., & Fan, H. (2020). A sensitive homogenous aptasensor based on tetraferrocene labeling for thrombin detection. *Anal. Chim. Acta*, 1111, 1–7. https://doi.org/10.1016/j.aca.2020.03.017
57. Zhang, Y., Xia, J., Zhang, F., Wang, Z., & Liu, Q. (2018). Ultrasensitive label-free homogeneous electrochemical aptasensor based on sandwich structure for thrombin detection. *Sens. Actuat. B Chem.*, 267, 412–418. https://doi.org/10.3390/bios12040253
58. Chen, Y., Song, X., Li, L., & Tang, B. (2020). A high-fidelity electrochemical platform Based on Au–Se interface for biological detection. *Anal. Chem.*, 92, 5855–5861. https://doi.org/10.1021/acs.analchem.9b05509
59. Lee, J., Yun, J.Y., Lee, W.C., Choi, S., Lim, J., Jeong, H., Shin, D.-S., & Park, Y.J. (2017). A reference electrode-free electrochemical biosensor for detecting MMP-9 using a concentric electrode device. *Sens. Actuat. B Chem.*, 240, 735–741. https://doi.org/10.1016/j.snb.2016.09.026
60. Ahirwar, R., Dalal, A., Sharma, J.G., Yadav, B.K., Nahar, P., Kumar, A., & Kumar, S. (2019). An aptasensor for rapid and sensitive detection of estrogen receptor alpha in human breast cancer. *Biotechnol. Bioeng.*, 116, 227–233. https://doi.org/10.1002/bit.26819
61. Liu, M., Ke, H., Sun, C., Wang, G., Wang, Y., & Zhao, G. (2019). A simple and highly selective electrochemical label-free aptasensor of 17β-estradiol based on signal amplification of bi-functional graphene. *Talanta,* 194, 266–272. https://doi.org/10.1016/j.talanta.2018.10.035
62. Rizwan, M., Hazmi, M., Lim, S.A., & Ahmed, M.U. (2019). A highly sensitive electrochemical detection of human chorionic gonadotropin on a carbon nano-onions/gold, nanoparticles/polyethylene glycol nanocomposite modified glassy carbon electrode. *J. Electroanal. Chem.*, 833, 462–470. https://doi.org/10.1016/j.jelechem.2018.12.031.
63. Mishra, G.K., Barfidokht, A., Tehrani, F., & Mishra, R.K. (2018). Food safety analysis using electrochemical biosensors. *Foods,* 7(9), 141. https://doi.org/10.3390/foods7090141
64. Lee, S., Reuveny, A., Reeder, J., Lee, S., Jin, H., Liu, Q., Yokota, T., Sekitani, T., Isoyama, T., Abe, Y., & Suo, Z. (2016, May). A transparent bending-insensitive pressure sensor. *Nat. Nanotechnol.*, 11(5), 472–478. https://doi.org/10.1038/nnano.2015.324
65. Tao, L.Q., Zhang, K.N., Tian, H., Liu, Y., Wang, D.Y., Chen, Y.Q., Yang, Y., & Ren, T.L. (2017, September 26). Graphene-paper pressure sensor for detecting human motions. *ACS Nano*, 11(9), 8790–8795. https://doi.org/10.1021/acsnano.7b02826
66. Bai, N., Wang, L., Wang, Q., Deng, J., Wang, Y., Lu, P., Huang, J., Li, G., Zhang, Y., Yang, J., & Xie, K. (2020). Graded intrafillable architecture-based iontronic pressure sensor with ultra-broad-range high sensitivity. *Nat. Commun.*, 11(1), 1–9. https://doi.org/10.1038%2Fs41467-019-14054-9
67. Wu, Q., Qiao, Y., Guo, R., Naveed, S., Hirtz, T., Li, X., Fu, Y., Wei, Y., Deng, G., Yang, Y., & Wu, X. (2020, July 15). Triode-mimicking graphene pressure sensor with positive resistance variation for physiology and motion monitoring. *ACS Nano*, 14(8), 10104–10114. https://doi.org/10.1021/acsnano.0c03294
68. Rai, V.K. (2007, July). Temperature sensors and optical sensors. *Appl. Phys. B*, 88(2), 297–303. https://doi.org/10.1007/s00340-007-2717-4
69. Zhang, S., & Yu, F. (2011, October). Piezoelectric materials for high temperature sensors. *J. Am. Ceramic Soc.*, 94(10), 3153–3170. https://doi.org/10.1111/j.1551–2916.2011.04792.x
70. Chen, H.Y., Chen, A., & Chen, C. (2020, January). Investigation of the impact of infrared sensors on core body temperature monitoring by comparing measurement sites. *Sensors,* 20(10), 2885. https://doi.org/10.3390/s20102885.
71. Du, Y., Yu, G., Dai, X., Wang, X., Yao, B., & Kong, J. (2020, November 3). Highly stretchable, self-healable, ultrasensitive strain and proximity sensors based on skin-inspired conductive film for human motion monitoring. *ACS Appl. Mater. Interfaces*, 12(46), 51987–51998. https://doi.org/10.1021/acsami.0c15578.
72. Moheimani, R., Pasharavesh, A., Agarwal, M., & Dalir, H. (2020, August 17). Mathematical model and experimental design of nanocomposite proximity sensors. *IEEE Access*, 8, 153087–153097. https://doi.org/10.1109/ACCESS.2020.3017144
73. Grantz, K.H., Cummings, D.A., Zimmer, S., Vukotich Jr., C., Galloway, D., Schweizer, M.L., Guclu, H., Cousins, J., Lingle, C., Yearwood, G.M., & Li, K. (2021, January 27). Age-specific social mixing of school-aged children in a US setting using proximity detecting sensors and contact surveys. *Sci. Rep.*, 11(1), 11. https://doi.org/10.1038%2Fs41598-021-81673-y
74. Chen, Z., & Lu, C. (2005, December 1). Humidity sensors: a review of materials and mechanisms. *Sens. Lett.*, 3(4), 274–295. https://doi.org/10.1166/sl.2005.045
75. Farahani, H., Wagiran, R., & Hamidon, M.N. (2014, May). Humidity sensors principle, mechanism, and fabrication technologies: a comprehensive review. *Sensors,* 14(5), 7881–7939. https://doi.org/10.3390/s140507881
76. Delipinar, T., Shafique, A., Gohar, M.S., & Yapici, M.K. (2021, March 25). Fabrication and materials integration of flexible humidity sensors for emerging applications. *ACS Omega*, 6(13), 8744–8753. https://doi.org/10.1021/acsomega.0c06106
77. Lebosse, C., Renaud, P., Bayle, B., & de Mathelin, M. (2011, April 5). Modeling and evaluation of low-cost force sensors. *IEEE Trans. Robot.*, 27(4), 815–822. https://doi.org/10.1109/TRO.2011.2119850
78. Vecchi, F., Freschi, C., Micera, S., Sabatini, A.M., Dario, P., & Sacchetti, R. (2000, June 17). Experimental evaluation

of two commercial force sensors for applications in biomechanics and motor control. In *5th Annual Conference of International Functional Electrical Stimulation Society*. Corpus ID: 36971223.
79. Shi, J.X., Lei, X.W., & Natsuki, T. (2021, January). Review on carbon nanomaterials-based nano-mass and nano-force sensors by theoretical analysis of vibration behavior. *Sensors,* 21(5), 1907. https://doi.org/10.3390/s21051907
80. Fan, Z., Chen, J., Zou, J., Bullen, D., Liu, C., & Delcomyn, F. (2002, June 24). Design and fabrication of artificial lateral line flow sensors. *J. Micromech. Microeng.*, 12(5), 655. https://doi.org/10.1088/0960-1317/12/5/322
81. Kuo, J.T., Yu, L., & Meng, E. (2012, September). Micromachined thermal flow sensors – a review. *Micromachines,* 3(3), 550–573. https://doi.org/10.3390/mi3030550
82. Ejeian, F., Azadi, S., Razmjou, A., Orooji, Y., Kottapalli, A., Warkiani, M.E., & Asadnia, M. (2019, August 15). Design and applications of MEMS flow sensors: a review. *Sens. Actuat. A Phys.*, 295, 483–502. https://doi.org/10.1016/j.sna.2019.06.020
83. Wang, C., Yin, L., Zhang, L., Xiang, D., & Gao, R. (2010, March). Metal oxide gas sensors: sensitivity and influencing factors. *Sensors,* 10(3), 2088–2106. https://doi.org/10.3390/s100302088
84. Lee, J., Jung, Y., Sung, S.H., Lee, G., Kim, J., Seong, J., Shim, Y.S., Jun, S.C., & Jeon, S. (2021). High-performance gas sensor array for indoor air quality monitoring: the role of Au nanoparticles on WO3, SnO2, and NiO-based gas sensors. *J. Mater. Chem.*, 9(2), 1159–1167. https://doi.org/10.1039/D0TA08743B
85. Repelianto, A.S.A., & Kasai, N. (2019, January). The improvement of flaw detection by the configuration of uniform eddy current probes. *Sensors,* 19(2), 397. https://doi.org/10.3390/s19020397
86. Sun, H., Ali, R., Johnson, M., & Bowler, J.R. (2005, April 9). Enhanced flaw detection using an eddy current probe with a linear array of Hall sensors. In *AIP Conference Proceedings* (Vol. 760, pp. 516–522). https://doi.org/10.1063/1.1916719
87. Hu, F., Liu, M., Gao, H., & Lu, Z. (2009, January 26). Flaw-detected coating sensors applied in aircraft R&M. In *2009 Annual Reliability and Maintainability Symposium*. IEEE. https://doi.org/10.1016/j.sintl.2021.100110
88. Gómez-Robledo, L., López-Ruiz, N., Melgosa, M., Palma, A.J., Capitán-Vallvey, L.F., & Sánchez-Marañón, M. (2013, November 1). Using the mobile phone as Munsell soil-color sensor: an experiment under controlled illumination conditions. *Comput. Electron. Agric.*, 99, 200–208. https://doi.org/10.1016/j.compag.2013.10.002
89. Othman, M., Bakar, M.N., Ahmad, K.A., Razak, T.R., & Fuzzy, E. (2014, April 1). Ripening mango index using RGB color sensor model. *Research. World,* 5(2), 1. https://doi.org/10.3390/foods13050793
90. Sakai, O., Kitagawa, T., Sakurai, K., Itami, G., Miyagi, S., Noborio, K., & Taguchi, K. (2021, January 14). In-vacuum active color sensor and wireless communication across a vacuum. https://doi.org/10.1016/j.sintl.2021.100110
91. Silberman, N., & Fergus, R. (2011). Indoor scene segmentation using a structured light sensor. In *2011 IEEE International Conference on Computer Vision Workshops (ICCV Workshops)* (pp. 601–608). IEEE. https://doi.org/10.1109/ICCVW.2011.6130298
92. Fogle, K.J., Parson, K.G., Dahm, N.A., & Holmes, T.C. (2011, March 18). Cryptochrome is a blue-light sensor that regulates neuronal firing rate. *Science,* 331(6023), 1409–1413. https://doi.org/10.1126/science.1199702
93. Shi, Z., Wang, T., & Lin, J. (2021, March). A simultaneous calibration technique of the extrinsic and turntable for structured-light-sensor-integrated CNC system. *Opt. Laser Eng.*, 138, 106451. https://doi.org/10.1016/j.optlaseng.2020.106451
94. Chen, G.Y., Chai, T.Q., Wang, J.L., & Yang, F.Q. (2023). Recent advances in the colorimetric and fluorescence analysis of bioactive small molecule compounds based on the enzyme-like activity of nanomaterials. *J. Pharm. Biomed. Anal.*, 236, 115695. https://doi.org/10.1016/j.jpba.2023.115695
95. Kricka, L.J., & Park, J.Y. (2012). Optical techniques. In C.A. Burtis, E.R. Ashwood, & D.E. Bruns (Eds.), *Tietz Textbook of Clinical Chemistry and Molecular Diagnostics* (5th ed., pp. 242–244). Philadelphia: Saunders. https://doi.org/10.1007%2Fs12291-012-0287-7
96. Wilson, K., & Walker, J. (Eds.). (2010). *Principles and Techniques of Biochemistry and Molecular Biology* (7th ed.). New York: Cambridge University Press. https://doi.org/10.1017/9781316677056
97. Upadhyay, A., Upadhyay, K.N., & Nath, N. (2014). *Biophysical Chemistry: Principles and Techniques* (4th ed.). Gurgaon; Mumbai. ISBN-10: 8183188656, ISBN-13: 978–8183188654.
98. Upadhyay, A., Upadhyay, K., & Nath, N. (2010). *Techniques in Biochemistry* (1st ed.). Keshav Puram, Delhi: Ahuja Publishing House. ISBN: 8183188656, 9788183188654.
99. Bishop, M.L., Fody, E.P., & Schoef, L.E. (2010). Clinical chemistry techniques, principles, correlations. In C.M.L. Bishop, E.P. Fody, & E.P. Schoeff (Eds.), *Clinical Chemistry Techniques, Principles, Correlations* (6th ed., pp. 130–140). ISBN-10: 1451109008; ISBN-13: 978-1451109009.
100. Shrestha, Y.K., & Shrestha, S.K. (Eds.). (n.d.). Fundamentals of colorimetry. In A.K. Samanta (Ed.), *Advances in Colorimetry*. https://doi.org/10.5772/intechopen.112099
101. Smirnova, S.V., & Apyari, V.V. (n.d.). Aqueous two-phase systems based on cationic and anionic surfactants mixture for rapid extraction and colorimetric determination of synthetic food dyes. *Sensors*. https://doi.org/10.3390/s23073519
102. Chai, T.Q., Chen, G.Y., Chen, L.X., Wang, J.L., Zhang, C.Y., & Yang, F.Q. (2023). Adenine phosphate-Cu nanozyme with multienzyme mimicking activity for efficient degrading phenolic compounds and detection of hydrogen peroxide, epinephrine and glutathione. *Anal. Chim. Acta*, 1279, 341771. https://doi.org/10.1016/j.microc.2021.106992
103. Stewart, M.E., Anderton, C.R., Thompson, L.B., et al. (2008). Nanostructured plasmonic sensors. *Chem. Rev.* 108, 494–521. https://doi.org/10.1021/cr068126n
104. Atwater, H.A. (2007). The promise of plasmonics. *Sci. Am.*, 296, 56–63. https://doi.org/10.1145/1859855.1859856
105. Ozbay, E. (2006). Plasmonics: merging photonics and electronics at nanoscale dimensions. *Science,* 311, 189–193. https://doi.org/10.1126/science.1114849
106. Maier, S.A., & Atwater, H.A. (2005). Plasmonics: localization and guiding of electromagnetic energy in metal/dielectric structures. *J. Appl. Phys.*, 98, 011101. https://doi.org/10.1063/1.1951057
107. Duyne, R.P.V. (2004). Molecular plasmonics. *Science,* 306, 985–986. https://doi.org/10.1126/science.1104976
108. Barnes, W.L., Dereux, A., & Ebbesen, T.W. (2003). Surface plasmon subwavelength optics. *Nature,* 424, 824–830. https://doi.org/10.1038/nature01937
109. Haynes, C.L., McFarland, A.D., Zhao, L., et al. (2003). Nanoparticle optics: the importance of radiative dipole

coupling in two-dimensional nanoparticle arrays. *J. Phys. Chem. B*, 107, 7337–7342. https://doi.org/10.1021/jp034234r
110. Maier, S.A., Kik, P.G., Atwater, H.A., et al. (2003). Local detection of electromagnetic energy transport below the diffraction limit in metal nanoparticle plasmon waveguides. *Nat. Mater.*, 2, 229–232. https://doi.org/10.1038/nmat852
111. Willets, K.A., & Van Duyne, R.P. (2007). Localized surface plasmon resonance spectroscopy and sensing. *Annu. Rev. Phys. Chem.*, 58, 267–297. https://doi.org/10.1146/annurev.physchem.58.032806.104607
112. Haes, A.J., Haynes, C.L., McFarland, A.D., Schatz, G.C., Van Duyne, R.P., & Zou, S. (2005). Plasmonic materials for surface-enhanced sensing and spectroscopy. *MRS Bull.*, 30, 368–375. https://doi.org/10.1557/mrs2005.100
113. Pyrak, E., Krajczewski, J., Kowalik, A., Kudelski, A., & Jaworska, A. (2019). Surface enhanced Raman spectroscopy for DNA biosensors – how far are we? *Molecules*, 24(24), 4423. https://doi.org/10.3390/molecules24244423
114. Zhang, J., Dong, Y., Zhu, W., et al. (2019). Ultrasensitive detection of circulating tumor DNA of lung cancer via an enzymatically amplified SERS-based frequency shift assay. *ACS Appl. Mater. Interfaces*, 11, 18145–18152. https://doi.org/10.1021/acsami.9b02953
115. Lee, J.U., Kim, W.H., Lee, H.S., Park, K.H., & Sim, S.J. (2019). Quantitative and specific detection of exosomal miRNAs for accurate diagnosis of breast cancer using a surface-enhanced Raman scattering sensor based on plasmonic head-flocked gold nanopillars. *Small,* 15(18), 1804968. https://doi.org/10.1002/smll.201970091
116. Han, Y., Qiang, L., Gao, Y., et al. (2021). Large-area surface-enhanced Raman spectroscopy substrate by hybrid porous GaN with Au/Ag for breast cancer miRNA detection. *Appl. Surf. Sci.*, 154. https://doi.org/10.1016/j.apsusc.2020.148456
117. Choi, N., Dang, H., Das, A., Sim, M.S., Chung, I.Y., & Choo, J. (2020). SERS biosensors for ultrasensitive detection of multiple biomarkers expressed in cancer cells. *Biosens. Bioelectron.*, 164, 112326. https://doi.org/10.1016/j.bios.2020.112326
118. Jaworska, A., Fornasaro, S., Sergo, V., & Bonifacio, A. (2016). Potential of surface enhanced Raman spectroscopy (SERS) in therapeutic drug monitoring (TDM). A critical review. *Biosensors,* 6(3), 47. https://doi.org/10.3390/bios6030047
119. Wallace, G.Q., & Masson, J.-F. (2020). From single cells to complex tissues in applications of surface-enhanced Raman scattering. *Analyst,* 145, 7162–7185. https://doi.org/10.1039/D0AN01274B
120. Jaworska, A., Malek, K., & Kudelski, A. (2021). Intracellular pH – Advantages and pitfalls of surface-enhanced Raman scattering and fluorescence microscopy – a review. *Spectrochim. Acta Part A Mol. Biomol. Spectrosc.*, 251, 119410. https://doi.org/10.1016/j.saa.2020.119410
121. Wen, C., Chen, H., Guo, X., et al. (2021). Lysosome-targeted gold nanotheranostics for in situ SERS monitoring pH and multimodal imaging-guided phototherapy. *Langmuir,* 37, 569–577. https://doi.org/10.1021/acs.langmuir.0c03290
122. Cialla, D., März, A., Böhme, R., et al. (2012). Surface-enhanced Raman spectroscopy (SERS): progress and trends. *Anal. Bioanal. Chem.*, 403(1), 27–54. https://doi.org/10.1007/s00216-011-5631-x
123. Terry, L.R., Sanders, S., Potoff, R.H., et al. Applications of surface-enhanced Raman spectroscopy in environmental detection. *Anal. Sci. Adv.* https://doi.org/10.1002/ansa.202200003
124. Xu, M.L., Gao, Y., Han, X.X., et al. Innovative application of SERS in food quality and safety: a brief review of recent trends. *Foods,* 11(14), 2097. https://doi.org/10.3390/foods11142097
125. Wilson, A.D., & Baietto, M. (2009). Applications and advances in electronic-nose technologies. *Sensors,* 9, 5099–5148. https://doi.org/10.3390/s90705099
126. Zhang, J., Liu, X., Neri, G., & Pinna, N. (2016). Nanostructured materials for room-temperature gas sensors. *Adv. Mater.*, 28, 795–831. https://doi.org/10.1002/adma.201503825
127. Meng, F.L., Guo, Z., & Huang, X.J. (2015). *Trends Anal. Chem.*, 68, 37–47. https://doi.org/10.1021/ac5039863
128. Yuan, S., & Zhang, S. (2019). *J. Semicond.*, 40, 111608. https://doi.org/10.1088
129. Tyagi, P., Sharma, A., Tomar, M., & Gupta, V. (2016). *Sens. Actuat. B*, 224, 282–289. https://doi.org/10.3390/chemosensors12030042
130. Joshi, N., Hayasaka, T., Liu, Y., Liu, H., Oliveira, O.N., & Lin, L. (2018). Microchim. *Acta,* 185, 213. https://doi.org/10.1007/s00604-018-2750-5
131. Kumar, S., Pavelyev, V., Mishra, P., Tripathi, N., Sharma, P., & Calle, F. (2020). *Mater. Sci. Semicond. Process.,* 107, 104865. https://doi.org/10.1016/j.mssp.2019.104865
132. Kannan, P.K., Late, D.J., Morgan, H., & Rout, C.S. (2015). *Nanoscale,* 7, 13293–13312. https://doi.org/10.1039/C5NR03633J
133. Fine, G.F., Cavanagh, L.M., Afonja, A., & Binions, R. (2010). Metal oxide semi-conductor gas sensors in environmental monitoring. *Sensors,* 10, 5469–5502. https://doi.org/10.3390/s100605469
134. Rheaume, J.M. (2010). *Solid State Electrochemical Sensors for Nitrogen Oxide (NOx) Detection in Lean Exhaust Gases*. Berkeley: University of California.
135. Mehmood, F., Ahmad, S., & Kim, D.H. (2019). Design and implementation of an interworking IoT platform and marketplace in Cloud of Things. *Sustainability,* 11, 5952. https://doi.org/10.3390/su11215952
136. Blank, T.A., Eksperiandova, L.P., & Belikov, K.N. (2016). Recent trends of ceramic humidity sensors development: a review. *Sens. Actuat. B*, 228, 416–442. https://doi.org/10.1016/j.snb.2016.01.015
137. Cho, S.Y., Yoo, X.W., Kim, J.Y., Jung, W.B., Jiu, M.L., Kim, J.S., Jeon, H.J., & Jung, H.T. (2016). High resolution p-type metal oxide semiconductor nanowire array as an ultrasensitive sensor for volatile organic compounds. *Nano Lett.*, 16, 4508–4515. https://doi.org/10.1021/acs.nanolett.6b01713
138. Konvalina, G., & Haick, H. (2014). Sensors for breath testing: from nano-materials to comprehensive disease detection. *Acc. Chem. Res.*, 47, 66–76. https://doi.org/10.1021/ar400070m
139. Choi, S.J., Jang, B.H., Lee, S.J.B., Min, K., Rotschield, A., & Kim, I.D. (2014). Selective detection of acetone and hydrogen sulfide for diagnosis of diabetes and halitosis using SnO2 nanofibers functionalized with reduced graphene oxide nanosheets. *ACS Appl. Mater. Interfaces*, 6, 2588–2597. https://doi.org/10.1021/am405088q
140. Xing, Y., Vincent, T.A., Cole, M., & Gardner, J.W. (2019). Real-time thermal modulation of high-bandwidth MOX gas sensors for mobile robot applications. *Sensors,* 19, 1180. https://doi.org/10.3390/s19051180
141. Garcia-Orellana, C.J., Macias-Macias, M., Gonzales-Verasco, H.M., Garcia-Manso, A., & Gallardo-Caballero, R. (2019). Low power and low cost environmental IoT

electronic nose using initial action periodic measurements. *Sensors,* 19, 3183. https://doi.org/10.3390/s19143183
142. Rossi, M., & Brunelli, D. (2016). Autonomous gas detection and mapping with unmanned aerial vehicles. *IEEE Trans. Instrument. Measure.*, 65, 765–775. https://doi.org/10.1109/TIM.2015.2506319
143. Burgues, J., Hernandez, V., Lilienthal, A.J., & Marco, S. (2019). Smelling nano aerial vehicle for gas source localization and mapping. *Sensors,* 19, 478. https://doi.org/10.3390/s19030478
144. Punnoose, A., Reddy, K.M., Thurber, A., Hays, J., & Engelhard, M.H. (2007). *Nanotechnology*, 18, 165502. https;//doi.org/10.1039/c7ra11106a
145. Gerber, A., Kopnov, G., & Karpovski, M. (2017). *Appl. Phys. Lett.,* 111, 143505. https://doi.org/10.3390/chemosensors10020049
146. Pugh, E.M. (1930). *Phys. Rev.*, 36, 1503. https://doi.org/10.1103/PhysRev.36.1503
147. Lin, J.Y., Xie, W.M., He, X.L., & Wang, H.C. (2016). *Appl. Phys. A*, 122, 801. https://doi.org/10.1039/d0na00826e
148. Ciprian, R., Baratto, C., Giglia, A., Koshmak, K., Vinai, G., Donarelli, M., . . . Sberveglieri, G. (2016). *RSC Adv.*, 6, 42517–42521. https://doi.org/10.1039/C6RA18213E
149. Liang, J.Y., Pai, Y.C., Lam, T.N., Lin, W.C., Chan, T.S., Lai, C.H., & Tseng, Y.C. (2017). *Appl. Phys. Lett.,* 111, 023503. https://doi.org/arXiv:2112.07296v1
150. Ciprian, R., Torelli, P., Giglia, A., Gobaut, B., Ressel, B., Stupar, M., . . . Casarin, B. (2016). *RSC Adv.*, 6, 83399–83405. https://doi.org/10.1039/C6RA14131E
151. Hamrle, J., Blomeier, S., Gaier, O., Hillebrands, B., Schneider, H., Jakob, G., . . . Felser, C. (2007). *J. Phys. D: Appl. Phys.*, 40, 1563. https://doi.org/10.1088/0022–3727/48/33/333001
152. Lederman, D., Wang, Y., Morales, E.H., Matelon, R.J., Cabrera, G.B., Volkmann, U.G., & Cabrera, A.L. (2004). *Appl. Phys. Lett.,* 85, 615–617. https://doi.org/10.1063/1.1737481
153. Göpel, W., Hesse, J., & Zemel, J.N. (1989). *Sensors, A Comprehensive Survey, Magnetic Sensors.* VCH Publishers Inc. ISBN: 978-3-527-62058-6.
154. Cantor, R. (1996). DC SQUIDs: design, optimization and practical applications. In H. Weinstock (Ed.), *SQUID Sensors: Fundamentals, Fabrication and Applications* (pp. 179–233). Dordrecht, The Netherlands: Kluwer Academic Publisher. https://doi.org/10.1007/978-94-011-5674-5
155. Ketchen, M.B. (1987). Integrated thin-film dc SQUID sensors. *IEEE Trans. Magnet.*, 23, 1650–1657.
156. Singh, S.P. (2014). Magnetoencephalography: basic principles. *Ann. Indian Acad. Neurol.*, 17(Suppl 1), S107–S112. https://doi.org/10.4103/0972–2327.128676
157. Eichler, C.E., Cheng, L.K., Paskaranandavadivel, N., Du, P., Bradshaw, L.A., & Avci, R. (2021). Effects of magneto gastrography sensor configurations in tracking slow wave propagation. *Comput. Biol. Med.*, 129, 104169. https://doi.org/10.1016/j.compbiomed.2020.104169
158. Bergstrand, M., Soderlind, E., Weitschies, W., & Karlsson, M.O. (2009). Mechanistic modeling of a magnetic marker monitoring study linking gastrointestinal tablet transit, in vivo drug release, and pharmacokinetics. *Clin. Pharmacol. Therap.*, 86(1), 77–83. https://doi.org/10.1038/clpt.2009.43
159. Black, R., & Wellstood, F.C. (Eds.). (2004). *The SQUID Handbook: Applications of SQUIDs and SQUID Systems* (Vol. II, pp. 391–440). https://doi.org/10.1002/9783527609956.ch12
160. Adolphi, N.L., Huber, D.L., Bryant, H.C., et al. (2010). Characterization of single-core magnetite nanoparticles for magnetic imaging by SQUID relaxometry. *Phys. Med. Biol.*, 55, 5985–6003. https://doi.org/10.1088/0031–9155/55/19/023
161. Flynn, E.R., & Bryant, H.C. (2005). A SQUID based system for in-vivo cancer imaging. *Phys. Med. Biol.*, 50, 1273–1293. https://doi.org/10.1088/0031–9155/50/6/016
162. Kirtley, J.R., & Wikswo, J.P., Jr. (1999). Scanning squid microscopy. *Annu. Rev. Mater. Sci.*, 29, 117–148. https://doi.org/10.1146/annurev.matsci.29.1.117.
163. Sheinker, A., Ginzburg, B., Salomonski, N., Dickstein, P.A., Frumkis, L., & Kaplan, B.Z. (2012). Magnetic anomaly detection using high-order crossing method. *IEEE Trans. Geosci. Remote Sens.*, 50(4), 1095–1103. https://doi.org/10.1109/TGRS.2011.2164086.
164. Weinstock, H. (1996). *SQUID Sensors: Fundamentals, Fabrication, and Applications*. https://doi.org/10.1007/978-94-011-5674-5
165. Moseley, P.T. (1992). *Sens. Actuat. B Chem.*, 6, 149–156. https://doi.org/10.1016/0925–4005(92)80047–2
166. Taylor, R.F., & Schultz, J.S. (1996). *Handbook of Chemical and Biological Sensors.* CRC Press. https://doi.org/10.1201/9780367802516
167. Punnoose, A., Reddy, K.M., Hays, J., Thurber, A., & Engelhard, M.H. (2006). *Appl. Phys. Lett.,* 89, 112509. https://doi.org/10.3390%2Fs17040714
168. Ciprian, R., Torelli, P., Giglia, A., Gobaut, B., Ressel, B., Vinai, G., . . . Casarin, B. (2016). *RSC Adv.*, 6, 83399–83405. https://doi.org/10.1039/C6RA18213E
169. Lueng, C., Lupo, P., Schefer, T., Metaxas, P.J., Adeyeye, A.O., & Kostylev, M. (2019). *Int. J. Hydrogen Energy*, 44, 7715–7724. https://doi.org/10.3390/chemosensors10020049
170. Ignatyeva, D.O., Knyazev, G.A., Kapralov, P.O., Dietler, G., Sekatskii, S.K., & Belotelov, V.I. (2016). *Sci. Rep.*, 6, 28077. https://doi.org/10.1038/srep28077
171. Hsu, C.C., Chang, P.C., Chen, Y.H., Liu, C.M., Wu, C.T., Yen, H.W., & Lin, W.C. (2018). *Sci. Rep.*, 8, 1–2. https://doi.org/10.1039%2Fd0na00826e
172. Matatagui, D., Kolokoltsev, O.V., Qureshi, N., Mejía-Uriarte, E.V., & Saniger, J.M. (2015). A magnonic gas sensor based on magnetic nanoparticles. *Nanoscale,* 7, 9607–9613. https://doi.org/10.1039/C5NR01499A

# 5 Metal Oxide Nanomaterial-Based Optoelectronic and Sensing Devices

*Roselin George[1], Riju K. Thomas[1], Fidha Fathima K. H.[2], Diya Sebastian[2], and Nayana S. George[3]*
[1]Department of Physics, Deva Matha College, Kuravilangad, Kottayam, Kerala, India
[2]Bharata Mata College (Autonomous), Thrikkakara, Kochi, Ernakulam, Kerala, India
[3]Institute of Child Health (ICH), Kottayam, Kerala, India

## 5.1 IMPORTANCE OF METAL OXIDE NANOMATERIALS IN OPTOELECTRONICS AND SENSING

Metal oxide nanoparticles are a diverse form of nanomaterials, with their main applications in the chemical, physical, and materials sciences [1]. Metal oxide nanomaterials are a particularly important class of nanomaterials for optoelectronics and sensing applications. Metal oxides have several unique properties that make them ideal for these applications, including:

- **High surface area:** Metal oxide nanomaterials have a high surface area, which means that they have a large number of atoms on their surface. This is important for optoelectronics and sensing applications because it allows the nanomaterials to interact with a large number of light photons or molecules.
- **Tunable optical properties:** The optical properties of metal oxide nanomaterials can be tuned by varying their composition, size, and shape. This makes them ideal for a wide range of optoelectronics applications.
- **Chemical sensitivity:** Metal oxide nanomaterials are sensitive to a wide range of gases and chemicals.
- **Mechanical strength:** Metal oxide nanomaterials are often mechanically strong, which makes them suitable for use in a variety of devices.
- **Biocompatibility:** Many metal oxide nanomaterials are biocompatible, meaning that they can be used in medical applications.
- **Good stability:** Metal oxide nanomaterials are often more stable than other materials, which makes them more suitable for long-term applications.
- **Low cost:** Metal oxide nanomaterials are often relatively inexpensive to produce, which makes them more cost effective than other materials.

## 5.2 DEFINITION AND CLASSIFICATION OF METAL OXIDE NANOMATERIALS

Metal oxide nanomaterials are materials that contain metal and oxygen atoms and have at least one dimension in the nanometer (nm) range, which is 1 to 100 nm. Metal oxide nanomaterials can be classified into two main categories, metal oxides and complex metal oxides. Simple metal oxides contain only one type of metal atom and one type of oxygen atom. Examples of simple metal oxides include titanium dioxide ($TiO_2$), zinc oxide (ZnO), and tin dioxide ($SnO_2$). Complex metal oxides are materials which contain two or more types of metal atoms and one or more types of oxygen atoms. Examples of complex metal oxides include perovskites, such as lead titanate ($PbTiO_3$) and barium titanate ($BaTiO_3$).

## 5.3 SYNTHESIS METHODS FOR METAL OXIDE NANOPARTICLES

There are a number of different methods for synthesizing metal oxide nanoparticles. Some of the most common methods include hydrothermal synthesis, sol-gel synthesis, and chemical vapor deposition (CVD). During hydrothermal synthesis, a sealed container is used to heat a mixture of metal salts and water under high pressure. This creates a supercritical fluid that can be used to synthesize nanoparticles with a variety of different shapes and sizes. The sol-gel method involves dissolving metal salts in a solvent and then adding a gelling agent. The gelling agent causes the solution to form a gel, which can then be heated to produce metal oxide nanoparticles. Chemical vapor deposition is the process of growing metal oxide nanoparticles on a substrate in a vacuum chamber. The substrate is heated and then exposed to a gas mixture that contains the metal precursor and oxygen. The metal precursor reacts with the oxygen to form metal oxide nanoparticles on the surface of the substrate.

The specific synthesis method that is used depends on the desired properties of the metal oxide nanoparticles (Table 5.1).


DOI: 10.1201/9781032660639-5

**TABLE 5.1**
**Synthesis Methods of Nanoparticles and Their Advantages**

| Synthesis Method | Advantages |
|---|---|
| Sol-gel | Can be used to produce a wide range of nanoparticle shapes and sizes. |
| Hydrothermal synthesis | Can be used to produce nanoparticles with a narrow size distribution. |
| Flame synthesis | Can be used to produce nanoparticles with a high surface area. |
| Laser ablation | Can be used to produce nanoparticles with a narrow size distribution and a high degree of crystallinity. |
| Electrochemical deposition | Can be used to produce nanoparticles with a high degree of purity. |

For example, hydrothermal synthesis is often used to synthesize nanoparticles with a variety of different shapes and sizes, while sol-gel synthesis is often used to synthesize nanoparticles with a narrow size distribution. A wide range of physical, chemical, and biological techniques have been developed to produce metal oxide nanoparticles of different sizes, shapes, and compositions [2].

Here are some additional factors to consider when choosing a synthesis method for metal oxide nanoparticles:

- **The desired particle size and shape:** Some synthesis methods are better at producing nanoparticles of a particular size or shape than others.
- **The desired surface area:** Some synthesis methods produce nanoparticles with a higher surface area than others. This can be important for applications where the nanoparticles are used for catalysis or sensing.
- **The desired crystallinity**: Some synthesis methods produce nanoparticles with a higher degree of crystallinity than others. This can be important for applications where the nanoparticles are used for optical or electronic devices.
- **The desired purity:** Some synthesis methods produce nanoparticles with a higher degree of purity than others. This can be important for applications where the nanoparticles are used for biomedical applications.

## 5.4 METAL OXIDE NANOPARTICLE OPTOELECTRONIC DEVICES

Metal oxide semiconductors (MOSs) are a class of materials with unique electronic and optical properties that make them ideal for use as photoelectrodes in dye-sensitized solar cells (DSCs). DSCs are a type of low-cost, high-efficiency solar cell that uses a dye to absorb sunlight and generate electricity. MOSs are used in DSCs as the photoelectrode, which is the material that absorbs light and converts it into electrical energy [3].

Metal oxide nanoparticle optoelectronic devices have emerged as a cutting-edge technology with significant potential in various applications. These devices harness the unique properties of metal oxide nanoparticles to create highly efficient and versatile optoelectronic systems. One of the key advantages is their tunable electronic properties, enabling precise control over the bandgap and charge transport characteristics. This tunability allows engineers to design devices with tailored light-emitting or light-detecting properties, making them well suited for use in solar cells, photodetectors, and light-emitting diodes. Furthermore, metal oxide nanoparticle optoelectronic devices often exhibit exceptional optical transparency and electrical conductivity, making them ideal for transparent electronics, such as transparent solar panels, displays, and touchscreens. Their compatibility with flexible substrates and low-temperature processing methods allows for the creation of bendable and even wearable optoelectronic devices, opening up possibilities for novel applications in areas like wearable technology and flexible electronics. In addition to their inherent versatility, the cost-effective fabrication of these devices, often achieved through solution-based processes, adds to their attractiveness for large-scale production. As a result, metal oxide nanoparticle optoelectronic devices represent a promising avenue for advancing modern technology by providing solutions that are energy-efficient, adaptable, and economically viable for a wide range of optoelectronic applications. Classifications of optoelectronic devices are shown in Figure 5.1.

### 5.4.1 Solar Cells

Metal oxide nanomaterials can be used to improve the efficiency of solar cells by increasing the amount of light that is absorbed and converted into electricity. For example, titanium dioxide ($TiO_2$) nanomaterials are often used in solar cells because they have a high band gap and can absorb a wide range of wavelengths of light.

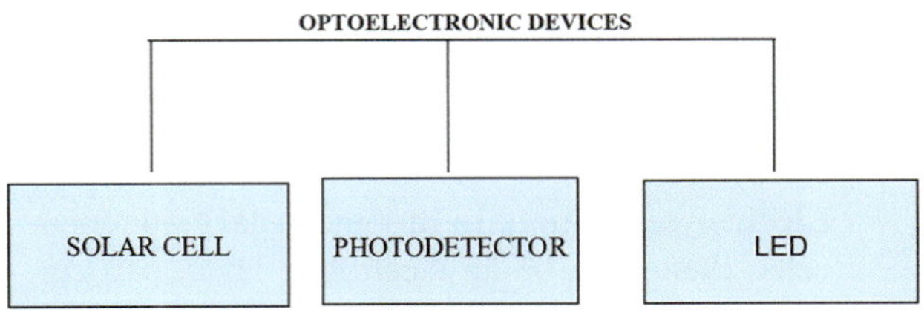


**FIGURE 5.1** A diagram illustrating the classification of optoelectronic devices, including solar cells, photodetectors, and light-emitting diodes (LEDs).

Developing solar cells with high power conversion efficiencies is critical for the sustainable development of human society but remains a challenge. Rapid charge separation, transport, and collection are essential for high-efficiency solar cells and can be achieved by optimizing the microscopic structures and morphologies of the materials used. One-dimensional nanomaterials have a large specific surface area and one-dimensional configuration, which favor charge transport and collection, making them ideal for use in solar cell fabrication. The impact of the one-dimensional structure on device performance is highlighted to elucidate the advantages of such nanomaterials [4]. Metal oxide nanomaterial solar cells are a class of photovoltaic devices that use metal oxide nanoparticles as a key component to capture and convert solar energy into electricity. These cells offer several advantages, including low cost, environmental friendliness, and the potential for high efficiency. A working description of metal oxide nanomaterial solar cells includes:

- **Light Absorption:** The basic principle of any solar cell is to absorb photons (particles of light) from sunlight. Metal oxide nanomaterial solar cells typically utilize metal oxide nanoparticles with a large surface area that can efficiently absorb a broad spectrum of sunlight. The choice of the metal oxide material can be tailored to absorb specific wavelengths or to enhance the absorption of light.
- **Generation of Electron-Hole Pairs:** Electron–hole pairs are developed by photon absorption and their excitations to the conduction band. This process is known as the photovoltaic effect.
- **Transport of Electrons and Holes:** Once electron–hole pairs are generated, the next step is to transport these charge carriers to the respective electrodes (anode and cathode) in the solar cell. Metal oxides are typically n-type semiconductors, meaning they have an excess of electrons. Electrons are transported through the metal oxide to the anode.
- **Electrode Materials:** The anode is typically made of a transparent conductive material such as indium tin oxide (ITO), while the cathode can be a metal or another suitable conductor. The electrons flow from the metal oxide nanoparticles to the anode through an external circuit, creating an electric current.
- **Electrolyte (Optional):** In some metal oxide solar cells, there may be an electrolyte layer, particularly in dye-sensitized solar cells (DSSCs). The electrolyte helps regenerate the electron-donating dye molecules by transferring electrons to them.
- **Reduction of Electron-Hole Recombination:** To improve the efficiency of the solar cell, it's essential to minimize the recombination of electrons and holes, which can reduce the current output. Metal oxide nanomaterials with a large surface area and appropriate energy band structures help to reduce recombination by providing more opportunities for charge carriers to reach the electrodes.
- **External Circuit:** The electric current generated by the flow of electrons from the anode to the cathode through the external circuit can be harnessed for various applications, including charging batteries or powering electrical devices.
- **Optical Management:** Metal oxide nanomaterial solar cells may also incorporate additional optical management layers, such as anti-reflection coatings or light-trapping structures, to enhance light absorption and improve overall efficiency.
- **Stability and Durability:** To ensure the long-term performance of metal oxide nanomaterial solar cells, materials, and device structures must be designed to withstand environmental factors like moisture, heat, and UV exposure.
- **Efficiency Improvement:** Ongoing research in the field of metal oxide nanomaterial solar cells focuses on improving their efficiency through innovations in materials, device architecture and fabrication techniques. Metal oxide nanomaterial solar cells hold promise for sustainable energy generation due to their cost-effectiveness and potential for high efficiency.

#### 5.4.1.1 Dye Sensitized Solar Cells

Organic dyes were shown to generate electricity at oxide electrodes in electrochemical cells when illuminated. A dye-sensitized solar cell contains a thick film of semiconductor nanoparticles, a monolayer of organic dye molecules adsorbed to the nanostructured layer, liquid electrolyte, and counter electrode. When the DSSC is illuminated, the dye molecules absorb photons and become excited. The excited electrons in the dye are then injected into the conduction band of the semiconductor material. These electrons diffuse through the semiconductor nanoparticles to the anode, where they are collected. The holes left behind in the dye molecules are refilled by electrons from the electrolyte [4].

In a p-type DSSC, the electrolyte contains a redox couple, which is a molecule that can exist in two different oxidation states. When the dye molecule injects an electron into the semiconductor, the redox couple in the electrolyte is reduced. The reduced redox couple then diffuses to the counter electrode, where it is oxidized. The recombination of electrons in the semiconductor with the electrolyte limits the efficiency of DSSCs. Recombination occurs when a hole, which is the absence of an electron, is injected into the semiconductor's conduction band as a result of photoexcitation. The hole then combines with an electron in the conduction band, releasing energy in the form of heat.

P-type DSSCs have traditionally had lower performance than n-type DSSCs because of the lack of suitable wide-bandgap semiconductors with good charge transport properties and sufficient transparency. However, recent advances in materials science have led to the development of new p-type semiconductors that have improved the performance of p-type DSSCs. The overall power conversion efficiency of DSSCs has now reached as high as 15.3% for n-type DSSCs and 11.3% for p-type DSSCs. DSSCs are a promising technology for solar energy conversion because they are relatively inexpensive to produce and can be made using a variety of materials. DSSCs are also flexible and lightweight, making them suitable for a variety of applications [5].

Metal oxide nanoparticles play a crucial role in dye-sensitized solar cells due to their significance in enhancing the cell's efficiency and functionality. These nanoparticles serve as key components in various parts of the DSSC structure.

**Electron Transport:** Metal oxide nanoparticles like titanium dioxide ($TiO_2$) or zinc oxide (ZnO) are commonly used as the semiconductor material in the photoanode of DSSCs. These nanoparticles facilitate the efficient transport of electrons generated from the dye upon light absorption.

**Charge Separation and Collection:** When light strikes the dye molecules, they get excited and release electrons. Metal oxide nanoparticles facilitate the separation of these charges, ensuring that the electrons are effectively collected and transported through the cell to produce an electric current. The configuration is shown in Figure 5.2.

##### *5.4.1.1.1 Advantages and Contributions of One-Dimensional Nanomaterials in Solar Cells*

The one-dimensional configuration and nanomaterial characteristics have made one-dimensional nanomaterials very promising as the building blocks to enhance the performance of solar cells. Transparent conductive electrodes are widely used in the fabrication of various solar cells, and transparency and electrical conductivity are two important factors determining the device's performance. A high transparency represents less loss of the incident light, and a low resistance means lower internal resistance of the solar cell [4].

**High specific surface area**: The interface between two phases has always been considered crucial for various systems, not limited to photovoltaic conversion. Larger specific areas enable sufficient charge generation, separation, and collection, which leads to high device performance. Compared with conventional bulk materials, materials at the nanoscale demonstrated much larger specific surface areas with higher interacting efficiencies and thus are favorable for performance enhancement [4].

**Oriented structure**: The assembly method of nanomaterials is considered very important to extend their physical properties from nanoscale to macroscopic scale. An oriented structure has been proved to efficiently improve the mechanical and electrical properties of nanomaterials in the oriented direction, which is favorable for the improvement of device performance [4].

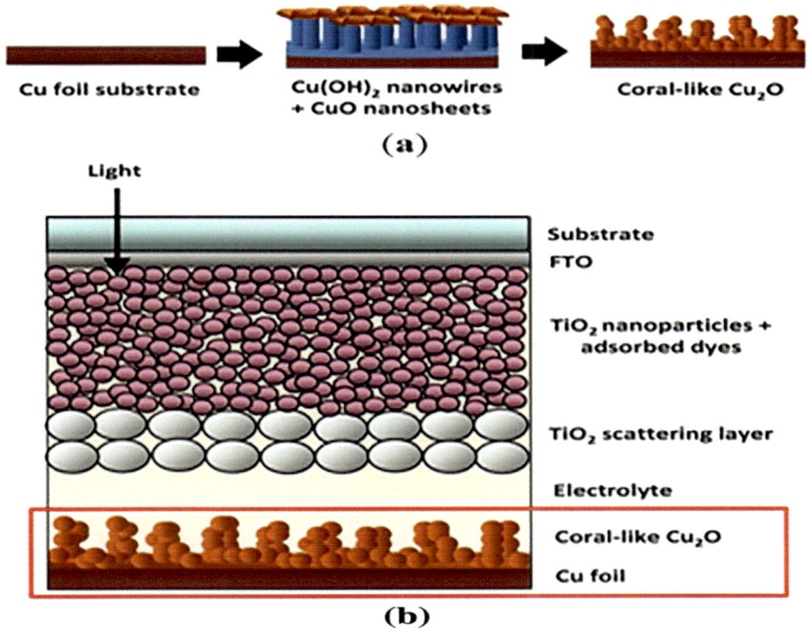


**FIGURE 5.2** A typical configuration of DSSC exhibiting a sandwich structure. (a) Fabrication of a coral-like $Cu_2O$ counter electrode – Shows the synthesis process of the $Cu_2O$ electrode, designed to enhance electron transfer in DSSCs. (b) Schematic structure of a DSSC with a $Cu_2O$ counter electrode – Illustrates how dye molecules on the nano-oxide layer capture solar energy, transferring electrons through the transparent conducting oxide (fluorine-doped tin oxide(FTO)) to the counter electrode. The nanostructured $TiO_2$ serves as a photoanode and the dye absorbs sunlight. The $Cu_2O$ counter electrode facilitates charge transfer, while the electrolyte replenishes electrons to maintain continuous energy conversion, improving DSSC efficiency.

**Source: (From Chih-Hung Tsai, Po-Hsi Fei, and Chih-Han Chen (2015, 8(9), 5715–5729 P 5718)).**

#### 5.4.1.2 Titanium Dioxide Nanofibers

Titanium dioxide ($TiO_2$) nanofibers operate based on their high surface area and unique photocatalytic properties. When exposed to light, $TiO_2$ nanofibers generate electron–hole pairs, which enable them to catalyze the decomposition of organic pollutants, making them effective in applications such as air and water purification, self-cleaning surfaces, and decontamination. Additionally, $TiO_2$ nanofibers can serve as highly sensitive sensing elements, with changes in electrical conductivity or optical properties indicating the presence of target analytes, making them valuable in environmental monitoring and medical diagnostics. They are also used in air filtration systems to capture and decompose airborne pollutants and are employed in photovoltaic devices, where their high surface area enhances light absorption and electron transport, contributing to the overall efficiency of solar cells. $TiO_2$ nanofibers are versatile nanomaterials with broad applications in various technological and environmental solutions [4].

#### 5.4.1.3 $Cu_2O$ Solar Cells

MO*x* compounds have a bandgap range between 1.4 and 2.2 eV. It can be covered from CuO*x* by tuning stable binary oxide phases of copper oxides $Cu_2O$, CuO and $Cu_4O_3$. Cuprous oxide ($Cu_2O$) is a *p*-type semiconductor attractive as photovoltaic material because of its low cost, high absorption coefficient, good mobility, great availability, and nontoxicity. In solar cells, $Cu_2O$ power conversion efficiency (PCE) was higher than ~8%, and it was used as a photocathode in photoelectrochemical cells. A Schottky barrier (heterojunction) solar cell can be fabricated as a back wall or front wall structure. A Schottky barrier solar cell indicates a major part of the potential drop at the metal/cuprous oxide interface, and only the electrons that could overcome this high-resistance barrier contribute to the external current of the cell [5]. Figure 5.3 illustrates a CuO/$Cu_2O$ heterojunction solar cell.

#### 5.4.1.4 Indium Tin Oxide

Indium tin oxide is a widely used transparent conducting oxide with high conductivity, hardness, chemical inertness, a high work function, and optical transparency. It is broadly applied in both research and industry as a thin film in photovoltaics, gas sensors, polymer-based electronics, antennas, and energy-saving glass windows [6]. Conventionally used ITO electrodes have the drawbacks of low resistivity, high transmittance, and high process temperatures for crystallization. Apart from the high cost and energy that vacuum deposition techniques consume, ITO production requires volatile indium incorporation. The film usually indicates a brittle nature and low thermal expansion coefficient [5].

#### 5.4.1.5 Nanoceria in Quantum Dot Solar Cells

Cerium oxide ($CeO_2$) nanoparticles, also known as nanoceria, have been investigated for use in quantum dot solar cells. Quantum dot solar cells utilize nanoscale semiconductor materials to absorb and convert light into electricity. Nanoceria, when incorporated as a layer, can enhance charge separation and collection within quantum dot solar cells, potentially leading to more efficient and cost-effective photovoltaic devices. The utilization of metal oxide nanoparticles in solar cells holds immense significance in the pursuit of clean and renewable energy sources. Here are several key points highlighting their importance:

- **Enhanced Efficiency:** Metal oxide nanoparticles improve the efficiency of solar cells by enabling efficient charge separation, electron transport, and reduced charge recombination. This results in higher power conversion efficiencies, making

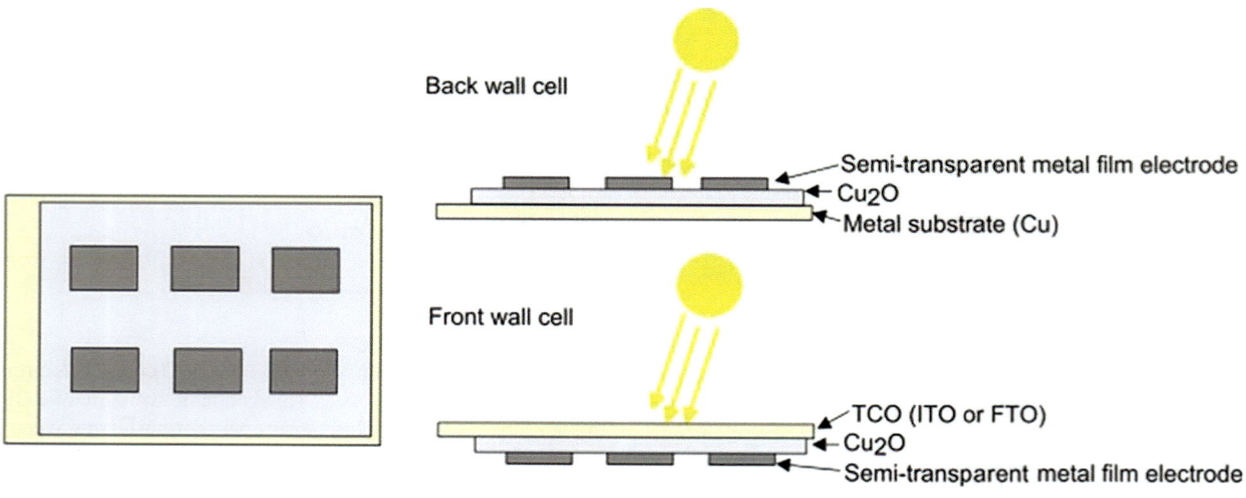


FIGURE 5.3 Illustration of CuO/$Cu_2O$ heterojunction solar cell (from Terence K. S. Wong, Siarhei Zhuk, Saeid Masudy-Panah, and Goutam K. Dalapati (2016, 9(4), 271 P 5)). Photovoltaic effect is due to the Schottky barrier, CuO/ZnO heterojunction with CuO/$Cu_2O$ nanopowder – enhances light absorption and charge transport.

solar energy a more competitive and accessible renewable energy source.

- **Low-Cost Materials:** Many metal oxide nanoparticles, such as titanium dioxide and zinc oxide, are abundant and relatively inexpensive, contributing to cost-effective solar cell manufacturing. This lowers the overall cost of solar energy production and promotes its widespread adoption.
- **Sustainability:** Solar cells employing metal oxide nanoparticles align with sustainable practices, as they rely on non-toxic and environmentally friendly materials. This is crucial for minimizing the environmental impact of solar cell production and disposal.
- **Versatility:** The versatility of metal oxide nanoparticles allows for their incorporation into various solar cell technologies, including both established ones like DSSCs and emerging ones like perovskite solar cells. This adaptability accelerates the development and adoption of advanced photovoltaic devices.
- **Reduction in Greenhouse Gas Emissions:** By harnessing solar energy, solar cells with metal oxide nanoparticles contribute to reducing greenhouse gas emissions associated with conventional energy sources. This transition to cleaner energy is vital in combating climate change and promoting sustainability.
- **Energy Independence:** Solar energy harnessed through these advanced solar cells promotes energy independence and security, reducing dependence on fossil fuels and non-renewable resources.

## 5.4.2 LEDs

Metal oxide nanoparticle LEDs (MON-LEDs) are a type of light-emitting diode (LED) that uses metal oxide nanoparticles as the active light-emitting material. It can be used to develop LEDs with different colors and brightness levels. For example, zinc oxide (ZnO) nanomaterials are often used in blue LEDs. MON-LEDs have a number of advantages over traditional LEDs. They have broader emission spectrum, so they can emit light across a wide range of wavelengths, from ultraviolet to infrared. This makes them suitable for a variety of applications, including lighting, displays, and sensors. They are more efficient and have a longer lifetime than traditional LEDs. They can last for longer periods of time without replacement. Examples of metal oxide nanoparticles that can be used in MON-LEDs are zinc oxide (ZnO), titanium dioxide ($TiO_2$), aluminum oxide ($Al_2O_3$), gallium nitride (GaN), and indium gallium nitride (InGaN). These nanoparticles can be synthesized in a variety of ways, including chemical vapor deposition, sol-gel synthesis, and hydrothermal synthesis. Once the nanoparticles have been synthesized, they can be incorporated into an LED device using a variety of methods. One common method is to spin-coat a thin film of nanoparticles onto a substrate. The substrate can be made of a variety of materials, such as glass, sapphire, or silicon. Another method for incorporating nanoparticles into an LED device is to embed them in a polymer matrix. This approach can be used to create flexible MON-LEDs.

**Applications in light-emitting devices:** As nanoparticles are prone to have high luminescence quantum yield, they can possibly be applied in light-emitting devices. They should also possess a direct band gap to emit light in the limited and desired spectral range [7]. Metal oxide nanoparticles have emerged as promising materials for light-emitting diodes due to their unique optical and electrical properties. MONPs can be used to fabricate LEDs with a wide range of colors, including white, blue, green, red, and yellow [7].

**White LEDs**: MONPs can be used to fabricate white LEDs by combining them with a phosphor material. The phosphor material absorbs blue light from the MONPs and emits green and red light, creating white light.

**Blue LEDs**: MONPs can be used to fabricate blue LEDs by exploiting their direct bandgaps. When excited by an electrical current, MONPs can emit blue light. MONPs such as zinc oxide and GaN are commonly used for blue LEDs. Here are some advantages of using metal oxide nanoparticles in blue LEDs:

- They can emit a wide range of blue colors, from deep blue to light blue.
- They are very efficient, meaning that they can convert a lot of electrical energy into light.
- They are very stable, meaning that they can be used in a variety of applications without degrading.
- They are relatively inexpensive to manufacture.

**Green LEDs**: MONPs can be used to fabricate green LEDs by doping them with specific impurities. Doping introduces energy levels within the bandgap, allowing for the emission of green light.

**Red LEDs:** MONPs can be used to fabricate red LEDs by using a combination of MONPs and a phosphor material. The MONPs emit orange light, and the phosphor material absorbs orange light and emits red light.

**Yellow LEDs:** MONPs can be used to fabricate yellow LEDs by using a combination of MONPs and a phosphor material. The MONPs emit green light, and the phosphor material absorbs green light and emits yellow light.

### 5.4.2.1 Metal Oxide Nanoparticle-Based LEDs

Metal oxide nanoparticle LEDs possess unique characteristics that make them promising for various applications. First and foremost, these LEDs offer tunable emission properties, allowing precise control over the emitted light's wavelength and color. By adjusting the size, composition, and doping of the metal oxide nanoparticles, engineers can create LEDs with specific emission spectra, making them adaptable for diverse uses, including displays, lighting, and optoelectronics. Another key characteristic is their high quantum efficiency, which indicates their ability to efficiently convert electrical energy into visible light. This efficiency results in energy-saving LED devices with reduced power consumption, contributing to sustainability and cost-effectiveness. Metal oxide nanoparticles, such as zinc oxide and tin dioxide, can exhibit excellent stability and durability, ensuring long-lasting LED performance even in challenging environments, making them suitable for applications where reliability is paramount. Moreover, the compatibility of metal oxide nanoparticles with thin-film and flexible substrates allows for the development of flexible and bendable LED displays, making them ideal for emerging technologies like wearable devices and curved screens. Additionally, their low toxicity and potential for multifunctionality, combined with their ability to serve as transparent conductive layers, enhance their versatility for a wide range of LED applications. These characteristics make metal oxide nanoparticle LEDs a compelling area of research and development in the field of optoelectronics and solid-state lighting.

#### 5.4.2.2 Zinc Oxide LEDs

Zinc oxide stands out as a highly promising material for UV light emission owing to its environmentally friendly nature. It possesses the advantage of being grown at low temperatures on inexpensive transparent substrates. Moreover, ZnO boasts a direct wide band gap of 3.3 eV and a notably large exciton binding energy of 60 MeV, essential characteristics for robust light emission applications. UV light emitters find extensive utility across various domains, including solid-state lighting, forensic applications, non-line-of-sight communications, biosensors, and air purification systems. These applications benefit from the unique properties of ZnO, making it an attractive choice for fulfilling the demands of UV-based technologies in diverse fields. Metal oxide quantum dots are employed in quantum dot light-emitting diodes (QLEDs) to generate efficient and tunable light emission. Their size-dependent bandgap allows precise control over the colors emitted, making them valuable in displays and lighting. The QLED technology is integrated into displays and lighting products to enhance color accuracy, brightness, and energy efficiency. They are used in TVs, monitors, and lighting solutions for improved visual quality.

##### *5.4.2.2.1 Advantages of Zinc Oxide LEDs*

ZnO is a wide-bandgap semiconductor with a band gap of 3.37 eV, making it well suited for UV and blue LED applications. ZnO LEDs have been demonstrated with efficiencies of over 20% for blue emission and over 50% for UV emission. ZnO LEDs are also relatively inexpensive to produce and can be processed using solution-based methods, making them a promising technology for low-cost and flexible LED displays. Zinc oxide LEDs, a class of light-emitting diodes that utilize ZnO as the active material, offer several unique characteristics that make them noteworthy in the realm of optoelectronics. First, ZnO LEDs are known for their wide bandgap, which enables them to emit ultraviolet (UV) or blue light. This characteristic is crucial for applications like data storage, sterilization, and the generation of high-quality white light. ZnO LEDs are particularly attractive for their ability to produce UV light efficiently, a feature often used in medical and industrial applications.

Another key characteristic is ZnO's excellent transparency and high electron mobility. This transparency allows ZnO LEDs to serve as transparent conductive layers in devices like flat-panel displays, solar cells, and touchscreens. Moreover, ZnO LEDs can be fabricated on flexible substrates, paving the way for flexible and bendable LED displays, offering new possibilities in the development of wearable technology and innovative lighting solutions. Furthermore, ZnO is a semiconductor material that is chemically and thermally stable, contributing to the durability and long-term performance of ZnO LEDs. Their stability and resistance to environmental factors make them suitable for various outdoor and industrial applications. Additionally, ZnO is considered a less toxic material compared to some alternatives. ZnO LEDs offer a combination of characteristics, including a wide bandgap for UV and blue emission, excellent transparency, flexibility, and stability, making them a valuable choice for applications ranging from UV sterilization and data storage to transparent displays and wearable technology. Their unique properties continue to drive research and development in the optoelectronics industry.

#### 5.4.2.3 Titanium Dioxide LEDs

$TiO_2$ is another wide-bandgap semiconductor with a band gap of 3.23 eV, making it well suited for UV LED applications. $TiO_2$ LEDs have been demonstrated with efficiencies of over 90% for UV emission. $TiO_2$ LEDs are also very stable and can be operated in harsh environments. $TiO_2$ nanomaterial LEDs work in a similar way to other semiconductor LEDs. When the $TiO_2$ nanomaterials are illuminated with light, the electrons in the valence band are excited into the conduction band. These excited electrons can then recombine with holes in the valence band, releasing energy in the form of photons. The wavelength of the emitted photons is determined by the band gap of the semiconductor material. In the case of $TiO_2$, the band gap is 3.23 eV, which corresponds to UV light. However, there are a few unique features of $TiO_2$ nanomaterial LEDs that make them attractive for UV light generation. First, $TiO_2$ nanomaterials have a very high surface area, which allows them to absorb more light and generate more electron–hole pairs. Second, $TiO_2$ nanomaterials are very stable and can be operated in harsh environments. Third, $TiO_2$ nanomaterials are relatively inexpensive to produce. The various steps involved in this procedure are

- $TiO_2$ nanomaterials are deposited on a substrate, such as sapphire or glass.
- A metal contact is deposited on top of the $TiO_2$ nanomaterials.

- A voltage is applied between the metal contact and the substrate.
- When the $TiO_2$ nanomaterials are illuminated with light, electrons in the valence band are excited into the conduction band.
- These excited electrons diffuse to the metal contact and are collected.
- Holes left behind in the valence band diffuse to the substrate.
- Electrons from the substrate recombine with the holes in the valence band, releasing energy in the form of photons.
- The wavelength of the emitted photons is determined by the band gap of the $TiO_2$ semiconductor material.

$TiO_2$ nanomaterial LEDs have a number of potential applications, including water and air sterilization, medical diagnostics, sensor applications, UV curing, and photocatalysis.

#### 5.4.2.4 Copper Oxide LEDs

CuO is a p-type semiconductor with a band gap of 1.2 eV, making it suitable for red and green LED applications. CuO LEDs have been demonstrated with efficiencies of over 10% for red and green emission. CuO LEDs are also relatively inexpensive to produce and can be processed using solution-based methods. Copper oxide nanomaterial LEDs work by emitting light when electrons and holes recombine in the CuO semiconductor. CuO is a p-type semiconductor, which means that it has more holes than electrons. When a CuO LED is illuminated, photons are absorbed by the CuO semiconductor and generate electron-hole pairs. The electrons and holes then diffuse through the CuO semiconductor and recombine, emitting light. The wavelength of the emitted light depends on the band gap of the CuO semiconductor. The band gap is the energy difference between the valence band and the conduction band. In CuO, the band gap is 1.2 eV, which corresponds to a wavelength of 1000 nanometers (nm). This means that CuO LEDs can emit red and green light.

There are two main types of CuO nanomaterial LEDs: thin film LEDs and nanowire LEDs. Thin film CuO LEDs are fabricated by depositing a thin film of CuO onto a substrate. Nanowire CuO LEDs are fabricated by growing CuO nanowires on a substrate.

#### 5.4.2.5 Iron Oxide LEDs

$Fe_2O_3$ is a p-type semiconductor with a band gap of 2.2 eV, making it suitable for red and yellow LED applications. $Fe_2O_3$ LEDs have been demonstrated with efficiencies of over 5% for red and yellow emission. $Fe_2O_3$ LEDs are also very stable and can be operated at high temperatures. Iron oxide nanomaterial LEDs work in a similar way to other semiconductor LEDs. When the $Fe_2O_3$ nanomaterial LED is illuminated, photons are absorbed by the $Fe_2O_3$ nanoparticles and excite electrons from the valence band to the conduction band. These excited electrons then diffuse through the $Fe_2O_3$ nanoparticles to the anode, where they are collected. The holes left behind in the $Fe_2O_3$ nanoparticles are refilled by electrons from the electrolyte.

The recombination of electrons in the $Fe_2O_3$ nanoparticles with the electrolyte limits the efficiency of $Fe_2O_3$ nanomaterial LEDs. Recombination occurs when a hole, which is the absence of an electron, combines with an electron in the conduction band, releasing energy in the form of heat. Researchers are working on ways to improve the efficiency of $Fe_2O_3$ nanomaterial LEDs by reducing recombination. One approach is to use surface treatments to modify the surface of the $Fe_2O_3$ nanoparticles and reduce the number of defects that act as recombination centers. Another approach is to develop new electrolytes that are less likely to recombine with electrons in the $Fe_2O_3$ nanoparticles. $Fe_2O_3$ nanomaterial LEDs have a number of advantages over traditional inorganic LEDs. $Fe_2O_3$ nanomaterials are relatively inexpensive to produce, and $Fe_2O_3$ nanomaterial LEDs are relatively easy to fabricate. $Fe_2O_3$ nanomaterial LEDs are also environmentally friendly and non-toxic. $Fe_2O_3$ nanomaterial LEDs are still under development, but they have the potential to be used in a variety of applications, such as lighting, displays, and sensors.

#### 5.4.2.6 Aluminum Oxide LEDs

$Al_2O_3$ is a transparent semiconductor with a band gap of 9.0 eV, making it suitable for white LED applications. $Al_2O_3$ LEDs have been demonstrated with efficiencies of over 20% for white emission. $Al_2O_3$ LEDs are also very stable and can be operated in harsh environments. Aluminum oxide nanomaterial LEDs work in a similar way to other metal oxide nanomaterial LEDs. When the LED is illuminated, the $Al_2O_3$ nanomaterial absorbs photons and becomes excited. The excited electrons in the $Al_2O_3$ are then injected into the conduction band of the LED. These electrons diffuse through the $Al_2O_3$ nanoparticles to the anode, where they are collected. The holes left behind in the$Al_2O_3$ nanomaterial are refilled by electrons from the electrolyte.

The main difference between $Al_2O_3$ nanomaterial LEDs and other metal oxide nanomaterial LEDs is that $Al_2O_3$ is a transparent semiconductor. This means that $Al_2O_3$ nanomaterial LEDs can be used to create white LEDs by combining them with phosphors. Phosphors are materials that absorb light at one wavelength and emit light at another wavelength. When $Al_2O_3$ nanomaterial LED is combined with a phosphor, the blue light emitted by the LED is absorbed by the phosphor and re-emitted as white light. $Al_2O_3$ nanomaterial LEDs have a number of advantages over other metal oxide nanomaterial LEDs, including:

**Higher efficiency:** $Al_2O_3$ nanomaterial LEDs have been demonstrated with efficiencies of over 20% for white emission, which is higher than the efficiency of most other metal oxide nanomaterial LEDs.

**Better transparency:** $Al_2O_3$ is a transparent semiconductor, which makes it suitable for white LED applications.

**Better stability**: $Al_2O_3$ is a very stable material, which makes $Al_2O_3$ nanomaterial LEDs suitable for a wide range of applications.

In addition to the types of metal oxide nanomaterial LEDs listed previously, there are also a number of other metal oxide nanomaterials that are being investigated for LED applications. Some of these materials include:

- Gallium nitride LEDs: GaN is a wide-bandgap semiconductor with a band gap of 3.4 eV, making it well suited for UV and blue LED applications. GaN LEDs are already widely used in commercial LED products, such as LED displays and headlights.
- Indium gallium nitride LEDs: InGaN is an alloy of GaN and indium nitride (InN), which has a tunable band gap from 0.7 eV to 3.4 eV. This makes InGaN LEDs suitable for a wide range of applications, from infrared to ultraviolet. InGaN LEDs are already widely used in commercial LED products, such as LED displays and lighting.
- Perovskite LEDs: Perovskites are a class of materials with a unique crystal structure. Perovskite LEDs have been demonstrated with efficiencies of over 25% for white emission. Perovskite LEDs are still under development, but they have the potential to revolutionize the LED industry.

Metal oxide nanomaterial LEDs have the potential to revolutionize the LED industry due to their advantages in cost, solution processability, and environmental friendliness. Metal oxide nanomaterial LEDs are still under development, but they are already being used in a variety of commercial products.

### 5.4.3 Photodetectors

Metal oxide nanoparticles, such as zinc oxide, titanium dioxide, and tin oxide, have demonstrated exceptional performance in photodetectors. These nanoparticles exhibit excellent photoconductivity, making them ideal for converting light into electrical signals. Quantum confinement effects in nanoparticles enhance their photo response. They are used in imaging sensors, solar cells, and optical communication devices. Metal oxide nanomaterials are a promising class of materials for photodetectors due to their unique properties, including high surface area, good photoconductivity, and tunable bandgap. MON photodetectors work by converting light into an electrical signal. MON photodetectors can be used to detect a wide range of wavelengths of light, from ultraviolet to infrared (IR). They are also highly sensitive and can be used to detect very low levels of light. MON photodetectors have a number of advantages over traditional photodetectors based on other materials, such as silicon and germanium. MON photodetectors are more flexible and can be fabricated on a variety of substrates. They are also more cost effective to manufacture.

Photodetectors can be used in optical communication systems to receive and transmit data, imaging, environmental monitoring, and medical diagnostics such as optical coherence tomography (OCT) and photoacoustic imaging. Some specific examples of MON photodetectors are discussed in the following subsections.

#### 5.4.3.1 Titanium Dioxide Photodetectors

Titanium dioxide photodetectors (PDs) are sensitive to ultraviolet light and are widely used in optical communication systems and UV sensors. $TiO_2$ nanostructure material is a highly prized metal oxide semiconductor material due to its key features such as high transparency, high refractive index, good chemical stability, relative hardness, and wide bandgap (see Figure 5.4). These material oxides exist in three distinct structural systems: rutile (tetragonal), anatase (trigonal), and brookite (orthorhombic). Various techniques are employed to synthesize $TiO_2$ nanoparticles, including sol-gel, chemical vapor deposition, sputtering, and hydrothermal methods. The nanosecond pulsed laser ablation method in a fluid environment is a particularly effective approach for producing $TiO_2$ nanoparticles with different sizes, shapes, and morphologies, depending on laser ablation parameters such as wavelength, fluence, number of laser pulses, and liquid medium. This method has recently emerged as a promising, simple, and cost-effective technique for synthesizing size-controlled nanoparticles in solution with high purity.

During the laser ablation process, irradiation of a target immersed in a liquid with a pulsed laser initiates the formation of plasma and material removal. These plasmas contain ions, atoms, and clusters, which collide and react with molecules in the surrounding medium to form new compounds, including atoms from both the target and the solution. Furthermore, due to the high-density conditions, this method enables the fabrication of unique materials that are not achievable using traditional methods. $TiO_2$ nanoparticles offer a range of desirable properties and can be synthesized using various techniques. The nanosecond pulsed laser ablation method in a fluid environment stands out as a promising approach for producing high-quality $TiO_2$ nanoparticles with tailored properties [6].

#### 5.4.3.2 Zinc Oxide Photodetectors

Zinc oxide photodetectors can detect both ultraviolet and visible light, making them well suited for imaging systems and environmental monitoring sensors.

Traditional ZnO UV photodetectors typically require an external driving force, such as a voltage or current source, to operate. When exposed to UV light, photogenerated electrons and holes are driven by the external driving force to generate a photocurrent. Consequently, the external driving force significantly influences the recombination, separation, and transport processes of photogenerated electrons and holes, which ultimately determine the overall photoresponse behavior of a conventional photodiode. ZnO photodetectors function by absorbing incident light and

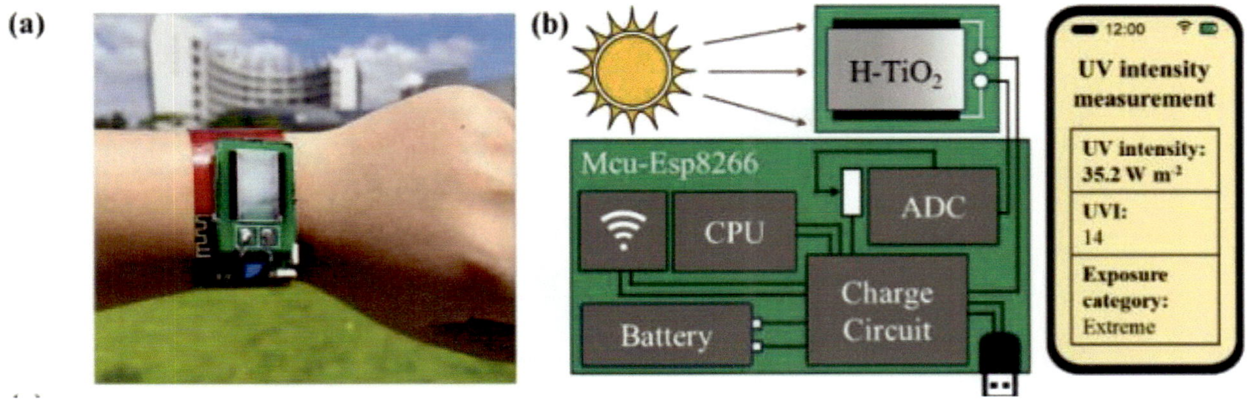


**FIGURE 5.4** An ultraviolet photodetector based on conductive hydrogenated $TiO_2$ (H-$TiO_2$) Film (from Y. Zhang (2024, p. 16)). (a) A wearable UV photodetector. (b) An illustration of PD device consisting of the H-$TiO_2$ PD, an analog-to-digital converter (ADC), battery, Wi-Fi, a charger, and multicontroller (MCU) assembled in a circuit board, whose data values can be accessed from mobile browser.

creating electron-hole pairs. These electron–hole pairs are then separated by an internal electric field, resulting in a photocurrent. The strength of the photocurrent is directly proportional to the intensity of the incident light, making ZnO nanomaterial photodetectors excellent for light sensing applications.

#### 5.4.3.3 Gallium Oxide Photodetectors

$Ga_2O_3$ photodetectors are sensitive to UV and visible light and are also sensitive to deep UV light, which is not detectable by most other photodetectors. $Ga_2O_3$ photodetectors are often used in applications such as flame detection and ozone monitoring. MON photodetectors are a promising new technology with the potential to revolutionize the way we communicate, image, and monitor our environment and our health. Gallium oxide photodetectors are semiconductor devices that exploit the unique properties of $Ga_2O_3$ to detect and convert incoming photons into electrical signals. These photodetectors exhibit several distinctive working principles and properties.

##### *5.4.3.3.1 Working Principles*

Gallium oxide photodetectors operate based on the intrinsic semiconductor properties of $Ga_2O_3$. When photons with energy greater than the bandgap of $Ga_2O_3$ strike the material, they generate electron–hole pairs. These charge carriers are then separated and collected, resulting in a measurable photocurrent. $Ga_2O_3$'s wide bandgap, typically around 4.5 to 4.9 eV, makes it particularly suitable for detecting ultraviolet light and even some shorter-wavelength visible light, as it allows efficient absorption of high-energy photons. $Ga_2O_3$ photodetectors are known for their exceptional sensitivity to UV light, making them valuable for applications in UV radiation monitoring and flame detection. Additionally, they offer fast response times due to the efficient charge carrier generation and collection processes, which are crucial in applications where rapid detection is required. $Ga_2O_3$ photodetectors benefit from the material's inherent properties, including its wide bandgap, high thermal stability, and excellent chemical resistance. These characteristics make $Ga_2O_3$ photodetectors suitable for harsh environmental conditions. Furthermore, they can be engineered as thin-film devices, which enhances their integration into various electronic and optoelectronic systems, making them promising candidates for UV sensing, optical communication and environmental monitoring applications. $Ga_2O_3$ photo detectors utilize the unique properties of gallium oxide, such as its wide bandgap, high sensitivity to UV light, and fast response times, to efficiently detect and convert photons into electrical signals. These detectors are particularly valuable for applications requiring UV light sensing and can withstand challenging environmental conditions due to $Ga_2O_3$'s exceptional thermal stability and chemical resistance.

## 5.5 SENSING DEVICES

Sensing devices are instruments designed to detect and measure physical or chemical properties in the environment and are given in Figure 5.5. They play a crucial role in various fields, including environmental monitoring, healthcare, and industrial processes. These devices can be based on a wide range of technologies, such as chemical sensors, biosensors, gas sensors, and environmental sensors. They have been intensively employed in diverse applications [8]. The structures of some oxides which have been commonly used for making sensing devices are shown in Figure 5.6.

Metal oxide nanomaterials are a promising class of materials for sensing devices due to their unique properties, such as high surface area, good electrical conductivity, and tunable band gap. MON sensing devices work by detecting changes in the electrical conductivity of the MON material when it is exposed to a target analyte.

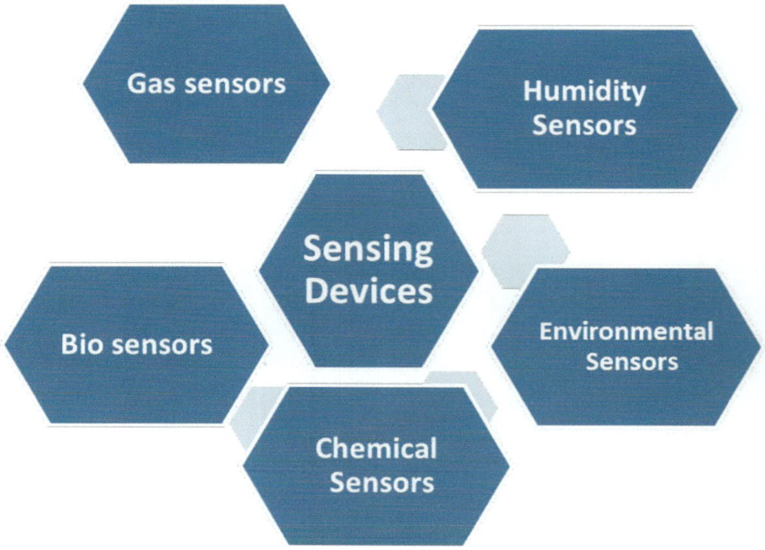


**FIGURE 5.5** Different types of sensors.

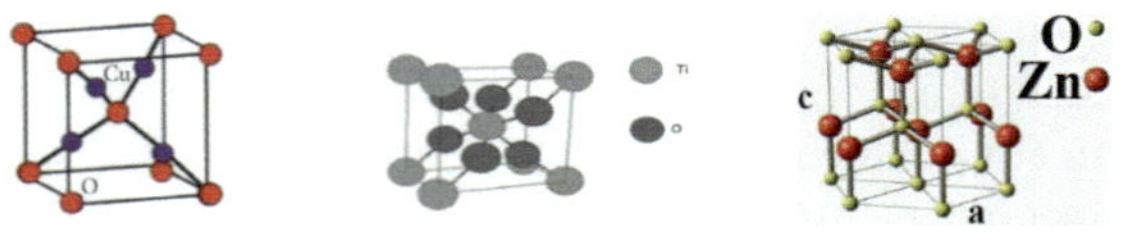


(a) $Cu_2O$ (b) $TiO_2$ (c) ZnO

**FIGURE 5.6** The structures of some oxides used for making sensing devices.

These oxides, including (a) Cu2O (from F. Chaudhary, I. Shoukat, D. Afsal, C. Park, M. Cancan, M R. Farahani(2021)), (b) $TiO_2$ (from Stephen L (2020)), and (c) ZnO (from Al-Hardan N H, Hamid M A A, Shamsudin R, Othman N K (2017)), are widely utilized in applications like gas sensors and biosensors because of their unique electronic properties.

There are two main types of MON sensing devices:

Resistive MON sensing devices: Resistive MON sensing devices measure the change in electrical resistance of the MON material when it is exposed to a target analyte. This type of sensing device is often used to detect gasses, such as carbon monoxide and nitrogen dioxide.

Capacitive MON sensing devices: Capacitive MON sensing devices measure the change in capacitance of the MON material when it is exposed to a target analyte. This type of sensing device is often used to detect biomolecules, such as DNA and proteins. MON sensing devices have a number of advantages over traditional sensing devices based on other materials, such as polymers and metal films. MON sensing devices are more sensitive, selective, and durable. They are also more cost effective to manufacture.

MON sensing devices are used in a wide range of applications such as environmental monitoring, food safety, and medical diagnostics. MON sensing devices are used to monitor air and water quality for a variety of pollutants, such as gases, heavy metals, and pesticides. They are used to detect leaks of toxic gases in industrial settings and to detect foodborne pathogens and toxins in food products. MON sensing devices are used to detect diseases by detecting biomarkers that are present in the blood, urine, or other bodily fluids. MON sensing devices are still under development, but there are a number of trends that are likely to shape the future of this technology.

### 5.5.1 Gas Sensors

Metal oxide nanomaterials can be used to develop sensors that are sensitive to a wide range of gases. For example, tin dioxide nanomaterials are often used in carbon monoxide sensors. Gas sensors are designed to detect and quantify the presence of specific gases in the atmosphere. Metal oxide nanomaterials have gained prominence in gas sensor development due to their high sensitivity, selectivity, and low-cost fabrication. They are widely used in applications like air quality monitoring, industrial safety, and automotive exhaust analysis. Conductometric semiconducting metal oxide gas sensors are currently one of the most actively researched types of gas sensors. Their ability to detect a wide range of gases under ambient conditions, ease of use, and low production costs have made them particularly attractive in the field of gas sensing [9]. Many studies have highlighted that conductometric semiconducting metal oxide gas sensors exhibit a distinct feature: the reversible interaction between the gas and the material's surface. This interaction, pivotal to the sensor's functionality, is subject to various influences stemming from both internal and external factors. Internal factors encompass inherent characteristics of the base materials used, the surface areas, and the microstructure of sensing layers. Meanwhile, external influences include factors such as surface additives, temperature fluctuations, and humidity levels, among others. Sensitivity stands out as a crucial parameter in gas sensors, prompting considerable attention and concerted efforts aimed at its enhancement. Researchers have dedicated substantial endeavors to augment the sensitivity of these gas sensors, recognizing its significance in optimizing their performance and reliability [9].

#### 5.5.1.1 Metal Oxides for Gas Sensors

A range of metal oxides demonstrate the capability to detect various types of gases – be it combustible, reducing, or oxidizing – through the measurement of their conductivity. The conductivity of specific metal oxides, including $Cr_2O_3$, $Mn_2O_3$, $Co_3O_4$, NiO, CuO, SrO, $In_2O_3$, $WO_3$, $TiO_2$, $V_2O_3$, $Fe_2O_3$, $GeO_2$, $Nb_2O_5$, $MoO_3$, $Ta_2O_5$, $La_2O_3$, $CeO_2$, and $Nd_2O_3$, is directly influenced by the presence of gases. These oxides exhibit changes in conductivity when exposed to different gas types, making them suitable candidates for gas-sensing applications. Leveraging the electronic structure of these metal oxides enables the strategic selection of materials ideal for specific gas sensors. By understanding and analyzing the electronic configurations and properties of these oxides, scientists and engineers can identify and utilize the most appropriate material for enhancing gas sensor performance and sensitivity in detecting the gas types.

The range of electronic structures of oxides is so wide that metal oxides are divided into two categories [9]. Transition-metal oxides such as $Fe_2O_3$, NiO, $Cr_2O_3$, alongside non-transition-metal oxides like pre-transition-metal oxides (e.g., $Al_2O_3$) and post-transition-metal oxides (e.g., ZnO, $SnO_2$), constitute a diverse array of materials used in gas-sensing applications. Notably, the amalgamation of graphene with metal oxide nanoparticles has proven effective in enhancing the selectivity and response time of graphene-based gas sensors, even when operating at room temperature. Specifically, our graphene transistors, adorned with $SnO_2$ NPs, demonstrate rapid response and swift recovery times, approximately 1 second each, at 50°C when exposed to a hydrogen concentration of 100 ppm. Employing X-ray photoelectron spectroscopy (XPS) and conductive atomic force microscopy (CAFM), we delve into investigating the interface properties between graphene and $SnO_2$ nanoparticles. This comprehensive characterization enables the elucidation of a mechanism grounded in charge transfer and band alignment. This mechanism elucidates the unique physical attributes of these graphene-based gas sensors, notably the high carrier mobility of graphene and the minimal energy barrier existing between graphene and $SnO_2$ NPs. This combination ensures the devices' remarkable attributes, including fast response, heightened sensitivity, and exceptional selectivity [10].

#### 5.5.1.2 Sensing Mechanism

Understanding the influential factors impacting the gas sensing capabilities of metal oxide sensors necessitates uncovering the underlying sensing mechanism. Fundamentally, this mechanism revolves around the trapping of electrons at molecules adsorbed onto the sensor's surface, along with the subsequent induction of band bending due to these charged molecules. This interplay leads to a significant alteration in conductivity. Specifically, the Schottky barrier existing between adjacent grains diminishes, facilitating the ease of electron conduction across various grains within the sensing layers. This modification allows for improved electron mobility and enhanced conductivity, contributing to the sensor's responsiveness and sensitivity to the gas molecules [9].

Nanomaterial-based sensors have gained prominence due to their high sensitivity and specificity in detecting a wide range of analytes, such as gases, chemicals, and biomolecules. The sensing mechanism of these sensors relies on the unique properties of nanomaterials at the nanoscale. When an analyte interacts with the nanomaterial's surface, it can induce changes in the material's electrical, optical, or mechanical properties. These changes are then transduced into a measurable signal, enabling the sensor to detect and quantify the presence of the analyte. For example, in a gas sensor using metal oxide nanoparticles, when the target gas molecules adsorb onto the nanoparticles' surface, they alter the electrical conductivity of the material. This change in conductivity is then converted into a signal that can be measured and correlated to the gas concentration. Similarly, in a biosensor employing nanomaterials like carbon nanotubes or quantum dots, the binding of biomolecules to the nanosurface can lead to fluorescence or electrical changes, allowing for the detection of specific biomarkers. The working of nanomaterial-based sensors typically involves a transduction mechanism that translates the interaction between the nanomaterial and analyte into a measurable response, such as a change in electrical current, voltage, or optical properties. This response is then quantified to determine the concentration or presence of the target analyte. The small size and high surface area of nanomaterials enhance their reactivity, resulting in improved sensitivity and rapid response times. As a result, nanomaterial-based sensors find applications in fields such as environmental monitoring, healthcare, and security, where precise and rapid detection of analytes is of paramount importance.

#### 5.5.1.3 Factors Affecting Sensitivity

Gas sensor applications rely on a spectrum of material properties encompassing adsorption capacity, catalytic activity, sensitivity, and thermodynamic stability, among others. Despite the favorable traits exhibited by numerous metal oxide materials for gas sensors, few meet all necessary requisites. For instance, composite ZnO–$SnO_2$ sensors have showcased notably heightened sensitivity compared to sensors solely comprising tin dioxide or zinc oxide under identical testing conditions. This superior performance arises from the synergistic catalytic reactions between both components, fostering an improved conductivity response. Generally, the conductivity response in gas sensors hinges on the efficacy of catalytic reactions occurring at the sensing material's surface. Hence, regulating the catalytic activity of gas sensor materials stands as a common strategy to augment sensor efficiency. The functionality of solid-state gas sensors hinges on both receptor and transducer functions, with the latter being particularly crucial in converting chemical interactions into electrical signals efficiently. Hence, synthesizing metal oxides with optimal morphology and crystallographic structure assumes paramount importance. Environmental humidity significantly impacts the performance of metal oxide gas sensors, leading to the development of numerous humidity-based gas sensors rooted in metal oxides. However, the sensing mechanisms for water vapor and other pollutants such as CO, $NO_2$, and $H_2S$ vary. Ionic-type sensors predominantly prevail in metal oxide humidity gas sensors, where the conduction mechanism relies on H+ or $H_3O$+ ions generated from the dissociation of adsorbed water. These ions then traverse between adjacent hydroxyl groups, shaping the sensor's response to environmental humidity levels and different gas pollutants. The gas sensing mechanism of $Sn_2O_3$-doped $SnO_2$ in the presence of (a) $C_2H_2$ and (b) $C_2H_2$ and humidity is illustrated in Figure 5.7 [9]. The gas-sensing process is intrinsically linked to surface reactions. Different metal oxide-based materials exhibit varying reaction activation energies for target gases. Composite metal oxides generally display enhanced gas responses compared to single-component materials if the catalytic actions of the components complement each other. Porous structures with high surface areas are considered the standard structure for metal oxide gas sensor layers. These structures are assembled from numerous small grains with voids and pores interspersed among them. It has also been demonstrated that small grain sizes contribute to the enhanced sensitivity [9].

Metal oxide nanoparticle gas sensors offer several advantages, making them valuable tools for detecting and monitoring various gases. These sensors are highly sensitive and can provide accurate measurements even at low gas concentrations, critical for early gas detection and safety applications. Their rapid response times enable real-time monitoring, ensuring quick detection of changes in gas concentrations. Additionally, metal oxide nanoparticle sensors can be engineered for selectivity, reducing the likelihood of false alarms and enhancing detection reliability. Their wide detection range, low power consumption, and compact size make them suitable for various applications, including portable devices and IoT systems. These sensors are also known for their long-term stability, environmental compatibility, and cost-effectiveness, which is crucial for large-scale and extended deployments. Overall, the versatility and reliability of metal oxide nanoparticle gas sensors contribute to their widespread adoption in fields such as industrial safety, air quality monitoring, and environmental protection. Metal oxide nanoparticle gas sensors hold great importance in a wide range of applications, owing to their ability to detect and monitor various gases with high sensitivity, reliability, and efficiency. These sensors have become indispensable tools in fields such as environmental monitoring, industrial safety, healthcare, and security. In this comprehensive exploration, we will delve into the significance of metal oxide nanoparticle gas sensors and their contributions to ensuring safety, environmental protection, and public health.

- **Environmental Monitoring:** Environmental monitoring is a critical application where metal oxide nanoparticle gas sensors play a pivotal role. They are extensively used to detect and quantify pollutants in the air and water. With their high sensitivity, these sensors can measure trace levels of gases, such as carbon monoxide (CO), nitrogen dioxide ($NO_2$), sulfur dioxide ($SO_2$), and volatile

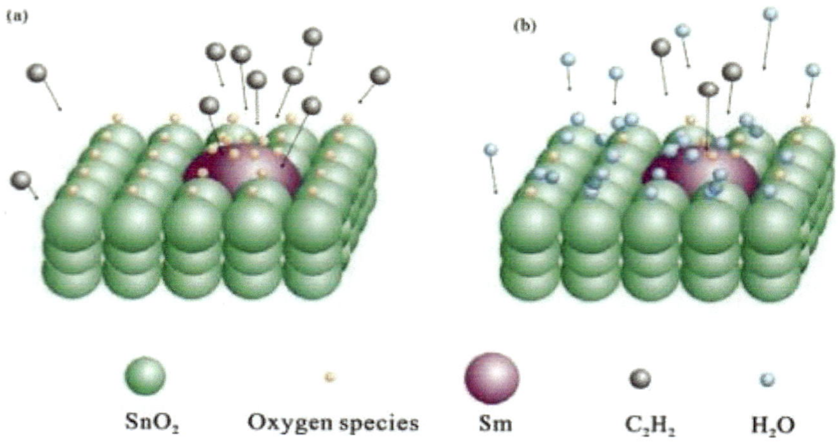


FIGURE 5.7 Schematic representation of the gas sensing mechanism of $Sn_2O_3$-doped $SnO_2$ in the presence of (a) $C_2H_2$ and (b) $C_2H_2$ with humidity. The diagrams illustrate how oxygen species interact with the sensor material, affecting the sensitivity and gas detection process (from Wang C, Yin L, Zhang L, Xiang D, Gao R (2010)).

organic compounds (VOCs), which are associated with adverse health effects and environmental degradation. In urban areas, where air quality is a significant concern, these sensors provide real-time data on pollutant levels, enabling authorities to implement timely actions to combat air pollution and protect public health. Furthermore, in water quality monitoring, metal oxide nanoparticle gas sensors help identify contaminants and ensure the safety of drinking water sources and aquatic ecosystems. By detecting and quantifying gases, these sensors contribute to preserving the environment and enhancing the quality of life.

- **Industrial Safety:** In industrial settings, the importance of metal oxide nanoparticle gas sensors cannot be overstated. Industries often deal with potentially hazardous gases, and the safety of workers is of paramount importance. These sensors are employed to detect gases like methane, hydrogen sulfide, and volatile organic compounds, ensuring that workers are protected from exposure to harmful substances. Industrial applications include chemical manufacturing, oil and gas production, and manufacturing processes involving solvents or volatile materials. By continuously monitoring gas concentrations and providing timely warnings or control signals, metal oxide nanoparticle gas sensors mitigate the risks associated with industrial activities, preventing accidents, and safeguarding human lives.
- **Healthcare:** In the healthcare sector, the significance of metal oxide nanoparticle gas sensors is particularly evident in the development of medical devices, such as breath analyzers and biosensors. Breath analyzers utilize these sensors to measure blood alcohol content (BAC) quickly and accurately, a critical function for law enforcement agencies in enforcing alcohol-related regulations. Additionally, in the management of health conditions like diabetes, metal oxide nanoparticle gas sensors, often integrated into portable glucometers, help individuals monitor blood glucose levels. By offering real-time data, these sensors empower patients and healthcare professionals to make informed decisions about treatment and lifestyle choices, ultimately contributing to improved health outcomes.
- **Environmental Protection:** The protection of the environment is of global concern, and metal oxide nanoparticle gas sensors are instrumental in achieving this goal. They are employed in the monitoring of greenhouse gases, enabling the assessment and mitigation of climate change. Sensors capable of detecting gases like carbon dioxide ($CO_2$) or methane ($CH_4$) contribute to the monitoring of emissions from various sources, such as power plants, transportation, and agriculture. By providing accurate data on greenhouse gas concentrations, these sensors support efforts to reduce emissions and combat climate change. Additionally, in the context of environmental conservation, sensors are used to detect gases emitted by industrial facilities and landfills, helping to identify and mitigate environmental pollution. Through their contribution to environmental protection, metal oxide nanoparticle gas sensors play a crucial role in preserving the planet for future generations.
- **Security and Defense:** The significance of metal oxide nanoparticle gas sensors extends to security and defense applications. They are essential tools for detecting chemical warfare agents and toxic industrial chemicals, providing protection to military personnel and civilian populations in the event of a chemical threat. Additionally, these sensors are used to identify explosive materials and illicit substances in security screening applications, including at airports and public transportation hubs. By ensuring the timely detection of potentially harmful substances, metal oxide nanoparticle gas sensors enhance security and protect the safety of individuals in high-risk environments.
- **Emerging Technologies:** Metal oxide nanoparticle gas sensors are also making significant contributions to emerging technologies. For example, they are integrated into wearable devices, providing individuals with real-time information about their immediate environment's air quality. This data empowers users to make informed decisions and take preventive measures to minimize exposure to air pollutants. Furthermore, these sensors find applications in the development of smart cities, where environmental parameters are continuously monitored, and data is used to optimize urban planning, transportation, and resource management. The importance of these sensors in emerging technologies lies in their ability to enhance the quality of life, promote sustainability, and create more resilient and efficient urban environments.
- **Scientific Research and Policy-Making:** In addition to their practical applications, metal oxide nanoparticle gas sensors contribute significantly to scientific research and policy-making. Data collected from these sensors are invaluable for studying air quality, tracking pollution trends, and investigating the effects of environmental pollutants on human health and ecosystems. Researchers and policymakers rely on this information to make informed decisions and implement strategies aimed at improving public health and environmental sustainability. The data generated by metal oxide nanoparticle gas sensors are critical for developing regulations and standards that protect air and water quality, as well as for assessing the effectiveness of environmental policies.

#### 5.5.1.4 Carbon Dioxide Sensors

Carbon dioxide sensors are used in a wide variety of fields, including air quality monitoring, hospital indoor environments, and the food packaging industry. The performance, linearity range, limit of detection, and price requirements for these sensors vary significantly depending on the field of application. $CO_2$ sensors are widely needed for many indoor offices, industrial applications, and environmental monitoring. Various semiconductor nanomaterials have been tested for $CO_2$ detection, but most of them have shown to be ineffective in improving sensor response through doping. This is why researchers have focused on developing new nanocomposite metal oxides ($MO_x$) to address this challenge. One example of a nanocomposite $MO_x$ that has been developed for $CO_2$ sensing is $BaTiO_3$–CuO. In summary, nanocomposite $MO_x$ sensors offer a promising new approach to developing $CO_2$ sensors with high sensitivity and selectivity. $BaTiO_3$–CuO is just one example of a nanocomposite MOx that has been shown to be effective for $CO_2$ sensing. Further research is needed to develop new nanocomposite MOx sensors with even better performance and to explore their potential applications in a wider range of fields [11].

#### 5.5.1.5 Ammonia Gas Sensors

Ammonia is a colorless gas recognized for its distinctively pungent odor. As the simplest pnictogen hydride, it is notably prevalent in nitrogenous waste, particularly among aquatic organisms. $NH_3$ assumes a significant role in sustaining surface-dwelling life forms, serving as a precursor to food production and the manufacturing of chemical fertilizers. Consequently, various metal oxide nanoparticles like ZnO, $SnO_2$, and $In_2O_3$ have undergone thorough investigation for their potential application as $NH_3$ gas sensors. Among these options, $SnO_2$ has emerged as the most promising material for $NH_3$ sensing applications.

An exemplar of a $SnO_2$-based $NH_3$ sensor involves a cluster matrix comprising thousands of extremely small-sized $SnO_2$ nanoparticles with interconnecting wormhole-like pores dispersed randomly throughout the oxide nanoparticles. The introduction of Pt activation on the surface of these clusters results in a significant enhancement in gas response to 500 ppm of $NH_3$, elevating it from 6.5 to an impressive 203.5. This augmentation in sensitivity showcases the potential of this modified $SnO_2$ structure as a highly effective $NH_3$ gas sensor, demonstrating considerable promise for practical application in detecting and quantifying ammonia gas concentrations. This significant improvement in gas response is attributed to the following factors:

- The small size of the $SnO_2$ NPs provides a large surface area for adsorption of $NH_3$ molecules.
- The wormhole-like pores facilitate the rapid diffusion of $NH_3$ molecules to the surface of the $SnO_2$ NPs.
- The Pt catalyst enhances the adsorption and dissociation of $NH_3$ molecules.

Hydrogen sulfide ($H_2S$) is a colorless, toxic, flammable gas known for its explosive nature, naturally found in petroleum rigs, as well as occurring in natural geological processes such as volcanic eruptions and hot-bath springs. Accurate measurement of hydrogen sulfide at low micromolar or nanomolar concentrations in real time has been a long-standing challenge. $H_2S$ sensors are metal oxide semiconductor devices that change their electrical resistance in response to the adsorption and desorption of $H_2S$ molecules on a thin film surface. The thin film can be made of a variety of materials, such as gold or tin oxide, each of which has a different sensitivity to $H_2S$. One of the key challenges in developing $H_2S$ sensors with high sensitivity and fast response time is the low concentration of $H_2S$ that is typically present in the environment. At these low concentrations, the adsorption and desorption of $H_2S$ molecules on the thin film surface is a slow process, which can lead to long response times. Recent advances in nanotechnology have led to the development of new $H_2S$ sensors with significantly improved sensitivity and response time. For example, one type of sensor uses a thin film of tin oxide nanoparticles decorated with platinum nanoparticles. The platinum nanoparticles act as catalysts, which accelerate the adsorption and desorption of $H_2S$ molecules on the tin oxide surface. This results in a sensor with a response time of less than 1 minute for $H_2S$ concentrations ranging from 25 ppb to 10 ppm [11].

### 5.5.2 Biosensors

The advent of nanomaterials with unique properties and tailored functionalities has paved the way for the development of miniaturized biosensors with exceptional sensitivity and selectivity for biomolecular detection. The interplay of distinct molecular recognition interactions at the nanoscale, coupled with the integration of nanoscale components, further enhances the sensitivity of these biosensors. Among the diverse array of nanomaterials, nanostructured metal oxides (NMOs) have emerged as promising immobilization matrices for biosensor development. Their exceptional optical and electrical properties, stemming from electron and phonon confinement, high surface-to-volume ratios, tunable surface work functions, high surface reaction activity, high catalytic efficiency, and strong adsorption capabilities make NMOs ideal candidates for this application. Enzyme-immobilized NMO-based biosensors, in particular, exemplify the potential of NMOs in biosensing. These biosensors rely on the biochemical reaction between an analyte and a biocatalyst immobilized on an NMO substrate [5].

Nanostructured metal oxides hold immense promise for revolutionizing biosensors by facilitating direct electron transfer between enzymes and NMO electrodes, leading to enhanced biosensing capabilities. These new-generation biosensors achieve direct coupling between active enzyme sites and the NMO electrode, improving biosensing properties. Nanostructured metal oxide-based electrodes offer a biocompatible electroactive surface for enzyme immobilization, optimizing enzyme conformation, orientation, and biological activity. Enhancing electrical communication between the active sites of biomolecules and the NMO electrode surface remains a crucial challenge in developing amperometric bioelectrodes [5]. Metal oxide nanomaterials are also important for biosensors, which are devices that use biological molecules to detect and measure specific substances. Biosensors are used in a wide range of applications, including medical diagnostics, environmental monitoring, and food safety. Biosensors are devices that incorporate biological elements (e.g., enzymes, antibodies, or DNA) to detect specific biological molecules. Metal oxide nanomaterials have found applications in biosensors, enhancing their performance by improving immobilization and electron transfer processes. These sensors are essential for medical diagnostics, environmental monitoring, and food safety.

Metal oxide nanomaterials are a promising class of materials for biosensors due to their unique properties, including high surface area, good biocompatibility, and good electrical conductivity. MON biosensors work by detecting the binding of a target biomolecule to the surface of a MON nanoparticle. This binding can cause a change in the electrical properties of the MON nanoparticle, such as its conductivity, capacitance, or impedance. This change in electrical properties can then be measured using an external circuit. MON biosensors can be classified into two main types: electrochemical and field-effect transistor (FET) biosensors. Electrochemical biosensors measure the change in electrical current or potential that occurs when a target biomolecule binds to the surface of a MON nanoparticle. This type of biosensor is often used to detect glucose, lactate, and other biomolecules that can be oxidized or reduced. FET biosensors measure the change in electrical conductance of a MON nanoparticle when a target biomolecule binds to its surface. This type of biosensor is often used to detect DNA, proteins, and other biomolecules that can bind to the surface of the MON nanoparticle. MON biosensors have a number of advantages over traditional biosensors based on other materials, such as enzymes and antibodies. Metal oxide nanomaterial biosensors are more stable and less susceptible to environmental factors, such as pH and temperature, than traditional biosensors based on other materials, such as enzymes and antibodies. This makes them ideal for use in a variety of applications where environmental conditions are not well controlled.

Additionally, MON biosensors are more scalable and can be manufactured at a lower cost, making them more accessible to a wider range of users. MON biosensors are still under development, but they have the potential to revolutionize the way we diagnose diseases, monitor food safety, and protect the environment. For example, MON biosensors are being developed to detect cancer biomarkers in the blood, glucose levels in diabetics, and infectious diseases in urine or other bodily fluids. MON biosensors can also be used to detect foodborne pathogens and toxins in food products. Here are some specific examples of how MON biosensors are being developed for use in medical diagnostics, food safety, and environmental monitoring:

- **Medical diagnostics**
  MON biosensors are being developed to detect cancer biomarkers in the blood, such as circulating tumor cells and exosomes. These biomarkers can be used to diagnose cancer early, when it is most treatable. MON biosensors are also being developed to monitor glucose levels in diabetics. These biosensors could be implanted in the body and provide real-time glucose readings to the patient. MON biosensors are also being developed to detect infectious diseases, such as HIV, malaria, and tuberculosis. These biosensors could be used to diagnose diseases quickly and accurately, even in remote or resource-limited settings.
- **Food safety**
  MON biosensors are being developed to detect foodborne pathogens, such as *E. coli* and Salmonella, in food products. These biosensors could be used to prevent foodborne illnesses, which are a major public health problem.

  MON biosensors are also being developed to detect toxins, such as mycotoxins and pesticides, in food products. These biosensors could help to ensure that food is safe to eat.
- **Environmental monitoring**
  MON biosensors are being developed to detect pollutants, such as heavy metals and pesticides, in water and air. These biosensors could be used to monitor environmental quality and identify potential health hazards.

  MON biosensors are also being developed to detect pathogens in environmental samples, such as water and soil. These biosensors could be used to identify and prevent the spread of infectious diseases.

For example, MON biosensors are being developed to detect *E. coli* and Salmonella in food products. Environmental monitoring, MON biosensors can be used to detect pollutants and other harmful substances in the environment. For example, MON biosensors are being developed to detect heavy metals, pesticides, and herbicides in water and air. Overall, MON biosensors are a promising new technology with the potential to revolutionize the way we diagnose diseases, monitor food safety, and protect the environment. Metal oxide nanoparticles intended for biomedical applications must fulfill specific criteria. For instance, MONPs used as drug carriers should exhibit release kinetics aligned with the treatment requirements for a particular infection. Additionally, they should be biodegradable to avoid the need for further surgical procedures. Despite the vast array of available MONPs, only titanium dioxide, zinc oxide, copper oxide, ferric oxide, and ferrous oxide have demonstrated relative safety in mammals.

- **Internal Tissue Therapy**
  The effectiveness of therapeutic interventions is often limited by the ability to modulate diverse molecular signaling pathways that regulate cell growth, division, differentiation, migration, and apoptosis. Nanomedicine offers a promising approach to overcoming these challenges by enabling the development of drug delivery systems with enhanced delivery efficiency and targeted delivery to specific organs or tissues. Nanocarriers can enter the body through various routes, including inhalation, ingestion, transdermal absorption, and injection. Once in the body, nanocarriers can translocate through the cell membrane and deliver their payload directly to the target cells. This ability to bypass cellular barriers is a major advantage of nanocarriers over traditional drug delivery systems. The development of nanocarriers as drug delivery systems offers a number of benefits, including:

- Enhanced pharmacological activity: Nanocarriers can protect drugs from degradation and enhance their absorption into the body, leading to increased efficacy.
- Sustained and targeted delivery: Nanocarriers can be designed to release their payload slowly over time or to target specific organs or tissues, reducing the need for frequent dosing and minimizing side effects.
- Co-delivery of multiple therapeutic agents: Nanocarriers can be used to deliver multiple therapeutic agents simultaneously, allowing for combination therapy strategies that are more effective than monotherapy.
- Improved stability and bioavailability: Nanocarriers can improve the stability and bioavailability of drugs, making them more effective and easier to administer.
- Reduced side effects: By delivering drugs directly to the target cells, nanocarriers can help to reduce the systemic exposure to drugs and minimize side effects.

Engineering materials can be used to improve the specificity of nanocarriers and further reduce side effects. For example, nanocarriers can be coated with ligands that bind to specific receptors on target cells, enabling targeted delivery. Additionally, nanocarriers can be designed to release their payload in response to specific stimuli, such as pH, temperature, or light. The toxicity of nanoparticles can vary depending on their size, composition, and surface properties. Some nanoparticles are non-toxic, while others can be toxic to certain types of cells. It is important to carefully evaluate the toxicity of nanoparticles before using them in drug delivery applications. Despite the challenges, nanomedicine has the potential to revolutionize the way we treat diseases. By enabling targeted delivery of drugs with enhanced efficacy and reduced side effects, nanocarriers can help to improve patient outcomes [12].

### 5.5.3 Chemical Sensors

Metal oxide nanomaterials can be used to develop sensors that are sensitive to a wide range of chemicals. For example, zinc oxide nanomaterials are often used in humidity sensors. Metal oxide nanomaterials are a class of materials with unique properties that make them ideal for chemical sensors. MONs have a high surface area, which allows them to interact with a large number of gas molecules. They are also good semiconductors, which means that their electrical conductivity changes when they are exposed to different gases. The MON material is typically deposited on a substrate in the form of a thin film or nanostructures. When the sensor is exposed to air, oxygen molecules adsorb on the surface of the MON material and form a depletion layer. The depletion layer is a region near the surface of the MON material where there are fewer free electrons, which reduces the electrical conductivity of the material. When the sensor is exposed to a target gas, the gas molecules interact with the adsorbed oxygen molecules and change the composition of the depletion layer. This can either increase or decrease the electrical conductivity of the MON material, depending on the type of gas. For example, reducing gases, such as hydrogen and carbon monoxide, donate electrons to the depletion layer, which increases the electrical conductivity of the MON material. Oxidizing gases, such as nitrogen dioxide and ozone, remove electrons from the depletion layer, which decreases the electrical conductivity of the MON material. The change in electrical conductivity of the MON material is measured by an external circuit. The magnitude of the change in electrical conductivity is proportional to the concentration of the target gas. This allows the sensor to be used to measure the concentration of a target gas in the air.

- **Carbon monoxide sensors**
  MON carbon monoxide sensors are widely used in homes and businesses to detect leaks of carbon monoxide, which is a toxic and odorless gas. MON nitrogen dioxide sensors are used to monitor air quality for nitrogen dioxide, which is a major air pollutant that can cause respiratory problems. Ozone sensors are used to monitor air quality for ozone, which is a major air pollutant that can cause respiratory problems and damage plants. MON VOC sensors are used to detect a wide range of VOCs, which are gases that are released by a variety of sources, including bacteria, plants, and industrial processes. MON VOC sensors are used in a variety of applications, including medical diagnostics, food safety, and environmental monitoring [13].

Metal oxide nanoparticle chemical sensors are an emerging technology with the potential to transform air quality monitoring. Conductometric chemical sensors have gained popularity among sensor researchers due to their low fabrication cost, simple preparation and operation, wide range of potential detection and application possibilities, and potential for device miniaturization. However, enhancing their sensitivity, selectivity, stability, and speed (response and recovery rates) remains a challenge. The interaction between gas molecules and the active oxide occurs on its surface, making the surface atoms critical for chemical sensing performance. Nanowires, in particular, offer a high surface-to-volume ratio, amplifying the effects of surface phenomena (which underpin the chemical sensing mechanism). Additionally, they may be single-crystalline and have well-defined crystal orientations, leading to controlled reactions and improved stability. Response dynamics should be faster compared to their polycrystalline counterparts due to the absence of a need for gas diffusion prior to surface reaction.

Finally, some interesting effects that can be exploited in chemical sensing, such as self-heating, exist only for the nanowire morphology. These effects can be used to detect leaks of toxic gases and to diagnose diseases [13]. Metal oxide nanoparticle chemical sensors find diverse applications across various fields due to their ability to interact with specific chemicals and provide measurable responses. For instance, carbon monoxide gas sensors employ metal oxide nanoparticles like tin dioxide or zinc oxide to detect CO in the environment by measuring changes in electrical conductivity upon CO adsorption. Ethanol breath analyzers utilize metal oxide nanoparticles to determine blood alcohol content by monitoring changes in electrical conductivity when ethanol molecules interact with the nanoparticle surface.

Environmental VOC sensors, on the other hand, employ metal oxide nanoparticles like tungsten oxide to detect volatile organic compounds in the surroundings by measuring changes in electrical resistance upon VOC adsorption. Glucose biosensors utilize metal oxide nanoparticles combined with enzymes to detect glucose levels through alterations in electrical or optical properties. Explosive vapor detectors employ functionalized metal oxide nanoparticles to interact with explosive compounds, causing changes in electrical conductivity upon exposure to explosive vapors. In the realm of pH sensors, metal oxide nanoparticles, such as indium oxide or ruthenium dioxide, determine the acidity or alkalinity of solutions by sensing changes in surface charge, which is detected through alterations in electrical potential. These examples illustrate the versatility and effectiveness of metal oxide nanoparticle chemical sensors in various applications by utilizing their unique interactions with specific target analytes.

### 5.5.4 Environmental Sensors

Environmental sensors are designed to monitor and detect various environmental parameters, including air and water quality. Metal oxide nanomaterials have proven effective in sensing and identifying pollutants, such as gases and heavy metals, in the environment. They are essential for tracking pollution levels and ensuring environmental safety. Metal oxide nanomaterials are a promising class of materials for environmental sensors due to their unique properties, including high surface area, good biocompatibility, and good electrical conductivity. MON environmental sensors work by detecting the presence or concentration of specific pollutants or other contaminants in the environment. MON environmental sensors can be used to detect a wide range of pollutants, including air pollutants.

Metal oxide gas sensors are versatile tools for monitoring air quality, effectively detecting various harmful pollutants such as carbon monoxide, nitrogen dioxide, ozone, and sulfur dioxide. MON sensors can be used to detect water and soil pollutants such as heavy metals, pesticides, and herbicides. MON environmental sensors are often used in fixed-site monitoring stations, but they can also be used in mobile and portable devices. This makes them well suited for a variety of applications, including: MON sensors can be used to monitor ambient air quality in urban and rural areas. This information can be used to inform public health advisories and air pollution control strategies. These sensors can be used to monitor emissions from industrial facilities to ensure compliance with environmental regulations. They can also be used to monitor soil contamination in agricultural areas, industrial sites, and other locations of concern, providing information that can guide soil remediation and protect human health and the environment.

#### Examples of MON environmental sensors

- **Metal oxide gas sensors:** Metal oxide gas sensors are versatile tools for monitoring air quality, effectively detecting various harmful pollutants such as carbon monoxide, nitrogen dioxide, ozone, and sulfur dioxide. These sensors are often used in fixed-site air quality monitoring stations and in mobile and portable air quality monitoring devices.
- **Metal oxide water quality sensors:** Metal oxide water quality sensors are used to detect a wide range of water pollutants, including heavy metals, pesticides, and herbicides. These sensors are often used in water quality monitoring stations and in portable water quality monitoring devices.
- **Metal oxide soil contamination sensors:** Metal oxide soil contamination sensors are used to detect heavy metals, pesticides, and other contaminants in soil. These sensors are often used in environmental site assessments and in the remediation of contaminated soil. MON environmental sensors are a promising new technology with the potential to revolutionize the way we monitor and protect the environment.

### 5.5.5 Humidity Sensors

The humidity sensor developed using 3D networks of AuNPs interconnected by oligoethylene glycol di-thiols (OEG) linkers possesses remarkable sensitivity and responsiveness, surpassing previous devices due to its unique design and the effective utilization of OEG linkers [14]. Metal oxide nanoparticle humidity sensors have gained significant attention in recent years due to their numerous advantages over traditional humidity sensors. These advantages include high sensitivity, fast response and recovery times, wide operating range, low cost, ease of fabrication, sustainability, tunable properties, small size and portability, compatibility with microfabrication techniques, long-term stability, and potential for self-cleaning properties [14].

## 5.6 CONCLUSION

Metal oxide nanomaterials are a promising platform for the development of new and improved optoelectronics and sensing devices. They have a number of unique properties and characteristics that make them ideal for these applications, including high surface area, tunable optical properties, chemical sensitivity, mechanical strength, and biocompatibility. There are a number of different methods for synthesizing metal oxide nanoparticles, and the specific method that is used depends on the desired properties of the nanoparticles.

Metal oxide nanoparticles have emerged as versatile materials in the realm of optoelectronic and sensing devices, showcasing their immense potential in various applications such as solar cells, LEDs, and sensors. The unique properties of metal oxide nanoparticles, including their tunable optical and electrical characteristics, make them particularly attractive for harnessing renewable energy and advancing sensor technologies. Their ability to facilitate charge transport and promote light absorption has significantly contributed to the development of more efficient and cost-effective solar cells. This breakthrough not only addresses the global energy demand but also underscores the importance of sustainable technologies in mitigating environmental challenges. Moreover, metal oxide nanoparticles have revolutionized the field of LEDs, offering a pathway towards more energy-efficient and high-performance lighting solutions. By exploiting the unique optical properties of these nanoparticles, researchers have been able to achieve improved color rendering and luminous efficacy in LED devices. This has implications not only for energy conservation but also for creating enhanced visual experiences in various applications.

In the realm of sensing devices, metal oxide nanoparticles have proven highly effective in detecting and responding to various environmental stimuli. Their sensitivity to changes in gas concentrations, temperature, and other factors makes them invaluable for the development of advanced sensor technologies. Whether applied in environmental monitoring, healthcare diagnostics, or industrial safety, these nanoparticles offer a versatile platform for creating sensitive and selective sensing devices. The integration of metal oxide nanoparticles into optoelectronic and sensing devices represents a significant step towards a more sustainable and technologically advanced future. The ongoing research and innovations in this field continue to unlock new possibilities, paving the way for further advancements in renewable energy, lighting, and sensing technologies. As we delve deeper into the potential of metal oxide nanoparticles, we anticipate continued breakthroughs that will shape the landscape of optoelectronics and sensing for years to come.

## Keywords

Biosensors
Chemical sensors
Humidity sensors
LEDs
Metal oxide nanoparticles
Optoelectronic devices
Sensing devices
Solar cells

## REFERENCES

1. Alhalili, Z. (2023, March 30). Metal oxides nanoparticles: general structural description, chemical, physical, and biological synthesis methods, role in pesticides and heavy metal removal through wastewater treatment. *MDPI.* https://www.mdpi.com/1420–3049/28/7/3086
2. Gebre, S. H., & Sendeku, M. G. (2019, July 27). New frontiers in the biosynthesis of metal oxide nanoparticles and their environmental applications: an overview. *SN Applied Sciences.* https://link.springer.com/article/10.1007/s42452-019-0931-4
3. Jose, R., Thavasi, V., & Ramakrishna, S. (2009). Metal oxides for dye-sensitized solar cells. *Journal of the American Ceramic Society*, 92(2), 289–301. https://doi.org/10.1111/j.1551–2916.2008.02870.x
4. Jose, R., Thavasi, V., & Ramakrishna, S. (2009). Metal oxides for dye-sensitized solar cells. *Journal of the American Ceramic Society*, 92(2), 289–301. https://doi.org/10.1111/j.1551–2916.2008.02870.x
5. Solanki, P., Kaushik, A., Agrawal, V., et al. (2011). Nanostructured metal oxide-based biosensors. *NPG Asia Materials* 3, 17–24. https://doi.org/10.1038/asiamat.2010.137
6. Bader, B. B., Hadi, I. H., Slewa, M. Y., Khashan, K. S., Abdulameer, F. A. Photodetector based on titanium oxide nanoparticles produced via pulsed laser ablation. *Advances in Condensed Matter Physics*, 2022, 8066167, 5. https://doi.org/10.1155/2022/8066167
7. Sosna-Głębska, A., Szczecińska, N., Znajdek, K., & Sibiński, M. (2019). Review on metallic oxide nanoparticles and their application in optoelectronic devices. *Acta Innovations*, (30), 5–15.
8. Yoon, Y. (2022). PhuocLoc Truong, Daeho Lee, and Seung Hwan Ko. *ACS Nanoscience Au*, 2(2), 64–92.

9. Wang, C., Yin, L., Zhang, L., Xiang, D., & Gao, R. (2010). Metal oxide gas sensors: sensitivity and influencing factors. *Sensors (Basel)*, 10(3), 2088–2106. https://doi.org/10.3390/s100302088. Epub 2010 Mar 15. PMID: 22294916; PMCID: PMC3264469.
10. Zhang, Z., Zou, X., Xu, L., Liao, L., Liu, W., Ho, J., Xiao, X., Jiang, C., & Li, J. (2015, May 5). Hydrogen gas sensor based on metal oxide nanoparticles decorated graphene transistor. *Nanoscale*. https://pubs.rsc.org/en/content/articlelanding/2015/nr/c5nr01924a
11. Chavali, M. S., & Nikolova, M. P. (2019). Metal oxide nanoparticles and their applications in nanotechnology. *SN Applied Sciences*, 1, 607. https://doi.org/10.1007/s42452-019-0592-3
12. Materials Today. (n.d.). Retrieved from https://www.sciencedirect.com/journal/materials-today/vol/19/issue/2
13. Materials Today. (n.d.). Retrieved from https://www.sciencedirect.com/journal/materials-today/vol/19/issue/2
14. Montes-García, V., Squillaci, M. A., Diez-Castellnou, M., Ong, Q. K., Stellacci, F., & Samorì, P. (n.d.). Retrieved from https://pubmed.ncbi.nlm.nih.gov/33290474/

# 6 Metal Oxide Nanomaterial-Based Biosensors for Cancer Diagnosis

*Aparna C. M. and Shibil S.*
School Pure and Applied Physics, Mahatma Gandhi University, Kottayam, India

## 6.1 INTRODUCTION

In 2015, approximately 9 million people died from various types of cancer, according to reports from the World Health Organization (WHO) [1]. The delayed detection of cancer is primarily due to conventional diagnostic methods, which depend on recognizing morphological changes in cancer cells that typically manifest only in advanced stages. This means that widely used methods like mammography, endoscopy, and MRI are not effective in catching cancer early on [2]. Additionally, cancer brings about various intricate changes in cells, like DNA or RNA mutations, shifts in protein structure, and alterations in expression levels. These complex transformations make it difficult to identify cancer in its early phases using conventional techniques [2]. Consequently, there is a growing need for more advanced (easy, fast, and cost-effective) diagnostic approaches to overcome these challenges and detect cancer at earlier, more treatable stages. Today, the implementation of biosensors, a type of chemical sensor, for cancer diagnosis has accelerated diagnostic activities [3]. In recent times, various biosensors have emerged, aiming to simplify, accelerate, and make disease detection more affordable. The International Union of Pure and Applied Chemistry (IUPAC) defined a biosensor in 1997 as a 'a device that transforms chemical information, ranging from the concentration of a specific sample component to total composition analysis into an analytically useful signal'. Updike and Hicks made the initial report on biosensors in 1967. A biosensor integrates three scientific disciplines – chemistry, biology, and engineering – within a single lab-on-chip (LOC) device [2].

Although biosensors offer a more advanced approach for early cancer diagnosis compared to conventional methods, they come with certain limitations. Unlike economical handheld devices such as blood-sugar testing kits, biosensors face challenges in terms of cost-effectiveness due to the inability to miniaturize the transduction principle. A Russian research group has highlighted issues with the durability of biorecognition elements on the transducer surface, noting their short lifespan and the difficulty in quick and easy assembly [4]. Additionally, conventional biosensors may lack accuracy when detecting biomarkers at minute concentrations in a patient's sample. The drawbacks of conventional biosensors are overcome by incorporation of nanomaterials, particularly nanoparticles, in their fabrication. These biosensors are called nanobiosensors. They operate on the same principles of conventional biosensors. Incorporating nanomaterials improves the functionality of biomolecules, enabling precise targeting [2]. This adaptability allows nanomaterials to be used in different ways within biosensors, advancing their applications in nanobiosensors.

Nanoparticles (NPs) have a small size, giving them a high surface-to-volume ratio with unique physical and chemical properties like strong surface plasmon resonance (SPR), high magnetization, and chemical reactivity. This small size also means a larger surface area, allowing more efficient immobilization of bioreceptors. This, in turn, maximizes the binding of biomarkers from the patient's sample and enhances the chemical signal of binding. Consequently, the biosensor's sensitivity and the quality of in vivo imaging can improve [4]. Various nanomaterials find applications in biosensor fabrication, including metal oxide-based semiconducting nanoparticles like ZnO and $TiO_2$, particularly promising for electrochemical biosensors. ZnO, characterized as a direct and wide band gap semiconductor material, stands out in these applications [2].

Numerous nanomaterials and nanocomposites are being explored for biosensing applications. The progress in developing nanomaterials as transducers is significantly advancing biosensing applications, especially in the context of cancer detection [5]. Despite these developments, it is essential to note that most biosensor development remains in the laboratory stage, indicating that there is still a considerable journey ahead before commercialization becomes a reality [5].

## 6.2 METAL OXIDE NANOMATERIALS

Metal oxide nanomaterials (MONs), situated at the forefront of nanotechnology, have emerged as resourceful and promising materials with a wide range of applications. These materials are compounds composed of metal cations and oxygen ions, which exhibit remarkable properties at the nanoscale. Examples of these materials include zinc oxide (ZnO) nanoparticles, titanium dioxide ($TiO_2$) nanotubes, and iron oxide ($Fe_3O_4$) nanoparticles [6]. These nanomaterials are employed in various fields such as electronics, photovoltaics, catalysis, and medicine. Their remarkable


DOI: 10.1201/9781032660639-6

properties and tunability at the nanoscale offer opportunities for innovation and have the potential to revolutionize industries ranging from renewable energy to healthcare. As research in nanomaterials advances, their role in addressing global challenges becomes increasingly significant.

## 6.3 CHARACTERISTICS

Metal oxide nanomaterials exhibit a unique set of characteristics that make them highly valuable in a wide range of applications.

### 6.3.1 High Surface Area

Metal oxide nanomaterials possess a special advantage due to their high surface area. The high surface area provides an extensive contact interface, making them highly effective in various applications [6]. In catalysis, for instance, a greater surface area allows for more active sites, enhancing catalytic reactions. This property is also beneficial in the field of sensors, as it increases the likelihood of interactions with target molecules, improving sensitivity and selectivity [6].

### 6.3.2 Tunability

The properties of metal oxide nanomaterials can be finely tuned by adjusting their size, shape, and composition, allowing customization for specific applications [7]. These materials can be precisely engineered at nanoscale to exhibit specific properties, making them adaptable in a range of applications. This tunability is particularly valuable in catalyst design, where tailored reactivity can improve efficiency [7]. Moreover, achieving stability and reproducibility in nanoscale tuning can be challenging [7].

### 6.3.3 Semiconducting Behavior

Many metal oxides, such as titanium dioxide and zinc oxide, exhibit semiconducting behavior, making them ideal for electronic and optoelectronic devices. One significant advantage is their adaptability for use in electronic devices, as they can function as semiconductors. This feature enables their incorporation into sensors, transistors, and photovoltaic cells, contributing to technological advancements [6]. Sensitivity to environmental conditions, such as humidity or oxygen levels, can affect the stability and reproducibility of their electronic properties. But with the careful engineering and surface modification, semiconducting metal oxide nanomaterials hold potential for a wide range of applications, from gas sensors to energy conversion devices, where their tunable semiconductor behavior is an asset [6].

### 6.3.4 Catalytic Activity

These materials have remarkable catalytic capability, primarily due to their high surface area and the availability of active sites, making them essential in applications ranging from environmental remediation to industrial chemical processes. Their catalytic efficiency and specificity can be designed through size, shape, and composition adjustments, which is a distinct advantage [6]. Achieving long-term stability and preventing sintering or agglomeration of these nanomaterials can be challenging, potentially affecting their catalytic performance over time. Additionally, the toxicity of certain metal oxides, like those containing heavy metals, can raise environmental and safety concerns. Therefore, while metal oxide nanomaterials exhibit impressive catalytic potential, their practical use necessitates careful design and consideration of these pros and cons [6].

### 6.3.5 Biocompatibility

Some metal oxide nanomaterials are bio-compatible, enabling their use in biomedical applications, including drug delivery and imaging. They can be used as drug delivery carriers, imaging agents, or in tissue engineering due to their controlled release properties and lack of cytotoxicity in certain formulations [8]. Some metal oxides can raise toxicity concerns, and their potential long-term effects on living organisms must be thoroughly assessed [8].

These unique characteristics make the way for innovative solutions in numerous scientific, industrial, and medical fields.

## 6.4 BIOSENSORS

Biosensors are sophisticated analytical devices that have gained immense importance in science and technology due to their remarkable ability to detect specific biological molecules or markers [4]. Biosensors are devices that integrate biological recognition elements with transducers to convert biological signals into detectable electronic signals. Enzymes, antibodies, DNA, and other biomolecules serve as bioreceptors, interacting with specific bioanalytes to produce measurable signals [9].

### 6.4.1 The Working Principles of Biosensors

The working principles of biosensors (Figure 6.1) typically involve the following steps.

#### 6.4.1.1 Recognition Element

The recognition element in biosensors is a crucial component that provides specificity by interacting selectively with the target analyte (the substance being detected). It is responsible for recognizing and binding to the analyte, initiating the signal transduction process. The choice of recognition element depends on the type of biosensor and the specific application [4]. Enzymes are often used as recognition elements in biosensors. They are highly specific and can catalyze reactions with their target analytes, resulting in the production of a measurable signal. For example,

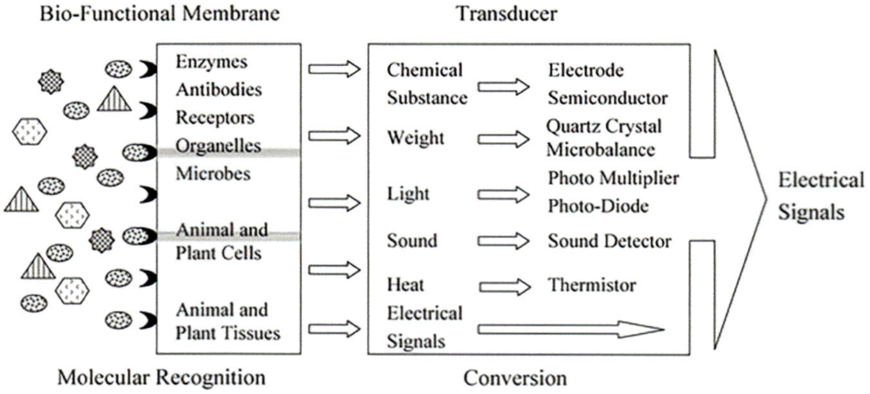


**FIGURE 6.1** Principles of a typical biosensor.

**Source: Reprinted with permission from Ref. [10]**

glucose biosensors use glucose oxidase as the recognition element to detect glucose levels [11]. Monoclonal or polyclonal antibodies can be used to recognize and bind to specific proteins or antigens. Immunosensors, for instance, use antibodies to detect pathogens, biomarkers, or other specific molecules in clinical diagnostics [11]. DNA or RNA strands can be employed as recognition elements in biosensors to bind to complementary sequences. DNA-based biosensors are used for genetic testing, identifying specific DNA or RNA sequences, and detecting pathogens [11]. Aptamers are short, single-stranded DNA or RNA molecules that can be engineered to bind to specific target molecules with high affinity. They are increasingly used as recognition elements in biosensors due to their specificity and ease of modification [12].

#### 6.4.1.2 Transducer

The transducer in biosensors is a fundamental component responsible for converting the biological or chemical recognition event, which occurs when the recognition element interacts with the target analyte, into a measurable signal. The type of transducer used in a biosensor depends on the application and the desired output [10]. Optical transducers depend on changes in light properties to detect and quantify the interaction between the recognition element and the analyte. Common types of optical transducers include fluorescence-based transducers, absorption-based transducers, and surface plasmon resonance transducers. Fluorescence-based transducers detect changes in fluorescence intensity or wavelength when the recognition element interacts with the analyte. For example, a fluorescent dye might be used to label the recognition element [13]. Absorption-based transducers measure changes in the absorption of light, often in the UV-vis or IR range, when the analyte binds to the recognition element. SPR biosensors measure changes in the angle or wavelength of light as it interacts with a thin metal film upon binding of the analyte to the recognition element. SPR is commonly used for studying biomolecular interactions [13].

An electrochemical transducer depends on changes in electrical properties, such as current or voltage, due to the electrochemical reactions that occur when the recognition element interacts with the analyte [14]. A typical example of electrochemical transducers is amperometric transducers, which measure the current produced by electrochemical reactions at an electrode surface. Potentiometric transducers measure the potential (voltage) difference generated during the recognition event. The potential varies with the concentration of ions or redox species in the electrochemical reaction [14]. In conductometric transducers, changes in electrical conductivity are monitored [14]. Piezoelectric biosensors measure changes in mass or surface stress because of the binding of the analyte to the recognition element. These changes affect the sensor's resonance frequency, which can be precisely detected [14].

The choice of transducer depends on the specific application and the desired characteristics of the biosensor, such as sensitivity, selectivity, and the type of output signal [4]. Different transducers offer different advantages and limitations, making them suitable for various applications in medical diagnostics, environmental monitoring, and beyond.

#### 6.4.1.3 Signal Output

Signal output in biosensors refers to the measurable response or signal generated as a result of the interaction between the recognition element and the target analyte [10]. The type of signal output can vary depending on the biosensor's design and the transducer used. Various types of biosensors utilize different signal outputs, such as electrical, optical, or electrochemical signals, depending on the specific design and intended applications.

Different signal outputs offer various advantages in terms of sensitivity, selectivity, speed, and ease of detection. Biosensors are valuable tools in fields such as medical diagnostics, environmental monitoring, and food safety, primarily because they provide precise and timely information about the presence or concentration of specific analytes [13].

#### 6.4.1.4 Signal Processing

Signal processing in biosensors is a critical step that involves the manipulation and analysis of the raw data generated by the transducer. This essential stage enhances the accuracy, sensitivity, and specificity of biosensor measurements [15]. It encompasses tasks such as amplification to strengthen weak signals, filtering to remove noise and unwanted interference, data preprocessing for standardization and calibration, feature extraction to characterize the recognition event, and quantification to convert processed data into analyte concentration measurements [15]. Statistical analysis and pattern recognition techniques may be applied for reliability assessment and discrimination of different analytes [4]. Real-time monitoring ensures timely feedback, and data storage and visualization aid in data management and presentation.

### 6.4.2 Types of Biosensors

Biosensors are classified based on their bioreceptors and transducers, each tailored to specific applications. Among the myriad types, immunosensors, calorimetric sensors, DNA biosensors, enzyme-based sensors, tissue-based sensors, optical biosensors, thermal sensors, piezoelectric sensors, and electrochemical biosensors prominently feature. The latter category further branches into conductimetric, amperometric, and potentiometric sensors [14]. A standout example among these is the amperometric glucose biosensor, renowned for its role in enabling diabetic patients to monitor blood glucose levels regularly [11]. Here are some common types of biosensors:

- **Electrochemical Biosensors [16]:**
  - Utilize electrochemical reactions to detect analytes.
  - Commonly employ enzymes, such as glucose oxidase, to catalyze reactions.
  - Used in blood glucose monitoring, DNA analysis, and environmental monitoring.
- **Optical Biosensors [17]:**
  - Rely on light interactions for analyte detection.
  - May involve fluorescence, luminescence, or absorbance changes.
  - Widely used in DNA sequencing, protein-protein interactions, and label-free detection.
- **Piezoelectric Biosensors [17]:**
  - Utilize changes in mass or density to measure analyte binding.
  - Commonly based on quartz crystal microbalances (QCMs).
  - Used in studies of biomolecular interactions and real-time monitoring of cellular events.
- **Enzyme-Based Biosensors [16]:**
  - These biosensors use enzymes as the biological recognition element.
  - Enzymes catalyze a reaction in the presence of the target analyte, and the resulting change in properties (e.g., electrical or optical) is measured [12].
- **Immunoassay Biosensors [17]:**
  - These biosensors rely on the specific binding between antibodies and antigens.
  - They are widely used for the detection of proteins, hormones, and various disease markers.
- **Microbial Biosensors [17]:**
  - These biosensors use whole cells or microorganisms to detect analytes, such as bacteria or pollutants.
  - They are valuable in environmental monitoring and bioremediation.
- **Thermal Biosensors [17]:**
  - Measure changes in heat associated with biochemical reactions.
  - Used for DNA hybridization studies, protein-protein interactions, and drug discovery.
- **Mass-Based Biosensors [17]:**
  - Detect analyte binding by monitoring mass changes.
  - Often used for studying molecular interactions, such as antibody-antigen binding.
- **Gas Biosensors [17]:**
  - Detect gases or volatile organic compounds using biological components.
  - Applied in environmental monitoring, detection of toxic gases, and breath analysis for disease diagnosis.
- **Magnetic Biosensors [18]:**
  - Employ magnetic nanoparticles and their interactions with analytes.
  - Used in medical diagnostics and biomolecular studies, including DNA and protein analysis [4].
- **Acoustic Biosensors [18]:**
  - Utilize acoustic wave properties to detect changes upon analyte binding.
  - Applied in monitoring various biomolecular interactions and cellular responses.
- **Surface Plasmon Resonance Biosensors [18]:**
  - Measure changes in the refractive index of a surface.
  - Widely used for studying molecular binding events and real-time monitoring of biomolecular interactions.
- **Nanoparticle-Based Biosensors:**
  - Utilize nanomaterials, such as nanoparticles and nanowires, to enhance sensitivity and specificity [12].
  - Applied in various fields, including disease diagnostics, environmental monitoring, and drug development.

These types of biosensors represent a diverse toolkit for detecting a wide range of analytes, making them invaluable in scientific research, medical diagnostics, environmental monitoring, and various industrial applications. Each type of biosensor has its own advantages and limitations, making them suitable for different analytical tasks.

### 6.4.3 Metal Oxide Nanostructure-Based Biosensors

Metal oxide nanostructures, such as zinc oxide, titanium dioxide, and iron oxide, exhibit unique properties due to their nanoscale dimensions. Metal oxide nanostructures can serve as transducers, facilitating signal transduction, and as substrates for bioreceptor immobilization, enhancing biosensor sensitivity and selectivity [10].

Bioreceptors, such as enzymes or antibodies, specifically interact with target bioanalytes, initiating a signal-generating event. Metal oxide nanostructures can be functionalized with bioreceptors to improve their stability and interaction efficiency. The interaction between the bioreceptor and the bioanalyte leads to a measurable signal that is then transduced by the transducer into an electronic or optical output [4].

Metal oxide nanostructure-based biosensors have a broad range of applications.

1. They are extensively used in medical diagnostics for the rapid and precise detection of diseases (cancer) and viruses. Figure 6.2 examines various specific capabilities of biosensors that are encompassed within the domains of healthcare and related services [19].
2. These biosensors also play a critical role in food quality assessment, monitoring fermentation processes, and environmental monitoring. Their ability to detect various bioanalytes, such as proteins, DNA, glucose, and toxins, makes them capable tools in multiple domains [20].
3. Analyzing pollutants and pathogens in air and water, ensuring environmental safety.
4. Monitoring bioprocesses, such as fermentation and enzymatic reactions, for pharmaceutical and biofuel production.

Metal oxide nanostructure-based biosensors offer several advantages, including enhanced sensitivity, rapid response times, and the potential for miniaturization [12].

## 6.5 ROLE OF BIOSENSORS IN CANCER DIAGNOSIS

Cancer remains a influential global health challenge, responsible for millions of lives lost each year. In Figure 6.3, we can see a significant decrease in the age-adjusted death rate for heart disease over the past five decades, whereas cancer has not shown a similar decline in this regard. The critical importance of early detection in effectively managing and treating cancer cannot be overstated. In recent years, biosensors have emerged as powerful tools in the field of oncology, offering innovative approaches to cancer diagnosis. These devices, which combine biological recognition elements with various physicochemical transducers, have opened new avenues for early detection, precision medicine, monitoring, and non-invasive testing [5].

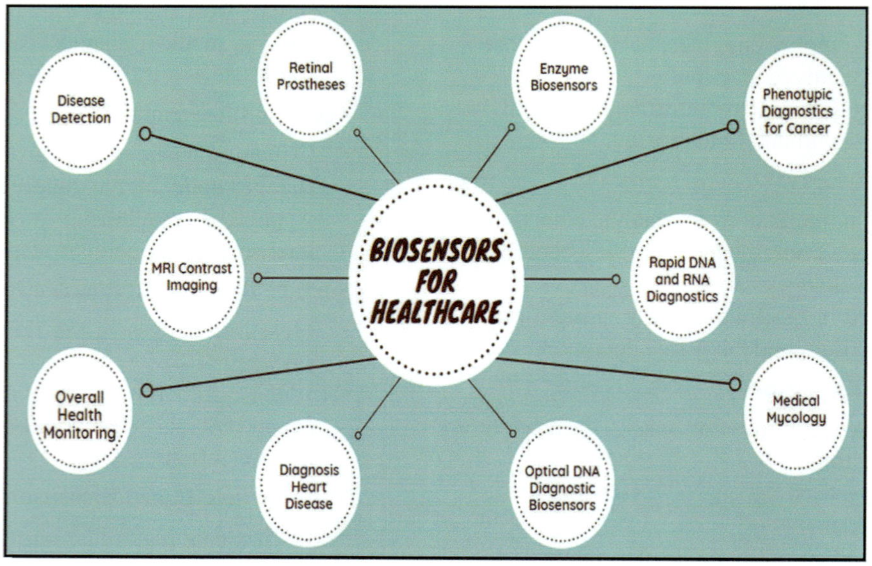


FIGURE 6.2 Specific capabilities of biosensors in health services.

**Source: Reprinted with permission from Ref. [19].**

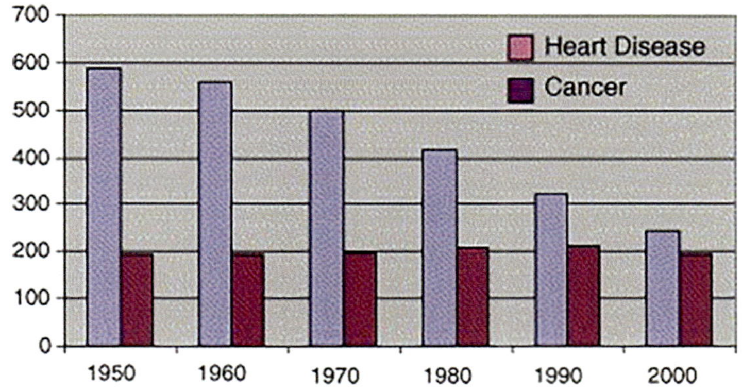


**FIGURE 6.3** Death rates from heart disease and all cancer-related diseases from 1950 to 2000.

**Source: Reprinted with permission from Ref. [21].**

### 6.5.1 Applications of Biosensors in Cancer Diagnosis

#### 6.5.1.1 Early Detection

The practical applications of biosensors in early cancer detection are vast and impactful. One of the most significant roles of biosensors is in the identification of cancer-specific biomarkers in bodily fluids or tissues. This enables early diagnosis, which is often a crucial determinant of successful treatment outcomes. Furthermore, biosensors have paved the way for liquid biopsies, a non-invasive approach to early cancer detection. By analyzing circulating tumor cells and cell-free DNA, biosensors are transforming the way we monitor and identify cancer, minimizing the need for painful procedures [2]. In addition, molecular biosensors play a pivotal role in tumor imaging by helping visualize tumor specific markers. Such imaging techniques facilitate early detection by revealing the presence of malignancies at their onset. Moreover, biosensors contribute to the concept of precision medicine by profiling the genetic and molecular characteristics of an individual's cancer. This enables personalized treatments, maximizing their effectiveness and minimizing adverse effects [22].

#### 6.5.1.2 Monitoring

Monitoring in biosensors for cancer detection is a pivotal aspect of their utility in the field of oncology. Once cancer is identified, biosensors continue to be indispensable tools for assessing the progression of the disease and the effectiveness of treatment. By continuously tracking specific biomarkers or cellular changes, biosensors enable healthcare professionals to make informed decisions about patient care [12]. Whether it is monitoring the levels of cancer-related biomarkers in bodily fluids or gauging the response to therapeutic interventions, biosensors offer real-time data that guides treatment strategies, ultimately leading to improved patient outcomes and individualized care.

#### 6.5.1.3 Non-Invasive Testing

Non-invasive testing using biosensors has changed the complexion of cancer detection. Traditionally, diagnosing cancer often involved invasive and sometimes uncomfortable procedures. Biosensors have changed this paradigm by offering innovative, patient-friendly alternatives. Through the analysis of bodily fluids, such as blood or urine, and even exhaled breath, biosensors can identify cancer-specific biomarkers or cellular changes without the need for painful biopsies or uncomfortable imaging procedures [21]. This non-invasive approach not only reduces patient discomfort and anxiety but also significantly enhances the likelihood of early cancer detection, as it encourages more individuals to undergo regular screening. Non-invasive biosensors hold the promise of making cancer diagnosis more accessible and less burdensome for patients, ultimately improving the chances of early intervention and successful treatment.

#### 6.5.1.4 Point-of-Care Testing

Point-of-care testing (POCT) using biosensors represents a transformative approach to cancer detection and management. These portable, user-friendly devices bring diagnostics directly to the patient, offering several advantages in the context of cancer [21]. In cancer detection, the ability to perform rapid tests at the patient's bedside or in remote locations has far-reaching implications. With biosensors, healthcare providers can swiftly analyze samples, such as blood or urine, to detect cancer-specific biomarkers or cellular changes. This immediate access to diagnostic information accelerates the early detection of cancer, allowing for timely interventions and improved patient outcomes. The convenience and speed of point-of-care biosensors make them indispensable tools in the fight against cancer, enhancing accessibility to diagnostics and reducing the time between testing and treatment decisions [21].

#### 6.5.1.5 Cost Reduction

Cost reduction is a significant benefit of employing biosensors for cancer detection. Traditional diagnostic methods, such as extensive laboratory tests and imaging procedures, can be costly for both healthcare systems and patients [12]. In contrast, biosensors offer a more cost-effective approach to cancer diagnosis. These devices often require smaller sample volumes, reducing the expenses associated with sample collection and transportation. Furthermore, biosensors can streamline the diagnostic process, delivering results rapidly and minimizing the need for extensive and expensive laboratory infrastructure. By enabling early cancer detection, biosensors contribute to more cost-effective treatment, as interventions at earlier stages of the disease tend to be less resource-intensive and more successful. Overall, the integration of biosensors in cancer detection helps reduce the financial burden on healthcare systems and patients, making diagnostics more accessible and affordable.

#### 6.5.1.6 Research and Drug Development

Biosensors have become invaluable tools in cancer research and drug development. Their impact extends beyond diagnosis to understanding the underlying mechanisms of cancer and testing potential therapies. In the domain of research, biosensors assist in clarifying the complex biology of cancer by enabling the real-time monitoring of biomarkers and cellular changes [14]. This contributes to a deeper understanding of the disease, its progression, and its response to various interventions.

Moreover, biosensors play a crucial role in drug development. They facilitate the testing of potential anticancer drugs by providing a means to assess their efficacy in real-time. Researchers can use biosensors to screen compounds, study their impact on cancer cells, and refine drug candidates, ultimately accelerating the development of new therapeutic approaches. Additionally, biosensors are integral to the study of drug resistance, helping researchers identify and address challenges in treatment.

By bridging the gap between research and practical applications, biosensors contribute to the development of more targeted and effective cancer therapies, with the potential to improve patient outcomes and reduce side effects. They have become fundamental tools in the quest to understand and combat cancer at a molecular level, making significant strides in the field of oncology research and drug development.

## 6.6 PRESENT EXAMPLES OF METAL OXIDE NANOMATERIAL-BASED BIOSENSORS FOR CANCER DETECTION

Nanomaterials, including zinc oxide, titanium dioxide, and tin oxide ($SnO_2$), offer unique properties that enhance the sensitivity and specificity of biosensors. For instance, ZnO nanomaterials can be functionalized to selectively detect specific cancer biomarkers, such as microRNA or proteins, with high precision. $TiO_2$ nanomaterials have been employed in the creation of electrochemical biosensors for cancer biomarker quantification, allowing for early diagnosis [2]. $SnO_2$ nanomaterials have demonstrated potential in gas sensing applications, which can be harnessed for detecting volatile organic compounds associated with cancer.

Sarcosine oxidase-targeted biosensors are an optimistic field of research in cancer diagnosis. Sarcosine is a metabolite that has been associated with prostate cancer, and biosensors that target this enzyme can potentially help in early cancer detection [12]. These biosensors typically work by detecting elevated levels of sarcosine in biological samples, which can be indicative of cancer. The biosensor technology involves immobilizing sarcosine oxidase on a substrate and measuring the enzymatic reaction when sarcosine is present. This can provide a quantitative assessment of sarcosine levels, which may help in cancer diagnosis and monitoring [12]. Ovarian cancer diagnosis has witnessed a significant breakthrough through the application of nanoparticle-based biosensors. Overall, the integration of metal oxide nanomaterials in biosensors holds great promise for more accurate and early cancer diagnosis. Here are a few examples of metal oxide nanomaterial-based biosensors that have been developed for cancer diagnosis:

### 6.6.1 Zinc Oxide Nanowire Biosensors

ZnO nanowires are nanostructured materials characterized by their high surface area, biocompatibility, and excellent electronic properties. These attributes position them as ideal candidates for biosensor development. In the context of cancer diagnosis, ZnO nanowire biosensors have been specifically designed to detect cancer-related biomolecules, such as prostate-specific antigen (PSA) for prostate cancer or specific microRNAs associated with various cancers [23]. The fundamental principle underlying ZnO nanowire biosensors involves the immobilization of selective receptors, such as aptamers or antibodies, onto the nanowire surface. These receptors bind selectively to cancer biomarkers of interest, inducing changes in the electrical properties of the ZnO nanowires. These changes are then transduced into measurable signals, offering the potential for highly sensitive and specific cancer biomarker detection [14].

ZnO nanowire biosensors offer unparalleled sensitivity, rapid detection, and the possibility of point-of-care testing, crucial for early cancer diagnosis.

#### 6.6.1.1 Functionalization and Selectivity

ZnO nanowires are functionalized with specific capture molecules, such as aptamers, antibodies, or DNA probes. These capture molecules are highly selective and can specifically bind to the target biomolecules of interest (PSA) [23]. The binding of PSA to the functionalized ZnO nanowires induces changes in their electrical properties, resulting in measurable signals.

#### 6.6.1.2 Transduction of Signals

The change in electrical conductance due to the binding event is transduced into measurable signals. This transduction can occur in various ways, including monitoring changes in resistance, current, voltage, or impedance. The extent of the change is correlated with the concentration of the target biomolecule [24].

#### 6.6.1.3 Sensitivity and Specificity

- ZnO nanowire biosensors exhibit exceptional sensitivity, enabling the detection of PSA even at trace levels, which is critical for the early diagnosis of prostate cancer.
- The specificity is achieved through the choice of highly selective capture molecules, reducing false-positive results.

#### 6.6.1.4 Applications and Challenges

These biosensors have the potential to revolutionize prostate cancer diagnosis. They can be applied to detect PSA in patient samples, facilitating early disease detection and monitoring. While ZnO nanowire biosensors offer significant promise, challenges include optimizing functionalization, ensuring long-term stability, and minimizing interference from other molecules in complex biological samples [24].

### 6.6.2 Iron Oxide Nanoparticles for MRI

Magnetic resonance imaging is a cornerstone of non-invasive medical diagnostics, providing detailed images of internal structures and physiological processes in the human body [25]. Advancements in MRI technology have led to the development of contrast agents that enhance the visibility of specific tissues and anomalies within the body. Among these agents, iron oxide nanoparticles, known as superparamagnetic iron oxide nanoparticles (SPIONs), have emerged as potential device for improving MRI contrast and facilitating early disease detection. SPIONs have attracted a lot of attention due to their widespread diagnostic and therapeutic applications in neuro-oncology [26].

Superparamagnetic NPs are unique inorganic NPs that were first introduced for biomedical applications in the late 1980s. The core of superparamagnetic NPs consists of metal molecules of nickel, cobalt, or iron oxide ($Fe_3O_4$ magnetite, which is the most commonly used metal). The surface of superparamagnetic NPs can be modified by coating the core in a few atomic layers of organic polymers. Superparamagnetic NPs have been also used successfully for cell labelling, drug delivery, MRI of tumor tissue and vasculature, hyperthermia induction, and magnetofection of cancer cells [26].

In this context, the application of iron oxide nanoparticles in MRI holds significant potential for the detection of cancer cells and tumors [27]. By leveraging their magnetic properties, these nanoparticles can enhance the contrast in MRI images, thereby enabling more precise and timely cancer diagnosis. Iron oxide nanoparticles are known for their biocompatibility and stability, making them suitable for in vivo applications [28].

#### 6.6.2.1 Enhancement of Cancer Detection

- **Improved Tumor Imaging:** Iron oxide nanoparticles enhance the visualization of cancer cells and tumors, aiding in their early detection.
- **Targeted Imaging:** Functionalized iron oxide nanoparticles can be directed toward cancer-specific receptors, enabling precise imaging of malignant tissues.

#### 6.6.2.2 Advantages of Iron Oxide Nanoparticles

- **High Sensitivity:** Iron oxide nanoparticles provide high sensitivity, allowing for the detection of subtle abnormalities [29].
- **Non-Invasive:** MRI with iron oxide nanoparticles is a non-invasive method that reduces patient discomfort [28].
- **No Ionizing Radiation:** Unlike some imaging techniques, MRI does not involve ionizing radiation, making it safer for repeated use.

#### 6.6.2.3 Applications and Case Studies

- **Breast Cancer Imaging:** Breast cancer is one of the most prevalent cancers and a significant cause of fatalities in women. Diagnosis of breast cancer faces substantial challenges, owing to the multi-factored nature of the illness [30]. $Fe_3O_4$ nanoparticles have been employed to improve breast cancer detection through enhanced MRI contrast [31]. Iron oxide nanoparticles can be efficaciously functionalized based on breast cancer molecular mechanisms for potential application in breast cancer drug delivery, gene delivery, immunotherapy, photo responsive therapy, and bio-imaging [32].
- **Brain Tumor Diagnosis:** SPIONs have been utilized to aid in the identification and characterization of brain tumors [33]. Brain tumors, especially aggressive glioblastoma, are challenging to treat due to limited options. The blood–brain barrier (BBB) poses a hurdle for effective theranostics, restricting the entry of imaging probes and drugs. To enhance brain tumor diagnosis and treatment, researchers explore multifunctional nanoparticles with size effects and diverse properties. Modified nanoparticles can overcome the BBB, accumulating at the tumor for precise imaging and drug release [26].
- **Liver cancer:** Hepatocellular carcinoma (HCC) is one of the most common cancers worldwide, leading to the second most likely cause of cancer-related deaths [34]. In clinical settings, medical

imaging is vital for HCC screening and diagnosis. Magnetic resonance imaging employs contrast agents based on magnetic nanoparticles, offering high spatial resolution and sensitivity. This technique shows promise for HCC imaging and staging, with significant potential for practical in vivo applications [34]. This detection strategy is based on the fact that hepatic macrophages, or Kupffer cells, can readily internalize SPIONs in normal hepatic tissue, while liver tumors are mostly devoid of such macrophages. The first such SPIONs clinically used in Europe were coated with dextran T10. Later, other SPIONs with different sizes and coatings were introduced into the market.

- **Prostate Cancer:** Prostate cancer is the most frequently diagnosed malignancy in men. Using magnetic nanoparticles to target drugs is a new way to treat prostate cancer. There's strong evidence that guiding nanoparticles with magnets to a specific area works when a magnetic field is applied. This method can boost the concentration of the drug's effect in that particular area, offering a potential improvement in treating prostate cancer [35].
- **Other Cancer Types:** Research is ongoing to expand the application of iron oxide nanoparticles to detect cancer in various organs, such as ovarian cancer and lung cancer.

#### 6.6.2.4 Challenges and limitations

- **Nanoparticle Functionalization:** One challenge is optimizing the functionalization of $Fe_3O_4$ nanoparticles to ensure their targeting specificity and biocompatibility.
- **Clinical Translation:** A key hurdle is moving from experimental settings to clinical applications [28]. Ensuring the safety, regulatory approval, and widespread use of $Fe_3O_4$ nanoparticles in MRI requires rigorous testing and validation.
- **Multimodal Imaging:** Integrating $Fe_3O_4$ nanoparticles with other imaging modalities (e.g., fluorescence imaging, PET) presents opportunities for comprehensive diagnostics but requires advancements in imaging integration and interpretation [31].
- **Long-term Stability:** Ensuring the long-term stability and biocompatibility of $Fe_3O_4$ nanoparticles in the human body is a priority.

### 6.6.3 Copper Oxide Nanoparticles in microRNA Detection

Copper oxide nanoparticles are sometimes used in microRNA detection due to their unique properties. They can be functionalized with specific molecules, such as aptamers or DNA probes, to selectively bind to microRNAs of interest [36]. When this binding occurs, it can lead to changes in the nanoparticles' properties, such as their optical or electrical characteristics, which can be detected and quantified [36].

This approach allows for the sensitive and specific detection of microRNAs, which are important in gene regulation and disease diagnostics.

#### 6.6.3.1 microRNAs in Cancer

microRNAs (miRNAs) are single-stranded noncoding RNA molecules that act as key regulators of mRNA expression and are emerging biomarkers for disease [37]. They are becoming important markers for diseases. Due to their small size (18–25 nucleotides), it is challenging to recognize, label, and generate signals for these molecules. miRNAs found in various biological fluids resist breakdown by ribonucleases and remain stable during storage [38].

In many cancers, the profiles of miRNAs undergo significant changes, impacting the cell's transcriptome. However, these small RNAs are also regulated by several proteins associated with cancer, like p53 and c-Myc. The way miRNAs are expressed depends on the type of tumor, its stage, and grade. miRNAs play a role in how tumors respond to treatment and how likely they are to be cured [37].

- **Relevance of microRNAs as Biomarkers for Cancer:**
  1. **Dysregulation in Cancer:** microRNAs are small RNA molecules that are involved in the post-transcriptional regulation of gene expression [39]. Dysregulation of miRNAs is commonly observed in various cancers. They can act as either oncogenes (promote cancer) or tumor suppressors (inhibit cancer) by targeting specific genes involved in cell proliferation, apoptosis, and metastasis [36].
  2. **Specificity to Cancer Types:** Certain microRNAs are uniquely associated with particular cancer types. For instance, miR-21 is often overexpressed in breast, lung, and pancreatic cancers [39]. Profiling the expression of these specific miRNAs can help identify the type of cancer a patient may have.
  3. **Prognostic and Predictive Value:** microRNA profiles can provide prognostic information by indicating the likelihood of disease progression or recurrence. They can also serve as predictive biomarkers, guiding treatment decisions and predicting responses to therapies [38].
  4. **Minimally Invasive Detection:** microRNAs can be detected in various biological samples, including blood, urine, and tissue, which makes their collection minimally invasive. This allows for regular monitoring and early cancer detection.

- **Need for Highly Sensitive and Specific Detection Methods:**
  1. **Early Detection:** Cancer is often most treatable when detected at an early stage. Highly sensitive methods are necessary to detect low levels of miRNAs, especially in the early stages of the disease [40].
  2. **Specificity:** Since different miRNAs are associated with different cancer types, it's crucial to have detection methods that are highly specific and can distinguish between miRNA variants [38].
  3. **Patient Stratification:** For personalized medicine and tailored cancer treatments, sensitive and specific detection methods are required to identify the unique miRNA profiles of individual patients.
  4. **Monitoring Treatment Response:** Sensitive methods allow for the monitoring of treatment responses, helping to assess the effectiveness of therapies and adjust treatment plans as necessary [39].
  5. **Reducing False Positives/Negatives:** Highly sensitive and specific methods can minimize the likelihood of false positive or false negative results, which can have significant clinical implications [41].

#### 6.6.3.2 Functionalization of Copper Oxide Nanoparticles

Functionalizing copper oxide nanoparticles for microRNA binding is a crucial step in developing sensitive and specific detection methods. Common methods for functionalization include chemical functionalization, where the surface of CuO NPs is modified with chemical ligands such as carboxylic acids, amines, or thiol groups [42]. These ligands provide binding sites for miRNAs through electrostatic interactions, hydrogen bonding, or covalent linkage [43]. Other methods are the attachment of molecules like aptamers, DNA probes, or peptides that have specific affinity for the target miRNA [12]. This functionalization imparts selectivity by ensuring that CuO NPs primarily interact with the desired miRNA sequences. Additionally, optimizing factors such as ligand density, pH, and ionic strength in the reaction environment can further enhance the selective binding. The specific design of ligands and their interaction with the CuO NP surface plays a vital role in achieving this selectivity [43]. These techniques not only allow for the detection of miRNAs but also pave the way for their precise identification and quantification in various applications, such as cancer diagnostics and personalized medicine [43].

#### 6.6.3.3 Detection Methods

Copper oxide nanoparticles have found extensive application in conjunction with various detection methods, such as electrochemical and fluorescence assays, for cancer-related microRNA detection, offering impressive sensitivity and specificity [44]. In electrochemical assays, CuO NPs are functionalized with miRNA-specific probes. The binding of miRNA to these probes results in alterations in the electrical properties of the CuO NPs. These electrochemical changes are precisely quantified, facilitating the highly sensitive detection of miRNAs [44]. Notably, research has demonstrated the remarkable sensitivity and specificity of these assays, with detection limits reaching at molar concentrations [14]. Furthermore, in fluorescence assays, CuO NPs are used as quenchers, and when miRNAs bind to their corresponding probes, a spatial separation occurs, leading to the generation of a fluorescence signal. These assays offer real-time monitoring capabilities and have proven to be highly sensitive and specific in differentiating various miRNA targets [44].

#### 6.6.3.4 Applications in Cancer Diagnosis

Copper oxide nanoparticles have demonstrated promise in the detection and monitoring of various cancer types, with relevance in early-stage diagnosis and tracking the progression of the disease. The CuO NPs, when functionalized with miRNA-specific probes, have been used to detect miRNA biomarkers associated with breast cancer [45]. These assays offer the sensitivity required to identify miRNAs at low concentrations, enabling early-stage diagnosis. Furthermore, they hold potential in monitoring the progression of breast cancer, aiding in the assessment of treatment effectiveness and relapse prediction [45].

Lung cancer is often diagnosed at advanced stages, but CuO NP-based miRNA detection methods can contribute to early diagnosis. Pancreatic cancer is notorious for its late diagnosis and poor prognosis [46]. CuO NPs have been utilized to detect specific miRNAs associated with pancreatic cancer. CuO NP-based miRNA detection can be employed to identify miRNAs linked to prostate cancer. Colorectal cancer is another condition where early detection is critical. CuO NP-enabled assays can detect miRNAs indicative of colorectal cancer, making it possible to identify the disease in its early stages and track changes in miRNA expression during disease progression [45].

#### 6.6.3.5 Challenges and Limitations

- **Clinical Translation:** One of the primary challenges is the transition from research to clinical applications. While CuO NPs show promise in laboratories, their clinical translation involves navigating complex regulatory pathways, safety assessments, and large-scale production challenges [47].
- **Specificity:** Achieving high specificity in miRNA detection remains a challenge, as miRNA sequences can be quite similar, and cross-reactivity with related miRNAs may occur.
- **Quantification:** Accurate quantification of miRNAs is essential for clinical applications, and

current methods may need refinement to provide precise measurements [42].

- **Sample Complexity:** Clinical samples, such as blood or tissue, can be complex and contain various interfering substances that may affect the accuracy and sensitivity of miRNA detection [47].
- **Biocompatibility:** Ensuring the biocompatibility and safety of CuO NPs for in vivo applications is crucial, as the nanoparticles' long-term effects need to be thoroughly evaluated [47].

### 6.6.4 Manganese Oxide Nanoparticles in Targeted Drug Delivery

Targeted drug delivery is a critical area of research and innovation in the field of healthcare and pharmaceuticals [48]. The primary goal of drug delivery systems is to transport therapeutic agents to specific sites in the body with the utmost precision and efficiency while minimizing collateral damage to healthy tissues [48]. Achieving this goal is essential in enhancing the efficacy of treatments and reducing adverse side effects associated with many conventional drug delivery methods [48].

Nanoparticles have emerged as promising carriers in drug delivery due to their unique physical and chemical properties. At the nanoscale, materials exhibit distinct behaviors that can be harnessed to overcome the limitations of traditional drug delivery approaches. Their size, surface modifiability, and controlled drug release capabilities make them a powerful tool for enhancing drug delivery precision. Yet, despite the tremendous potential, numerous research challenges persist, including the optimization of targeting strategies, the assessment of safety and toxicity, and the navigation of regulatory requirements.

Manganese oxide (MnO) nanoparticles have gained attention in the field of targeted drug delivery due to their unique properties. They can serve as promising carriers for delivering drugs to specific sites in the body.

#### 6.6.4.1 Manganese Oxide Nanoparticles

Manganese oxide (MnO) nanoparticles have garnered significant attention in the field of drug delivery, owing to their distinctive characteristics. These nanoparticles, typically within the nanoscale range, offer remarkable properties such as excellent biocompatibility, which is essential for their safe use in medical applications [49]. Their synthesis methods encompass a variety of techniques, including sol-gel processes, co-precipitation, and hydrothermal methods. Moreover, MnO nanoparticles can be readily surface-modified by attaching ligands, antibodies, or coatings, allowing for active targeting of specific cells or tissues [49]. Such surface engineering empowers these nanoparticles to play a pivotal role in enhancing drug delivery precision. The advantages of MnO nanoparticles in drug delivery are multifaceted, from controlled drug release and minimized toxicity to their potential for diagnostic imaging, making them a versatile and promising candidate for revolutionizing drug delivery systems [49].

#### 6.6.4.2 Targeted Drug Delivery

Targeted drug delivery represents a sophisticated approach designed to transport therapeutic agents precisely to their intended sites within the body [50]; see Figure 6.4.

This concept arises from the recognition that conventional systemic drug administration often results in unintended side effects, suboptimal treatment efficacy, and the potential for toxicity in healthy tissues. Targeted drug delivery, in contrast, strives to improve these aspects. Its core advantages are rooted in its ability to enhance treatment outcomes while minimizing adverse effects [50]. Two key targeting strategies underpin this concept: passive and active targeting [50].

Passive targeting relies on the inherent characteristics of the target tissue. For example, tumors tend to have leaky vasculature, allowing nanoparticles to accumulate more within them than in healthy tissues. Active targeting, on the other hand, involves modifying drug carriers, like nanoparticles, with specific molecules that can recognize and bind to molecular markers on the target tissue. This approach enhances the selectivity of drug delivery [50].

Minimizing off-target effects is of paramount importance in drug delivery. Unintended distribution of drugs throughout the body can lead to systemic toxicity, a key concern in traditional therapies [52]. By using targeted drug delivery, the risk of drug exposure to non-targeted tissues is significantly reduced. This precision not only enhances treatment effectiveness but also reduces the potential for side effects, improving patient safety and the overall quality of medical care [52].

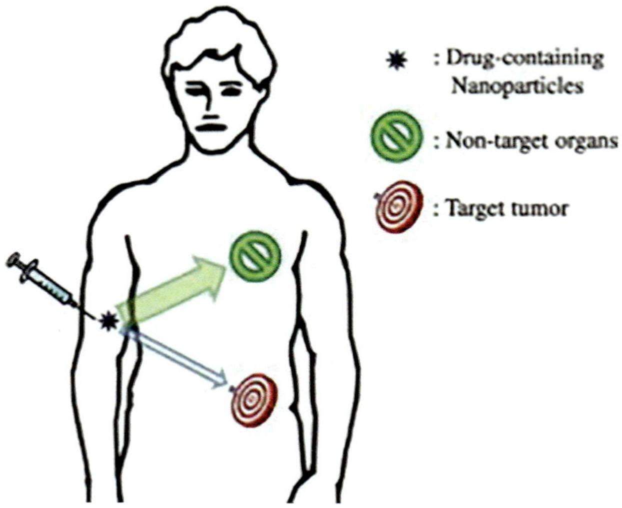


FIGURE 6.4 Diagram of targeted drug delivery.

Source: Reprinted with permission from Ref. [51].

#### 6.6.4.3 Functionalization of MnO Nanoparticles for Targeted Drug Delivery

MnO nanoparticles can be engineered to achieve specific drug delivery objectives, and one key strategy involves active targeting through ligand conjugation. Active targeting seeks to enhance the nanoparticles' selectivity for particular cells or tissues by attaching ligands or antibodies that can recognize and bind to specific cell surface receptors. This approach is pivotal in improving drug delivery precision [53].

- **Ligand Conjugation for Active Targeting:** Ligand conjugation involves the attachment of ligands, such as antibodies, peptides, or small molecules, to the surface of MnO nanoparticles [53]. These ligands are carefully selected to recognize overexpressed receptors on the target cells, such as those found in cancer cells. When MnO nanoparticles are administered, they circulate through the body and actively seek out these target cells due to the specific interaction between the ligands on the nanoparticles and the receptors on the cell surface. The advantages of ligand-conjugated MnO nanoparticles are substantial [53]. They allow for:
    1. **Enhanced Selectivity:** Ligand conjugation increases the nanoparticles' specificity to the target cells, ensuring that drug payloads are delivered primarily to the diseased tissue [53].
    2. **Minimized Off-Target Effects:** By actively homing in on the target cells, the risk of off-target drug delivery is significantly reduced, thereby minimizing systemic toxicity and side effects [53].
    3. **Improved Therapeutic Efficacy:** Targeted drug delivery enhances drug accumulation at the site of action, increasing treatment efficacy while potentially lowering the required drug dosage [54].
- **Coating and Encapsulation Techniques to Control Drug Release:** Beyond active targeting, another essential aspect of functionalization is the ability to control drug release from MnO nanoparticles. This is crucial in achieving sustained therapeutic effects and mitigating potential side effects [55].

MnO nanoparticles can be coated with various materials or encapsulated within a shell that governs the release of the loaded drug. These coating and encapsulation techniques can be tailored to control the drug release kinetics, achieving the desired duration and profile of drug delivery [55].

**Coating Techniques:** MnO nanoparticles can be coated with polymers or other materials that act as barriers to drug release. These coatings can be designed to respond to specific environmental cues, such as changes in pH, temperature, or the presence of certain enzymes, which trigger drug release when and where needed [56].

**Encapsulation Techniques:** MnO nanoparticles can also be encapsulated within microspheres or liposomes. These structures can provide a reservoir for the drug payload, allowing for controlled and sustained release over time [55]. The functionalization of MnO nanoparticles is instrumental in advancing targeted drug delivery. Ligand conjugation for active targeting enhances specificity, reduces off-target effects, and improves therapeutic outcomes.

#### 6.6.4.4 In Vitro and In Vivo Studies

In vitro and in vivo studies involving MnO nanoparticles for drug delivery have yielded compelling results, demonstrating their potential in improving drug delivery efficiency and reducing toxicity [57, 58]. These studies have encompassed various drug formulations and medical applications.

In vitro experiments using MnO nanoparticles loaded with anticancer drugs, such as doxorubicin, revealed significantly enhanced drug delivery efficiency [57]. MnO nanoparticles, with their tailored surface functionalization, actively targeted cancer cells overexpressing specific receptors, increasing drug accumulation within tumor tissues. This was evidenced by quantitative assays showing a substantial uptake of the drug payload by cancer cells when delivered via MnO nanoparticles compared to free drug administration [57].

In vivo studies in animal models, such as murine xenograft models, further emphasized the potential of MnO nanoparticles in minimizing off-target effects and reducing drug toxicity. Animals treated with MnO nanoparticle-based drug delivery systems exhibited notably less damage to healthy tissues compared to those receiving conventional systemic drug administration. This was corroborated by histological examinations and assessments of organ function, which consistently indicated lower levels of toxicity in animals treated with MnO nanoparticles [59].

#### 6.6.4.5 Application in Cancer Therapy

Manganese oxide nanoparticles possess remarkable imaging capabilities, particularly in magnetic resonance imaging [59]. The unique magnetic properties of MnO make them excellent contrast agents for visualizing tissues and pathological conditions. In MRI, these nanoparticles generate contrasting signals that enhance the clarity and specificity of images. Their capability in MRI imaging arises from their ability to modulate relaxation times (T1 and T2), which can be exploited to provide detailed anatomical and functional information [59].

In oncology, MnO nanoparticles have demonstrated their potential as both imaging agents and carriers for cancer therapy. In an illustrative example, MnO nanoparticles functionalized with cancer-targeting ligands were used to image and treat a breast cancer xenograft model in mice. The MRI images clearly delineated the tumor, and the same nanoparticles, loaded with an anticancer drug, facilitated targeted therapy, resulting in reduced tumor size and enhanced survival [59].

One notable case study focused on using MnO nanoparticles for the targeted delivery of chemotherapeutic agents to solid tumors [60]. In a murine model of breast cancer, MnO nanoparticles with surface-bound antibodies specific to cancer cell receptors demonstrated remarkable efficacy. They significantly inhibited tumor growth and prolonged survival compared to the control group receiving free chemotherapy, all while reducing systemic toxicity [60].

#### 6.6.4.6 Challenges and Limitations

- **Biocompatibility and Safety:** One of the primary concerns with MnO nanoparticles is their biocompatibility and potential toxicity. Ensuring that these nanoparticles do not induce adverse effects in the body, such as inflammation or organ damage, is a critical challenge [48].
- **Stability and Aggregation:** MnO nanoparticles are prone to aggregation, which can affect their performance and drug delivery efficiency [48].
- **Drug Loading and Release Control:** Achieving optimal drug loading in MnO nanoparticles and controlling the release kinetics are ongoing challenges. These factors are critical in ensuring that the right amount of drug is delivered to the target site at the right rate [61].
- **Tissue Penetration:** MnO nanoparticles, like many nanoparticles, face limitations in deep tissue penetration [61].

## 6.7 CONCLUSION

Metal oxide nanomaterial-based biosensors represent a critical advancement in the field of cancer diagnosis. Their remarkable sensitivity and specificity empower the detection of cancer biomarkers with unparalleled precision, laying the foundation for early cancer diagnosis. This is of immense significance, as the key to successful cancer treatment often hinges on catching the disease at its earliest stages. These biosensors have the potential to revolutionize cancer detection by offering non-invasive, cost-effective, and rapid testing methods. In conclusion, they hold the promise of saving lives through the early detection and timely intervention in cancer cases, ultimately improving patient outcomes and reducing the burden of this devastating disease.

### Keywords

- Biomarkers
- Biosensors
- Cancer diagnosis
- Metal oxide nanomaterials
- Targeted drug delivery

## REFERENCES

1. Christina Fitzmaurice, Christine Allen, Ryan M Barber, Lars Barregard, Zulfiqar A Bhutta, Hermann Brenner, Daniel J Dicker, Odgerel ChimedOrchir, Rakhi Dandona, Lalit Dandona, et al. Global, regional, and national cancer incidence, mortality, years of life lost, years lived with disability, and disability-adjusted life-years for 32 cancer groups, 1990 to 2015: a systematic analysis for the global burden of disease study. *JAMA Oncology*, 3(4):524–548, 2017.
2. Majid Sharifi, Mohammad Reza Avadi, Farnoosh Attar, Fariba Dashtestani, Hedayatollah Ghorchian, Seyed Mahdi Rezayat, Ali Akbar Saboury, and Mojtaba Falahati. Cancer diagnosis using nanomaterials based electrochemical nanobiosensors. *Biosensors and Bioelectronics*, 126:773–784, 2019.
3. Jaise Mariya George, Arun Antony, and Beena Mathew. Metal oxide nanoparticles in electrochemical sensing and biosensing: a review. *Microchimica Acta*, 185:1–26, 2018.
4. Goutam Ghosh. Early detection of cancer: focus on antibody coated metal and magnetic nanoparticle-based biosensors. *Sensors International*, 1:100050, 2020.
5. Nasuha Rohaizad, Carmen C Mayorga-Martinez, Michaela Fojt, Naziah M Latiff, and Martin Pumera. Two-dimensional materials in biomedical, biosensing and sensing applications. *Chemical Society Reviews*, 50(1):619–657, 2021.
6. Lionel Vayssieres. On the design of advanced metal oxide nanomaterials. *International Journal of Nanotechnology*, 1(1–2):1–41, 2004.
7. Michael A Carpenter, Sanjay Mathur, and Andrei Kolmakov. *Metal oxide nanomaterials for chemical sensors*. Springer Science & Business Media, 2012.
8. Enza Fazio, Salvatore Spadaro, Carmelo Corsaro, Giulia Neri, Salvatore Gianluca Leonardi, Fortunato Neri, Nehru Lavanya, Chinnathambi Sekar, Nicola Donato, and Giovanni Neri. Metal-oxide based nanomaterials: synthesis, characterization and their applications in electrical and electrochemical sensors. *Sensors*, 21(7):2494, 2021.
9. Veeradasan Perumal and Uda Hashim. Advances in biosensors: principle, architecture and applications. *Journal of Applied Biomedicine*, 12(1):1–15, 2014.
10. Hideaki Nakamura and Isao Karube. Current research activity in biosensors. *Analytical and Bioanalytical Chemistry*, 377:446–468, 2003.
11. Saraju P Mohanty and Elias Kougianos. Biosensors: a tutorial review. *IEEE Potentials*, 25(2):35–40, 2006.
12. Baljit Singh, Shiliang Ma, Tony O Hara, and Sargun Singh. Nanomaterialsbased biosensors for the detection of prostate cancer biomarkers: recent trends and future perspective. *Advanced Materials Technologies*, 2201860, 2023.
13. Baoyi Shao and Zhongdang Xiao. Recent achievements in exosomal biomarkers detection by nanomaterials-based optical biosensors-a review. *Analytica Chimica Acta*, 1114:74–84, 2020.
14. Bansi D Malhotra, Saurabh Kumar, and Chandra Mouli Pandey. Nanomaterials based biosensors for cancer biomarker detection. In *Journal of Physics: Conference Series*, 704, 012011. IOP Publishing, 2016.
15. Pierre R Coulet. What is a biosensor? In *Biosensor principles and applications*, CRC Press, 1–6, 2019.

16. Parikha Mehrotra. Biosensors and their applications–a review. *Journal of Oral Biology and Craniofacial Research*, 6(2):153–159, 2016.
17. Nuggehalli M Ravindra, Camelia Prodan, Shanmugamurthy Fnu, Ivan Padronl, and Sushil K Sikha. Advances in the manufacturing, types, and applications of biosensors. *Jom*, 59:37–43, 2007.
18. RK Arora and RP Saini. Biosensors: way of diagnosis. *International Journal of Pharmaceutical Sciences and Research*, 4(7):2517–2527, 2013.
19. Abid Haleem, Mohd Javaid, Ravi Pratap Singh, Rajiv Suman, and Shanay Rab. Biosensors applications in medical field: a brief review. *Sensors International*, 2:100100, 2021.
20. Sara Eskandarinezhad, Irshad Ahmad Wani, Mohammad Nourollahileilan, Ajit Khosla, and Tokeer Ahmad. Metal and metal oxide nanoparticles/nanocomposites as electrochemical biosensors for cancer detection. *Journal of The Electrochemical Society*, 169(4):047504, 2022.
21. Steven A Soper, Kathlynn Brown, Andrew Ellington, Bruno Frazier, Guillermo Garcia-Manero, Vincent Gau, Steven I Gutman, Daniel F Hayes, Brenda Korte, James L Landers, et al. Point-of-care biosensor systems for cancer diagnostics/prognostics. *Biosensors and Bioelectronics*, 21(10):1932–1942, 2006.
22. Yuqi Yang, Qiong Huang, Zuoxiu Xiao, Min Liu, Yan Zhu, Qiaohui Chen, Yumei Li, and Kelong Ai. Nanomaterial-based biosensor developing as a route toward in vitro diagnosis of early ovarian cancer. *Materials Today Bio*, 13:100218, 2022.
23. Soyoun Jung and Taeksoo Ji. Protein biosensors based on polymer nanowires, carbon nanotubes and zinc oxide nanorods. *Sensors*, 11:5087–5111, 2011.
24. Sahar Saleem Bhat, Ahsanulhaq Qurashi, and Firdous Ahmad Khanday. Zno nanostructures based biosensors for cancer and infectious disease applications: perspectives, prospects and promises. *TrAC Trends in Analytical Chemistry*, 86:1–13, 2017.
25. Soudabeh Arsalani, Eder J Guidelli, Matheus A Silveira, Carlos EG Salmon, Jefferson FDF Araujo, Antonio C Bruno, and Oswaldo Baffa. Magnetic fe3o4 nanoparticles coated by natural rubber latex as mri contrast agent. *Journal of Magnetism and Magnetic Materials*, 475:458–464, 2019.
26. Xiaoli Wu, Han Yang, Weitao Yang, Xingmeng Chen, Junxiao Gao, Xiaoqun Gong, Hanjie Wang, Yue Duan, Daohe Wei, and Jin Chang. Nanoparticle-based diagnostic and therapeutic systems for brain tumors. *Journal of Materials Chemistry B*, 7(31):4734–4750, 2019.
27. Seyed Mohammadali Dadfar, Karolin Roemhild, Natascha I Drude, Saskia von Stillfried, Ruth Knüchel, Fabian Kiessling, and Twan Lammers. Iron oxide nanoparticles: diagnostic, therapeutic and theranostic applications. *Advanced Drug Delivery Reviews*, 138:302–325, 2019.
28. Roberto Gonzalez-Rodriguez, Elizabeth Campbell, and Anton Naumov. Multifunctional graphene oxide/iron oxide nanoparticles for magnetic targeted drug delivery dual magnetic resonance/fluorescence imaging and cancer sensing. *PLoS One*, 14(6):e0217072, 2019.
29. Bo Zheng, Minghua Xue, Xinyi Zhang, Ning Tian, and Dongmei Wang. Breast cancer diagnosed by mri using mesoporous tio2-coated (fe3o4) nanoparticles. *Journal of Nanoscience and Nanotechnology*, 20(10):6561–6567, 2020.
30. Samira Sadeghi and Mahdi Rahaie. Design and fabrication of a dna-copper nanocluster-based biosensor for multiple detections of circulating mirnas in early screening of breast cancer. *Journal of Fluorescence*, 32(6):2297–2307, 2022.
31. Ali Hossein Rezayan, Majid Mousavi, Somayyeh Kheirjou, Ghasem Amoabediny, Mehdi Shafiee Ardestani, and Javad Mohammadnejad. Monodisperse magnetite (fe3o4) nanoparticles modified with water soluble polymers for the diagnosis of breast cancer by mri method. *Journal of Magnetism and Magnetic Materials*, 420:210–217, 2016.
32. Jitu Halder, Deepak Pradhan, Prativa Biswasroy, Vineet Kumar Rai, Biswakanth Kar, Goutam Ghosh, and Goutam Rath. Trends in iron oxide nanoparticles: a nano-platform for theranostic application in breast cancer. *Journal of Drug Targeting*, 30(10):1055–1075, 2022.
33. Marc-André Weber, Frederik Lars Giesel, and Bram Stieltjes. Mri for identification of progression in brain tumors: from morphology to function. *Expert Review of Neurotherapeutics*, 8(10):1507–1525, 2008.
34. Massoud Amanlou, Seyed Davar Siadat, Dariush Norouzian, Seyed Esmaeil Sadat Ebrahimi, Mohammad Reza Aghasadeghi, Masoud Ghorbani, Mohammad Shafiee Alavidjeh, Davoud Nouri Inanlou, Ali Jabbari Arabzadeh, and Mehdi Shafiee Ardestani. Magnetic resonance contrast media sensing in vivo molecular imaging agents: an overview. *Current Radiopharmaceuticals*, 4(1):31–43, 2011.
35. Pallabita Chowdhury, Allison M Roberts, Sheema Khan, Bilal B Hafeez, Subhash C Chauhan, Meena Jaggi, and Murali M Yallapu. Magnetic nanoformulations for prostate cancer. *Drug Discovery Today*, 22(8):1233–1241, 2017.
36. Ragini Singh, Santosh Kumar, Feng-Zhen Liu, Cheng Shuang, Bingyuan Zhang, Rajan Jha, and Brajesh Kumar Kaushik. Etched multicore fiber sensor using copper oxide and gold nanoparticles decorated graphene oxide structure for cancer cells detection. *Biosensors and Bioelectronics*, 168:112557, 2020.
37. Despina P Kalogianni, Panagiota M Kalligosfyri, Iraklis K Kyriakou, and Theodore K Christopoulos. Advances in microrna analysis. *Analytical and Bioanalytical Chemistry*, 410:695–713, 2018.
38. Mohadese Mahdie, Mahdi Rahaie, and Fateme Shateri. A colorimetric biosensor based on peroxidase-like activity of cuo nanoparticles for simultaneous detection of micrornas. *Chemical Papers*, 76(12):7305–7312, 2022.
39. Josie Hayes, Pier Paolo Peruzzi, and Sean Lawler. Micrornas in cancer: biomarkers, functions and therapy. *Trends in Molecular Medicine*, 20(8):460–469, 2014.
40. Maliana El Aamri, Ghita Yammouri, Hasna Mohammadi, Aziz Amine, and Hafsa Korri-Youssoufi. Electrochemical biosensors for detection of microrna as a cancer biomarker: pros and cons. *Biosensors*, 10(11):186, 2020.
41. Kamini Sundarbose, Reena V Kartha, and Subbaya Subramanian. Micrornas as biomarkers in cancer. *Diagnostics*, 3(1):84–104, 2013.
42. Nishant Verma and Nikhil Kumar. Synthesis and biomedical applications of copper oxide nanoparticles: an expanding horizon. *ACS Biomaterials Science & Engineering*, 5(3):1170–1188, 2019.
43. Arokia Vijaya Anand Mariadoss, Kandasamy Saravanakumar, Anbazhagan Sathiyaseelan, Karthikkumar Venkatachalam, and Myeong-Hyeon Wang. Folic acid functionalized starch encapsulated green synthesized

copper oxide nanoparticles for targeted drug delivery in breast cancer therapy. *International Journal of Biological Macromolecules*, 164:2073–2084, 2020.
44. Peng Miao, Tian Zhang, Jianhua Xu, and Yuguo Tang. Electrochemical detection of mirna combining t7 exonuclease-assisted cascade signal amplification and dna-templated copper nanoparticles. *Analytical Chemistry*, 90.
45. Fatemeh Hakimian and Hedayatollah Ghourchian. Ultrasensitive electrochemical biosensor for detection of microrna-155 as a breast cancer risk factor. *Analytica Chimica Acta*, 1136:1–8, 2020.
46. Mahdi Norouzi, Saman Yasamineh, Maryam Montazeri, Mehdi Dadashpour, Roghayeh Sheervalilou, Mozhgan Abasi, and Younes PilehvarSoltanahmadi. Recent advances on nanomaterials-based fluorimetric approaches for micrornas detection. *Materials Science and Engineering: C*, 104:110007, 2019.
47. Hasna Mohammadi, Ghita Yammouri, and Aziz Amine. Current advances in electrochemical genosensors for detecting microrna cancer markers. *Current Opinion in Electrochemistry*, 16:96–105, 2019.
48. Ziyue Wang, Jinren Liu, Zihao Zhou, Junli Yu, Liguo Hao, Shunxin Hu, Zhongzhong Lu, Li Wang, and Yan Ding. Cell membrane-coated hollow manganese dioxide drug delivery nanosystem for targeted liver cancer diagnosis and treatment. *ACS Applied Nano Materials*, 7(16):18262–18272, 2023.
49. Binbin Ding, Pan Zheng, Ping'an Ma, and Jun Lin. Manganese oxide nanomaterials: synthesis, properties, and theranostic applications. *Advanced Materials*, 32(10):1905823, 2020.
50. Jeffrey K Mills and David Needham. Targeted drug delivery. *Expert Opinion on Therapeutic Patents*, 9(11):1499–1513, 1999.
51. Il Keun Kwon, Sang Cheon Lee, Bumsoo Han, and Kinam Park. Analysis on the current status of targeted drug delivery to tumors. *Journal of Controlled Release*, 164(2):108–114, 2012.
52. Gupta Manish and Sharma Vimukta. Targeted drug delivery system: a review. *Research Journal of Chemical Sciences*, 1(2):135–138, 2011.
53. Madduri Srinivasarao and Philip S Low. Ligand-targeted drug delivery. *Chemical Reviews*, 117(19):12133–12164, 2017.
54. Madduri Srinivasarao, Chris V Galliford, and Philip S Low. Principles in the design of ligand-targeted cancer therapeutics and imaging agents. *Nature Reviews Drug Discovery*, 14(3):203–219, 2015.
55. Simon D Anderson, Vanessa V Gwenin, and Christopher D Gwenin. Magnetic functionalized nanoparticles for biomedical, drug delivery and imaging applications. *Nanoscale Research Letters*, 14:1–16, 2019.
56. Ki Hyun Bae, Kyuri Lee, Chunsoo Kim, and Tae Gwan Park. Surface functionalized hollow manganese oxide nanoparticles for cancer targeted sirna delivery and magnetic resonance imaging. *Biomaterials*, 32(1):176–184, 2011.
57. Cheong A Choi, Jung Eun Lee, Zihnil Adha Islamy Mazrad, Insik In, Ji Hoon Jeong, and Sung Young Park. Redox-and ph-responsive fluorescent carbon nanoparticles-mno2-based fret system for tumor-targeted drug delivery in vivo and in vitro. *Journal of Industrial and Engineering Chemistry*, 63:208–219, 2018.
58. Zuzanna Sobańska, Joanna Roszak, Kornelia Kowalczyk, and Maciej Stepnik. Applications and biological activity of nanoparticles of manganese and manganese oxides in in vitro and in vivo models. *Nanomaterials*, 11(5):1084, 2021.
59. Xiaoxia Cai, Qingxia Zhu, Yun Zeng, Qi Zeng, Xueli Chen, and Yonghua Zhan. Manganese oxide nanoparticles as mri contrast agents in tumor multimodal imaging and therapy. *International Journal of Nanomedicine*, 8321–8344, 2019.
60. Manli Song, Ting Liu, Changrong Shi, Xiangzhong Zhang, and Xiaoyuan Chen. Bioconjugated manganese dioxide nanoparticles enhance chemotherapy response by priming tumor-associated macrophages toward m1-like phenotype and attenuating tumor hypoxia. *ACS Nano*, 10(1):633–647, 2016.
61. Michele Mauri, Veronica Collico, Lucia Morelli, Pradip Das, Isabel Garcia, Jesus Penaranda Avila, Michela Bellini, Rany Rotem, Marta Truffi, Fabio Corsi, et al. Mno nanoparticles embedded in functional polymers as t 1 contrast agents for magnetic resonance imaging. *ACS Applied Nano Materials*, 3(4):3787–3797, 2020.

# 7 Synthesis and Biomedical Applications of Biofunctionalized Metal Nanoparticles and Metal Oxide Nanoparticles

*Divya Jose*[1], *Preethy Treesa Paul*[2], *Baiju Koonathan Pappachan*[1,3], *Anjaly Jose*[4,5], *Nimya Ann Mathews*,[1] and *Jiya Jose*[6]

[1]Post Graduate, Department of Chemistry and Research Centre, Bharata Mata College (Autonomous), Thrikkakara, Ernakulam, Kerala, India

[2]Research Scholar, Biochemistry and Nutrition Division, ICAR-Central Institute of Fisheries Technology, Cochin, Kerala, India

[3]Post Graduate and Research Department of Chemistry, Maharaja's College, Ernakulam, Kerala, India

[4]Department of Physics, Nirmala College (Autonomous), Muvattupuzha, Ernakulam, Kerala, India

[5]Department of Physics, Newman College, Thodupuzha P.O., Ernakulam, Kerala, India

[6]Division of Microbiology, Department of Biosciences, Rajagiri College of Social Sciences (Autonomous), Kalamassery, Ernakulam, Kerala, India

## 7.1 INTRODUCTION

Metal nanoparticles (NPs) are metallic elements with dimensions less than 100 nm. The synthesis of inorganic NPs, including metal and metal oxide NPs, can be categorized into two main types: synthesis in a hydrophobic environment and direct formulation in aqueous media, yielding hydrophobic and hydrophilic NPs. The properties of metal NPs are influenced by their composition, size, shape, and structure. Advanced synthesis strategies are essential for customizing metal NPs optical, catalytic, electronic, and magnetic properties (1). Metal oxides are particularly intriguing due to their varied structural properties, which differ significantly from those of pure metals. Transition metals can form a wide array of oxides with distinct properties. These metal oxides have extensive applications in various technologies, aiding in developing magnetic ferrites, ferroelectric oxides, superconductors, ionic conductors, phosphors, and photocatalysts. Several synthetic methods have been developed to control metal oxides' composition, structure, and properties, including coprecipitation, thermal decomposition, reduction, micelle synthesis, and hydrothermal synthesis (2). Each method provides a unique approach to tailoring the characteristics of metal oxides, offering researchers and engineers flexible tools for designing materials with specific functionalities.

The solubility of hydrophobic NPs in water is often limited, which impedes their further functionalization (3). Addressing these challenges is essential for broadening the applications of NPs, especially in targeted drug delivery and biomedical research. Functionalizing NPs enhances colloidal stability, water solubility, and versatility (4). For metal and metal oxide NPs in biomedical applications, achieving stability, biocompatibility, and functionality necessitates effective surface functionalization. Various techniques have been developed for this purpose, including coating, conjugation strategies, surface encapsulation, in situ synthesis, self-assembly, and core-shell NP creation (5). Functionalization materials include silica, synthetic polymers, biopolymers, dendrimers, and small molecules. The literature extensively documents established protocols for functionalizing various NPs, such as gold, quantum dots, polymeric, and magnetic NPs (6, 7).

In addition to the well-known coating strategies using polymers and silica, various other methods have been developed to coat or functionalize metal oxide NPs. These methods offer diverse ways to modify and improve NP properties for specific uses. Bioconjugation is another common technique for functionalizing metal oxide NPs, especially in biomedical applications. This process involves attaching biomolecules to the surfaces of NPs, enabling specific interactions within biological systems (8). Bioconjugation often

DOI: 10.1201/9781032660639-7

involves covalent binding with primary amines, carboxylic acids, and thiols. It is commonly used with NPs such as quantum dots (QDs), gold nanoparticles (AuNPs), up conversion nanoparticles (UCNPs), and magnetic nanoparticles (MNPs) (9). For example, NPs functionalized with carboxyl groups through bioconjugation can interact more specifically with biomolecules containing amine groups, such as avidin, peptides, antibodies, and nucleic acids. AuNPs, in particular, show high efficiency in conjugation with thiol groups or thiol-containing molecules (10). Bioconjugation is especially important in drug delivery and cancer therapy, as conjugating NPs with targeting molecules such as antibodies, aptamers, and folate can enhance their ability to target cancer cells, thereby improving the effectiveness of drug delivery and diagnostic applications (11, 12).

## 7.2 SYNTHESIS OF BIOFUNCTIONALIZED METAL NANOPARTICLES

Metallic NPs have been extensively studied for various biological purposes due to their impressive inertness and nanoscale structures, which are comparable in size to many biological molecules. These properties make them highly attractive for biomedical applications. Their intrinsic electronic, optical, physicochemical, and surface plasmon resonance characteristics have enabled various uses in biomedicine. These characteristics can be modified by altering specific particle attributes, such as size, shape, environment, aspect ratio, synthesis ease, and functionalization capabilities. Applications include photothermal and photodynamic therapy, sensing, imaging, and targeted drug delivery (13). For biomedical applications, it is crucial that NPs exhibit predictable and controllable effects to achieve the desired outcomes with minimal harm. Tailoring the ligand shell allows for stabilization, targeted identification of biological species, and specific targeting. Silver NPs, for instance, are used in medical imaging and diagnosis due to their optoelectronic properties, particularly surface plasmon resonance (SPR). The size-dependent localized SPR enhances surface-enhanced Raman signals, enabling the detection of individual molecules. This has led to significant interest in plasmonic sensor-based biomolecular analyte detection. Surface-enhanced Raman spectroscopy (SERS) provides high signal intensities primarily due to the plasmonic properties of silver, making it effective for detecting low analyte concentrations in biosensing applications (14).

Silver nanoparticles (AgNPs) exhibit antibacterial properties against various viruses, fungi, and Gram-positive and Gram-negative bacteria. Their effectiveness as a biocidal agent is attributed to their ability to oxidize and release $Ag+$ ions, which are associated with their antibacterial characteristics (15). Additionally, their small size allows them to penetrate cell membranes, impacting intracellular activities. The extensive surface area of these NPs enhances their environmental interaction, contributing to their strong antibacterial effects. AgNPs disrupt cell membranes and hinder proper budding, demonstrating significant antifungal activity. Due to their small size, nanocrystalline silver dressings are effective in wound healing and are used in commercial antimicrobial dressings (16).

Similarly, AuNPs have gained attention due to their ability to interact with light through SPR (17). Their diverse applications in treatments, detection, imaging, and surface modification have led to their extensive use in chemistry, material sciences, physics, medicine, and life sciences (18). Gold and other metallic NPs are frequently employed for their versatile roles in cancer treatment. They are utilized in the biomedical field for sensitive biomolecular screening, photothermal therapy-assisted cancer cell destruction, protein labelling, and cellular drug delivery.

## 7.3 BIOFUNCTIONALIZED METAL NANOPARTICLES

Functionalizing metal NPs with biomolecules is crucial for developing biocompatible platforms with low toxicity for various biopharmaceutical applications. The preservation of the biological function of biomolecules conjugated to NPs depends significantly on the surface chemistry of the NPs. Surface modification, or "biofunctionalization", involves attaching suitable organic groups to the NPs. This process not only promotes self-organization but also stabilizes the NPs against agglomeration. Surface-modified AgNPs have improved biocompatibility and intracellular absorption for drug delivery. The high stability, excellent biocompatibility, and strong affinity for biomolecules make biomolecule-functionalized NPs highly attractive for clinical diagnostics and drug delivery. Controlled functionalization of NPs with biomolecules ensures that the bulk mechanical properties remain unchanged while altering the surface composition, structure, and morphology. Additionally, surface modification is essential as the surface layer enhances the long-term stability of the particles by preventing agglomeration and reducing surface energy (19).

The bioconjugation or biofunctionalization process can be viewed as the "natural" advancement of previously discussed concepts of ligand exchange and chemical biomolecule functionalization. Besides potentially biocompatible polymers like polyethylene glycol (PEG), metal NPs can be functionalized with various functional groups such as peptides, antibodies, RNA, and DNA for targeting specific cells. In biological treatments involving active targeting, NPs are attached to targeting agents known as ligands, which selectively bind to overexpressed receptors on target cells. Many metal NPs, particularly silver and gold, are produced through chemical reduction and biosynthetic methods. This involves using different reducing or antioxidant agents in plant/leaf extracts or other compounds to reduce silver and gold salts (20, 21). These biosynthesized metal NPs maintain their NP properties even with multiple functional groups on their surface. The modified surfaces of metal NPs can feature a variety of functional groups. These NPs, equipped with multiple functional groups, can be conjugated to the surface of biomolecules with different functional groups. The primary chemical modifications for many metal NPs include thiols, amines, nitriles, and carboxylic acids.

There are four primary methods for attaching biomolecules to NPs:

- Ligand-mediated binding of the biomolecule, often through chemisorption, such as thiol groups, to the surface of the inorganic NP core.
- Electrostatic charge exchange between negatively charged NPs and positively charged macromolecules, or the reverse.
- Covalent binding using conjugation chemistry utilizes the functional groups of particles and biomolecules.
- Affinity-based non-covalent receptor-ligand systems.

Thiols are organic sulfur analogues of alcohols or R-CH and can strongly bind to the surfaces of various metal NPs. The thiol group easily chemisorbs onto metal surfaces through two primary methods. In the first method, the thiol group can be adsorbed on the surface of metal nanoparticles (22). In the second method, thiols or disulfides chemisorb onto metal surfaces to form SR groups, enabling the biofunctionalization of metal NPs (23). During chemisorption, the SH bond cleaves as the thiols attach to the metal surface. Functionalizing metal NPs with DNA facilitates target-specific biosensing, bioimaging, and drug delivery. The initial and crucial step in the bioconjugation of NPs involves thiolate chemisorption on silver, gold, and other metals. Significant research has also explored thiolate chemisorption on Ag, Pd, and Pt planar substrates. These studies and comparisons to thiolate-AuNP systems provide a valuable foundation for preparing and applying thiolate-coated Ag, Pt, and Pd NPs.

DNA is an ideal biomolecule for modifying NP surfaces due to its excellent physicochemical stability and robust mechanical rigidity of its double helices. The interactions between metal ions and DNA or nucleotides are very important and have been thoroughly researched. Metal ions can interact with DNA covalently or noncovalently, acting as counterions. DNA-based probes are especially attractive due to their non-toxicity and biocompatibility. Metal NPs can be biofunctionalized with DNA through covalent conjugation and electrostatic interactions. The DNA functionalization with gold NPs via thiol-Au coupling is essentially used for biosensing applications (24). The conjugation of thiolated DNA to gold NPs is possible using a "salt aging" method by adding NaCl (25). Researchers have reported the attachment of DNA to citrate-capped gold and silver NPs (26).

Amines are employed to enhance the bio functionality of metal NPs, which increases their stability. Compared to thiol groups, the interaction between the amino group and metal NPs is weaker. By adsorbing an ammonium group onto their surface, metal NPs can become hydrophilic, a feature important for biomedical uses. Peptides are crucial for identifying antibodies, developing new medications and vaccines, and diagnosing infectious diseases. They are also used to coat NPs, enhancing their stability, biocompatibility, and targeting accuracy. Various methods exist for creating peptide-functionalized gold NPs, such as ligand exchange, chemical reduction, and chemical conjugation. For instance, Levy et al. (27) developed citrate-stabilized gold NPs with peptide coatings, where the thiol group of a cysteine-containing peptide covalently bonds with the gold NPs, while the amino group of the peptide aids in the conjugation to the metal surface. Peptide-gold NP complexes have been synthesized employing two methods. The first method involves the spontaneous attachment of a peptide with biological activity to the AuNP surface through a thiol group. The second method entails the activation and functionalization of a peptide after a functionalized NP has been capped with a linker molecule. This linker, which contains a functional group like a carboxyl group, has a thiol that facilitates binding to the gold surface (28).

Carboxylate groups can be introduced to NPs to modify their surface properties. Carboxylate ions can strongly adhere to NP surfaces, which can then be conjugated to functional groups using techniques like chemisorption or adsorption. As the size of NPs decreases, adsorption tends to diminish while the extent of chemisorption increases. Antibodies, polypeptides categorized into heavy and light chains, play a critical role in the immune system. Targeting ligands with antibodies is key for biofunctionalization. The IgG1 molecule features carboxyl (-COOH) groups on the side chains of aspartic or glutamic acid, along with amine (-$NH_2$) groups, particularly the ε-amino group of lysine and the α-amino group of polypeptide chains. Antibodies or their fragments can functionalize NPs via adsorption, covalent attachment, or adapter molecules (29). Effective conjugation should ensure the correct orientation and quantity of biomolecules on each NP while immobilizing the antibodies. In physical adsorption, weak interactions such as electrostatic forces, hydrogen bonding, hydrophobic interactions, and van der Waals forces hold the antibody to the NP's surface. The ionic binding arises from surface interactions with opposite charges on antibodies and NPs (30).

Biofunctionalized AgNPs can also be prepared through electrostatic interactions between biomolecules and NPs. This approach allows for the detection of biofunctionalized NPs. For instance, mercaptosuccinic acid (MSA)-functionalized AgNPs have been employed to attach anti-ngn1 antibodies, resulting in a stable Ag-antibody conjugate (31). This attachment is facilitated by electrostatic interactions and hydrogen bonding between the anionic carboxyl groups (–COO) on the NP surface and the positively charged amine groups (–NH3+) on the antibody residues. In physisorption, ligands adhere to the NP surface through non-covalent interactions such as hydrogen bonding, electrostatic forces, and hydrophobic interactions. AgNPs are stabilized with bovine serum albumin (BSA), a major circulatory protein, through chemisorption. Stabilizing agents like citrate or thiols are typically used in the wet-chemical synthesis of AgNPs, as they readily bind to exposed atoms on the NP surface. Functionalizing NPs with amino acids or peptides has enhanced the selectivity and effectiveness of NP-based delivery systems (32). Additionally, proteins such as enzymes and antibodies are particularly suited for molecular recognition and surface modification. Enzymes are specialized molecules that catalyze essential biological processes and regulate metabolism, while antibodies (immunoglobulins) specifically bind to antigens through molecular recognition.

## 7.4 COVALENT BIOCONJUGATION TO SURFACE COATINGS

Carbodiimide coupling is one of the simplest and most widely used methods for creating bioconjugates with Au and AgNPs. A common approach involves coating the AuNPs with thioalkyl amines. Protein-NP conjugation has diverse applications, including catalysis, delivery, sensing, imaging, treatment, and protein structure and function regulation. Various techniques exist for attaching proteins to NPs. One such technique is non-specific adsorption, where proteins are incubated with NPs and, if they carry opposite charges, adhere to the NP surface through electrostatic attraction. In these cases, interactions such as van der Waals forces, hydrogen bonds, silver-thiol bonds (from cysteine residues), or hydrophobic forces help hold the molecules onto the particle surface. Numerous mechanisms for NP adhesion, both covalent and non-covalent, have been identified. Non-covalent interactions generally involve physisorption, where ligands bind to the NP surface through forces like hydrophobic interactions, hydrogen bonding, and electrostatic attraction. Steric stabilization is another method for attaching capping agents, such as polymers or surfactants, to NPs via non-covalent means. In these cases, more effective binding ligands can facilitate the direct attachment of biomolecules to the particles through exchange processes.

Csapó et al. (33) reported the preparation of cysteine-functionalized AgNPs. Their study examined how pH levels influence the cross-linking of these NPs, which were stabilized using citrate. Raman spectroscopy revealed Ag-S, Ag-O, and Ag-N bands, indicating that L-cysteine interacts with the AgNPs. At pH 3 and 7, the electrostatic interactions between the deprotonated carboxylate and protonated amino groups on the surface facilitate cross-linking, leading to a predominance of chain-like networks. However, at pH levels below approximately 3, these interactions are disrupted due to the protonation of the carboxylate group. Conversely, at pH values above 7.0, L-cysteine's ammonium groups lose dissociable protons, resulting in negatively charged molecules. This change prevents the formation of electrostatic interactions between carboxylate and amino groups, increases interparticle distances, and maintains a stable dispersion without aggregation. Due to their size similarity to biological molecules, these NPs show considerable potential for biomedical uses, including enhanced biocompatibility and intracellular absorption for drug delivery and other biomedical applications.

## 7.5 BIOMEDICAL APPLICATIONS OF METAL NANOPARTICLES

Nanoparticles, ranging in size from 10 to 1000 nm, are solid colloidal particles with several advantages, including an enhanced surface-to-volume ratio and superior magnetic properties compared to larger particles (34, 35). Their unique characteristics have sparked increased interest in their application across various biomedical fields, such as targeted drug delivery, hyperthermia, photoablation therapy, bioimaging, and biosensing (36). Metal NPs have revolutionized biomedical science, catalysis, sensors, and optoelectronic materials. Their widespread application is due to their physical and electrochemical properties, which arise from their high surface area and small size. The distinctive features of metal NPs, such as their high surface-to-volume ratio, fixed particle mass, and surface reactivity, make them suitable for numerous applications. However, this increased surface-to-volume ratio also enhances their reactivity, which can lead to the generation of free radicals (37, 38). Additionally, their nanoscale size improves their ability to enter cells and interact with tissues. Jeng and Swanson noted that metal-based NPs, through Fenton-type reactions, can generate free radicals that interact with cellular macromolecules, causing oxidative stress (39). Metal-based NPs offer various therapeutic benefits, including antimicrobial (e.g., antiviral), antioxidant, and anti-inflammatory effects. They hold promise as drug carriers for delivering therapeutic agents or as diagnostic tools for non-invasive imaging and antigen detection (40, 41).

NPs are mainly divided into two categories: (1) inorganic, which encompass metals, metal oxides, magnetic NPs, and quantum dots, and (2) organic, which include polymeric NPs, liposomes, carbon-based NPs, and dendrimers (42). The development and fabrication of metallic NPs involve various methods of blending and functionalization. Surface functionalization can be achieved using surfactants, polymers, antibodies, drugs, oligonucleotides, or peptides to improve their use in theragnostic applications and biotechnological drug delivery. Numerous metal NPs, such as platinum (PtNPs), silver, gold, and copper (CuNPs), have been thoroughly studied and optimized to enhance their performance as therapeutic and diagnostic agents (Figure 7.1).

### 7.5.1 Silver Nanoparticles

Silver nanoparticles, ranging in size from 1 to 100 nm, are a subset of metal NPs known for their excellent thermal conductivity and exceptional optical and electrical characteristics (43). In nanomedicine, they hold considerable promise, being used as coatings for medical devices, imaging probes, antimicrobial agents, drug delivery systems, diagnostic

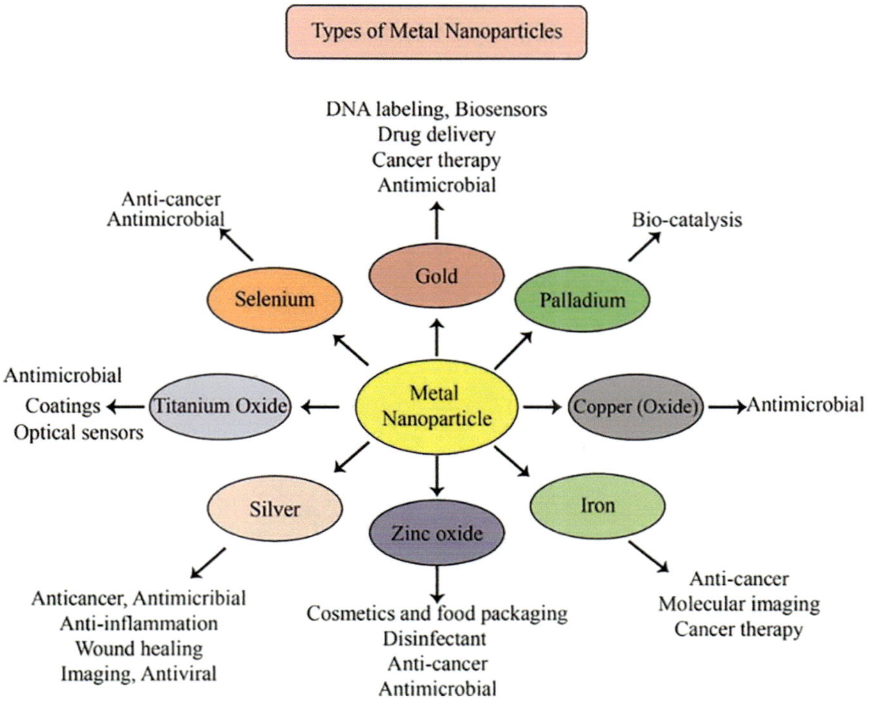


**FIGURE 7.1** Theragnostic applications of metallic nanoparticles (41).

platforms (Figure 7.2), and their applications in optoelectronics (44). Their antimicrobial effectiveness stems from their ability to release Ag+ ions into aqueous environments. Factors such as NP size, capping agents, and shape influence the release of these ions. The interaction of Ag+ ions with thiol groups on bacterial cell surfaces, particularly those containing sulfur-rich proteins, impacts bacterial cell viability by disrupting these proteins within the bacterial membrane (45). Research indicates that AgNPs act as antibiotic inhibitors, demonstrating significant inhibitory effects against wild strains of *K. pneumoniae, S. ebony,* and *E. coli.* They also exhibit a noticeable, though less pronounced, effect on drug-resistant strains of *S. aureus* and inhibit the growth of *K. pneumoniae, S. ebony*, and *E. coli* at concentrations of 4, 2, and 64 μg/ml, respectively (46).

The large surface area of AgNPs and their release of metallic ions enhance their antifungal properties (47) AgNPs notably inhibit biofilm formation by *S. ebony* and suppress the growth of the *S. aureus* strain. The Ag+ ions replace vital metal ions like $Ca^{2+}$ and $Zn^{2+}$ in essential bacterial enzymes and proteins, disrupting cellular respiration and leading to cell death. Additionally, AgNPs can bind to bacterial cell walls and cause membrane damage or increased permeability, contributing to cell death. The Ag+ ions released from AgNPs can interact with DNA phosphorus groups, disrupting DNA replication (45, 48). The cytotoxic effects of AgNPs, which include antibacterial, antifungal, antiviral, and anti-cancer activities, are largely due to their ability to generate reactive oxygen species (ROS) and free radicals, such as hydroxyl radicals, superoxide anions, hydrogen peroxide, singlet oxygen, and hypochlorous acid (49, 50). In their antibacterial action, AgNPs create pores in bacterial cell walls through free radical-mediated mechanisms, leading to cell death. Research has shown that AgNPs have strong inhibitory and fungicidal effects on in vitro fungal colony growth, creating an inhibition zone around the treated area. Moreover, a higher concentration of AgNPs correlates with a reduction in colony formation.

Extracellular polymeric substance (EPS) production is a key characteristic of pathogenic bacteria, as it supports bacterial colonization and cohesion within biofilms. NPs may potentially counteract this adhesive material vital for biofilm formation. For instance, AgNPs have been found to penetrate and accumulate within bacterial cells after binding to the plasma membrane. They then interfere with EPS synthesis by forming complexes with phosphorus and sulfur, inhibiting related enzymes' activity (51). Additionally, AgNPs have been shown to influence gene expression related to biofilm growth and maturation, leading to their investigation for biofilm-related infection prevention and treatment. This suggests their potential for application in coating pharmaceutical products and combating antibiotic-resistant biofilms (52, 53). A promising strategy in biofilm treatment involves combining AgNPs with conventional antibiotics.

AgNPs produced by *B. megaterium* have demonstrated enhanced efficacy against *S. aureus* biofilms, effectively curbing biofilm formation. This indicates their potential in neutralizing adhesive substances and preventing biofilm

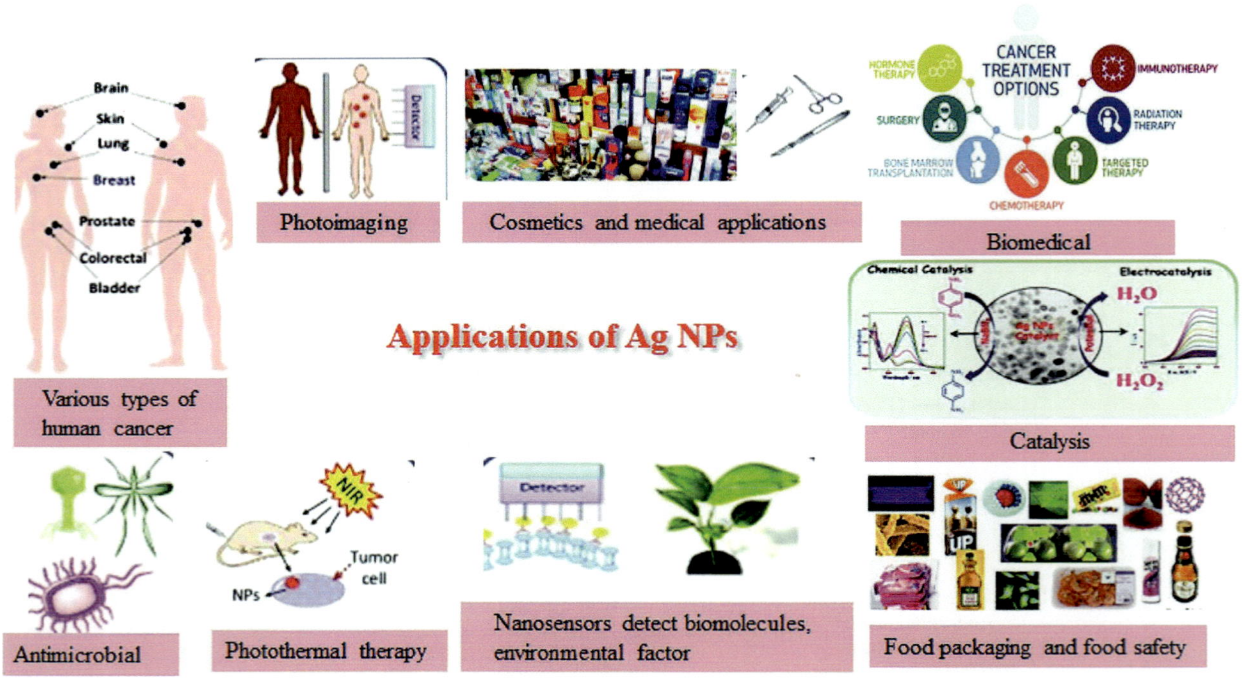


FIGURE 7.2 Applications of silver NPs (56).

development (54). Recent research by da Silva et al. has highlighted the relationship between AgNPs, localized surface plasmon resonance (LSPR), reactive oxygen species, and antimicrobial activity. LSPR occurs when small conductive NPs interact with light at a specific wavelength due to electromagnetic interactions between the metal and light. By utilizing this phenomenon, researchers have enhanced the antimicrobial effectiveness of AgNPs through increased ROS generation when exposed to visible light (55).

Additionally, AgNPs and Ag+ have the potential to trigger the production of ROS, which can adversely impact the cell cytoskeleton and lead to the oxidation of nucleic acids. This process may result in chromosomal abnormalities and cell death (57). This development offers promising opportunities for exploring the cytotoxic effects of AgNPs against cancer cells (58).

### 7.5.2 Gold Nanoparticles

Gold (Au) emerges as a noteworthy noble metal with significant potential and promising applications in the medical field, as discussed by Wang et al. (59). Gold nanoparticles have revealed anti-cancer properties employing diverse mechanisms, including drug delivery, anti-angiogenic effects, and photothermal and photodynamic effects, as Bhattacharyya et al. highlighted (60). The photothermal abilities of AuNPs stem from the combined impacts of locally boosted laser radioactivity absorption at near-infrared frequencies through localized surface plasmon resonance. This results in hyperthermia within cancer tissue. Yang et al. elaborate on this groundbreaking method, signifying a shift from traditional laser hyperthermia in tissues (61).

Brazzale and colleagues introduced a compelling technique wherein targeted gold AuNPs could be activated for cancer cell eradication (62). This innovative *in vitro* study delves into using targeted AuNPs as sonosensitizers, presenting a novel approach to combating cancer. The strategy involves employing a non-toxic molecule to induce oxidative damage by generating ROS, leading to the demise of cancerous cells. Tachibana et al. and Mišík and Riesz discussed this approach (63, 64). The authors suggest that an incidence of 1.866 MHz, with a whole exposure time of 5 minutes facilitated by the physical phenomenon of sonoluminescence (65), could harness the plasmonic effect of AuNPs resulting from LSPR (63). This process converts photon energy to heat, causing cell damage through ROS synthesis. However, an HCT-116 and KB cell line analysis revealed that cells raised with AuNPs showed no ROS production growth. Combining 14 nm AuNPs with green laser light can be utilized to irradiate hyperthermia via LSPR. However, the effectiveness of this method in cancer treatment is hindered by the limited penetration of light through tissue, as highlighted by Ash et al. and Mendes et al. (66, 67). Au-based nanotherapeutics absorb radiofrequency and generate heat. This provides a promising approach for addressing intensely localized tumors through strategies involving Au and hyperthermia, as discussed by Abadeer and Murphy (68).

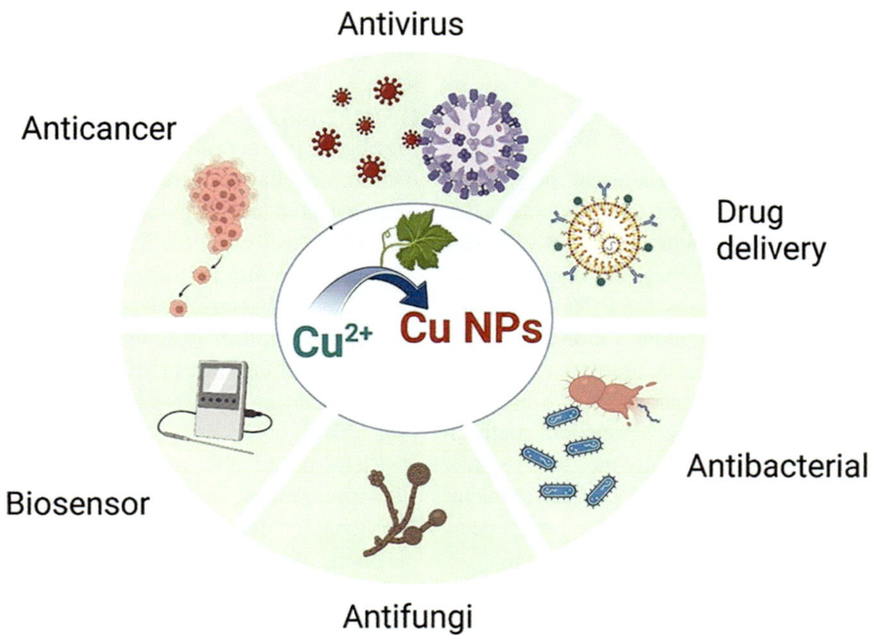


**FIGURE 7.3** Biomedical applications of biologically synthesized CuNPs (72).

### 7.5.3 Copper Nanoparticles

Copper nanoparticles exhibit diverse applications (Figure 7.3), encompassing numerous therapeutic and innovative attributes. Copper transporters have a crucial role in maintaining copper homeostasis inside the body. These particles find utility in various fields, such as electronics, biomedicine, agriculture, photocatalysis, and thermodynamics. The increasing focus on CuNPs is attributed to their cost-effectiveness and notable antibacterial and antifungal properties. Notably, CuNPs demonstrate heightened efficacy against Gram-positive bacteria, primarily due to their interaction with bacterial membranes, leading to intracellular signaling cascades that induce oxidative stress. Additionally, $Cu^{2+}$ ions prompt disturbances in DNA helical structure by eventually combining with it (69, 70). In specific scenarios, CuNPs generate ROS, which degrades chromosomal DNA. As emphasized by Sistemática et al., this underscores the release of metallic ions (71). Previous research suggests that CuNPs display significant repressing and antimycotic effects on *in vitro* colonization. They impede fungal growth, creating an inhibition zone from place to place in the central cavity. A reciprocal relationship was also noted, where increased NP concentration led to a drop in colony development.

CuNPs are stored to facilitate the controlled release of copper particles, thereby mitigating their toxicity (73). They have been created to be incorporated into active bandages, bedding, wound dressings, and medical fabrics. When applied to these surfaces, CuNPs led to a 90% reduction in *E. coli* compared to pristine reverse osmosis membranes. The quantity of CuNPs decreases as the reaction of these NPs progresses (74). Copper sulfide NPs exhibit absorbance in NIR-II light, making them appealing for biomedical applications. Their absorbance under NIR light is advantageous, as the maximum light induces only minimal temperature increase, causing minimal harm to normal cells. This property positions them for potential use in biomedicine. The potential of CuNPs to interfere with "contact-mediated killing" against different pathogenic fungi is evident through their effects on intracellular protein expression and bacterial denitrification. Cellular copper ion homeostasis mechanisms regulate the destruction of pathogenic microorganisms upon contact (75). Remarkably, copper surfaces have demonstrated the ability to act as a self-sanitizing substance, as evidenced by the absence of viable microorganisms recovered upon exposure (76).

CuNPs have displayed effective antibacterial properties against bacterial strains, with notable effectiveness even at higher concentrations of 0.15 and 0.25 μg/ml for *K. pneumoniae* and *S. aureus*, respectively (46). The susceptibility of bacteria to NP-induced biofilm formation implies that several signaling pathways are involved in both cell survival and the production of biofilms. The development of EPS, crucial for bacterial cell colonization and cell cohesion in biofilm formation, is a significant characteristic of pathogenic bacteria. NPs have the potential to neutralize this adhesive material essential for biofilm formation. CuNPs, through the release of $Cu^{2+}$ ions and the induction of significant damage via communication-mediated mechanisms, exhibited a promising capability to combat metal-tolerant biofilm bacteria (77, 78). A comprehensive examination of biofilm-specific gene expression indicates that NPs down-regulate genes associated with DNA replication, biofilm matrix formation, membrane lipoprotein synthesis, and motility. Ssekatawa et al. stated NPs regulate metal tolerance and biofilm dispersion genes (79).

### 7.5.4 Platinum Nanoparticles

PtNPs showcase exceptional physicochemical properties and hold significant promise in biomedical applications. To overcome the constraints connected with conservative physical and chemical synthesis approaches, environmentally friendly and cost-effective approaches for PtNP synthesis have been developed. Utilizing biogenic sources such as plant extracts, algae, fungi, and bacteria, along with biological derivatives like honey, bovine serum albumin protein, sheep milk, and egg yolk, has become a prevalent method for synthesizing PtNPs. These eco-friendly methods reduce the required reaction time and energy and provide ambient conditions for fabrication. PtNPs exhibit notable anti-cancer activity against cervical (HeLa) cancer cells ($IC_{50}$ – 19.83 μg/mL) and human breast (MDA-MB-231) ($IC_{50}$ – 36.86 μg/mL), respectively (80, 81).

Furthermore, these NPs exhibit remarkable bactericidal action against both Gram-negative (100 μg/mL) and Gram-positive bacteria (500 μg/mL) (80). Specifically, PtNPs display potent cytotoxicity against cervical human cancer cells (SiHa), with $IC_{50}$ and $IC_{75}$ values of 18.34 μg/mL and 11.4 μg/mL, respectively, triggering apoptosis (82). PtNPs synthesized through biological methods demonstrate important cytotoxicity counter to MCF-7 cancer cells, with an $IC_{50}$ of 17.84 μg/mL subsequently 48 hours of incubation (83). Furthermore, these NPs show notable antioxidant and neuroprotective activities, indicating their potential in treating Parkinson's disease (84). Additionally, PtNPs exhibit anti-cancer properties against human cervical (HeLa) cancer cells, displaying practical free radical scavenging abilities against superoxide, 2,2-diphenyl-1-picrylhydrazyl, nitric oxide, and hydroxyl radicals (85).

Employing PtNPs for treating human colon cancer cell lines (SW620 and SW480) and low-intensity exposure to 650 and 808 nm lasers for 5 minutes resulted in substantial cytotoxicity. The percentage of viable cancer cells decreased to 18% and 21% for SW620 and SW480 under a 650-nm laser and 25% and 22% under an 808-nm laser (86). In cancer cell lines (A549), PtNPs exhibited supreme cell death of 8.8% after 72 hours at a concentration of 100 μg/mL (87). Moreover, the cytotoxic activity against A549 cancer cell lines over 24 hours showed diminished cell viability, altered cellular morphology, and a dose-dependent decrease in cell number. Importantly, these PtNPs demonstrated biocompatibility with normal cells, suggesting their possibility for solid and selective anti-cancer drug economic progress (88).

PtNPs displayed notable anti-cancer activity compared to the colon cancer cell line (HCT116), showing a significant reduction in tumor cell feasibility with an $IC_{50}$ value of 20 μg/mL (89). PtNPs synthesized through environmentally friendly methods were evaluated for their bactericidal efficacy against Gram-negative bacteria (*Pseudomonas aeruginosa*) and Gram-positive bacteria (*Bacillus subtilis*). The outcomes indicated substantial antibacterial effectiveness, with zones of inhibition measuring *P. aeruginosa* (15±0.5 mm) and *B. subtilis* (18±0.8 mm), respectively. This highlights the potential of PtNPs as antibiotics capable of overcoming bacterial resistance (90). Additionally, PtNPs demonstrated antifungal activity against *Cladosporium fulvum* (8 μg/well) and *Colletotrichum acutatum* (4 μg/well), measuring 15 mm and 18 mm, respectively (91). These NPs also showed bactericidal action against *E. coli, L. lactis*, and *K. pneumoniae* (92), anti-cancer effects against human breast cancer cell lines (MCF-7) in various studies (93–98) and anti-cancer activity against A549 lung carcinoma cells (99, 100).

Metal-based NPs are in high demand for their versatile applications in healthcare, cosmetics, and industry. This chapter delves into the emerging biomedical uses of platinum nanoparticles, gold nanoparticles, silver nanoparticles, copper nanoparticles, and various therapeutic fields. Remarkably, diverse physical actuators can trigger these metal NPs to produce reactive oxygen species, enabling targeted elimination of cancer cells and bacteria. The crucial part of these NPs as activated agents for therapeutic ROS generation is underscored, especially under light or ultrasound exposure. While promising, the potential of metal NPs requires a thorough evaluation before clinical implementation. Consequently, forthcoming investigations on metal NPs must prioritize pre-clinical solid studies, focusing on expediting their scientific translation for diverse biomedical applications. Incorporating bioactive compounds in these metal NPs enhances their efficacy as antibacterial, anti-cancer, and antifungal agents.

## 7.6 SYNTHESIS OF BIOFUNCTIONALIZED METAL OXIDE NANOPARTICLES

Developing nanomaterials for biomedical applications is very important in this era due to the different properties of nanomaterials, such as optical, magnetic, electronic, and electrical properties (101). Different nanomaterials are made, which depends on their synthesis methods and surface functionalization with the desired properties. During synthesis, their physical and chemical features determine the morphology, size, and structural properties (102). Because of these features, NPs, especially metal oxide and metal NPs, are used in biomedical applications. Metal oxide particles such as $TiO_2$, ZnO, CuO, $Fe_2O_3$, and metal oxides doped with different elements have various applications in photocatalytic, water treatment, semiconductor devices, and magnetic properties. To utilize these metal oxides for biomedical applications, their physico-chemical behavior must be modified by functionalization to make them biocompatible, with less enzyme activity, and increase their distribution in the biological substrate (103). Here, we mainly discuss the synthesis and biomedical applications of zinc oxide, titanium oxide, copper oxide, and ferric oxide NPs.

### 7.6.1 Zinc Oxide Nanoparticles

Zinc oxide NPs have garnered significant interest because of their 60meV and wide band gap of 3.3eV, enabling them for various applications such as optoelectronic devices, gas sensors, pharmaceutical industries, and cosmetics. Their nontoxic and biocompatible nature enables them to be used as content in sunscreen cosmetics and medical applications. To synthesize NPs of zinc oxide to be useful for biomedical applications requires an environmentally beneficial approach, such as using plant extracts, bacteria, fungi, algae, and biopolymers, making the ZnO NPs biocompatible for medical applications (Figure 7.4). Sangeetha et al. utilized leaf extract from *Aloe barbadensis miller* to create ZnO NPs using chemical and biological synthetic routes (104). The variation in concentrations of zinc nitrate and *aloe vera* leaf extract resulted in ZnO NPs with an average size of 25–40 nm. The variation in leaf extract concentration was essential in size reduction (104, 105).

Fakhari et al. compared the role of *Laurus nobilis* leaf extract in synthesizing AgNPs and zinc oxide nanoparticles using metal ion precursors such as zinc acetate and zinc nitrate. Phytochemicals such as alcohols, polyphenols, and amines present in the leaf extract enable it to be an excellent stabilizing and capping agent for synthesizing ZnO NPs. The average size of ZnO NPs was 21.49 and 25.26 nm with bullet and flower-like structures when zinc acetate and zinc nitrate were used as the precursors, respectively. Leaf extract assisted the slow nucleation of metal ions in zinc acetate, while the same extract enabled zinc nitrate to result in a flower-like appearance (106). In a previous study, researchers utilized water extract of *Trifolium pretense* to synthesize ZnO NPs, resulting in a hexagonal and crystalline structure with an average size range of 60–70 nm (107). Selim et al. used an aqueous extract of aerial parts of the *Deverra tortuosa*, resulting in ZnO NPs with cytotoxic activity (108). The extract is able to produce ZnO NPs because it contains phytochemicals such as flavonoids, terpenoids, alkaloids, and other compounds. The extract's ability to function as a reductant and stabilizer is due to these phytochemicals present in it. The color change of zinc nitrate to ZnO NPs is attributed to excitations due to vibrations of surface plasmon resonance.

Recently, algae have influenced metal ions to convert them into metal or metal oxide NPs. *Sargassum muticum*, an alga, is the capping agent for synthesizing zinc oxide NPs. The presence of surface moieties and hydroxyl groups in algae polysaccharides is responsible for their capping ability. This results in a hexagonal wurtzite structure of

**FIGURE 7.4** Some of the plants used for the synthesis of ZnO-NPs are (a) *Cassia auriculata*, (b) *Parthenium hysterophorous*, (c) *Aloe vera*, (d) *Acalypha indica*, (e) *Calotropis gigantean*, and (f) *Abrus precatorius* (105).

ZnO NPs with 30–75-nm particle size, with some agglomeration. Azizi et al., using *Sargassum muticum* algal extract, synthesized ZnO NPs with an average particle size of 42 nm and a hexagonal wurtzite structure. The algal extract contains phytochemicals such as sulfated polysaccharides responsible for NP formation. Also, FTIR spectra of algal extract, ZnO NPs with extract content, and pure ZnO NPs formed after annealing at 450°C reveal that the presence of NH=CO bond plays a vital role in stabilizing NPs. (109, 110). Jayaseelan et al. reported *Aeromonas hydrophila*, a bacteria utilized for the biosynthesis of ZnO NPs and as-synthesized ZnO NPs, exhibiting antimicrobial activity (111). Boroumand Moghaddam et al. reported synthesizing ZnO NPs using the *Pichia kudriavzevii* yeast strain (112). Here, fungal cell-free filtrate helps form ZnO NPs when the zinc acetate precursor is treated with filtrate and after shaking incubation at 35°C with different durations. The color of the solution changed from light brown to pale and finally white, signifying the development of NPs. Similar reports were obtained using different algal extracts in the synthesis of zinc oxide nanoparticles (113).

### 7.6.2 Titanium Oxide Nanoparticles

$TiO_2$ NPs were synthesized using leaf extract of *Azadirachta indica* with a particle size of 41.9 nm with crystalline anatase and rutile structures. The formation of NPs is due to the extract's terpenoids, flavonoids, and proteins. These phytochemicals act as stabilizing and reducing agents of titanium oxide NPs (114). *Annona squamosa* peel's aqueous extract was used to synthesize $TiO_2$ NPs. It has a rutile crystal structure, and particle size was 23 nm with spherical morphology. The aqueous peel extract containing water-soluble hydroxy functional groups is mainly responsible for serving as a reducing agent (115).

Jayaseelan et al. reported synthesizing $TiO_2$ NPs using *Aeromonas hydrophila* bacteria (116). The main compound, glycyl-L-proline, acts as a capping, reducing, and stabilizing agent for NP formation. GC-MS analysis of bacterial culture showed the presence of glycyl-L-proline as the principal constituent and uric acid, glycyl-L-glutamic acid, and l-Leucyl-d-leucine; most of the compounds contained -COOH and -C=O groups. TiO $(OH)_2$ was used as the precursor, added to the bacterial culture, and kept in a shaking incubator at room temperature for 24 hours until white precipitation at the bottom of the reaction flask appeared. Then it was air-dried. During the 24 hours, lone-pair electrons of the oxygen molecule in the titania precursor picked up protons of glycyl-L-proline, were protonated, and then lost a water molecule to give a $Ti^{3+}$ ion, and later deprotonation occurred, resulting in $TiO_2$. The resulting NPs were regular, spherical, and had an uneven distribution with a size of 40–50 nm (116).

Jha et al. described the synthesis of $TiO_2$ NPs using the *Lactobacillus* strain (derived from buttermilk) and *Saccharomyces cerevisiae* (baker's yeast) (117). Lactobacilli have negative electrokinetic potential and can attract cations, which is the crucial step in this biosynthesis. Lactobacilli can grow even in the presence of oxygen, but the presence of glucose-like compounds reduces oxidation-reduction potential. Energy-giving material glucose and pH conditions also affect the binding capacity of *Lactobacillus* bacteria to the cations, which is investigated in the biosynthesis method. An acidic environment and lower oxidation-reduction potential are better for activating oxidoreductases to form $TiO_2$ NPs. In the biosynthesis route, yeast oxidoreductases and quinones are essential in NP formation and reduction. A lower pH activates the oxidases, while a higher pH activates the reductases. Simple hydroxy or methoxy derivatives of various quinones facilitate redox reactions by tautomerization in lower fungi (117).

### 7.6.3 Copper Oxide Nanoparticles

Copper oxide nanoparticles at 5–45 nm were synthesized using *Bifurcaria bifurcate* algal extract. This extract causes the formation of Copper (I) and Copper (II) oxide by the reduction property of algae on 1 mM $CuSO_4$ solution. UV-visible spectral analysis indicates the presence of diterpenoids, which results in the reduction of copper sulfate ($Cu^{2+}$) ions to Cu (I) and then to Cu (II) oxide. The color change of algae extracts to red blood color showing absorbance at 260 and 650 nm shows the formation of copper oxide NPs (118). Soureshjani et al. synthesized copper oxide using different brown algae extracts, *Cystoseira myrica* and *Sargassum latifolium*, a fast biosynthetic procedure that resulted in NPs of average size 12–24 nm. Copper sulphate solution was taken, sodium citrate and sodium carbonate solution were added, and pH was varied from 4, 7, 9, 12, and so on. Then the extract was added and stirred for 1 hour at 100°C. It was found that the reaction medium of pH 12 resulted in CuO NPs as the final product with a reddish-orange color. At first, sodium citrate acts as a ligand, stabilizing metal ions and promoting capping and reducing the effect of the algal extract on forming NPs.

The addition of algal extract followed by 1 hour of stirring transformed the blue-colored copper solution into an orange-red product, providing a visual indication of nanoparticle formation, which was subsequently confirmed by X-ray Diffraction (XRD) analysis. In XRD, only the peaks of $Cu_2O$ were obtained and were in good agreement with JCPDS values. No peaks corresponding to Cu or CuO were obtained. Brown algae contain biocomponents such as phenolics, sulphated polysaccharides, fucoidan, laminarin, fucoxanthin, alginic acid (alginate), and mannitol. The presence of hydroxyl groups in these biocomponents and released enzymes from the cells in the extracts enable the algal extract to act as the reducing agent. In the FTIR spectra of algal extract and NPs, it was determined that fucoxanthin is the main component present in algal extract, stabilizing and capping the metal ions to form NPs. The actual reduction mechanism of the extract could not be explained because of the diversity of biocomponents present in the algal extract. The peaks corresponding to algal extract in their FTIR spectra indicate C=O, O-H, C-H, and C-O stretching vibrations. In the FTIR spectrum of NPs, peaks corresponding to biostructures were shifted, and intensity was decreased, indicating that these biocomponents were responsible for the capping mechanism of metal ions and stabilizing the NPs formed (119).

Nabila et al. reported actinomycete-mediated synthesis, where the soil-isolated strain VITBN4 was capable of producing nanoparticles, with enzymes, proteins, peptides, reducing cofactors, and other organic materials from the microorganism acting as both reducing and capping agents. Actinomycetes have the potential to generate specific proteins that interact with nanocrystals, leading to the formation of NPs with complex morphology (120). CuO NP synthesis using plant extract has overcome the limitations of microbes, such as phytochemicals, which are non-toxic, and there is no need for incubation or culturing procedures for extract development and synthesis of NPs. Synthesis using plant extract has several advantages: it is a cost effective, safe, and easy to use synthetic procedure; has low energy; has several biomolecules; has good stability; and the absence of toxic chemicals. Biomolecules in plants, such as tannins, terpenoids, flavonoids, and polyphenols, act as good reducing and stabilizing agents for synthesizing CuO NPs (121).

Kerour et al. reported the synthesis of CuO NPs using *aloe vera* extract with different concentrations, and they had octahedron structures with crystallite sizes ranging between 24 and 61 nm (122). The morphology and size of NPs vary depending on the concentration of *Aloe vera* extract. *Aloe vera* extract contains alkaloids, flavonoids, terpenoids, and polyphenols responsible for the formation of NP and stability. The FTIR spectra of the synthesized NPs exhibited characteristic organic absorption bands, including C–H stretching, C=O stretching, and C–O out-of-plane bending vibrations, which can be attributed to organic compounds present in the *aloe vera* extract. Other plant extracts have also been used for the synthesis of CuO NPs, such as *Ferulago angulata* (schlecht) boiss extract (123) and *Caloropis procera* (124). In forming CuO NPs, phytochemicals such as alkaloids, triterpenoids, flavonoids, and polyphenols play an essential role as capping and reducing agents, thereby minimizing agglomeration. The oxygen atoms of the ester group and phenolic hydroxyl groups in phytochemicals influence metal ions to be chelated and to undergo decomposition to CuO on annealing to release NPs (121, 124).

### 7.6.4 Iron Oxide Nanoparticles

The contrast MRI technique is the most noninvasive anatomy and tissue function assessment methodology. Gadolinium-based contrast agents are more frequently used as contrast agents, which are toxic to the kidneys. To overcome this disadvantage, biocompatible superparamagnetic iron oxide nanoparticles (SPIONPs) are used, engulfed by macrophages in the liver and excreted by hepatobiliary processing. SPIONPs available are about 20 nm, and surface functionalization is required to decrease their size to less than 20 nm. For that, capping, ligands include silica (125), dextran derivative (126), chitosan derivative (127), poly-L-lysine (128, 129), and poly (ethyleneimine) (130). The most widely used synthesis methods for SPIONPs are coprecipitation and thermal decomposition. However, control over size is impossible via coprecipitation, and thermal decomposition will obtain a narrow size. To overwhelm this, combined synthetic approaches taking advantage of both co-precipitation and thermal decomposition, with hydrophilic nature playing an essential role in controlling size and geometry, are reported in a few works (131, 132). Water-soluble SPIONs are reported by the thermal decomposition of ferric acetate in PEG, poly (ethylene imine) PEI, and poly (vinyl pyrrolidone) as high boiling solvents (133). Zhang et al. reported a one-pot synthesis of water-dispersible PEG-based SPIONPs (134). They are colloidally stable across a broad pH range and display MRI contrasting ability due to their superparamagnetic behavior. Here, polyethylene glycol acts as a stabilizing agent, ferric acetate as the iron precursor, and tetra ethylene glycol as the solvent. PEG. SPIONPs are well dispersed without much aggregation.

$Fe_2O_3$ NPs synthesized are hydrophobic to make them water dispersible; an alcohol of high boiling point was used in thermal decomposition, as synthesized NPs have a hydrodynamic diameter of 11.7 nm. The size-determining factors are polymer length, conformation on the surface, and coverage on the surface. The presence of polyethylene glycol covers these vital parameters. FTIR and XRD patterns confirm the formation of $Fe_2O_3$ NPs and the role of PEG as the stabilizing and capping agent. The peaks of PEG and PEG.SPIONPs are similar in FTIR, and the carboxyl band is moved to a lower wavenumber in SPIONs, indicating the capping ability. XRD peaks reveal the high crystalline and monodispersed nature of SPIONPs due to adopting an alcohol-based thermal decomposition method (133). $Fe_2O_3$ NPs were synthesized taking iron oleate as the precursor by a solvothermal process in an organic medium, oleic acid-coated NPs were formed, and an oleic acid-di mercapto succinic acid (DMSA) ligand exchange reaction was conducted in water medium. Then methoxy amine polyethylene glycol with two different chain lengths was added, forming an amide linkage between the amine PEG and the carboxyl group of DMSA. This led to random coil conformation of PEG around the $Fe_2O_3$ NPs, which was confirmed by hydrodynamic size measurement. The newly formed NPs of $Fe_2O_3$ have increased water dispersibility. The PEG coating helps to develop SPIONs with biocompatibility, water solubility, anti-fouling ability, and wide circulation for in vivo reactions (125, 126).

Shameli et al. synthesized $Fe_2O_3$ NPs using red seaweed *Kappaphycus alvarezii* extract. It is commonly found in Malaysia, and dried seaweed powder is used for the extract preparation. $Fe^{2+}$ and $Fe^{3+}$ solutions were added to the extract in a 2:1 ratio, followed by NaOH solution dropwise, and pH was maintained at 11. As synthesized, $Fe_3O_4$ NPs were

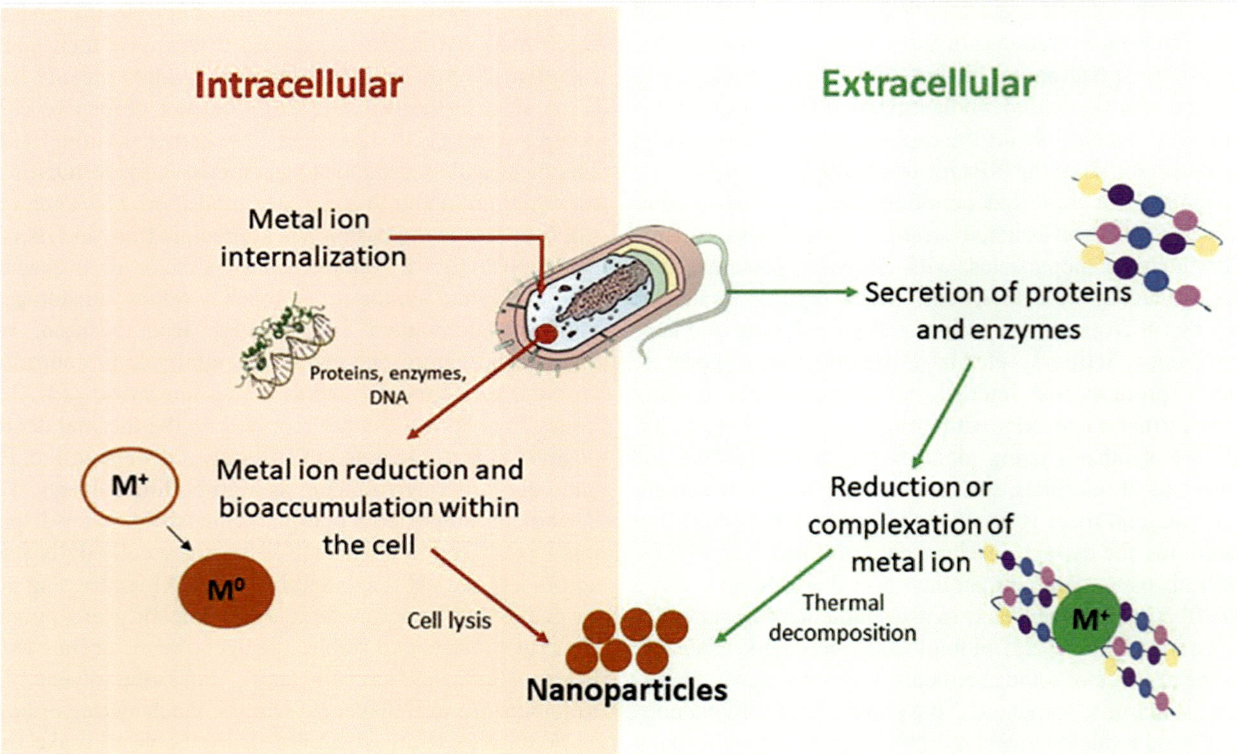


FIGURE 7.5 Green synthesis of nanoparticles using bacterial cultures (136).

separated by a bar magnet. One hour after adding the NaOH solution, it turned from light brown to black, indicating NP formation. The FTIR spectra of extract and synthesized $Fe_3O_4$ particles reveal that seaweed extract contains sulfated hydroxyl groups, C-O stretching vibrations, $CH_2$ bending, and C-H stretching vibrations. The peaks were shifted, indicating the capping and stabilizing property of extract, and peaks confirming the presence of $Fe_3O_4$ were observed at 556 $cm^{-1}$ due to tetrahedral site vibrations of the Fe-O bond, and 423 $cm^{-1}$ corresponds to the octahedral Fe (135).

Iron oxide NPs were synthesized from $FeCl_3$ solution using *Sargassum muticum* brown algae extract. $FeCl_3$ is converted to $Fe_2O_3$ NPs with cubic morphology at 25°C with particle size 18+/-4 nm due to the influence of polysaccharides in the algae (109, 137). Biosynthesis methods involving microbes such as bacteria and fungi release extracellular enzymes and proteins to the reaction environment, stabilizers or reducing agents in forming metal ions in the reduced state or metal oxide NPs (Figure 7.5). These enzymes and proteins also avoid NP agglomeration (136, 138). The advantages of fungi over bacteria are that they release more metabolites to the reaction medium and can withstand high pressure and flow rate, enabling large-scale NP synthesis (136). Plants have many phytochemicals such as methylxanthines, saponins, polyphenols, flavonoids, and saponins, which are known to be antioxidants and have radical scavenging properties as well as the ability to reduce metal chelates or metal ions to form metal oxide NPs. Moreover, plant extracts are not as toxic as the microbes. However, the only problem is that researchers cannot determine the number of phytochemicals released to the reaction medium for converting metal ions to metal oxide NPs from plant extract in different solvents. In different temperatures or processes, the release of several phytochemicals varies. In algae, the hydroxyl and carboxyl groups act as antioxidants for the conversion, like plant extract (125–127, 130–134, 136, 139–141).

## 7.7 BIOMEDICAL APPLICATIONS OF BIOFUNCTIONALIZED METAL OXIDE NANOPARTICLES

Biofunctionalized metal oxide NPs are mainly synthesized to improve their environmental sustainability and make them capable of biomedical applications (Figure 7.6).

### 7.7.1 Antimicrobial Applications

Biofunctionalization enhances biomedical applications by improving the performance and effectiveness of metal oxide NPs. Many pathogenic fungi and bacteria threaten human health, agriculture, and the environment, leading to extensive research into the antimicrobial activity of metal oxides (143). When the outer portion of the bacterial cells makes contact with a metal oxide, especially ZnO, CuO,

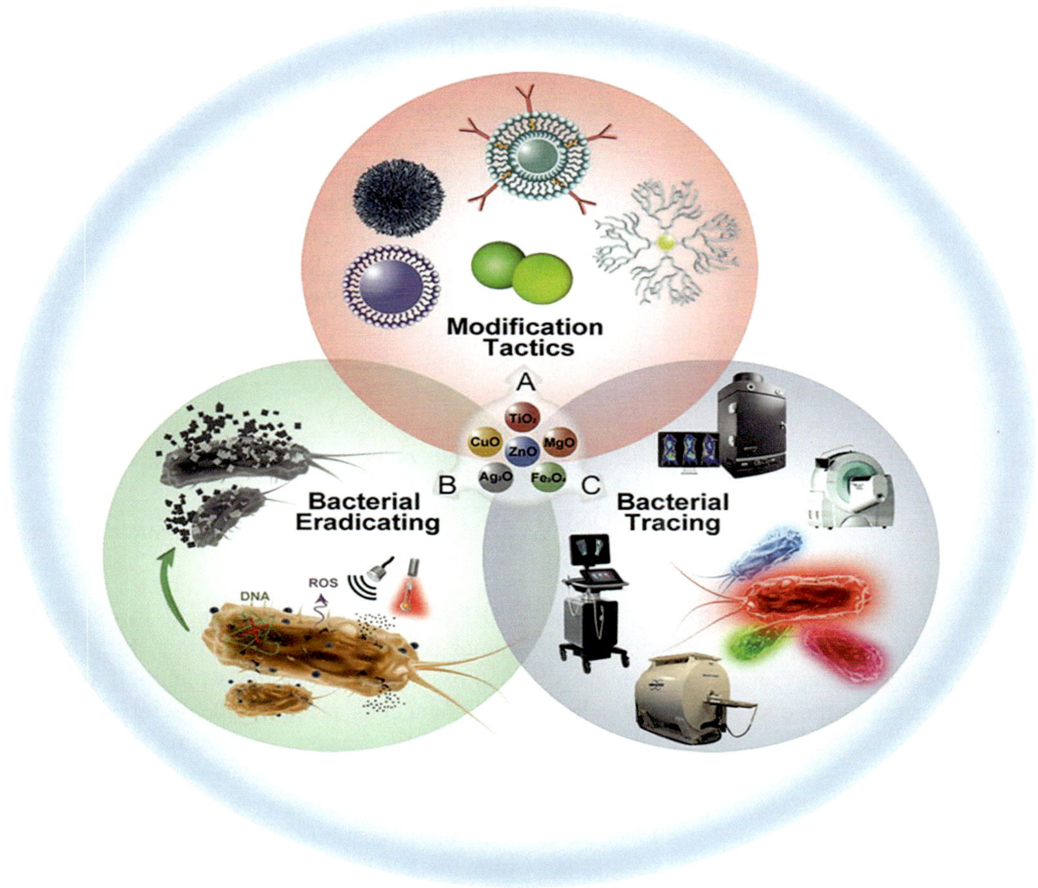


FIGURE 7.6 Antibacterial activity of biofunctionalized metal oxide nanoparticles. (A) Various strategies have been formulated to minimize the biotoxicity and immunogenicity of metal oxide nanoparticles through surface coating and hybridization, enhancing their ability to eradicate and trace bacteria. (B) Metal oxide nanoparticles target the cell wall, membrane, and cytoplasmic contents to disrupt cellular balance and produce highly cytotoxic ROS effective against nearly all microorganisms without resistance concerns. (C) Metal oxide nanoparticles leverage their inherent imaging properties and easy surface functionalization with targeted imaging agents, and they hold significant potential for tracing bacteria (142).

and $TiO_2$ NPs, they disrupt the structure and the permeability of the cell changes. Metal oxide NPs move in the cytoplasm and generate hydrogen peroxide and reactive oxygen species, which will cause cell death by inhibiting the cell's natural functioning (113, 144). The zone of inhibition can confirm the antimicrobial activity. The greater the zone of inhibition, the more significant the antimicrobial activity, which is associated with larger inhibition zones and smaller NP sizes. Rajiv et al. reported anti-fungal activity of ZnO NPs, which are size dependent, and found that the smaller the size of NPs, the greater the zone of inhibition (145). Also, the antifungal effect of NPs varied from fungal strains. Another critical factor that influences the zone of inhibition is the concentration of ZnO NPs. Patil and Taranath explained the dependence on the zone of inhibition in their work and the possible mechanism of the antimicrobial activity of ZnO on *Mycobacterium tuberculosis* (146). The zone of inhibition increases with the concentration of ZnO NPs.

ZnO NPs synthesized using zinc acetate as the precursor and different capping agents such as carboxymethyl cellulose, soluble starch, urea, lactose, and polyvinylpyrrolidone have antimicrobial activity. Antimicrobial action is due to the ability of ZnO NPs to release reactive oxygen species, hydroxy free radicals, and hydrogen peroxide towards the reaction medium, which causes cell death. Also, $Zn^{2+}$ enters the cytoplasm of bacteria, disrupting microbial growth and functioning, finally leading to death.

ZnO NPs exhibit inhibition zones against the bacteria *Bacillus subtilis* (Gram-positive) and *Escherichia coli* (Gram-negative) and fungi *Candida albicans* (unicellular) and *Aspergillus niger* (filamentous). ZnO NPs synthesized using soluble starch, urea, and PVP showed greater antibacterial and antifungal activity than those synthesized using lactose. This is because as the size of the NPs reduces, the antimicrobial activity will be more significant. They can enter the cytoplasm more quickly and cause cell death (116, 147). Titanium dioxide and copper oxide NPs also show similar antimicrobial activity against *Escherichia coli, Bacillus subtilis, Salmonella typhi,* and *Klebsiella pneumonia* (114, 119, 148–154).

### 7.7.2 Drug Delivery

Due to their lower cytotoxicity and biocompatibility, IONPs are used for drug delivery. By drug delivery, we can deliver drugs to the specific site that is damaged or affected. Due to their small size and biocompatibility, IONPs are widely used in in vivo studies. Due to the hydrophobic nature of the surface of the bare IONPs, they are biofunctionalized with organic or polymer compounds, making them hydrophilic (155, 156). Different endocytosis mechanisms are there for the entrance of NPs into cells, and these pathways or the mechanisms are influenced by the size, morphology, and surface charges of the NPs. PEG-coated IONPs are expected to adopt two pathways based on TEM micrographs: caveolae-dependent endocytosis and micropinocytosis. If the NPs are smaller than 200 nm, they will adopt the caveolae pathway; micropinocytosis mechanisms are adapted to larger aggregates. The caveolae-dependent pathway can directly be targeted to the endoplasmic reticulum or the Golgi apparatus (103, 157).

### 7.7.3 Anti-Tumor Activity

In vitro anticancer studies were done on HeLa cells by CuO NPs synthesized using black bean extract. HeLa cells were treated with CuO NPs at various concentrations and durations. Fluorescence studies were done using fluorescent probe DCFH-DA. The studies indicate that CuO NPs enhanced reactive oxygen species generation in a dose-dependent manner. Mito Tracker Red and Hoechst 33258 were used to assess mitochondrial and nuclear morphology in response to NP exposure. Mitochondria changed their structure after exposure to CuO NPs, and the number of colonies of cancer-affected cells was also reduced after treatment with NPs. The NPs, by necrosis apoptosis mechanisms, inhibit cell growth. Cytotoxic studies show that CuO NPs have a great influence on HeLa cells (158). Cytotoxic studies by CuO NPs synthesized using *Pterolobium hexapetalum* leaf extract on human breast cancer cells MDA-MB-231 and HBL 100 normal cells were carried out by MTT assay (dimethyl thiazolyl tetrazolium bromide). It was observed that CuO NPs inhibited the proliferation of cancer cells more effectively than normal cells at an $IC_{50}$ value of 30 μg/ml. The studies also indicated that as the concentration of NPs increases, their accumulation of cancer cells inhibits cell viability and finally leads to cell death. Here also, reactive oxygen species generation plays a vital role in the apoptosis mechanism; CuO NPs can cause more reactive oxygen species generation, which plays a vital role in anticancer activity (159). Similarly, CuO NPs synthesized using different plant extracts show anticancer activity by apoptosis mechanism through reactive oxygen species generation (160, 161). Zinc oxide shows anticancer activity (162–164). Because of the electrostatic force of attraction on the NP surface, they are attached to tumor cells or normal cells. The metal oxide NPs split into ionic states of metal and oxygen radicals because of the presence of phytochemicals or biomolecules on the capping agent. The cytotoxic effect on tumor cells occurs due to the generation of reactive oxygen species when metal oxide NPs enter the cells. When NPs enter normal cells, the metal ions will function as metal ion supplements for the growth of the cells, and radicals will be converted into water and oxygen molecules by superoxidase dismutase, which is absent in tumor cells. So, in tumor cells, more and more reactive oxygen species enter and form free radicals, triggering the generation of hydrogen peroxide, inhibiting cell proliferation, and finally causing cell death (165).

### 7.7.4 Magnetic Resonance Imaging Contrast Agents

MRI is an essential instrument in diagnosing cancer or tumor cells. The most important thing is the administration of contrasting agents, which will accumulate over the malignant cells, revealing the presence of tumorous or cancerous cells. Various contrasting agents have been developed by researchers worldwide, but biofunctionalized iron oxide NPs have attracted considerable attention because of their biocompatibility. and unique magnetic properties. In immunodeficient mice, PEG-coated iron oxide NPs and commercially available contrasting agents were compared by injecting them into MDA-MB-231 xenograft tumors (~1 $cm^3$). Longer images could develop using PEG-coated iron oxide NPs even after 24 hours of injection. The presence of commercially available NPs indicates that NPs escaped from the inoculation point, passed through the bloodstream, and accumulated over the surface of the cancerous cells. However, in the case of PEG-coated IONPs, nothing is accumulated on the periphery of cancerous cells. Instead, they remain at the initial inoculation point. This matter helps with longitudinal imaging and anticancer drug delivery (103, 166).

## 7.8 CONCLUSION

Metal NPs have broad application prospects in all walks of life because of their unique performance. These NPs, whether metallic particles or metal oxides, can be tailored with advanced preparations to control their size, shape, and composition. From a distinct perspective, metal oxides exhibit various structural properties indispensable in applications like magnetic ferrites and photocatalysts. Metal and metal oxide NPs should be stable, biocompatible, and functional in biomedical applications. The improvement in colloidal stability, water solubility, and overall versatility is attributed to surface functionalization. Although polymer and silica coatings are commonly used, bioconjugation has emerged to attach biomolecules, including antibodies and aptamers, onto NP surfaces to increase specificity for biological interactions. Targeting strategies for efficient drug delivery are exemplified by NP bioconjugation of moieties to NPs, allowing them to couple drugs with targeting ligands to enhance drug delivery or engage specific cells/media for cancer therapy. This principle has further improved the efficiency of the drug and reduced the adverse effects by delivering drugs to selective sites. To summarize, the synthesis and functionalization of metal and metal oxide NPs provide many different routes to customized materials with intriguing properties, which contribute to the state of the art in such diverse fields as biomedical and high-tech applications. Ongoing investigation bodes well for enhanced advancement and the production of new applications in the near future.

## Keywords

Biofunctionalized Nanoparticles
Biomedical Applications
Cancer Treatment
Drug Delivery Systems
Green Synthesis
Targeted Therapy

## REFERENCES

1. Tiwari, A.; Verma, N. C.; Singh, A.; Nandi, C. K.; Randhawa, J. K.; Carbon coated core–shell multifunctional fluorescent SPIONs. *Nanoscale,* 2018, 10(22), 10389–10394.
2. Medhi, R.; Marquez, M. D.; Lee, T. R.; Visible-light-active doped metal oxide nanoparticles: review of their synthesis, properties, and applications. *ACS Appl. Nano Mater.*, 2020, 3(7), 6156–6185.
3. Paramasivam, G.; Kayambu, N.; Rabel, A. M.; Sundramoorthy, A. K.; Sundaramurthy, A.; Anisotropic noble metal nanoparticles: synthesis, surface functionalization and applications in biosensing, bioimaging, drug delivery and theranostics. *Acta Biomater.*, 2017, 49, 45–65.
4. Tiwari, P. M.; Vig, K.; Dennis, V. A.; Singh, S. R.; Functionalized gold nanoparticles and their biomedical applications. *Nanomaterials,* 2011, 31–63.
5. Erathodiyil, N.; Ying, J. Y.; Functionalization of inorganic nanoparticles for bioimaging applications. *Acc. Chem. Res.*, 2011, 44(10), 925–935.
6. Kumar, R.; Kumar, V. B.; Gedanken, A.; Sonochemical synthesis of carbon dots, mechanism, effect of parameters, and catalytic, energy, biomedical and tissue engineering applications. *Ultrason. Sonochem.*, 2020, 64, 105009.
7. Lingeshwar, R. K.; Balaji, R.; Kumar, A.; Krishnan, V.; Lanthanide doped near infrared active upconversion nanophosphors: fundamental concepts, synthesis strategies, and technological applications. *Small*, 2018, 14(37), 1801304.
8. Sperling, R. A.; Parak, W. J.; Surface modification, functionalization and bioconjugation of colloidal inorganic nanoparticles. *Philos. Trans. R. Soc. A Math. Phys. Eng. Sci.*, 2010, 368(1915), 1333–1383.
9. Sinha, R.; Kim, G. J.; Nie, S.; Shin, D. M.; Nanotechnology in cancer therapeutics: bioconjugated nanoparticles for drug delivery. *Mol. Cancer Ther.*, 2006, 5(8), 1909–1917.
10. Gao, J.; Huang, X.; Liu, H.; Zan, F.; Ren, J.; Colloidal stability of gold nanoparticles modified with thiol compounds: bioconjugation and application in cancer cell imaging. *Langmuir*, 2012, 28(9), 4464–4471.
11. Farokhzad, O. C.; Karp, J. M.; Langer, R.; Nanoparticle–aptamer bioconjugates for cancer targeting. *Expert Opin. Drug Deliv.*, 2006, 3(3), 311–324.
12. Zhou, J.; Rossi, J.; Aptamers as targeted therapeutics: current potential and challenges. *Nat. Rev. Drug Discov.*, 2017, 16(3), 181–202.
13. Ravindran, A.; Chandran, P.; Khan, S. S.; Biofunctionalized silver nanoparticles: advances and prospects. *Coll. Surf. B.*, 2013, 105, 342–352.
14. Fang, Y.; Jing C.; Experimental (SERS) and theoretical (DFT) studies on the adsorption behaviors of l-cysteine on gold/silver nanoparticles. *Chem. Phys.*, 2007, 332(1), 27–32.
15. Stellacci, F.; Ouay, B. L.; Antibacterial activity of silver nanoparticles: a surface science insight. *Nano Today*, 2015, 10(3), 339–354.
16. Tang, S.; Zheng, J.; Antibacterial activity of silver nanoparticles: structural effects. *Adv. Healthc. Mater.*, 2018, 7(13), e1701503.
17. Joseph, V.; Matschulat, A.; Polte, J.; Rolf, S.; Emmerling, F.; Kneipp. J.; SERS enhancement of gold nanospheres of defined size. *J. Raman Spectrosc.*, 2011, 42(9), 1736–1742.
18. Sztandera, K.; Gorzkiewicz, M.; Klajnert-Maculewicz, B.; Gold nanoparticles in cancer treatment. *Mol. Pharm.*, 2019, 16(1), 1–23.
19. Guerrini, L.; Alvarez-Puebla, R. A.; Pazos-Perez. N.; Surface modifications of nanoparticles for stability in biological fluids. *Materials,* 2018, 11(7), 1154.
20. Gopal, K.; Dwivedi, A. D.; Biosynthesis of silver and gold nanoparticles using Chenopodium album leaf extract. *Coll. Surf. A,* 2010, 369(1), 27–33.
21. Dubey, S. P.; Lahtinen, M., Sillanpää, M.; Green synthesis and characterizations of silver and gold nanoparticles using leaf extract of Rosa rugosa. *Coll. Surf. A.*, 2010, 364(1–3), 34–41.
22. Xu, B.; Gonella, G.; DeLacy, B. G.; Hai-Lung, D.; Adsorption of anionic thiols on silver nanoparticles. *J. Phys. Chem. C,* 2015, 119(10), 5454–5461.
23. Fontana, L.; Bassetti, M.; Battocchio, C.; Venditti, I.; Fratoddi, I.; Synthesis of gold and silver nanoparticles functionalized with organic dithiols. *Coll. Surf. A,* 2017, 532, 282–289.
24. Peng, H.; Xing-Fang, L.; Zhang, H.; Le, X. C.; A microRNA-initiated DNAzyme motor operating in living cells. *Nat. Commun.*, 2017, 8(1), 14378.
25. Elghanian, R.; Storhoff, J. J.; Mucic, R. C.; Letsinger, R. L.; Mirkin, C. A.; Selective colorimetric detection of polynucleotides based on the distance-dependent optical properties of gold nanoparticles. *Science*, 1997, 277(5329), 1078–1081.
26. Zhang, X.; Servos, M. R.; Liu, J.; Fast pH-assisted functionalization of silver nanoparticles with monothiolated DNA. *Chem. Commun.*, 2012, 48(81), 10114–10116.
27. Lévy, R.; Thanh, N. T., Doty, R. C.; Hussain, I.; Nichols, R. J.; Schiffrin, D. J.; Brust, M.; Fernig, D. G.; Rational and combinatorial design of peptide capping ligands for gold nanoparticles. *J. Am. Chem. Soc.*, 2004 126(32), 10076–10084.
28. Kogan, M. J.; Olmedo, I.; Hosta, L.; Guerrero, A. R.; Cruz, L. J.; Albericio, F.; Peptides and metallic nanoparticles for biomedical applications. *Nanomedicine*, 2007 2(3), 287–306.
29. Montenegro, J.; Grazu, V.; Sukhanova, A.; Agarwal, S.; de la Fuente, J. M.; Nabiev, I.; Greiner, A.; Parak, W. J.; Controlled antibody/(bio-) conjugation of inorganic nanoparticles for targeted delivery. *Adv. Drug Deliv. Rev.*, 2013, 65(5), 677–688.

30. Arruebo, M.; Valladares, M.; González-Fernández, Á.; Antibody-conjugated nanoparticles for biomedical applications. *J. Nanomater.*, 2009, 1–24.
31. Ścigalski, F.; Kaczmarek, H.; Wolnicka, M.; Preparation and properties of silver nanoparticles stabilized by mercaptosuccinic acid with a controllable size. *Adv. Sci. Eng. Med.*, 2013, 5(4), 309–313.
32. Zong, J.; Cobb, S. L.; Cameron, N. R.; Peptide-functionalized gold nanoparticles: versatile biomaterials for diagnostic and therapeutic applications. *Biomater. Sci.*, 2017, 5(5), 872–886.
33. Csapó, E.; Patakfalvi, R.; Hornok, V.; Tóth, L. T.; Sipos, Á.; Szalai, A.; Csete, M.; Dékány, I.; Effect of pH on stability and plasmonic properties of cysteine-functionalized silver nanoparticle dispersion. *Coll. Surf. B.*, 2012, 98, 43–49.
34. Kreuter, J.; Nanoparticles – a historical perspective. *Int. J. Pharm.*, 2007, 331(1), 1–10.
35. Issa, B.; Obaidat, I. M.; Albiss, B. A.; Haik, Y.; Magnetic nanoparticles: surface effects and properties related to biomedicine applications. *Int. J. Mol. Sci.*, 2013, 14(11), 21266–21305.
36. McNamara, K.; Tofail, S. A.; Nanoparticles in biomedical applications. *Adv. Phys. X*, 2017, 2(1), 54–88.
37. Naqvi, S.; Samim M.; Abdin M. Z.; Ahmad, F. J.; Maitra, A. N.; Dinda, A. K.; Prashant, C. K.; Concentration-dependent toxicity of iron oxide nanoparticles mediated by increased oxidative stress [Retraction]. *Int. J. Nanomed.*, 2022, 17, 1459–1460.
38. Pujalté, I.; Passagne, I.; Brouillaud, B.; Tréguer, M.; Durand, E.; Ohayon-Courtès, C.; L'Azou, B.; Cytotoxicity and oxidative stress induced by different metallic nanoparticles on human kidney cells. *Part. Fibre Toxicol.*, 2011, 8, 1–16.
39. Jeng, H. A.; Swanson, J.; Toxicity of metal oxide nanoparticles in mammalian cells. *J. Environ. Sci. Health A*, 2006, 41(12), 2699–2711.
40. Fouad, G. I.; A proposed insight into the anti-viral potential of metallic nanoparticles against novel coronavirus disease-19 [COVID-19]. *Bull. Natl. Res. Cent.*, 2021, 45, 1–22.
41. Roshani, M.; Rezaian-Isfahni, A.; Lotfalizadeh, M. H.; Khassaf, N.; Abadi, M. H. J. N.; Nejati, M.; Metal nanoparticles as a potential technique for the diagnosis and treatment of gastrointestinal cancer: a comprehensive review. *Cancer Cell Int.*, 2023, 23(280), 1–23.
42. Khalid, K.; Tan, X.; Zaid, H. F. M.; Tao, Y.; Chew, C. L.; Din-Thoi, C.; Lam, M. K.; Yeek-Chia, H.; Lim, J. W.; Wei, L. C.; Advanced in developmental organic and inorganic nanomaterial: a review. *Bioengineered*, 2020, 11(1), 328–355.
43. Nel, A.; Xia, T.; Madler, L.; Li, N.; Toxic potential of materials at the nanolevel. *Science*, 2006, 311(5761), 622–627.
44. Lee, S. H.; Jun, B. H.; Silver nanoparticles: synthesis and application for nanomedicine. *Int. J. Mol. Sci.*, 2019, 20(4), 865.
45. Feng, Q. L.; Wu, J.; Chen, G. Q.; Cui, F. Z.; Kim, T. N.; Kim, J. O.; A mechanistic study of the antibacterial effect of silver ions on Escherichia coli and Staphylococcus aureus. *J. Biomed. Mater. Res.*, 2000, 52(4), 662–668.
46. Fatima, F.; Wahid, I.; Eco-friendly synthesis of silver and copper nanoparticles by Shizophyllum commune fungus and its biomedical applications. *Int. J. Environ. Sci. Technol.*, 2022, 19(8), 7915–7926.
47. Soliman, A. M.; Abdel-Latif, W.; Shehata, I. H.; Fouda, A.; Abdo, A. M.; Ahmed, Y. M.; Green approach to overcome the resistance pattern of Candida spp. using biosynthesized silver nanoparticles fabricated by Penicillium chrysogenum F9. *Biol. Trace Elem. Res.*, 2021, 199, 800–811.
48. Anh, N. T. P.; Loan, P. T. T.; Khang, N. V. D.; Khoa, V. N. D.; Synthesis of graphene oxide-$TiO_2$ nanotubes-silver nanoparticles nanocomposite by gamma irradiation for antibacterial and post-harvest preservation purposes. *J. Biotechnol.*, 2019, 17(4), 719–727.
49. Azizi, M.; Ghourchian, H.; Yazdian, F.; Bagherifam, S.; Bekhradnia, S.; Nyström, B.; Anti-cancerous effect of albumin coated silver nanoparticles on MDA-MB 231 human breast cancer cell line. *Sci. Rep.*, 2017, 7(1), 5178.
50. Shahverdi, A. R.; Fakhimi, A.; Shahverdi, H. R.; Minaian, S.; Synthesis and effect of silver nanoparticles on the antibacterial activity of different antibiotics against Staphylococcus aureus and Escherichia coli. *Nanomedicine.*, 2007, 3(2), 168–171.
51. Ahmed, B.; Hashmi, A.; Khan, M. S.; Musarrat, J.; ROS mediated destruction of cell membrane, growth and biofilms of human bacterial pathogens by stable metallic AgNPs functionalized from bell pepper extract and quercetin. *Adv. Powder Technol.*, 2018, 29(7), 1601–1616.
52. Rokaya, D.; Srimaneepong, V.; Qin, J.; Siraleartmukul, K.; Siriwongrungson, V.; Graphene oxide/silver nanoparticle coating produced by electrophoretic deposition improved the mechanical and tribological properties of NiTi alloy for biomedical applications. *J. Nanosci. Nanotechnol.*, 2019, 19(7), 3804–3810.
53. Mokabber, T.; Cao, H. T.; Norouzi, N.; Van Rijn, P.; Pei, Y. T.; Antimicrobial electrodeposited silver-containing calcium phosphate coatings. *ACS Appl. Mater. Interfaces*, 2020, 12(5), 5531–5541.
54. Kahzad, N.; Salehzadeh, A.; Green synthesis of $CuFe_2O_4$@Ag nanocomposite using the Chlorella vulgaris and evaluation of its effect on the expression of norA efflux pump gene among Staphylococcus aureus strains. *Biol. Trace Elem. Res.*, 2020, 198(1), 359–370.
55. Da Silva, R. T.; Petri, M. V.; Valencia, E. Y.; Camargo, P. H.; de Torresi, S. I.; Spira, B.; Visible light plasmon excitation of silver nanoparticles against antibiotic-resistant Pseudomonas aeruginosa. *Photodiag. Photodyn. Ther.*, 2020, 31, 101908.
56. Tarannum, N.; Gautam, Y. K.; Facile green synthesis and applications of silver nanoparticles: a state-of-the-art review. *RSC Adv.*, 2019, 9(60), 34926–34948.
57. Lee, W.; Keuk-Jun, K.; Lee, D. G.; A novel mechanism for the antibacterial effect of silver nanoparticles on Escherichia coli. *Biometals*, 2014, 27, 1191–1201.
58. Vasanth, K.; Ilango, K.; MohanKumar, R.; Agrawal, A.; Dubey, G. P.; Anticancer activity of Moringa oleifera mediated silver nanoparticles on human cervical carcinoma cells by apoptosis induction. *Coll. Surf. B.*, 2014, 117(3), 354–359.
59. Hsueh-Hsiao, W.; Cheng-Huang, S.; Yih-Jer, W.; Cheng-An, J. L.; Chih-Hsien, L.; Ji-Lin, S.; Wen-Hsiung, C.; Hung-I, Y.; Application of gold in biomedicine: past, present and future. *Int. J. Gerontol.*, 2012, 6(1), 1–4.
60. Bhattacharyya, S., Kudgus, R. A., Bhattacharya, R., Mukherjee, P.; Inorganic nanoparticles in cancer therapy. *Pharm. Res.*, 2011, 28, 237–259.
61. Yang, W., Liang, H., Ma, S., Wang, D., Huang, J.; Gold nanoparticle based photothermal therapy: development and application for effective cancer treatment. *Sustain. Mater. Technol.*, 2019, 22, e00109.
62. Brazzale, C.; Canaparo, R.; Racca, L.; Foglietta, F.; Durando, G.; Fantozzi, R., Caliceti, P.; Salmaso, S.; Serpe, L.; Enhanced selective sonosensitizing efficacy of

ultrasound-based anticancer treatment by targeted gold nanoparticles. *Nanomedicine.*, 2016, 11(23), 3053–3070.
63. Tachibana, K.; Feril Jr., L. B.; Ikeda-Dantsuji, Y.; Sonodynamic therapy. *Ultrasonics*, 2008, 48(4), 253–259.
64. Mišík, V., Riesz, P.; Free radical intermediates in sonodynamic therapy; *Ann. N. Y. Acad. Sci.*, 2000, 899(1), 335–348.
65. Canaparo, R.; Foglietta, F.; Giuntini, F.; Francovich, A.; Serpe, L.; The bright side of sound: perspectives on the biomedical application of sonoluminescence. *Photochem. Photobiol. Sci.*, 2020, 19, 1114–1121.
66. Ash, C.; Dubec, M.; Donne, K.; Bashford, T.; Effect of wavelength and beam width on penetration in light-tissue interaction using computational methods. *Lasers Med. Sci.*, 2017, 32, 1909–1918.
67. Mendes, R.; Pedrosa, P.; Lima, J. C.; Fernandes, A. R.; Baptista, P. V.; Photothermal enhancement of chemotherapy in breast cancer by visible irradiation of gold nanoparticles. *Sci. Rep.*, 2017, 7(1), 10872.
68. Abadeer, N. S.; Murphy, C. J.; Recent progress in cancer thermal therapy using gold nanoparticles. *Nanomater. Neoplasms*, 2021, 143–217.
69. Chen, C. W.; Hsu, C. Y.; Lai, S. M.; Syu, W. J.; Wang, T. Y.; Lai, P. S.; Metal nanobullets for multidrug resistant bacteria and biofilms. *Adv. Drug Deliv. Rev.*, 2014, 78, 88–104.
70. Lee, H. J.; Song, J. Y.; Kim, B. S.; Biological synthesis of copper nanoparticles using Magnolia kobus leaf extract and their antibacterial activity. *J. Chem. Technol. Biotechnol.*, 11, 2013, 88(11), 1971–1977.
71. Sistemática, D.; Gabriela, S. S.; Daniela, F. R.; Helia, B. T.; Copper nanoparticles as potential antimicrobial agent in disinfecting root canals. A systematic review. *Int. J. Odontostomat.*, 2016, 10(3), 547–554.
72. Murugan, B.; Md Zillur, R.; Fatimah, I.; Lett, J. A.; Annaraj, J.; Kaus, N. H. M.; Al-Anber, M. A.; Sagadevan, S.; Green synthesis of CuO nanoparticles for biological applications. *Inorg. Chem. Commun.*, 2023, 111088.
73. Kanninen, P.; Johans, C.; Merta, J.; Kontturi, K.; Influence of ligand structure on the stability and oxidation of copper nanoparticles. *J. Colloid Interface Sci.*, 2008, 318(1), 88–95.
74. Jain, S.; Nagar, N.; Devra, V.; Synthesis and characterization of highly efficient copper nanoparticles and their catalytic application in oxidative kinetic study. *Adv. Appl. Sci. Res.*, 2015, 171–180.
75. Quaranta, D.; Krans, T.; Santo, C. E.; Elowsky, C. G.; Domaille, D. W.; Chang, C. J.; Grass, G.; Mechanisms of contact-mediated killing of yeast cells on dry metallic copper surfaces; *Appl. Environ. Microbiol.*, 2011, 77(2), 416–426.
76. Liu, S.; Zhang, X. X.; Small colony variants are more susceptible to copper-mediated contact killing for Pseudomonas aeruginosa and Staphylococcus aureus; *J. Med. Microbiol.*, 2016, 65(10), 1143–1151.
77. Chaudhary, P.; Fatima, F.; Kumar, A.; Relevance of nanomaterials in food packaging and its advanced future prospects. *J. Inorg. Organomet. Polym.*, 2020, 30, 5180–5192.
78. Aldhameer, A.; Synthesizing and characterizations of one-dimensional Cu-based antibiofilm surface protective coating. *J. Nanoparticle Res.*, 2020, 22(5), 120.
79. Ssekatawa, K.; Byarugaba, D. K.; Kato, C. D.; Ejobi, F.; Tweyongyere, R.; Lubwama, M.; Kirabira, J. B.; Wampande, E. M.; Nanotechnological solutions for controlling transmission and emergence of antimicrobial-resistant bacteria future prospects, and challenges: a systematic review. *J. Nanoparticle Res.*, 2020, 22, 1–30.
80. Aygun, A.; Gülbagca, F.; Ozer, L. Y.; Ustaoglu, B., Altunoglu, Y. C.; Baloglu, M. C., Atalar, M. N.; Alma, M. H.; Sen, F.; Biogenic platinum nanoparticles using black cumin seed and their potential usage as antimicrobial and anticancer agent. *J. Pharm. Biomed. Anal.*, 2020, 179, 112961.
81. Guleria, A.; Sachdeva, H.; Saini, K.; Gupta, K.; Mathur, J.; Recent trends and advancements in synthesis and applications of plant-based green metal nanoparticles: a critical review; *App. Organomet. Chem.*, 2022, 36(9), e6778.
82. Alshatwi, A. A.; Athinarayanan, J.; Subbarayan, P. V.; Green synthesis of platinum nanoparticles that induce cell death and G2/M-phase cell cycle arrest in human cervical cancer cells. *J. Mater. Sci. Mater. Med.*, 2015, 26, 1–9.
83. Şahin, B.; Aygün, A.; Gündüz, H.; Şahin, K.; Demir, E.; Akocak, S.; Şen, F.; Cytotoxic effects of platinum nanoparticles obtained from pomegranate extract by the green synthesis method on the MCF-7 cell line. *Coll. Surf. B.*, 2018, 163, 119–124.
84. Nellore, J.; Pauline, C.; Amarnath, K.; Bacopa monnieri phytochemicals mediated synthesis of platinum nanoparticles and its neurorescue effect on 1-methyl 4-phenyl 1, 2, 3, 6 tetrahydropyridine-induced experimental parkinsonism in zebrafish. *J. Neurodegener. Dis.*, 2013, 972391.
85. Ghosh, S.; Nitnavare, R.; Dewle, A.; Tomar, G. B.; Chippalkatti, R.; More, P.; Kitture, R.; Kale, S.; Bellare, J.; Chopade, B. A.; Novel platinum–palladium bimetallic nanoparticles synthesized by Dioscorea bulbifera: anticancer and antioxidant activities. *Int. J. Nanomedicine*, 2015, 10, 7477–7490.
86. Depciuch, J.; Stec, M.; Maximenko, A.; Drzymała, E.; Pawlyta, M.; Baran, J.; Parlinska-Wojtan, M.; Synthesis method-dependent photothermal effects of colloidal solutions of platinum nanoparticles used in photothermal anticancer therapy. *Appl. Organomet. Chem.*, 2020, 34(3), e5401.
87. Dobrucka, R.; Romaniuk-Drapała, A.; Kaczmarek, M.; Evaluation of biological synthesized platinum nanoparticles using Ononidis radix extract on the cell lung carcinoma A549. *Biomed. Microdevices*, 2019, 21(3), 1–10.
88. Ullah, S.; Ahmad, A.; Wang, A.; Raza, M.; Jan, A. U.; Tahir, K.; Rahman, A. U.; Qipeng, Y.; Bio-fabrication of catalytic platinum nanoparticles and their in vitro efficacy against lungs cancer cells line (A549). *J. Photochem. Photobiol. B*, 2017, 173, 368–375.
89. Yang, C.; Wang, M.; Zhou, J.; Chi, Q.; Bio-synthesis of peppermint leaf extract polyphenols capped nano-platinum and their in-vitro cytotoxicity towards colon cancer cell lines (HCT 116). *Mater. Sci. Eng. C*, 2017, 77, 1012–1016.
90. Tahir, K.; Nazir, S.; Ahmad, A.; Li, B.; Khan, A. U.; Khan, Z. U. H.; Khan, F. U.; Khan, Q. U.; Khan, A.; Rahman, A. U.; Facile and green synthesis of phytochemicals capped platinum nanoparticles and in vitro their superior antibacterial activity. *J. Photochem. Photobiol. B*, 2017, 166, 246–251.
91. Velmurugan, P.; Shim, J.; Kim, K.; Oh, B. T.; Prunusxyedoensis tree gum mediated synthesis of platinum nanoparticles with antifungal activity against phytopathogens. *Mater. lett.*, 2016, 174, 61–65.
92. Ramkumar, V. S.; Pugazhendhi, A.; Prakash, S.; Ahila, N. K.; Vinoj, G.; Selvam, S.; Kumar, G.; Kannapiran, E.; Rajendran, R. B.; Synthesis of platinum nanoparticles using seaweed Padina gymnospora and their catalytic activity as PVP/PtNPs nanocomposite towards biological applications. *Biomed. Pharmacother.*, 2017, 92, 479–490.
93. Nadaroglu, H.; Gungor, A. A.; Ince, S.; Babagil, A.; Green synthesis and characterisation of platinum nanoparticles

using quail egg yolk. *Spectrochim. Acta Part A Mol. Biomol. Spectrosc.*, 2017, 172, 43–47.
94. Baskaran, B.; Muthukumarasamy, A.; Chidambaram, S.; Sugumaran, A.; Ramachandran, K.; Rasu Manimuthu, T.; Cytotoxic potentials of biologically fabricated platinum nanoparticles from Streptomyces sp. on MCF-7 breast cancer cells. *IET Nanobiotechnol.*, 2017, 11(3), 241–246.
95. Gaidhani, S. V.; Yeshvekar, R. K.; Shedbalkar, U. U.; Bellare, J. H.; Chopade, B. A.; Bio-reduction of hexachloroplatinic acid to platinum nanoparticles employing Acinetobacter calcoaceticus. *Proc. Biochem.*, 2014, 49(12), 2313–2319.
96. Gholami-Shabani, M.; Gholami-Shabani, Z.; Shams-Ghahfarokhi, M.; Akbarzadeh, A.; Riazi, G.; Razzaghi-Abyaneh, M.; Biogenic approach using sheep milk for the synthesis of platinum nanoparticles: the role of milk protein in platinum reduction and stabilization. *Int. J. Nanosci. Nanotechnol.*, 2016, 12(4), 199–206.
97. Venu, R.; Ramulu, T. S.; Anandakumar, S.; Rani, V. S.; Kim, C. G.; Bio-directed synthesis of platinum nanoparticles using aqueous honey solutions and their catalytic applications. *Coll. Surf. A.*, 2011, 384(1–3), 733–738.
98. Chen, L.; Wang, N.; Wang, X.; Ai, S.; Protein-directed in situ synthesis of platinum nanoparticles with superior peroxidase-like activity, and their use for photometric determination of hydrogen peroxide. *Microchim. Acta*, 2013, 180, 1517–1522.
99. Shiny, P. J.; Mukherjee, A.; Chandrasekaran, N.; DNA damage and mitochondria-mediated apoptosis of A549 lung carcinoma cells induced by biosynthesised silver and platinum nanoparticles. *RSC Adv.*, 2016, 6(33), 27775–27787.
100. Syed, A.; Ahmad, A.; Extracellular biosynthesis of platinum nanoparticles using the fungus Fusarium oxysporum. *Coll. Surf. B.*, 2012, 97, 27–31.
101. Fratila, R. M.; Mitchell, S. G.; del Pino, P.; Grazu, V.; and de la Fuente, J. M.; Strategies for the biofunctionalization of gold and iron oxide nanoparticles. *Langmuir*, 2014, 30(50), 15057–15071.
102. Gambhir, R. P.; Rohiwal, S. S.; Tiwari, A. P.; Multifunctional surface functionalized magnetic iron oxide nanoparticles for biomedical applications: a review. *Appl. Surf. Sci. Adv.*, 2022, 11, 100303.
103. Lazaro-Carrillo, A.; Filice, M.; Guillén, M. J.; Amaro, R.; Viñambres, M.; Tabero, A.; Paredes, K. O.; Villanueva, A.; Calvo, P.; Morales, M. D. P.; Marciello, M.; Tailor-made PEG coated iron oxide nanoparticles as contrast agents for long lasting magnetic resonance molecular imaging of solid cancers. *Mater. Sci. Eng. C* 2020, 107, 110262.
104. Sangeetha, G.; Rajeshwari, S.; Venckatesh, R.; Green synthesis of zinc oxide nanoparticles by aloe barbadensis miller leaf extract: structure and optical properties. *Mater. Res. Bull.*, 2011, 46(12), 2560–2566.
105. Mirzaei, H.; Darroudi, M.; Zinc oxide nanoparticles: biological synthesis and biomedical applications. *Ceramics Int.*, 2017, 43(1), 907–914.
106. Fakhari, S.; Jamzad, M.; Fard, H. K.; Green synthesis of zinc oxide nanoparticles: a comparison. *Green Chem. Lett. Rev.*, 2019, 12(1), 19–24.
107. Dobrucka, R.; Długaszewska, J.; Biosynthesis and antibacterial activity of ZnO nanoparticles using Trifolium pratense flower extract. *Saudi J. Biol. Sci.*, 2016, 23(4)517–523.
108. Selim, Y. A.; Azb, M. A.; Ragab, I.; El-Azim, M. H. M. A.; Green synthesis of zinc oxide nanoparticles using aqueous extract of deverra tortuosa and their cytotoxic activities. *Sci. Rep.*, 2020, 10, 3445.
109. Siddiqi, K. S.; Husen, A.; Fabrication of metal and metal oxide nanoparticles by algae and their toxic effects. *Nanoscale Res. Lett.*, 2016, 11.
110. Azizi S, Ahmad M. B.; Namvar, F.; Mohamad, R.; Green biosynthesis and characterization of zinc oxide nanoparticles using brown marine macroalga. *Mater. Lett.*, 2014, 116, 275–277.
111. Jayaseelan, C.; Rahuman, A. A.; Kirthi, A. V.; Marimuthu, S.; Santhoshkumar, T.; Bagavan, A.; Gaurav, K.; Karthik, L.; Rao, K. V. B.; Novel microbial route to synthesize ZnO nanoparticles using Aeromonas hydrophila and their activity against pathogenic bacteria and fungi. *Spectrochim. Acta Part A Mol. Biomol. Spectrosc.*, 2012, 90, 78–84.
112. Moghaddam, A. B.; Moniri, M.; Azizi, S.; Rahim, R. A.; Ariff, A. B.; Saad, W. Z.; Namvar, F.; Navaderi, M.; Biosynthesis of ZnO nanoparticles by a new Pichia kudriavzevii yeast strain and evaluation of their antimicrobial and antioxidant activities. *Molecules*, 2017, 22(6), 872.
113. Basnet, P.; Chanu, T. I.; Samanta, D.; Chatterjee, S.; A review on bio-synthesized zinc oxide nanoparticles using plant extracts as reductants and stabilizing agents. *J. Photochem. Photobiol. B*, 2018, 183, 201–221.
114. Thakur, B. K.; Kumar, A.; Kumar, D.; Green synthesis of titanium dioxide nanoparticles using Azadirachta. *S. Afr. J. Bot.*, 2019, 124, 223–227.
115. Roopan, S. M.; Bharathi, A.; Prabhakarn, A.; Rahuman, A. A.; Velayutham, K.; Rajakumar, G.; Padmaja, R. D.; Lekshmi, M.; Madhumitha, G.; Efficient phyto-synthesis and structural characterization of rutile TiO2 nanoparticles using Annona squamosa peel extract. *Spectrochim. Acta Part A Mol. Biomol. Spectrosc.*, 2012, 98, 86–90.
116. Jayaseelan, C.; Rahuman, A. A.; Roopan, S. M.; Kirthi, A. V.; Venkatesan, J.; Se-Kwon, K.; Iyappan, M.; Siva, C.; Biological approach to synthesize $TiO_2$ nanoparticles using Aeromonas hydrophila and its antibacterial activity. *Spectrochim. Acta Part A Mol. Biomol. Spectrosc.*, 2013, 107, 83–89.
117. Jha, A. K.; Prasad, K.; Kulkarni, A. R.; Synthesis of TiO2 nanoparticles using microorganisms. *Coll. Surf. B.*, 2009, 71, 226–229.
118. Abboud, Y.; Saffaj, T.; Chagraoui, A.; El Bouari, A.; Brouzi, K.; Tanane, O.; Ihssane, B.; Biosynthesis, characterization and antimicrobial activity of copper oxide nanoparticles (CONPs) produced using brown alga extract (Bifurcaria bifurcata). *Appl. Nanosci.* 2014, 4, 571–576.
119. Soureshjani, P. T.; Shad, A.; Mohammadsaleh, F.; Algae-mediated route to biogenic cuprous oxide nanoparticles and spindle-like CaCO3: a comparative study, facile synthesis, and biological properties. *RSC Adv.*, 2021, 11, 10599–10609.
120. Akintelu, S. A.; Folorunso, A. S.; Folorunso, F. A.; Oyebamiji, A. K.; Green synthesis of copper oxide nanoparticles for biomedical application and environmental remediation. *Heliyon*, 2020, 6(7), e04508.
121. Nabila, M. I.; Kannabiran, K.; Biosynthesis, characterization and antibacterial activity of copper oxide nanoparticles (CuO NPs) from actinomycetes. *Biocat. Agri. Biotech.*, 2018, 15, 56–62.
122. Kerour, A.; Boudjadar, S.; Bourzami, R.; Allouche, B.; Eco-friendly synthesis of cuprous oxide ($Cu_2O$) nanoparticles and improvement of their solar photocatalytic activities. *J. Solid State Chem.*, 2018, 263, 79–83.

123. Mehr, S. E.; Sorbiun, M.; Ramazani, A.; Plant-mediated synthesis of zinc oxide and copper oxide nanoparticles by using ferulago angulata (schlecht) boiss extract and comparison of their photocatalytic degradation of Rhodamine B (RhB) under visible light irradiation. *J. Mater. Sci. Mater. Electron.*, 2018, 29, 1333–1340.
124. Reddy, K. R.; Green synthesis, morphological and optical studies of CuO nanoparticles. J. *Mol. Struct.*, 2017, 1150, 553–557.
125. Alwi, R.; Telenkov, S.; Mandelis, A.; Leshuk, T.; Gu, F.; Oladepo, S.; Michaelian, K.; Silica-coated super paramagnetic iron oxide nanoparticles (SPION) as biocompatible contrast agent in biomedical photoacoustics. *Biomed. Opt. Express*, 2012, 3, 2500–2509.
126. Khot, V. M.; Salunkhe, A. B.; Thorat, N. D.; Ningthoujam, R. S.; Pawar, S. H.; Induction heating studies of dextran coated $MgFe_2O_4$ nanoparticles for magnetic hyperthermia. *Dalton Trans.*, 2013, 42(4), 1249–1258.
127. Qu, J. B.; Shao, H. H.; Jing, G. L.; Huang, F.; PEG-chitosan-coated iron oxide nanoparticles with high saturated magnetization as carriers of 10-hydroxycamptothecin: preparation, characterization and cytotoxicity studies. *Coll. Surf. B.*, 2013, 102, 37–44.
128. Babic, M.; Horák, D.; Trchová, M.; Jendelová, P.; Glogarová, K.; Lesný, P.; Herynek, V.; Hájek, M.; Syková, E.; Poly(L-lysine)-modified iron oxide nanoparticles for stem cell labeling. *Bioconjug. Chem.*, 2008, 19(3), 740–750.
129. Wang, X.; Wei, F.; Liu, A.; Wang, L.; Jian-Chun, W.; Ren, L.; Liu, W.; Tu, Q.; Li, L.; Wang, J.; Cancer stem cell labeling using poly(L-lysine)-modified iron oxide nanoparticles. *Biomaterials,* 2012, 33(14), 3719–3732.
130. Guo, C.; Wang, J.; Hui-zhou, L.; Jia-yong, C.; Hydration and conformation of temperature-dependent micellization of PEO–PPO–PEO block copolymers in aqueous solutions by FT-Raman. *Langmuir,* 1999, 15(8), 2703–2708.
131. Laurent, S.; Forge, D.; Port, M.; Roch, A.; Robic, C.; Luce, V. E.; Muller, R. N.; Magnetic iron oxide nanoparticles: synthesis, stabilization, vectorization, physicochemical characterizations, and biological applications. *Chem. Rev.*, 2008, 108, 2064–2110.
132. Sun, S.; Zeng, H.; Robinson, D. B.; Raoux, S. Rice, P. M.; Wang, S. X.; Li, G.; Monodisperse MFe2O4 (M = Fe, Co, Mn) nanoparticles. *J. Am. Chem. Soc.*, 2004, 126, 273–279.
133. Dai, L.; Liu, Y.; Wang, Z.; Guo, F.; Shi, D.; Zhang, B.; One-pot facile synthesis of PEGylated superparamagnetic iron oxide nanoparticles for MRI contrast enhancement. *Mater. Sci. Eng. C*, 2014, 41, 161–167.
134. Zhang, B.; Tu, Z.; Zhao, F.; Wang, J.; Superparamagnetic iron oxide nanoparticles prepared by using an improved polyol method. *Appl. Surf. Sci.*, 2013, 266, 375–379.
135. Yew, Y. P.; Shameli, K.; Miyake, M.; Kuwano, N.; Bt Ahmad, N. B. K.; Bt Mohamad, S. E.; Lee, K. X.; Green synthesis of magnetite (Fe3O4) nanoparticles using seaweed (Kappaphycus alvarezii) extract. *Nanoscale Res. Lett.*, 2016, 11(1), 276.
136. Bandeira, M.; Giovanela, M.; Roesch-Ely, M.; Devine, D. M.; da Silva C. J.; Green synthesis of zinc oxide nanoparticles: a review of the synthesis methodology and mechanism of formation. *Sustain. Chem. Pharm.*, 2020, 15, 100223.
137. Mahdavi, M.; Namvar, F.; Ahmad, M. B.; Mohamad, R.; Green biosynthesis and characterization of magnetic iron oxide (Fe3O4) nanoparticles using Seaweed (Sargassum muticum) Aqueous Extract. *Molecules*, 2013, 18(5), 5954–5964.
138. Tripathi, R. M.; Bhadwal, A. S.; Gupta, R. K.; Singh, P.; Shrivastav, A.; Shrivastav, B. R.; ZnO nanoflowers: novel biogenic synthesis and enhanced photocatalytic activity. *J. Photochem. Photobiol. B*, 2014, 141, 288–295.
139. Santos-Martinez, M. J.; Rahme, K.; Corbalan, J. J.; Faulkner, C.; Holmes, J. D.; Tajber, L.; Medina, C.; Radomski, M. W.; Pegylation increases platelet biocompatibility of gold nanoparticles. *J. Biomed. Nanotechnol.*, 2014, 10(6), 1004–1015.
140. Wang, L. J.; Wu, Q; Tang, S.; Zeng, J. F.; Qiao, R. R.; Zhao, P.; Zhang, Y.; Hu, F. Q.; Gao, M. Y.; Facile non-hydrothermal synthesis of oligosaccharide coated sub-5 nm magnetic iron oxide nanoparticles with dual MRI contrast enhancement effects. *J. Mater. Chem. B*, 2014, 2, 5344–5351.
141. Vllasaliu, D.; Fowler, R.; Stolnik, S.; PEGylated nanomedicines: recent progress and remaining concerns. *Expert Opin. Drug Deliv.*, 2014, 11, 139–154.
142. Ren, E.; Zhang, C.; Li, D.; Pang, X.; Liu, G.; Leveraging metal oxide nanoparticles for bacteria tracing and eradicating. *View*, 2020, 1, 20200052.
143. Elumalai, K.; Velmurugan, S.; Ravi, S.; Kathiravan, V.; Raj, G. A.; Bio-approach: plant mediated synthesis of ZnO nanoparticles and their catalytic reduction of methylene blue and antimicrobial activity. *Adv. Power Technol.*, 2015, 26(6), 1639–1651.
144. Jamdagni, P.; Khatri, P.; Rana, J. S.; Green synthesis of zinc oxide nanoparticles using flower extract of Nyctanthes arbor-tristis and their antifungal activity. *J. King Saud Univ. Sci.*, 2018, 30, 168–175.
145. Rajiv, P.; Rajeshwari, S.; Venckatesh, R.; Bio-Fabrication of zinc oxide nanoparticles using leaf extract of Parthenium hysterophorus L. and its size-dependent antifungal activity against plant fungal pathogens. *Spectrochim. Acta Part A Mol. Biomol. Spectrosc.*, 2013, 112, 384–387.
146. Taranath, T. C.; Patil, B. N.; Limonia acidissima L. leaf mediated synthesis of zinc oxide nanoparticles: a potent tool against Mycobacterium tuberculosis. *Int. J. Mycobacteriol.,* 2016, 5(2), 197–204.
147. Ibrahim, N. A.; Nada, A. A.; Hassabo, A. G.; Eid, B. M; El-Deen, A. M. N.; Abou-Zeid, N. Y.; Effect of different capping agents on physicochemical and antimicrobial properties of ZnO nanoparticles. *Chem. Papers*, 2017, 71(7), 1365–1375.
148. Sunkar, S.; Nachiyar, C. V.; Lerensha, R.; Renugadevi, K.; Biogenesis of $TiO_2$ nanoparticles using endophytic Bacillus cereus. *J. Nanopart. Res.*, 2014, 16(11), 1–11.
149. Devasena, T.; Kumar, P. S. M.; Francis, A. P.; Biosynthesized and chemically synthesized titania nanoparticles: comparative analysis of antibacterial activity. *J. Environ. Nanotechnol.*, 2014, 3(3), 73–81.
150. Santhoshkumar, T.; Rahuman, A. A.; Jayaseelan, C.; Rajakumar, G.; Marimuthu, S.; Kirthi, A. V.; Velayutham, K.; Thomas, J.; Venkatesan, J.; Se-Kwon, K.; Green synthesis of titanium dioxide nanoparticles using Psidium guajava extract and its antibacterial and antioxidant properties. *Asian Pac. J. Trop. Med.*, 2014, 7(12), 968–974.
151. El-Din, S. H.; Fouda, A.; Radwan, A. A.; Salem, S. S.; Barghoth, M. G.; Awad, M. A.; Abdo, A. M.; El-Gamal, M. S.; Endophytic actinomycetes Streptomyces spp mediated biosynthesis of copper oxide nanoparticles as a promising tool for biotechnological applications-. *J. Biol. Inorg. Chem.*, 2019, 24(3), 377–393
152. Černík, M.; Padil, V. V. T.; Green synthesis of copper oxide nanoparticles using gum karaya as a biotemplate and

their antibacterial application. *Int. J. Nanomed.*, 2013, 8, 889–898.
153. Rabiee, N.; Bagherzadeh, M.; Kiani, M.; Ghadiri, A. M.; Etessamifar, F.; Jaberizadeh, A. H.; Shakeri, A.; Biosynthesis of copper oxide nanoparticles with potential biomedical applications. *Int. J. Nanomed.*, 2020, 15, 3983–3999.
154. Nabila, M. I.; Kannabiran, K.; Biosynthesis, characterization and antibacterial activity of copper oxide nanoparticles (CuO NPs) from actinomycetes. *Biocat. Agri. Biotech.*, 2018, 15, 56–62.
155. Xiao, J.; Zhang, G.; Xu, R.; Chen, H.; Wang, H.; Tian, G.; Wang, B.; Yang, C.; Bai, G.; Zhang, Z.; Yang. H.; Zhong, K; Zou, D.; Wu, Z; A pH-responsive platform combining chemodynamic therapy with limotherapy for simultaneous bioimaging and synergistic cancer therapy. *Biomaterials*, 2019, 216, 119254
156. Mirza, M.; Ahmad, S.; Shah, M. S.; Ateeq, M. I. A.; Magnetic nanoparticles: drug delivery and bioimaging applications. In *Metal Nanoparticles for Drug Delivery and Diagnostic Applications*: *Micro Nano Technologies.* Elsevier, 2020, 1, 189–213.
157. Ma, Y.; Wang, X.; Chen, H.; Miao, Z.; He, G.; Zhou, J.; Polyacrylic acid functionalized $Co0_{.85}Se$ nanoparticles: an ultrasmall pH-responsive nanocarrier for synergistic photothermal-chemo treatment of cancer. *ACS Biomater. Sci. Eng.*, 2018, 4, 547–557.
158. Nagajyothi, P. C.; Muthuraman, P.; Sreekanth, T. V. M.; Kim, D. H.; Shim. J; Green synthesis: in-vitro anticancer activity of copper oxide nanoparticles against human cervical carcinoma cells. *Arab. J. Chem.*, 2017, 10, 215–225.
159. Nagaraj, E.; Karuppannan, K.; Shanmugam, P.; Venugopal, S.; Exploration of bio-synthesized copper oxide nanoparticles using pterolobium hexapetalum leaf extract by photocatalytic activity and biological evaluations. *J. Clust. Sci.*, 2019, 30, 1157–1168.
160. Sankar, R.; Maheswari, M.; Karthik, S.; Shivashangari, K. S.; Ravikumar, V.; Anticancer activity of Ficus religiosa engineered copper oxide nanoparticles; *Mater. Sci. Eng. C*, 2014, 44, 234–239.
161. Sulaiman, G. M.; Tawfeeq, A. T.; Jaaffer, M D.; Biogenic synthesis of copper oxide nanoparticles using olea europaea leaf extract and evaluation of their toxicity activities: an in vivo and in vitro study. *Biotechnol. Prog.*, 2018, 34, 218–230.
162. Prashanth, G. K.; Prashanth, P. A.; Bora, U.; Gadewar, M.; Nagabhushana, B.; Ananda, S.; Krishnaiah, G.; Sathyananda, H.; In vitro antibacterial and cytotoxicity studies of ZnO nanopowders prepared by combustion assisted facile green synthesis; *Karbala Int. J. Mod. Sci.*, 2015, 1, 67–77.
163. Suresh, J.; Pradheesh, G.; Alexramani, V.; Sundrarajan, M.; Hong, S. I.; Green synthesis and characterization of zinc oxide nanoparticle using insulin plant (Costus pictus D. Don) and investigation of its antimicrobial as well as anticancer activities. *Adv. Nat. Sci.: Nanosci. Nanotechnol.*, 2018, 9, 015008 (8 pages).
164. Ji, W.; Zhu, D.; Chen, Y.; Hu, J.; Li. F.; In-vitro cytotoxicity of biosynthesized Zinc oxide nanoparticles towards cardiac cell lines of Catla catla. *Biomed. Res.*, 2017, 28, 2262–2266.
165. Andra, S.; Balu, S. K.; Jeevanandham, J.; Muthalagu, M.; Vidyavathy, M.; Chan, Y. S.; Danquah, M. K.; Phytosynthesized metal oxide nanoparticles for pharmaceutical applications. *Naunyn-Schmiedeberg's Arch. Pharmacol.* 2019, 392, 755–771.
166. Kim, E. H.; Lee, H. S.; Kwak, B. K.; Kim, B. K.; Synthesis of ferrofluid with magnetic nanoparticles by sonochemical method for MRI contrast agent. *J. Magn. Magn. Mater.*, 2005, 289, 328–330.

## LIST OF ABBREVIATIONS

| | |
|---|---|
| **Ag** | Silver nanoparticles |
| **Au** | Gold nanoparticles |
| **BSA** | Bovine serum albumin |
| **-COOH** | Carboxyl groups |
| **Cu** | Copper nanoparticles |
| **CuO** | Copper oxide |
| **DMSA** | Di mercapto succinic acid |
| **DNA** | Deoxyribonucleic acid |
| **$eNH_2$** | Amine |
| **EPS** | Extracellular polymeric substances |
| **$Fe_2O_3$** | Iron oxide |
| **FTIR** | Fourier-transform infrared |
| **HCT-116** | Human colorectal carcinoma cell line |
| **IgG1** | Immunoglobulin G |
| **KB** | Human epithelial carcinoma cells |
| **LSPR** | Localized surface plasmon resonance |
| **MCF-7 cancer cells** | Michigan Cancer Foundation-7 |
| **MNPs** | Magnetic nanoparticles |
| **MRI** | Magnetic resonance imaging |
| **NPs** | Nanoparticles |
| **Pb** | Lead |
| **PEG** | Polyethylene glycol |
| **Pt** | Platinum nanoparticles |
| **QDs** | Quantum dots |
| **ROS** | Reactive oxygen species |
| **SERS** | Surface-enhanced Raman spectroscopy |
| **SPIONPs** | Superparamagnetic iron oxide nanoparticles |
| **SPR** | Surface plasmon resonance |
| **TEM** | Transmission electron microscopy |
| **$TiO_2$** | Titanium oxide |
| **UCNPs** | Up conversion nanoparticles |
| **VITBN4** | Vellore Institute of Technology actinomycetes non-motile |
| **XRD** | X-ray diffraction |
| **ZnO** | Zinc oxide |

# 8 Colorimetric Sensing Applications of Metal and Metal Oxide Nanoparticles

*Jincemon Cyriac*[1,2], *Tom Cherian*[3], *Gouri G. Nair*[1], *Niranjana A. R.*[1], *Sashank S.*[1], *Riju K. Thomas*[4], *and Sheeja V. N.*[1]

[1]Department of Physics, Marian College (Autonomous), Kuttikanam, Idukki, Kerala, India

[2]Department of Physics, Christ College (Autonomous), Calicut University, Irinjalakuda, Thrissur, Kerala, India

[3]Department of Chemistry, Christ College (Autonomous), Calicut University, Irinjalakuda, Thrissur, Kerala, India

[4]Department of Physics, Bharata Mata College, Thrikkakara, Idappalli, Kochi, Kerala, India

## 8.1 INTRODUCTION

Nanotechnology is the field of science that deals with materials with at least one of their dimensions in the nanoscale (0–100 nm). They exhibit unique physio-chemical properties such as optical, mechanical, magnetic, conductive and physical qualities that differ from the same chemical compounds at larger sizes.[1] Also, small changes in the size or shape of nanoparticles cause appropriate changes in their properties.

Colorimetry is an efficient analytic technique which has gained significance due to its simple and cost-efficient procedures. The incorporation of nanotechnology in colorimetry has resulted in the field being an important area of research. The use of nanoparticles as colorimetric sensors has led to colorimetry being one of the most preferred analytic methods. Colorimetry using nanoparticles is an analytic method with relatively low detection limits and high sensitivity, hence achieving real rate accuracy.[2]

Some of the noteworthy properties of nanoparticle colorimetric sensors are their high sensitivity and selectivity. Hybrid sensors are found to achieve a remarkable detection limit in the parts per billion by volume (ppbV) range. It is worth noting that with the noise level of the detection system, the estimated detection limit is even lower, reaching the level of a few tens of parts per trillion by volume (pptV). Further enhancements in detection limits can be achieved through the use of more advanced optical and electrochemical detection systems.[2]

The sensors were observed to demonstrate a high degree of selectivity in the presence of common interference, including changes in humidity, perfumes, mouthwash vapors, cleaners and petroleum products. The RGB color patterns provide a distinct fingerprint for explosives, allowing for discriminative detection even in complex environments.

### 8.1.1 Colorimetric Sensors and Classification

Colorimetric sensors are optically sensitive materials which, on changes in their environment, respond with a change in color. They exhibit a change in color of the substrate it is added to, providing elementary details of the chemical composition of the substrate. The sensors operate by displaying a color change, undergoing a bathochromic or hypsochromic shift visible to the naked eye, thus amplifying details at a molecular level, such as the presence of a functional group, to a macroscopic scale. Thus, colorimetric sensors open the door to qualitative or quantitative analysis of the materials. The unique localized surface plasmon resonance (LSPR) of metal nanoparticles makes them suitable candidates for colorimetric sensing. Metal oxides, graphene, DNA and some polymers are also found to make efficient colorimetric sensors.[1]

## 8.2 METAL-BASED SENSORS

The discovery of $Fe_3O_4$ mimicking the catalytic activity of horseradish peroxidase has opened new pathways in the development of artificial enzymes. These nanoparticles, which were capable of replacing natural enzymes, were named nanozymes. Further research led way to the discovery that nanoparticles of Au, Ag, Cu and Pt after processing also showed similar enzyme catalytic activity. Moreover, these nanozymes were found to be more efficient than naturally found enzymes. The colorimetric methods, closely related to nanozymes, were hence developed to prepare colorimetric materials optically sensitive to various analytes. Nanoparticles of alkali and noble metals also have broad

DOI: 10.1201/9781032660639-8

absorption bands in the visible region, which also aids in them being good colorimetric sensors.[2,3]

### 8.2.1 Gold Nanoparticles

Gold nanoparticles, due to their unique optical properties in various aggregated states, ease of preparation, high selectivity and sensitivity and good biological affinity, are the most commonly preferred nanoparticle for colorimetric sensing.[1] The high sensitivity and absorption coefficient properties of gold nanoparticles are exploited to produce nanozymes of relatively high catalytic activity. The colorimetric application of AuNPs can be attributed to their significant color change due to aggregate formation and morphology change due to changes in their chemical environment.

One of the primary applications of AuNPs is in food and water quality and safety testing. AuNPs can be used for the detection of heavy metal ions in samples, which are possible contaminants (toxins) of water; hence, they help to ensure the quality of water. They are also used to test the safety of food samples. An example is their use in the detection of kanamycin, an antibiotic present in milk, which in greater concentrations can result in ototoxicity and renal toxicity. The glycosides present in kanamycin are specific to AuNPs; hence, the interaction of both results in the formation of an aggregate.[1] This results in a change of the properties of the AuNPs so that kanamycin can be quantitatively determined by calculation.

On modification (phosphorus with citric acid, $Au^{3+}$-hexadecyl trimethyl ammonium bromide, fatty acid, cysteine, single-stranded DNA and other substances), AuNPs can be used to detect organophosphorus pesticides (OPs).[3] OPs are pollutants with adverse effects on the environment and human health. The presence of OPs in soil is not easily detected. However, with the help of highly sensitive and selective AuNPs, they are quite easily detected.

AuNPs are also widely used to prepare highly selective biosensors for the detection of biomolecules such as proteins, sugars, lipids and neurotransmitters. They have also been used for the detection of nucleic acid and biomarkers.

### 8.2.2 Copper Nanoparticles

Copper nanoparticles (CuNPs) also exhibit enzymatic catalytic properties, which contribute to their application as colorimetric sensors. CuNPs, a noble metal, apart from showing catalytic activity, also interact with other substances. This property can be used for the analysis of some systems. For example, the catalytic activity of CuNPs is reduced in the presence of pyrophosphate. Hence, activity changes of CuNPs can be used to detect the presence of pyrophosphate in a system.[4] CuNPs can also be used in food safety monitoring. An interesting example is the monitoring of fish spoilage using CuNPs. Spoilage of fish results in the release of hydrogen sulfide. It is a volatile sulfur gas released during the degradation of sulfur, containing amino acids. CuNPs can be modified for the detection of hydrogen sulfide, which indicates spoilage of fish. They can also be modified for the detection of trace amounts of metals and metal ions. Modified CuNPs (Rano-CuNPs) can be used as colorimetric sensors to detect the presence of $As^{3+}$ ions (which are toxic) in ground water, hence ensuring water safety. CuNPs are employed in the detection of several other heavy metals and ions as well such as mercury and sulfide ions.[1]

### 8.2.3 Iron Nanoparticles

Iron (Fe3O4) nanoparticles were one of the earliest colorimetric sensors and paved the way for research on similar properties of other nanoparticles. Iron nanoparticles (Fe NPs), along with several other NPs, were found to have enzyme catalytic activity similar to $Fe_3O_4$. Fe NPs were found to be efficient in detecting $H_2O_2$, glucose and biomolecules. An example of their application is the use of gold-coated iron oxide nanoparticles for the detection of coagulation factor thrombin.[1]

### 8.2.4 Other Nanoparticles

Other commonly used nanomaterials include platinum, palladium and silver.[1] Pt NPs are used to detect the presence of $Hg^{2+}$ ions through peroxidase activity simulation of metal NPs. Pd nanocubes whose enzymatic activity (peroxidase) was inhibited due to the presence of sulfide ions were used to detect the presence of those sulfide ions. Pd NPs with similar catalytic activity could be used to detect the presence of cysteine and homocysteine. Pt-Pd NPs (bimetallic) have also been used to detect the presence of the herbicide Atrazine. Another significant metal NP that is used is silver nanoparticles for colorimetric detection of water pollutants.

## 8.3 METAL OXIDE–BASED SENSORS

Apart from metals, some metal oxides have also been found to be efficient colorimetric sensors. One of the advantages of the use of metal oxides as colorimetric sensors is that they can be obtained using green synthesis, hence eliminating any toxicities during preparation of the sensor.[5]

### 8.3.1 Zinc Oxide

Zinc oxide nanoparticles (ZnO NPs) are used in the detection of Aflatoxin B1 in rice. Aflatoxin, produced as metabolites of the fungus Aspergillus, which often contaminates food crops and animal feed products, is a toxin and a Group 1 carcinogen.[6] Colorimetric detection of aflatoxin is carried out using a combination of ZnO NPs and curcumin composite solution.

ZnO NPs are also used in the detection of UVA (ultraviolet-A [315–400 nm]). They are also efficiently used in the detection of pesticidal residue in water. ZnO NPs are also used in the detection of heavy metal ions in water.[7]

### 8.3.2 Copper Oxide

Copper oxide nanoparticles (CuO NPs) are an important class of NPs due to their low band gap energy and easy and cheap synthesis. They are used in quantitative detection of nitrites in food due to their surface plasmon properties. Modified CuO NPs are efficiently used in the detection of hydrogen peroxide and glucose. They have been widely used in the detection of proteins and nucleic acids. Using this familiar technique, we can employ CuO NPs for the detection of sickle cell anemia disease. Sickle cell anemia is a genetic disorder due to a point mutation in the beta-globin gene.[8] The use of CuO NPs as colorimetric biosensors is based on their peroxidation property and ability to bind to double-stranded DNA. They can also be used to detect single-base mutations in the DNA with high sensitivity at low concentrations.[9]

## 8.4 GRAPHENE

Graphene and its derivatives possess remarkable optical, electrical and physicochemical properties that have demonstrated considerable potential for sensing applications. With a combination of its flat, two-dimensional nanostructure and high surface area, graphene can bind analytes with an exceptionally high density of surface-active sites, improving the sensitivity of colorimetric sensors.[10] The atomic-thick two-dimensional (2D) layered honeycomb lattice structure of carbon, referred to as graphene, has generated a great deal of interest. Excellent physical, chemical, optical and electrical capabilities set 2D nanomaterials and their nanocomposites apart from traditional bulk materials. These properties find widespread applications in biosensing, administration of drugs, reaction catalysis, energy storage and biological therapy.

Graphene and its derivatives, including graphene oxide (GO), reduced graphene oxide (rGO) and graphene quantum dots (GQDs), have numerous beneficial characteristics, which makes them widely used in biosensing applications. It has many useful features, such as a large surface area with plenty of active sites, a surface that can be easily modified because it contains oxygen groups, the ability to transfer energy efficiently, and good compatibility with living systems.

As graphene and its derivatives have strong polarization-dependent effects and broad band light absorption, they have been successfully used to build a visual detection platform that can identify a variety of analytes, including small molecules, DNA, pollutants and biomolecules.[11] Considering nanozyme sensors are more stable and less expensive than natural enzymes, they are becoming increasingly popular.[12] Researchers have demonstrated that, either employed alone or in combination with other nanomaterials, nanoparticles possess catalytic activity contributing to their development. For instance, in the presence of $H_2O_2$, 3,3′, 5,5′-tetramethylbenzidine (TMB) can react with graphene and its derivatives with peroxidase-like activity, resulting in color changes.[13,14] For instance, graphene nanoribbon (GNR)–based colorimetric sensors for dopamine detection have been reported by Ali et al.[15] The catalytic activity of GNR, GO and multi-walled carbon nanotubes (MWCNTs) was examined. The findings showed that GNR had greater intrinsic peroxidase-like activity than GO and MWCNT. This might be due to the smaller size and higher rate of electron transfer of nanoribbons than graphene sheets, which speeds up the redox interaction between TMB and $H_2O_2$.[16] Heteroatom-doped graphene-based nanozyme sensor arrays for the detection of aromatic pesticides were created by Wei's group.[17] When adsorption masks the active sites on graphene and the blue color is lowered, the peroxidase-mimicking activity of the nanozyme may be inhibited in the presence of pesticides.

In order to conduct a colorimetric analysis of trace mercury ions and enrichment simultaneously, graphene oxide nanosheets (CGO) functionalized with L-cysteine were developed. As CGO has higher S- and N-containing active sites than pure GO, it has been found to have stronger peroxidase-like catalytic activity, and it could be attributed to CGO's increased catalytic activity. In addition to this, $Hg^{2+}$ might be loaded onto CGO with a great affinity for the S-containing groups, which would significantly increase CGO's peroxidase-like activity. This was a result of the sensing principle's reliance on TMB and $Hg^{2+}$'s competitive adsorption over CGO. As a result of $Hg^{2+}$ occupying the active sites and preventing TMB from binding to CGO, there was a greater amount of free TMB in the solution, which ·OH could oxidize to form colored oxidation products. Due to their favorable optical characteristics, GQDs with a side length of less than 100 nm have been used as probes and nanozymes because of their inherent peroxidase-like activity.[18]

Researchers discovered that modified graphene and its derivatives with hemin and metal or metal oxide nanoparticles had intrinsic peroxidase-like catalytic activity, acting as synthetic enzymes for the detection of metal ions, tiny molecules and biological molecules. For instance, Jing et al.[19] synthesized molecular imprinting polymers (MIPs) on a hemin–graphene nanosheet (H-GN) surface in order to fabricate a colorimetric paper-based sensor for the detection of large molecular weight proteins. H-GNs served as catalysts for the synthesis of MIP film, supporters for the enrichment of high molecular weight proteins and catalytic enzymes for the oxidation of TMB. The functional paper-based colorimetric sensor exhibited a highly selective identification ability for thyroglobulin due to its higher adsorption capacity and imprinting features. Lu et al. developed histidine-capped gold nanoclusters (His@AuNCs) with intrinsic oxidase-like activity.[20] In the absence of $H_2O_2$, these nanoclusters could directly oxidize TMB to form blue-colored ox-TMB when His@Au NCs and rGO were further mixed, with the catalytic activity of the His@AuNCs/rGO nanocomposites significantly enhanced. Additionally, using nitrite and TMB as substrates, the oxidase-like activity of the as-prepared His@AuNCs/rGO was assessed. Because TMB and nitrite may have similar catalytic active sites, the results showed that nitrite might limit the catalytic activity of His@AuNCs/rGO in the oxidation of TMB. This result suggests that His@AuNCs/rGO were employed as an oxidase imitator for spectrophotometric sensor nitrite detection.

## 8.5 POLYMERS

Embedding pH-responsive organic molecules[21] into polymers allows the creation of smart materials, mainly visual sensors and, to a lesser extent, polymer-based chemo-sensors. The latter strategy has been shown to be more effective for immobilization in the design of pH-responsive materials because it creates a chemically anchored sensor motif on the polymer backbone by designing a chemo-sensing monomer and then polymerizing it. This is an advanced method to prevent sensor probe migration, leaching and hysteresis and to achieve unparalleled long-term stability.

The polymer membranes that Pablos's group[22] created are controllable films for colorimetric sensor materials that are used for the visual detection of water acidity in the low pH range (below 4) and in the high pH range (above 11.6 M of perchloric acid). Two monomers with weakly basic groups and up to four protonizable groups in the previously stated acidic regions were synthesized. The membranes in water, which demonstrated gel behavior, had pKa (solution's acid dissociation constant) values that varied from 2.7 to −6.5. As a result, visible color changes qualitatively revealed the acidity of water containing perchloric acid, and UV-vis spectroscopy allowed for a cost-effective way of measurement. Practical uses for titration with additional acids are anticipated. Moreover, smart fabrics that reacted colorimetrically to the acidity of the aqueous media were produced by polymer coatings of high-tech p-aramid fibers and traditional cotton utilizing the sensor monomers that were created. Additionally, both smart fibers and membranes showed color changes in acidic environments, indicating the potential for the development of completely sensory clothing and smart tags with intended uses in the fields of health and safety. Furthermore, the sensor materials can be reused because they are highly stable for extended periods of time in acidic environments and can be rinsed with water after protonation. Designing polymers with conjugated moieties that include weakly basic groups has allowed for the successful development of reusable colorimetric sensor materials for visual sensing of very acidic conditions. To create handleable films, coated cotton and high-tech p-aramidbres, monomers with up to four protonizable groups were synthesized and copolymerized with other commercial comonomers. This resulted in cross-linked membranes. With pKa ranging from 2.7 to −6.5 in aqueous solution, these materials allowed for the quantification of acidity using UV-vis spectroscopy titration curves and the visual control of acidity below pH = 4 and beyond the pH scale up to a concentration of 11.6 M in perchloric acid. These substances react colorimetrically to acids in gas phase as well as in acidic environments, such as those found in water media. Therefore, applications such as acidity control in environmental and industrial processes are anticipated.

## 8.6 DNA

As a highly stable, programmable and easily modifiable biomolecule, DNA has found widespread application as a sensing probe for target identification. Additionally, the cost of chemically synthesizing DNA oligonucleotides is inexpensive, and batch-to-batch variance is minimal.[23] The affinity of DNA structures with targets, which may result in the alteration of DNA nanostructures, is the basis for all of the mechanisms of DNA nanomaterial-based colorimetric sensors. Typical examples include DNAzymes (deoxyribozomes), ion-bridged complexes created between bases (such T-Hg-T and C-Ag-C bridges), G-quadruplexes generated by metal ions (like $K^+$ and $Na^+$) with G-rich sequences and so on.[24] Thus far, reports and applications of DNA-based nanomaterials for the detection of metal ions, viruses, pathogenic microorganisms and biomacromolecules have been made.

### 8.6.1 G-Quadruplex-Based Colorimetric Sensors

Biosensors have employed G-quadruplex (G4), a unique DNA structure with a variable binding capacity, for signal transduction and target recognition. G4-based colorimetric sensors have garnered significant interest in the identification of analytes because of their affordability, high sensitivity and simple operation. G4 has a number of intriguing characteristics, including the ability to bind dyes, metal ions and porphyrin. Furthermore, G4 may bind hemin to function as a DNAzyme mimicking peroxidase.[25] G4-based colorimetric sensors have therefore been widely used for the detection of metal ions, radioactive gas, small molecules, viruses and protein.[26]

Depending on their size, metal ions may form bonds with the carbonyl oxygens in the G4 plane or between two planes. Numerous metal ion detection sensors have been created based on this feature. After screening a few sequences in 2016, Liu's group[27] created a $Ti^+$ (titanium monoxide ion) biosensor based on a G4 sequence. $Ti^+$ has the ability to cause DNA folding, as is well known, while AuNPs exhibited very high extinction coefficients and color changes brought on by aggregation. The color turned blue because the $Ti^+$-folded DNA prevented AuNPs from adhering to it, allowing them to cluster more readily.

Hemin@G4 DNAzymes were created in order to improve the sensitivity and specificity of G4-based colorimetric sensors. Hemin@G4 DNAzymes could function as a mimic peroxidase to catalyze various colorless substrates to create a color shift such as ABTS (2,2-azino-bis-3-ethylbenothiazoline-6-sulphonic acid) and TMB.[28] G4/hemin DNAzymes (G4 DNAzymes) were developed by Ju and colleagues[29] for the sensitive colorimetric measurement of

$Pb^{2+}$, based on the oxidation of ABTS. A d [TC] (DNA [thymine cytosine] sequence) sequence was added to the 3′ end of the G4 DNAzymes, and $K^+$ was used to stabilize them. To maintain the parallel topology, $Pb^{2+}$ was introduced in place of $K^+$ in the parallel G4. However, the addition of $Pb^{2+}$ reduced the affinity between topology and hemin, which in turn reduced DNAzyme activity for catalyzing the oxidation of ABTS to produce a green dye. Without any amplification, this straightforward visual sensor produced a good linearity for $Pb^{2+}$ concentration ranging from 0.01 to 10 μM and a limit of detection (LOD) of 7.1 nM. Fei et al.[30] created an exonuclease-assisted multicolor apt sensor for ochratoxin A (OTA) visual detection based on the oxidation of TMB. In this colorimetric sensor, exonuclease I from the 3′-end of AG4-OTA could stop OTA's digestion when it bound to a DNA sequence (AG4-OTA) made up of a hemin aptamer and an OTA aptamer. The residual hemin aptamer exhibited peroxidase-like activity and was capable of binding to hemin to form a G4-hemin DNAzyme, which in turn could catalyze the oxidation of TMB by $H_2O_2$ to yield $TMB^{2+}$ in an acidic solution. $TMB^{2+}$ has the ability to oxidize Au (0) into Au (I), etching gold nanorods and producing hues similar to a rainbow. A multi-colored platform for the visual detection of OTA was made possible by this colorimetric sensor.

microRNA (miRNA) and other biomolecules have, as we know, evolved into biomarkers for a wide range of illnesses. Scientists are thus paying more and more attention to the creation of miRNA sensors that are inexpensive, easy to use and very sensitive. It was discovered in Lee's prior research[31] that GO could be joined with ssDNA through hydrogen bonding and π–π interactions, which could readily enrich the DNA probe. This discovery led Lee and colleagues[32] to describe the first GO-based paper sensor for colorimetric miRNA detection using the target-recycled signal amplification technique. In order to create the paper-based miRNA sensor, two steps were involved: first, G4 DNAzyme (Dz) was made and used to produce signals in colorimetric assays; second, GO as a net (GONET) was utilized to capture single-stranded DNA (ssDNA) probes through hydrogen bonding and π–π interactions; this allowed the Dz/GO composite to easily accumulate onto a desired spot on the paper, which could then be amplified in the presence of hemin, ABTS and $H_2O_2$.

### 8.6.2 DNA Tetrahedron-Based Colorimetric Sensors

The one-step building of the DNA tetrahedron, a classical and basic polyhedron, was first described by Turberfield and associates.[33] A stable, uniformly proportioned DNA tetrahedron is created by synthesizing four equal moles of carefully crafted 55-base oligonucleotides. DNA tetrahedrons have a great degree of programmability, which allows for creative design and material combinations. In particular, to address the needs of sensors, biomedical and other disciplines, functional DNA tetrahedrons were designed, with a primary focus on the tetrahedron's four vertices and six duplex-helix edges.[34] Thus far, colorimetric metal ion detection has been carried out worldwide using DNA tetrahedron nanostructures.[35]

Certain metal ions have drawn a lot of attention globally due to their high toxicity and bioaccumulation, which can lead to serious health and environmental issues. Thus, it is crucial to design a highly sensitive and selective monitoring technique for trace levels of metal ions. More critically, the shape of the hard and stable framework known as a DNA tetrahedron nanostructure can be changed by mechanically rearranging the DNA strand at the tetrahedron's arm in response to outside stimuli.[36] Furthermore, carrying an aptamer within the DNA tetrahedron structure might improve the sensitivity and specificity of DNA tetrahedron-based sensors. One such example is the DNA tetrahedron functionalized paper-based colorimetric sensor for $Hg^{2+}$ detection that Chen and colleagues developed. It is based on the strong and highly selective binding ability of thymine–thymine base pairs toward $Hg^{2+}$. The DNA tetrahedron's well-defined spacing, minimal steric hindrance effect and controlled particular orientation made it an excellent scaffold for anchoring aptamer and increasing the assay platform's capture efficiency.[37] In the absence of $Hg^{2+}$, biotinylated aptamer might bind with streptavidin-labeled HRP. However, biotinylated aptamer preferentially attached to $Hg^{2+}$ in the presence of $Hg^{2+}$ because T-Hg-T has a higher binding affinity than T-A (thymine-adenine base pair). This led to the biotinylated aptamer's release from the DNA tetrahedron. As a result, the DNA tetrahedron functionalized paper-based sensor was unresponsive to streptavidin-labeled horseradish peroxidase (HRP). Following the introduction of TMB–$H_2O_2$, a clear shift in the color signal was detected.

## 8.7 APPLICATION OF COLORIMETRIC SENSORS

Colorimetric sensors have applications in various fields, including chemistry, biology, environmental science and industry, as they are simple, quick and cost effective compared to alternatives.[24] Some of the applications include the following.

### 8.7.1 Water Pollution Detection

These days, the discussion of water pollution is imperative, as it has an impact on both human health and the environment. Pollutants present in water harm important organic mechanisms such as transport of nutrients into cells and cause many diseases. In water pollutants, we focus our attention on heavy metal ions such as Fe(III), Cu(II), Ni(II), Hg(II), Mn(II), Pb(II), Cd(II), Co(II) and Cr(III/V), as they are harmful for human health. Colorimetric sensors can be employed to detect these pollutants. Many silver nanoparticles have been employed to detect these pollutants. They differ mainly in the stabilizing agents used, which increases the protection of the nanoparticle core and provides the ability to interact selectively with contaminants.

Since AgNPs have a higher extinction coefficient and are less expensive than AuNPs, they have been utilized as an alternative to colorimetric sensing systems.[24] Based on the difference in optical brightness and improved sensitivity of UV-visible spectroscopic detection, AgNPs' high molar extinction coefficient improves visibility.[38] AgNP colorimetric sensing for copper ion by PADs has not been reported,[39] despite the fact that colorimetric applications of AgNP aggregation offer high sensitivity and selectivity for the determination of analytes such as histidine, homocysteine, cysteine, mercury ions and nickel ions using UV–visible spectroscopy.[40] AgNPs have only been used analytically once on cellulose paper, and that is to use Raman spectroscopy to determine the presence of Rhodamine 6 G dye.

## 8.7.2 Food Safety Screening

The method of systematically assessing and maintaining a check on food products to make sure they correspond to safety regulations and are free of substances that could be harmful to human health is known as food safety screening. Finding and removing possible risks along the whole food supply chain – from production and processing to distribution and consumption – is the aim of food safety screening.

### 8.7.2.1 Gold Nanoparticles

The most stable metal nanoparticles are gold nanoparticles, also known as gold colloids, which typically have sizes between 1 and 100 nm. They have special qualities that aren't present in bulk materials. From a sensing standpoint, the inherent qualities of gold nanoparticles present a number of advantages for the development of smart sensors.[3] Essentially, producing stable gold nanoparticles is really simple, and varying the experimental settings may precisely control the size and form of the particles to produce a unique optical property. In addition, due to their high surface-to-volume ratio, gold nanoparticles are a great scaffold for immobilizing large amounts of particular functional groups, which enables the targeted analyte to respond quickly and with high sensitivity. Furthermore, the distinctive optical properties of gold nanoparticles are derived from the collective oscillation of electrons at their surface. These properties can be optimized by manipulating factors like as size, composition, sharpness and chemistry. Practically, chemically and physiologically active compounds show great compatibility with gold nanoparticles. This indicates that the functional activity of gold nanoparticles is not significantly affected by their immobilization. Finally, gold nanoparticles can serve as a variety of platforms for detection. Stated differently, a target analyte can be identified using multiple detection methodologies. At present, colorimetric, fluorometric, spectroscopic and electrochemical approaches are the most widely used detection platforms that are aided by gold nanoparticles. As a result, gold nanoparticles can be created as powerful analytical instruments that are useful in a variety of applications when used properly. Significant efforts have been devoted over the past four decades to the fabrication of monodispersed gold nanoparticles with controlled size and shape. So far, gold nanoparticles with diameters varying from 2 nm to 150 nm and different shapes, including nanospheres, nanorods, nanoshells and nanocages, can be quickly synthesized through different chemical and biological protocols.

A thorough description of gold nanoparticles and their possible uses in colorimetric sensing-based food safety screening is given in the Xue et al. (2023) paper. AuNPs are special nanomaterials with a variety of characteristics, including as optical qualities, biocompatibility, and programmable size and shape. These characteristics make them perfect for application in colorimetric sensing, a quick and easy way to find a variety of analytes, including toxins in food. This section discusses the latest developments in gold nanoparticle (AuNP)-based colorimetric sensors for detecting infections, heavy metals, pesticides, and antibiotics in food. It also shows examples of how these sensors have been used to identify contaminants in a variety of food matrices including dairy, meat, fruits, and vegetables.

### 8.7.2.2 Silver Nanoparticles

AgNPs,[40] or silver nanoparticles, are a potential material for colorimetric water pollution monitors. They are superior to traditional sensors in a number of ways, such as high sensitivity, selectivity, simplicity and low cost.

AgNPs interact with contaminants in a way that modifies their surface plasmon resonance (SPR) characteristics, thereby functioning as colorimetric sensors. The AgNP SPR is extremely responsive to modifications in their size, shape, and surroundings. AgNPs may aggregate, undergo size changes or experience other changes in their SPR characteristics when exposed to a pollutant. As a result, the AgNPs' color changes, which can be utilized to determine the pollutant's concentration and presence. Colorimetric sensors for a range of water contaminants have been developed using AgNPs, including:

- Heavy metals: AgNPs can be functionalized to detect heavy metals like arsenic, mercury and lead in a selective manner.
- Organic pollutants: Pharmaceuticals, herbicides and insecticides are examples of organic pollutants that AgNPs may identify.

AgNPs are also useful in the detection of biological pollutants, including viruses and bacteria. It has been demonstrated that AgNP-based colorimetric sensors are useful for identifying water pollutants in a range of real-world samples, such as surface water, wastewater and drinking water. They represent a promising new paradigm in pollution prevention and water quality monitoring.

Examples of AgNP-based colorimetric water pollution sensors include:

- Lead detection: Thiol-containing compounds with a strong affinity for lead ions can be functionalized onto AgNPs. AgNPs undergo a color change from yellow to red upon exposure to lead. Lead at as low as 10 parts per billion can be found using this method.
- Mercuric ion detection: AgNPs can also be functionalized for the detection of mercuric ions. AgNPs are exposed to mercury and combine with the ions to produce a complex that turns them from yellow to white. As little as 1 ppb of mercury can be found using this method.
- Pesticide detection: Antibodies specific to particular pesticides can be functionalized onto AgNPs. AgNPs attach to antibodies and aggregate when exposed to pesticides. As a result, the hue turns from yellow to red. Pesticides at concentrations as low as 10 parts per billion can be found using this method.
- Bacteria detection: High-affinity compounds for bacteria can be functionalized onto AgNPs. They bind to bacteria when they come into contact with them and group together. As a result, the hue turns from yellow to red. Bacteria at concentrations as low as 100 cells/mL can be found with this method.

Although they are still in the early stages of development, AgNP-based colorimetric sensors have the potential to completely change pollution control and water quality monitoring. They can be used to identify a wide range of contaminants in real-world samples since they are straightforward, affordable and sensitive.[41]

## 8.8 FOOD PRESERVATIVES

The technique of preserving food involves increasing its shelf life without sacrificing its nutritional content, safety or flavor. Food can be preserved using a variety of processes, each with its own set of guidelines.

Food preservation is not only about preventing food from spoiling but also has far-reaching implications for economic, environmental, nutritional and safety considerations. It contributes to food security, reduces waste and ensures that a diverse range of foods is available and safe to consume.

### 8.8.1 Microbial Contamination

Contamination due to microbial toxins produced by microorganisms is a concern for public health. Colorimetric biosensors can be implemented as a quick and easy way to detect them at even trace levels. Nanomaterial-based transduction systems can be used for this purpose.

#### 8.8.1.1 Gold Nanoparticles

The selectivity of colorimetric biosensors can be achieved by the combination of aptamers and nanoparticles. Aptamers have several advantages: they are small and have high binding affinity; reproducible chemical synthesis; and modification, stability and specificity.[24]

The sample to be tested for microbial toxins is first incubated with AuNPs and aptamers. The aptamers then bind to the toxin, causing the AuNPs to aggregate, which changes the color of AuNPs from red to blue. This is a rapid and sensitive method with potential application in food safety and water quality monitoring.[42]

#### 8.8.1.2 Silver Nanoparticles

Silver nanoparticles and enzyme urease can be employed for colorimetric detection of pathogens. AgNPs preferentially bind to the bacterial surface, while urease is an enzyme that catalyzes the hydrolysis of urea to ammonia and carbon dioxide, which raises the pH of the solution. If pathogens are present, the AgNPs will bind to the bacterial surface, leaving the urease (protein) free to catalyze the hydrolysis of urea to ammonia. The increase in pH causes a color change in a pH-responsive dye, such as phenol red. This color change can be visualized by the naked eye or by using a spectrophotometer. This method is rapid and sensitive and can detect bacterial contamination levels as low as 102 cells $mL^{-1}$. It is also simple and inexpensive to implement, making it suitable for use in field settings.[43]

## 8.9 HEAVY METAL DETECTION USING AUNPS

The presence of heavy metals above certain threshold levels in our environment is a cause for concern, as these substances can be highly toxic to humans, leading to a range of life-threatening diseases. There is an urgent need to monitor and control the levels of toxic metals in water sources to ensure a safe and clean water supply. One of the significant challenges in this endeavor is the timely and accurate analysis of metal contamination in aqueous samples. Various approaches, including chemical, physical and biological methods, have been employed to detect these contaminants. While highly sophisticated instruments like atomic absorption spectrometry (AAS), inductively coupled plasma mass spectrometry (ICP-MS) and inductively coupled plasma atomic emission spectroscopy (ICP-AES) have been utilized for monitoring and detecting metal ions, they come with limitations such as lengthy protocols, extensive pre-concentration steps and expensive equipment.[44,45] Therefore, the development of a practical method for metal ion detection has become a major necessity.[46]

Nanotechnology has emerged as a significant milestone in addressing this need, with the creation of various nanoprobes designed for the detection of metal ions. These tiny particles, particularly gold nanoparticles, have gained attention for their exceptional properties, which offer advantages over conventional methods.[47] AuNPs are being used as sensors in waste management for the detection of toxic metal ions and contaminants due to their high surface-to-volume ratio and unique optical properties.[48] The optical properties of AuNPs are highly dependent on their size, shape and aggregation state, which can be fine-tuned by selecting appropriate synthesis and stabilization methods. In colloidal form, AuNPs are typically red or pink, but aggregation leads to a change in color to purple-blue. This phenomenon forms the basis of colorimetric detection, where binding of a target analyte to AuNP surfaces causes aggregation, resulting in a color change accompanied by a broadening and shifting of the surface plasmon resonance peak. SPR is a unique property of metal nanoparticles that arises when the oscillations of conduction band electrons resonate with electromagnetic radiation, leading to the fascinating optical properties associated with NPs.

### 8.9.1 Characteristics of Gold Nanoparticles and Their Attributes

Gold nanoparticles have gained significant importance in a wide range of applications, encompassing fields such as catalysis and biomedicine. This is primarily attributed to their remarkable physical and chemical characteristics, making them exceptional instruments for detection.[48,49] The suitability of AuNPs for biosensing applications stems from their numerous advantages over other materials. Various physio-chemical and biological routes can be utilized to synthesize them, as they are known to be highly inert and stable against oxidation.[50] The synthesis of highly stable gold nanoparticles, with an approximate size of 20 nm, can be achieved using the classical Turkevich method. In this method, the nanoparticle surfaces are coated with citrate molecules, which give them a net negative charge. This negative charge effectively prevents the particles from aggregating.[51] In addition to the Brust Schiffrin method, alternative techniques, including seed-mediated growth and the utilization of biological agents such as bacteria, algae and fungi, offer a wide range of shapes and sizes for gold nanoparticles.[52]

The optical characteristics of this AuNP colloidal solution are a result of the surface plasmon resonance phenomenon. Any alterations in the particle size, shape or geometry are directly apparent in the absorption maxima of SPR and subsequently influence the color of the solution.[53] When anisotropy is induced, such as in the case of nanorods, the properties of AuNPs undergo remarkable changes. This leads to enhanced sensitivity compared to nanospheres. Furthermore, the surfaces of AuNPs can be easily altered to link them with diverse organic and biological ligands that possess an attraction towards the analyte molecule under examination. The increasing use of AuNPs in detection, specifically for the development of AuNP-based sensors to detect metal ions in water, can be attributed to their useful properties and advantages. The functioning of these sensors depends on alterations in the optical properties influenced by particle size and aggregation. Consequently, the sensors exhibit modifications in color and shifts in plasmon absorption peaks.[54,55]

### 8.9.2 Functionalization of Gold Nanoparticles

Functionalization of gold nanoparticles is a crucial step in their use as colorimetric probes for specific metal ion detection. This process involves attaching target-specific ligand molecules to the AuNP surfaces, enabling precise binding to the desired analyte. Functionalization can be achieved through three main methods. The first method, physical adsorption, relies on rapid interactions driven by electrostatic or hydrophobic forces, making it a straightforward process without intricate synthesis. However, it exhibits sensitivity to variations in physical parameters like pH and ionic strength, which can result in the detachment of ligand molecules. The second method, covalent bonding, establishes robust bonds between gold (Au) and sulfur (S) groups, predominantly employing thiol-containing compounds such as alkanethiolates, glutathione, and other sulfur-based molecules to create functionalized AuNPs. The third technique, specific recognition, hinges on the precise binding of ligand molecules to the target analyte. In this scenario, AuNPs can be labeled with antibodies, enabling the identification of specific antigenic molecules. These diverse methods offer versatility in customizing AuNPs for various applications, enhancing their sensitivity and specificity in detection processes. Functionalization plays a vital role in enhancing the sensitivity of AuNP-based sensors, as demonstrated by AuNPs conjugated with mercaptoalkyl oligonucleotides, which showed increased sensitivity in detecting oligonucleotides down to femtomoles. However, it is worth noting that the functionalization process has some disadvantages, including a lengthy and expensive two-step synthesis and conjugation procedure, which may affect sensor reactivity. In contrast, unmodified or biologically synthesized AuNPs offer advantages because they are naturally stabilized by agents present in biological extracts. This eliminates the need to conjugate foreign ligand molecules, providing high specificity and stability compared to chemically synthesized AuNPs. These biologically synthesized AuNPs are a favorable choice for various applications.

As mentioned earlier, the aggregation of AuNPs induced by specific metal particles results in a color change and a shift in the plasmon absorption peak. This unique property of AuNPs has been harnessed for the colorimetric detection of heavy metal ions in aqueous solutions.

### 8.9.3 Arsenic Detection Using Gold Nanoparticles

Arsenic (As) is a hazardous metalloid responsible for various health issues in humans. AuNP-based colorimetric sensors for detecting $As^{3+}$ ions primarily rely on the strong binding affinity of $As^{3+}$ to sulfur-containing compounds. Kalluri et al. demonstrated the use of AuNPs functionalized with glutathione (GSH), dithiothreitol (DTT) and cysteine (Cys) for colorimetric detection of $As^{3+}$ ions. The minimum detection limits for these probes were reported, with DTT-conjugated AuNPs having a lower detection limit of 5 ppb.[56] Dynamic light scattering (DLS) analysis further improved the detection limit to 10 ppt, showcasing the method's sensitivity. Aptamer-based colorimetric detection of $As^{3+}$ ions was also demonstrated, where aptamers were used in conjunction with a cationic polymer, poly(diallyldimethylammonium) (PDDA), for specific detection. The presence of $As^{3+}$ ions induced the aggregation of AuNPs, resulting in a color change. Furthermore, poly(allylamine hydrochloride) (PAH)–conjugated AuNPs were used for detecting $As^{3+}$ ions, where arsenate formed electrostatic complexes with the amine groups of the conjugated molecule, causing changes in Rayleigh scattering profiles of AuNPs.[57,58] Laser-induced desorption/ionization mass spectrometric (LDI-MS) methods coupled with AuNP-modified mixed cellulose ester membrane were employed for the efficient detection of $As^{3+}$ ions, offering a rapid, highly selective and sensitive method with a low minimum detection limit. A novel approach for $As^{3+}$ ion detection used unmodified AuNPs and a phytochelatin-like peptide, $PC_3R$.[59] This method relied on preventing AuNP aggregation in the presence of $As^{3+}$ ions, as $As^{3+}$ ions have a strong binding affinity for thiol-containing molecules, and their complex formation with $PC_3R$ kept AuNPs stabilized in their original red state.

These methods exemplify the versatile nature of AuNP-based colorimetric sensors for detecting heavy metal ions in aqueous solutions. Each approach showcases the unique properties of AuNPs and their sensitivity to specific metal ions, allowing for the development of robust and effective detection methods. Additionally, they highlight the significance of functionalization in tuning the selectivity and sensitivity of AuNPs for target analytes, making AuNP-based sensors a promising tool for heavy metal detection in water. The development of these sensors is part of ongoing research and innovation in the field, with a focus on improving their performance and applicability for real-world water quality monitoring.

In summary, AuNPs have emerged as a powerful tool for detecting heavy metal ions in water, making them a promising platform for addressing the critical issue of heavy metal contamination in water sources. Researchers continue to refine and expand these sensors, improving their sensitivity, selectivity and scalability while working towards the development of portable and field-ready sensor systems for real-time water quality monitoring.

## 8.10 PLASMONIC COPPER SELENIDE NANOPARTICLES FOR COLORIMETRIC DETECTION OF GLUTATHIONE

Doped semiconductor nanocrystals, known for their ability to dynamically manipulate carrier densities, have gained considerable attention within the fields of sensing, catalysis, and photonics technologies. These materials exhibit plasmonic optical responses akin to those observed in noble metals like silver (Ag) and gold (Au) nanoparticles. Unlike metal nanoparticles, the tuning of their plasmonic spectroscopic signature can be achieved through the modulation of doping, making them particularly versatile.[60–65]

### 8.10.1 Focus on Cost-Effective Semiconductor Nanocrystals

Recent research has emphasized semiconductor nanocrystals based on earth-abundant transition metals, with copper being a notable choice due to its cost-effectiveness. Nonstoichiometric copper selenide nanoparticles ($Cu_{2-x}Se$) are exemplary self-doped representatives, offering a high carrier concentration that leads to the evolution of near-infrared localized surface plasmon resonance (NIR LSPR). The coexistence of Cu (I) and Cu (II) in these nanocrystals plays a pivotal role in their optical properties, which can be effectively adjusted by the presence of Cu defects, resulting in shifts in the absorption spectrum within the NIR range.[66–69]

The plasmonic properties of $Cu_{2-x}Se$ are intrinsically linked to its nonstoichiometric composition, with the introduction of Cu (II) being a key factor. This phenomenon has been extensively studied and validated, showing that an excess of $Cu^+$ ions reduces the NIR LSPR band, while the addition of an oxidizer such as Ce(IV) enhances the plasmon band by increasing copper vacancies.[57,70,71]

$Cu_{2-x}Se$, with its high copper vacancies, acts as both electron donor and acceptor, endowing these nanocrystals with remarkable catalytic capabilities. This facilitates electron transfer, making them valuable in catalytic reactions. Notably, they exhibit peroxidase-like activity and have been harnessed in colorimetric assays for various applications.[57–59,61,70,71]

### 8.10.2 Significance of Glutathione Quantification

Glutathione, a vital biological thiol, serves numerous physiological functions in biological systems, making it a critical component in medicinal and functional foods. However, excessive GSH intake can lead to health issues, emphasizing the need for accurate GSH quantification. Such quantification is essential for guiding GSH supplementation for disease prevention and treatment.[57,60–68,70,71]

### 8.10.3 Experimental Details and Analysis

The catalytic activity of $Cu_{2-x}Se$ is assessed using colorimetry, with the color change and absorbance at 650 nm serving as indicators of the catalytic process. GSH inhibits this process, leading to a decrease in the absorbance signal. This phenomenon is harnessed to design a colorimetric method for quantitative GSH detection. Optimal experimental conditions, including pH, TMB concentration,

$Cu_{2-x}Se$ concentration and $H_2O_2$ concentration, are determined to ensure accurate GSH quantification.[72] Interference studies are conducted to evaluate the platform's resistance to interference. The feasibility of the proposed method for real sample analysis is demonstrated by measuring GSH contents in tomatoes and cucumbers.

The presence of GSH hinders the peroxidase-like activity of $Cu_{2-x}Se$ and results in a decrease in its NIR LSPR.[49] Beneath the ideal response conditions, diverse concentrations of GSH were included to the sensing systems. With the increment within the GSH concentration, the absorbance of OxTMB at 650 nm slowly diminished, and the color of the solution changed from blue to light blue and after that to colorless. There is a strong linear relationship between the change within the absorbance signal and the GSH concentration. A linear equation can be expressed. The alter within the absorbance with the increment in GSH concentration.

## 8.11 COLORIMETRIC TECHNIQUES FOR DETECTING AND IDENTIFYING CHEMICAL WEAPONS AND EXPLOSIVES

The development of reliable, sensitive and cost-effective methods for detecting explosives is crucial for security and safety.[60,61,71] Recent advancements have led to the creation of a novel hybrid electrochemical-colorimetric (EC-C) sensing platform that addresses the challenges associated with explosives detection. This platform combines electrochemical reactions with colorimetric detection and harnesses the unique properties of explosives in an ionic liquid (IL), 1-butyl-3-methylimidazolium hexafluorophosphate ($BMIM\text{-}PF_6$).

### 8.11.1 Electrochemical Reactions and Colorimetric Detection

The hybrid sensor's core operation involves a synergistic combination of electrochemical reactions and colorimetric detection. Unlike traditional electrochemical methods that directly detect explosives, this approach primarily aims to generate reaction products. These products exhibit distinctive red colors with an absorption peak around 450 nm and a high extinction coefficient. The compounds responsible for these colors were identified as azo and azoxy derivatives through Fourier transform infrared spectroscopy (FTIR) and liquid chromatography/mass spectrometry (LC/MS).

### 8.11.2 The Role of Ionic Liquid

The use of a thin layer of 1-butyl-3-methylimidazolium hexafluorophosphate ($BMIM\text{-}PF_6$) plays a pivotal role in this sensing platform.[69,73,74] $BMIM\text{-}PF_6$ offers several remarkable properties, like selective preconcentration, stable electrolytes and colored reduction products. The system incorporates selective preconcentration to efficiently gather explosives and rapidly transport them to the electrodes, thereby significantly improving sensitivity in detection. It utilizes $BMIM\text{-}PF_6$ as a stable electrolyte, maintaining its reliability under ambient conditions for electrochemical reactions. Additionally, this medium facilitates the generation of colored reduction products, a crucial factor for subsequent colorimetric detection processes.

### 8.11.3 Explosive Detection and Preconcentration

The sensor is particularly effective in detecting nitroaromatic explosives such as 2,4,6-trinitrotoluene (TNT), picric acid (PA) and 2,4-dinitrotoluene (DNT). Notably, the IL provides an ideal environment for generating distinct colors during the electrochemical reduction of these explosives.[66–68] The system's preconcentration capabilities are outstanding, with the relative preconcentration capacities of $BMIM\text{-}PF_6$ varying according to the number of nitro groups in the analyte's aromatic structure. This preconcentration significantly contributes to the sensor's sensitivity and selectivity.

The innovative hybrid electrochemical-colorimetric sensing platform, with $BMIM\text{-}PF_6$ as a critical component, offers a cost-effective, sensitive and highly selective solution for explosives detection. Its ability to distinguish explosives from common interferences and provide quantifiable results makes it a valuable tool for enhancing security and safety in various applications. Further research and development may lead to even more sensitive and efficient detection systems for various hazardous substances.

## 8.12 CONCLUSION

This chapter offers a thorough and perceptive examination of the cutting-edge topic of nanotechnology in analytical chemistry. It draws attention to the exceptional sensitivity and adaptability of metal and metal oxide nanoparticles in identifying a variety of analytes, including biological molecules and ions. The colorimetric sensing methods covered in this chapter provide easy-to-use and reasonably priced options for a range of uses, such as food safety, medical diagnostics and environmental monitoring.

This chapter highlights how important nanomaterials are to solving today's problems and how they have the power to completely change how we think about and approach chemical analysis, which certainly opens the door for more developments in the creation of sensors based on nanomaterials, which provide improved specificity, sensitivity and selectivity. Through the use of these nanoparticles' distinct optical characteristics, scientists and researchers are expanding the potential of sensing technologies. In addition, the chapter clarifies fundamental ideas and provides several real-world examples and case studies that demonstrate the usefulness and efficiency of these colorimetric sensors. Also, the information presented here undoubtedly paves the way for advancements in the field of nanomaterial-based sensors in the future, which will lead to increased specificity, sensitivity and selectivity. The incorporation of metal and metal oxide nanoparticles into colorimetric sensing platforms is clearly going to be a major factor in the future development of analytical chemistry.

## REFERENCES


1. Xu, N.; Jin, S.; Wang, L.; Metal nanoparticles-based nanoplatforms for colorimetric sensing: a review: *Rev Anal Chem*, 2020;40(1):1–11. https://doi.org/10.1515/revac-2021-0122
2. Kaur, N.; Kumar, S.; Colorimetric metal ion sensors: *Tetrahedron*, 2011;67(48):9233–9264. https://doi.org/10.1016/j.tet.2011.09.003
3. Chen, H.; Zhou, K.; Zhao, G.; Gold nanoparticles: from synthesis, properties to their potential application as colorimetric sensors in food safety screening: *Trends Food Sci Technol*, 2018;78:83–94. https://doi.org/10.1016/j.tifs.2018.05.027
4. Teymouri, Z.; Shekarchizadeh, H.; A colorimetric indicator based on copper nanoparticles for volatile sulfur compounds to monitor fish spoilage in intelligent packaging: *Food Packag Shelf Life*, 2022;33:100884. https://doi.org/10.1016/j.fpsl.2022.100884
5. Ali, Z.; Ullah, R.; Tuzen, M.; Ullah, S.; Rahim, A.; Saleh, T. A.; Colorimetric sensing of heavy metals on metal doped metal oxide nanocomposites: a review: *Trends Environ Anal Chem*, 2023;37:e00187. https://doi.org/10.1016/j.teac.2022.e00187
6. Khansili, N.; Rattu, G.; Kumar, A.; Krishna, P. M.; Development of colorimetric sensor with zinc oxide nanoparticles for rapid detection of Aflatoxin B1 in rice: *Mater Today Proc*, 2020;21:1846–1855. https://doi.org/10.1016/j.matpr.2020.01.240
7. Mendoza, D. V. M.; Colorimetric detection of pesticide residues in water using zinc oxide (ZnO) nanoparticles: *Orient J Chem*, 2022;38(6):1476–1482.
8. Cuong, H. N.; Pansambal, S.; Ghotekar, S.; New frontiers in the plant extract mediated biosynthesis of copper oxide (CuO) nanoparticles and their potential applications: a review: *Environ Res*, 2022;203:111858. https://doi.org/10.1016/j.envres.2021.111858
9. Kasiri, M.; Rahaie, M.; A visible and colorimetric nanobiosensor based on DNA-CuO nanoparticle for detection of single nucleotide polymorphism involved in sickle cell anemia disease: *Mater Today Commun*, 2021;27:102423. https://doi.org/10.1016/j.mtcomm.2021.102423
10. Ji, G.; Tian, J.; Xing, F.; Feng, Y.; Optical biosensor based on graphene and its derivatives for detecting biomolecules: *Int J Mol Sci*, 2022;23(18):10838. https://doi.org/10.3390/ijms231810838
11. Wang, L.; Wu, A.; Wei, G.; Graphene-based aptasensors: from molecule–interface interactions to sensor design and biomedical diagnostics: *Analyst*, 2018;143(7):1526–1543. https://doi.org/10.1039/C8AN00081F
12. Kim, H.; Heme cofactor-resembling Fe–N single site embedded graphene as nanozymes to selectively detect H2O2 with high sensitivity: *Adv Funct Mater*, 2020. Accessed October 27, 2023. https://onlinelibrary.wiley.com/doi/abs/10.1002/adfm.201905410
13. Jin, S.; Wu, C.; Ye, Z.; Ying, Y.; Designed inorganic nanomaterials for intrinsic peroxidase mimics: a review: *Sens Actuators B Chem*, 2019;283:18–34. https://doi.org/10.1016/j.snb.2018.10.040
14. Sun, H.; Zhou, Y.; Ren, J.; Qu, X.; Carbon nanozymes: enzymatic properties, catalytic mechanism, and applications: *Angew Chem Int Ed*, 2018;57(30):9224–9237. https://doi.org/10.1002/anie.201712469
15. Rostami, S.; Mehdinia, A.; Jabbari, A.; Intrinsic peroxidase-like activity of graphene nanoribbons for label-free colorimetric detection of dopamine: *Mater Sci Eng C*, 2020;114:111034. https://doi.org/10.1016/j.msec.2020.111034
16. Liang, A.; Li, C.; Li, D.; Luo, Y.; Wen, G.; Jiang, Z.; A facile and sensitive peptide-modulating graphene oxide nanoribbon catalytic nanoplasmon analytical platform for human chorionic gonadotropin: *Int J Nanomed*, 2017;12:8725–8734. https://doi.org/10.2147/IJN.S149536
17. Zhu, Y.; Wu, J.; Han, L.; Nanozyme sensor arrays based on heteroatom-doped graphene for detecting pesticides: *Anal Chem*, 2020;92(11):7444–7452. https://doi.org/10.1021/acs.analchem.9b05110
18. Hai, X.; Li, Y.; Yu, K.; Synergistic in-situ growth of silver nanoparticles with nanozyme activity for dual-mode biosensing and cancer theranostics: *Chin Chem Lett*, 2021;32(3):1215–1219. https://doi.org/10.1016/j.cclet.2020.09.013
19. Wang, X.; Huang, K.; Zhang, H.; Zeng, L.; Zhou, Y.; Jing, T.; Preparation of molecularly imprinted polymers on hemin-graphene surface for recognition of high molecular weight protein: *Mater Sci Eng C*, 2019;105:110141. https://doi.org/10.1016/j.msec.2019.110141
20. Liu, L.; Du, J.; Liu, W.; Enhanced His@AuNCs oxidase-like activity by reduced graphene oxide and its application for colorimetric and electrochemical detection of nitrite: *Anal Bioanal Chem*, 2019;411(10):2189–2200. https://doi.org/10.1007/s00216-019-01655-y
21. Korostynska, O.; Arshak, K.; Gill, E.; Arshak, A.; State Key Laboratory of Nonlinear Mechanics (LNM), Institute of Mechanics, Chinese Academy of Sciences, Beijing 100080, China: *Sensors*, 2007;7(12):3027–3042. https://doi.org/10.3390/s7123027
22. Pablos, J. L.; Estévez, P.; Muñoz, A.; Polymer chemosensors as solid films and coated fibres for extreme acidity colorimetric sensing: *J Mater Chem A*, 2015;3(6):2833–2843. https://doi.org/10.1039/C4TA06009A
23. Yang, Y.; Li, W.; Liu, J.; Review of recent progress on DNA-based biosensors for Pb2+ detection: *Anal Chim Acta*, 2021;1147:124–143. https://doi.org/10.1016/j.aca.2020.12.056
24. Piriya, V. S. A.; Joseph, P.; Daniel, S. C. G. K.; Lakshmanan, S.; Kinoshita, T.; Muthusamy, S.; Colorimetric sensors for rapid detection of various analytes: *Mater Sci Eng C*, 2017;78:1231–1245. https://doi.org/10.1016/j.msec.2017.05.018
25. Li, R.; Liu, Q.; Jin, Y.; Li, B.; G-triplex/hemin DNAzyme: an ideal signal generator for isothermal exponential amplification reaction-based biosensing platform: *Anal Chim Acta*, 2019;1079:139–145. https://doi.org/10.1016/j.aca.2019.06.002
26. Wu, Y.; Feng, J.; Hu, G.; Zhang, E.; Yu, H. H.; Colorimetric sensors for chemical and biological sensing applications: *Sensors*, 2023;23(5):2749. https://doi.org/10.3390/s23052749
27. Hoang, M.; Huang, P. J. J.; Liu, J.; G-Quadruplex DNA for fluorescent and colorimetric detection of Thallium(I): *ACS Sens*, 2016;1(2):137–143. https://doi.org/10.1021/acssensors.5b00147
28. Wu, C.; Gao, G.; Zhai, K.; Xu, L.; Zhang, D.; A visual Hg2+ detection strategy based on distance as readout by G-quadruplex DNAzyme on microfluidic paper: *Food Chem*, 2020;331:127208. https://doi.org/10.1016/j.foodchem.2020.127208

29. Chen, J.; Zhang, Y.; Cheng, M.; Highly active G-quadruplex/hemin DNAzyme for sensitive colorimetric determination of lead(II): *Microchim Acta*, 2019;186(12):786. https://doi.org/10.1007/s00604-019-3950-3
30. Yu, X.; Lin, Y.; Wang, X.; Xu, L.; Wang, Z.; Fu, F.; Exonuclease-assisted multicolor aptasensor for visual detection of ochratoxin A based on G-quadruplex-hemin DNAzyme-mediated etching of gold nanorod: *Microchim Acta*, 2018;185(5):259. https://doi.org/10.1007/s00604-018-2811-9
31. Lee, J.; Kim, Y. kwan; Lee, S.; Yoon, S.; Kim, W. keun; Graphene oxide-based NET strategy for enhanced colorimetric sensing of miRNA: *Sens Actuators B Chem*, 2019;282:861–867. https://doi.org/10.1016/j.snb.2018.11.149
32. Lee, J.; Na, H. K.; Lee, S.; Kim, W. K.; Advanced graphene oxide-based paper sensor for colorimetric detection of miRNA: *Microchim Acta*, 2021;189(1):35. https://doi.org/10.1007/s00604-021-05140-1
33. Goodman, R. P.; Schaap, I. A. T.; Tardin, C. F.; Rapid chiral assembly of rigid DNA building blocks for molecular nanofabrication: *Science*, 2005;310(5754):1661–1665. https://doi.org/10.1126/science.1120367
34. Yang, X.; Wen, Y.; Wang, L.; PCR-free colorimetric DNA hybridization detection using a 3D DNA nanostructured reporter probe: *ACS Appl Mater Interfaces*, 2017;9(44):38281–38287. https://doi.org/10.1021/acsami.7b11994
35. Guan, H.; Yang, S.; Zheng, C.; DNAzyme-based sensing probe protected by DNA tetrahedron from nuclease degradation for the detection of lead ions: *Talanta*, 2021;233:122543. https://doi.org/10.1016/j.talanta.2021.122543
36. Pei, H.; Liang, L.; Yao, G.; Li, J.; Huang, Q.; Fan, C.; Innenrücktitelbild: reconfigurable three-dimensional DNA nanostructures for the construction of intracellular logic sensors (Angew. Chem. 36/2012): *Angew Chem*, 2012;124(36):9321–9321. https://doi.org/10.1002/ange.201206132
37. Fu, X.; Lin, H.; Qi, J.; A tetrahedral DNA nanostructure functionalized paper-based platform for ultrasensitive colorimetric mercury detection: *Sens Actuators B Chem*, 2022;362:131830. https://doi.org/10.1016/j.snb.2022.131830
38. Leesutthiphonchai, W.; Dungchai, W.; Siangproh, W.; Ngamrojnavanich, N.; Chailapakul, O.; Selective determination of homocysteine levels in human plasma using a silver nanoparticle-based colorimetric assay: *Talanta*, 2011;85(2):870–876. https://doi.org/10.1016/j.talanta.2011.04.041
39. Ngo, Y. H.; Li, D.; Simon, G. P.; Garnier, G.; Paper surfaces functionalized by nanoparticles: *Adv Colloid Interface Sci*, 2011;163(1):23–38. https://doi.org/10.1016/j.cis.2011.01.004
40. Ratnarathorn, N.; Chailapakul, O.; Henry, C. S.; Dungchai, W.; Simple silver nanoparticle colorimetric sensing for copper by paper-based devices: *Talanta*, 2012;99:552–557. https://doi.org/10.1016/j.talanta.2012.06.033
41. Prosposito, P.; Burratti, L.; Venditti, I.; Silver nanoparticles as colorimetric sensors for water pollutants: *Chemosensors*, 2020;8(2):26. https://doi.org/10.3390/chemosensors8020026
42. Geleta, G. S.; A colorimetric aptasensor based on gold nanoparticles for detection of microbial toxins: an alternative approach to conventional methods: *Anal Bioanal Chem*, 2022;414(24):7103–7122. https://doi.org/10.1007/s00216-022-04227-9
43. Singh, P.; Kakkar, S.; Bharti.; Kumar, R.; Bhalla, V.; Rapid and sensitive colorimetric detection of pathogens based on silver–urease interactions: *Chem Commun*, 2019;55(33):4765–4768. https://doi.org/10.1039/C9CC00225A
44. D'Ilio, S.; Petrucci, F.; D'Amato, M.; Di Gregorio, M.; Senofonte, O.; Violante, N.; Method validation for determination of arsenic, cadmium, chromium and lead in milk by means of dynamic reaction cell inductively coupled plasma mass spectrometry: *Anal Chim Acta*, 2008;624(1):59–67. https://doi.org/10.1016/j.aca.2008.06.024
45. Parham, H.; Pourreza, N.; Rahbar, N.; Solid phase extraction of lead and cadmium using solid sulfur as a new metal extractor prior to determination by flame atomic absorption spectrometry: *J Hazard Mater*, 2009;163(2):588–592. https://doi.org/10.1016/j.jhazmat.2008.07.007
46. Kim, T. H.; Kim, S. H.; Van Tan, L.; Dong, Y.; Kim, H.; Kim, J. S.; Diazo-coupled calix[4]arenes for qualitative analytical screening of metal ions: *Talanta*, 2008;74(5):1654–1658. https://doi.org/10.1016/j.talanta.2007.10.033
47. Wang, C.; Yu, C.; Detection of chemical pollutants in water using gold nanoparticles as sensors: a review: *Rev Anal Chem*, 2013;32(1):1–14. https://doi.org/10.1515/revac-2012–0023
48. Eustis, S.; El-Sayed, M. A.; Why gold nanoparticles are more precious than pretty gold: noble metal surface plasmon resonance and its enhancement of the radiative and nonradiative properties of nanocrystals of different shapes: *Chem Soc Rev*, 2006;35(3):209–217. https://doi.org/10.1039/B514191E
49. Bunz, U. H. F.; Rotello, V. M.; Gold nanoparticle–fluorophore complexes: sensitive and discerning "noses" for biosystems sensing: *Angew Chem Int Ed*, 2010;49(19):3268–3279. https://doi.org/10.1002/anie.200906928
50. Syed, M. A.; Bokhari, S. H. A.; Gold nanoparticle based microbial detection and identification: *J Biomed Nanotechnol*, 2011;7(2):229–237. https://doi.org/10.1166/jbn.2011.1281
51. Turkevich, J.; Stevenson, P. C.; Hillier, J.; A study of the nucleation and growth processes in the synthesis of colloidal gold: *Discuss Faraday Soc*, 1951;11(0):55–75. https://doi.org/10.1039/DF9511100055
52. Brust, M.; Walker, M.; Bethell, D.; Schiffrin, J.; Whyman, R.; Synthesis of thiol-derivatised gold nanoparticles in a two-phase Liquid–Liquid system: *J Chem Soc Chem Commun*, 1994;0(7):801–802. https://doi.org/10.1039/C39940000801
53. Jain, P. K.; Lee, K. S.; El-Sayed, I. H.; El-Sayed, M. A.; Calculated absorption and scattering properties of gold nanoparticles of different size, shape, and composition: applications in biological imaging and biomedicine: *J Phys Chem B*, 2006;110(14):7238–7248. https://doi.org/10.1021/jp057170o
54. Li, Y.; Schluesener, H. J.; Xu, S.; Gold nanoparticle-based biosensors: *Gold Bull*, 2010;43(1):29–41. https://doi.org/10.1007/BF03214964
55. Englebienne, P.; Use of colloidal gold surface plasmon resonance peak shift to infer affinity constants from the interactions between protein antigens and antibodies specific for single or multiple epitopes: *Analyst*, 1998;123(7):1599–1603. https://doi.org/10.1039/A804010I
56. Kalluri, J. R.; Arbneshi, T.; Afrin Khan, S.; Use of gold nanoparticles in a simple colorimetric and ultrasensitive dynamic light scattering assay: selective detection of arsenic in groundwater: *Angew Chem*, 2009;121(51):9848–9851. https://doi.org/10.1002/ange.200903958

57. Jain, P. K.; Manthiram, K.; Engel, J. H.; White, S. L.; Faucheaux, J. A.; Alivisatos, A. P.; Doped nanocrystals as plasmonic probes of redox chemistry: *Angew Chem*, 2013;125(51):13916–13920. https://doi.org/10.1002/ange.201303707
58. Zhan, S.; Yu, M.; Lv, J.; Colorimetric detection of trace Arsenic(iii) in aqueous solution using arsenic aptamer and gold nanoparticles: *Aust J Chem*, 2014;67(5):813–818. https://doi.org/10.1071/CH13512
59. Xia, N.; Shi, Y.; Zhang, R.; Zhao, F.; Liu, F.; Liu, L.; Simple, rapid and label-free colorimetric assay for arsenic based on unmodified gold nanoparticles and a phytochelatin-like peptide: *Anal Methods*, 2012;4(12):3937–3941. https://doi.org/10.1039/C2AY25803J
60. Li, K.; Xu, K.; He, Y.; Oxygen self-generating nanoreactor mediated ferroptosis activation and immunotherapy in triple-negative breast cancer: *ACS Nano*, 2023;17(5):4667–4687. https://doi.org/10.1021/acsnano.2c10893
61. Dorfs, D.; Härtling, T.; Miszta, K.; Reversible tunability of the near-infrared valence band plasmon resonance in Cu2–xSe nanocrystals: *J Am Chem Soc*, 2011;133(29):11175–11180. https://doi.org/10.1021/ja2016284
62. Cai, R.; Xiang, H.; Yang, D.; Plasmonic AuPt@CuS heterostructure with enhanced synergistic efficacy for radiophotothermal therapy: *J Am Chem Soc*, 2021;143(39):16113–16127. https://doi.org/10.1021/jacs.1c06652
63. Sun, M.; Fu, X.; Chen, K.; Wang, H.; Dual-plasmonic Gold@Copper sulfide core–shell nanoparticles: phase-selective synthesis and multimodal photothermal and photocatalytic behaviors: *ACS Appl Mater Interfaces*, 2020;12(41):46146–46161. https://doi.org/10.1021/acsami.0c13420
64. Luther, J. M.; Jain, P. K.; Ewers, T.; Alivisatos, A. P.; Localized surface plasmon resonances arising from free carriers in doped quantum dots: *Nat Mater*, 2011;10(5):361–366. https://doi.org/10.1038/nmat3004
65. Huang, Q.; Zhang, S.; Zhang, H.; Boosting the radiosensitizing and photothermal performance of Cu2–xSe nanocrystals for synergetic radiophotothermal therapy of orthotopic breast cancer: *ACS Nano*, 2019;13(2):1342–1353. https://doi.org/10.1021/acsnano.8b06795
66. Li, X.; Yang, L.; Men, C.; Photothermal soft nanoballs developed by loading plasmonic Cu2–xSe nanocrystals into liposomes for photothermal immunoassay of Aflatoxin B1: *Anal Chem*, 2019;91(7):4444–4450. https://doi.org/10.1021/acs.analchem.8b05031
67. Xu, W.; Liu, H.; Zhou, D.; Localized surface plasmon resonances in self-doped copper chalcogenide binary nanocrystals and their emerging applications: *Nano Today*, 2020;33:100892. https://doi.org/10.1016/j.nantod.2020.100892
68. Hu, Q.; Sun, H.; Zhou, X.; Bright-yellow-emissive nitrogen-doped carbon nanodots as a fluorescent nanoprobe for the straightforward detection of glutathione in food samples: *Food Chem*, 2020;325:126946. https://doi.org/10.1016/j.foodchem.2020.126946
69. Sharma, H.; White, J. M.; Lin, J.; New, E. J.; Pfeffer, F. M. A rhodamine-nitronaphthalimide Hg(II) complex for the simultaneous detection of oxidised and reduced glutathione: *Sens Actuators B Chem*, 2019;300:126825. https://doi.org/10.1016/j.snb.2019.126825
70. Kriegel, I.; Jiang, C.; Rodríguez-Fernández, J.; Tuning the excitonic and plasmonic properties of copper chalcogenide nanocrystals: *J Am Chem Soc*, 2012;134(3):1583–1590. https://doi.org/10.1021/ja207798q
71. Xie, Y.; Riedinger, A.; Prato, M.; Copper sulfide nanocrystals with tunable composition by reduction of covellite nanocrystals with Cu+ ions: *J Am Chem Soc*, 2013;135(46):17630–17637. https://doi.org/10.1021/ja409754v
72. Yan, G.; Ni, H.; Li, X.; Qi, X.; Yang, X.; Zou, H. Plasmonic Cu2–xSe mediated colorimetric/photothermal dual-readout detection of glutathione: *Nanomaterials*, 2023;13(11):1787. https://doi.org/10.3390/nano13111787
73. Li, J.; Jiao, L.; Xu, W.; Cobalt oxyhydroxide nanosheets integrating with metal indicator enable sensitive detection of glutathione: *Sens Actuators B Chem*, 2021;329:129247. https://doi.org/10.1016/j.snb.2020.129247
74. Samiec, P. S.; Drews-Botsch, C.; Flagg, E. W.; Glutathione in human plasma: decline in association with aging, age-related macular degeneration, and diabetes: *Free Radic Biol Med*, 1998;24(5):699–704. https://doi.org/10.1016/S0891-5849(97)00286-4


## LIST OF ABBREVIATIONS

| | |
|---|---|
| **2-D** | Two dimensions |
| **AAS** | Atomic absorption spectrometry |
| **ABTS** | 2,2-azino-bis-3-ethylbenothiazoline-6-sulphonic acid |
| **BMIM – PF6** | 1-butyl-3-methylimidazolium hexafluorophosphate |
| **CGO** | Chemically Modified Graphene Oxide |
| **Cys** | Cysteine |
| **d[TC]** | DNA [Thymine-Cytosine] |
| **DTT** | Dithiothreitol |
| **Dz** | DNA enzyme |
| **EC-C** | Electrochemical-colorimetric |
| **FTIR** | Fourier transform infrared spectroscopy |
| **G4** | G-quadruplex |
| **GNR** | Graphene Nano Ribbons |
| **GO** | Graphene Oxide |
| **GONET** | GO as a net |
| **GQD** | Graphene Quantum Dots |
| **GSH** | Glutathione |
| **Hemin@G4** | Hemin capped G4 |
| **H-GNs** | Hemin-Graphene Nanosheets |
| **His@AuNCs** | Histidine capped gold nanostructures |
| **HRP** | Horseradish Peroxidase |
| **ICP-AES** | Inductively coupled plasma mass spectrometry |
| **LOD** | Limit of Detection |
| **LSPR** | Localized Surface Plasmon Resonance |
| **MIP** | Molecule Imprinted Polymer |
| **miRNA** | Micro RNA |
| **MWCNT** | Multi Walled Carbon Nano Tubes |
| **NP** | Nano particles |
| **OP** | Organophosphorous Pesticides |
| **OTA** | Ochratoxin A |

| | |
|---|---|
| **PAH** | Poly (allylamine hydrochloride) |
| **PDDA** | Poly(diallyldimethylammonium) |
| **ppb** | Parts per billion |
| **ppbV** | Parts per billion volume |
| **pptV** | Parts per trillion volume |
| **RGB** | Red Green Blue |
| **rGO** | Reduced Graphene Oxide |
| **SPR** | Surface Plasma Resonance |
| **ssRNA** | Single strand RNA |
| **T-A** | Thymine – Adenine base |
| **TMB** | 3,3',5,5'-Tetramethyl benzidiene |
| **UV-A** | Ultraviolet-A |

# 9 Biolabeling and Bioimaging Applications of Functionalized Metal and Metal Oxide Nanoparticles

*Jincemon Cyriac[1, 6*], L. R. Sriramachandru[1], Linta P. Joseph[1], Gayathri M. J.[1], Riju K. Thomas[2], L. S. Archana[3], K. Sreelatha[4], Bindu R. G.[5], and Tom Cherian[7]*

[1]Department of Physics, Marian College (Autonomous), Kuttikkanam, Peermade, Idukki, India

[2]Department of Physics, Bharath Matha College, Trikkakkara, Kochi, India

[3]Department of Renewable Energy, St Alberts College (Autonomous), Kochi, Ernakulam, Kerala, India

[4]PG Department of Physics, Sree Narayana College, Chengannur, Alappuzha, Kerala, India

[5]Department of Physics, NSS College, Nilamel, Kollam, Kerala, India

[6]Department of Physics, Christ College (Autonomous), Calicut University, Irinjalakuda, Thrissur, Kerala, India

[7]Department of Chemistry, Christ College (Autonomous), Calicut University, Irinjalakuda, Thrissur, Kerala, India

## 9.1 INTRODUCTION

Diagnosis and treatment have advanced significantly as a result of modern biomedical imaging technology. Researchers still struggle to see and comprehend disease processes precisely and in real time because the majority of them happen at both the cellular and molecular levels. Since most biological activities occur on this length scale, imaging resolution in nanometers would be optimal. Consequently, there is a lot of interest in the functionalization of nanoparticles (NPs) as well as their usage in therapeutic and diagnostic applications. When used in imaging techniques such as fluorescence imaging, magnetic resonance imaging (MRI), positron emission tomography (PET), or multimodal imaging, specially designed nanoparticles can bind very specifically to their biological targets, enabling accurate visualization.

### 9.1.1 Bioimaging

Bioimaging is a non-invasive technique that allows for real-time visualization of biological processes. Since it is a non-invasive process, it causes less disturbance to biological processes, which is often desirable. It can also be used to obtain 3D structures of biological materials [1]. It collects functional data such as motion, magnetic field, electric field, and metabolism to provide precise anatomical structure. This technique is an effective method for capturing comprehensive images of the internal structure and abnormalities within the body. Hence, based on the indicated inherent qualities of this technology, it has a wide variety of applications in the medical field. including computed tomography (CT), MRI, optical and ultrasound imaging, PET imaging, and photoacoustic (PA) imaging [2]. Nanoparticles are promising tools in the field of bioimaging. They have particular properties such as their surface chemistry, magnetic characteristics, adjustable emission, and absorption properties, which show promising results for early detection of diseases such as cancer [3, 4].

### 9.1.2 Computed Tomography

Computed tomography, or more precisely X-ray computed tomography, is a vital tool for obtaining three-dimensional structural details. It is able to provide comprehensive insights into the composition and characteristics of the material under examination in addition to displaying its structural and morphological qualities in tridimensional. CT captures images of the examined area from 360°, which are then combined by computer software to create a 3D image. During CT imaging processes, contrasting agents are employed to enhance the visualization of specific tissues and organs. Differentially attenuated materials provide different scan contrasts. Furthermore, the contrast in the final image is influenced by the density of materials [5]. The contrast is improved by materials with higher densities and/or better X-ray attenuation characteristics. Medical procedures often require contrast agents for enhanced visibility, commonly using iodine-based ones. However, these pose environmental risks and may not be suitable for iodine-intolerant patients. Scientists are exploring metallic nanoparticles as alternatives due to their high X-ray

DOI: 10.1201/9781032660639-9

attenuation and density. Coating these nanoparticles with organic molecules reduces toxicity and enables targeted imaging of specific organs or tissues. Blending different nanoparticles into varied structures like core-shell or alloys provides them with dual or multiple properties. In addition to enhancing contrast, certain metallic nanoparticles possess photothermal or hyperthermia therapy capabilities, allowing therapeutic interventions on target organs or tissues during the imaging process, commonly known as theragnostic. Here we evaluate nanoparticles using distinct categories: (1) gold nanoparticles (GNPs) and (2) bismuth nanoparticles [5]. Recently, new CT contrast agents based on nanoparticles have been emerging to address the limitations of iodine agents, such as polymer-coated bismuth sulfide ($Bi_2S_3$) nanoparticles. They have vivo effectiveness and safety comparable to or better than iodinated agents. However, challenges arise in controlling the size and shape of $Bi_2S_3$ nanoparticles, and there is a lack of methods to modify their surface chemically, potentially hindering their clinical application, prompting research into alternative nanomaterial-based CT contrast agents. A novel CT contrast substance utilizing gold nanoparticles has been developed to address these challenges. Due to gold's higher atomic number and X-ray absorption capabilities compared to iodine, researchers anticipated that GNPs could serve as effective CT contrast agents [6]. Numerous studies have shown that GNPs can provide up to 2.5 times superior contrast at equivalent concentrations compared to the commercially available contrast agents utilized in clinical applications [5].

### 9.1.3 Magnetic Resonance Imaging

MRI is a bioimaging method that takes advantage of nuclear magnetic resonance (NMR) to provide a picture of internal anatomy and physiological processes. It contains a superconducting magnet that generates a static electromagnetic field; the nuclear spins of hydrogen protons inside water resonate at a specific frequency determined by the strength of the applied magnetic field of hydrogen proton inside water molecules in the test sample. After the specimen achieves equilibrium magnetization, the scanner inside the MRI machine emits a radiofrequency pulse, which is absorbed by the hydrogen protons and reemits a radio frequency, causing a shift in the overall magnetization. Once the radio frequency pulse ceases and the spins return to their initial equilibrium, they reemit an electromagnetic signal, which is measured by receiving coil inside MRI machine. 3D images of biomaterial can be obtained by analyzing these electromagnetic signals and their alterations in the presence of linear field gradients [7–10].

Metal oxide nanoparticles can be used to create targeted contrast agents that can deliver drugs to specific tissues or organs. Paramagnetic metal oxide nanoparticles are profoundly interesting to create contrast agents, as they can indirectly enhance the visibility of certain tissues and organs on the MRI. Paramagnetic contrast agents work by reducing the T1 and/or T2 relaxation times of water protons in their vicinity [11, 12]. Among the diverse range of metals and compounds, gadolinium-III is a superior contrast agent because of its substantial magnetic moment. Presently, the most commonly utilized MRI contrast agent is Gd-DTPA [diethyl triamine penta acetic acid], where DTPA is a chelating group that binds with the Gd III ion. The rapid spread of Gd-DTPA into the surrounding tissues and blood vessels limits its use as an MRI contrast agent. Due to the enhanced appearance of veins and arteries brought on by this non-specific distribution, image clarity is decreased and the target area is more difficult to highlight. Additionally, Gd, even in complex form, poses toxicity risks, with a relatively long half-life. Factors like temperature, pH, and surrounding ions affect its stability, and endogenous metals like Zn and Cu can displace Gd ions from the complex. Research has demonstrated in vivo demetallation, depositing free Gd III ions in bones and the liver after prolonged residence of Gd-DTPA in the body [13]. Iron oxide nanoparticles are a popular nanoparticle-based contrast agent for liver and spleen imaging. Superparamagnetic iron oxide (SPIO) or ultra-small paramagnetic iron oxide (USPIO, <50 nm) are the two types of iron oxide magnetic nanoparticles (IONPs) that are used in MRI. They act as $T_2$ contrast agents, enabling negative contrast enhancement and producing darker images of specific areas. They have been assessed as MRI contrast agents, especially for liver and spleen imaging. However, IONPs can be toxic to cells and tissues depending on their size, shape, surface coating, and concertation [10, 14, 15].

### 9.1.4 Optical Imaging

Light serves as a versatile form of imaging radiation, as it is non-invasive and can generate contrast through intensity, wavelength, polarization, and nonlinear effects [10]. Optical coherence tomography was modified to enable rapid visualization of living animal tissue using a catheter-endoscope measuring just 1 millimeter in diameter [16]. Among the optical imaging techniques, fluorescence microscopy has become highly influential. The phenomenon of absorption and nearly simultaneous re-radiation of light by substances is known as optical fluorescence [17]. It relies on the intrinsic properties of the fluorophore, which is typically an organic dye or lanthanide compound. Electrons in a fluorophore are raised from their ground state to a higher-energy singlet state when stimulated by particular energy quanta. Since electrons have different energies for vibration and rotation, a change in vibrational or rotational energy at a higher electronic state is required for the transition to a singlet state. A lower-energy photon is released as a result of the electron releasing some of its energy during this process, which frequently happens through thermal decay (also known as non-radiative energy decay). The Stokes shift, which is the wavelength difference between the excitation and emitted light, is the basic idea behind all fluorescence techniques. The key challenges in optical tissue imaging include significant light scattering, autofluorescence, and substantial absorption by hemoglobin (Hb) within the mid-visible band. Organic fluorescent dyes, such as fluorescein isothiocyanate (FITC) and carboxyfluorescein diacetate succinimidyl ester (CFSE), are widely utilized as fluorophores. These dyes find application in diverse biological contexts, including fluorescent-labelled antibodies and molecules used for staining cells or organelles. The limitations of using organic dyes include their vulnerability to rapid photobleaching, making them unsuitable for prolonged bioimaging observations. Rapid photobleaching represents a challenge associated with organic fluorescent dyes. Nanoparticles serve as the source of fluorescence. Fluorescent dyes can be efficiently trapped within silica particles (dye-doped silica nanoparticles), preserving the spectral characteristics of the dye molecules. Silica encapsulation forms a protective layer around the dye molecules, preventing oxygen molecule penetration, which commonly leads to the photodegradation of dyes in both air and aqueous environments (where dissolved oxygen is present). Consequently, the photo-stability of dye molecules significantly improves compared to bare dyes in solution [10]. Silica nanoparticles (SiNPs) and their aggregates larger than 300 nm result in significant light scattering. While SiNPs are generally biocompatible and not harmful to cells and tissues, their cell permeability relies on their zeta potential; negatively charged SiNPs struggle to pass cell membranes. They are mostly non-toxic and can be excreted easily unless their size exceeds 50 nm. However, they are rapidly coated by intracellular proteins and can be targeted by the immune system. Their surface is easily modifiable using various coatings and sol-gel techniques [18].

### 9.1.5 Ultrasound Imaging

Ultrasound imaging is a bioimaging technique widely used in medical diagnostics imaging. It is a low-cost, portable, and non-invasive real-time imaging technique. Nanoparticle-based contrast agents have the potential to significantly increase imaging resolution when compared to traditional contrast agents of size 1–8 μm. Additionally, they have a higher circulation lifetime and more stability than commercially available ones due to their small size. However, to ensure acoustic reflection, nanoparticles are generally of a size ranging from 100 to 1000 nm. These nanoparticles can be tagged with different surface ligands to facilitate targeted molecule binding [19].

There are three main types of nanoparticle contrast agents for ultrasound imaging. The first is gas-generating solid nanoparticles. These generate micro-bubbles upon activation, allowing for robust acoustic reflection and enhanced contrast in ultrasound imaging. These microbubbles include sulfur hexafluoride, carbon dioxide, nitrogen, and perfluorocarbon. Nanoparticles such as liposomes, silica nanoparticles, and polymer particles are employed in this context [10]. Azide-glycol chitosan-coated gold nanoparticles (AzGC-AuNPs) are an example of this kind of contrast agent. They are made up of gold nanoparticles (AuNPs) coated with azide compounds. They work by generating gas bubbles readily when exposed to pulsed laser light. These produced microbubbles are highly echogenic, causing the ultrasound to scatter very effectively. They have many advantages including smaller size (they are smaller than 100 nm), allowing them to escape from vasculature; on-demand gas generation, as they can generate gas bubbles on demand upon pulsed laser irradiation; and the higher echogenicity of nitrogen gas bubbles [20]. Then there are solid-core ultrasound contrast agents (USCAs), which are solid-based nanoparticles with strong scattering acoustic signals. Their small size, multifunctionality, favorable stability, ease of fabrication, highly targeted turnover, high abundance, and adaptable structures make them a promising class of contrast agent for ultrasound imaging. Through cellular absorption, the extravasation permeation retention (EPR) effect, or particular receptor binding, they can build up in target lesions. One example of a solid-core contrast agent is silica-based nanoparticles. Si NPs have unique properties such as biocompatibility, high specific surface area, controllable particle size, and unique porosity structure. They can be easily surface-modified to allow for many different compositions. Rattle-type mesoporous silica nanoparticles are an example of this. They are particularly promising for ultrasound detection of tumors. Finally, there are liquid-core nanoparticles like perfluorooctyl bromide (PFOB) nanoparticles, generating an acoustic signal due to their lower sound transmission speed compared to water [10, 21].

### 9.1.6 Positron Emission Tomography

Positron emission tomography and single positron emission computed tomography (SPECT), both widely used nuclear medicine technologies, offer real-time quantitative imaging analysis, high tissue penetration, and sensitivity. PET, in particular, provides molecular-level biological information through nuclide tracking, in addition to anatomical details. SPECT and PET can detect abnormal biochemical function before any anatomical changes occur. However, these advanced imaging techniques come with limitations, including their high cost and the exposure to radioactive elements. Long-lasting nuclides are essential for SPECT and PET imaging tracers using nanoparticles. Fluorine-18, typically used in clinical PET, has a half-life of 109.8 minutes, too short for nanoparticle applications due to preparation and cellular uptake time. Hence, nuclides with longer half-lives such as copper-64, with a half-life of 12.7 hours, or indium-111, with a half-life of 2.8 days are necessary for different nanoparticle preparations. The structure of nanoparticles used in SPECT imaging closely resembles that of those employed in PET imaging. Nanoparticles in SPECT and PET primarily find application in tumor detection, achieved through specific receptor binding, the EPR effect, or active and passive accumulation in target lesions. Various imaging methods, such as vector tracking, monitoring pharmacokinetic and atherosclerotic plaque imaging, and changes of particles, are utilized. Radiopharmaceutical nuclides are crucial in nuclear medicine imaging. However, the detachment of radio-labelled probes from nanoparticles poses a significant challenge. This issue has been addressed through direct conjugation of nuclides to stable nanoparticles synthesized using poly (4-vinyl phenol) (PVPh) [22].

### 9.1.7 Photoacoustic Imaging

Photoacoustic imaging, a swiftly advancing biomedical imaging technique, relies on the photoacoustic effect initially discovered by Alexander G. Bell in 1880. Photoacoustic imaging enables real-time analysis of diseased tissues, providing detailed insights into their anatomical, functional, and molecular aspects without the harmful effects of ionizing radiation. Unlike fluorescence optical imaging, PA imaging offers exceptional spatial resolution (as fine as 5 μm) and deep penetration (up to 5–6 cm) by minimizing the scattering of ultrasonic signals, as opposed to light in tissues. Overcoming the limitations of ultrasound imaging related to tissue's mechanical properties, PA imaging enhances tissue contrast based on the optical characteristics of different tissues. Moreover, PA imaging serves as a safer alternative to computed tomography, PET, and SPECT, eliminating the use of ionizing radiation while delivering comprehensive and high-quality imaging results, introducing contrast in PA imaging that can provide both functional and anatomical information. At present, there is a wide array of PA contrast agents available, but gold nanoparticles stand out as a highly promising exogenous PA contrast agent among them. The strong and adjustable optical absorption of AuNPs due to the surface plasmon resonance (SPR) effect has made them commonly chosen as PA contrast agents. The surface plasmon resonance effect occurs when free charges on the surface of gold nanoparticles oscillate in response to the electromagnetic field, leading to optical absorption levels that are significantly higher, by several orders of magnitude, than those observed in organic dyes [10]. In PA imaging utilizing contrast agents, a pulsed laser's optical energy is absorbed by the contrast agents, converting it into thermal energy and causing a temperature increase. This rapid temperature rise leads to thermoelastic expansion and subsequent ultrasonic emission, detectable by an ultrasound (US) transducer for PA image reconstruction. The effectiveness of PA imaging relies primarily on its optical absorption qualities and its ability to efficiently convert optical energy into acoustic signals. The amplitude of the resulting PA signal is directly proportional to the absorption of light and the thermoelastic properties of the absorbing substance. In PA imaging, the use of metallic nanomaterials is prominent, and it has exceptional light absorption and efficient photothermal conversion capabilities, which are attributed to the localized surface plasmon resonance (LSPR) effect. These materials (metallic nanoparticles) can be classified into two types: gold nanocrystals and their assemblies and transition-metal chalcogenide/MXene-based nanomaterials. PA imaging, when integrated with metal nanoparticles, finds applications in biosensing. This encompasses the detection of metal ions, monitoring pH levels, in vivo generation of reactive oxygen/nitrogen species (ROS/RNS), enzymatic activities, temperature variations, and the measurement of blood oxygen content [23].

### 9.1.8 Biolabeling

Various biosensors hold potential for diverse applications in science and technology. However, challenges such as portability, selectivity, sensitivity, and multi-analyte detection persist. Recent methodological advances integrating nanomaterials or nanoparticles/metals with biomarkers or bi labels as sensors offer promising solutions, overcoming these limitations and enabling the development of highly efficient detection devices. After the introduction of biosensors, the subsequent wave of technological advancement focuses on creating biolabels to serve as sensor detectors in both pathological and technical contexts. These biolabels employ fluorescent or bioluminescent materials as tags, paving the way for innovative research and practical applications. The integration of biolabels and nanoparticles has led to significant advancements in drug delivery technology, particularly in the realm of targeted therapy in biomedicine. In recent times, label-free sensors have emerged, utilizing nanomaterials/particles without the need for additional tools or combinations. These modern biosensors, integrating biological and physicochemical or mechanical transducers, enable the creation of highly accurate single or multiple analyte detectors. These detectors hold promise for applications in healthcare, the food industry, agriculture, and biodefense [24].

## 9.2 METAL AND METAL OXIDE NANOPARTICLES IN BIOIMAGING

### 9.2.1 Gold Nanoparticles

#### 9.2.1.1 Properties and Synthesis

Metal nanoparticles, with a specific emphasis on gold nanoparticles, have garnered significant attention across diverse scientific disciplines owing to their distinct attributes. These attributes include their exceptional X-ray absorption capabilities; the simplicity with which they can be synthesized and tailored; their ability to provide precise control over their physical and chemical characteristics [25]; a strong attraction to amines, disulfides, and thiols [26]; the capacity to fine-tune their optical properties [27]; and their unique electronic characteristics. When gold is in its bulk, solid state, it usually has a metallic yellow color and is inert. AuNPs are recognized for their many special qualities, one of which is their characteristic wine-red color when dissolved in a solution. The interaction of light with the nanoparticles at the nanoscale causes a process called surface plasmon resonance, which is responsible for this color shift. The color of gold nanoparticles in colloidal form can be directly affected by changing their size, and colorimetric detection techniques rely on this property. When suspended in an aqueous solution, spherical AuNPs exhibit a spectrum of colors, ranging from orange and red to purple, which corresponds to the increase in the core size, ranging from 1 to 100 nanometers [28].

There are different kind of methods used for the synthesis of functionalized gold nanoparticles, such as the Turkevich method and Brust-Schiffrin method, which serves as in-situ synthesis of AuNPs. The Turkevich method uses chloroauric acid with trisodium citrate as reducing agent, and about 10–20 nm of size-varying AuNPs can be synthesized [29]. The Brust-Schiffrin method is used for synthesis of thiolate-stabilized AuNPs. This two-phase method produces highly stable particles ranging in size from 1 to 6 nm [29]. Another method is the seed-mediated growth synthesis method, which uses AuNPs as a small seed in a growth solution containing $HAuCl_4$ and ascorbic acid, which act as reducing and stabilizing agents. From this technique, we can obtain AuNPs with size varying from 5–40 nm [30]. Methods such as UV radiation, γ-irradiation method, thermolytic and sonochemical method, photochemical method, laser ablation, and microwaves are categorized as physical methods [31]. Each of these physical methods differs from the others according to the required type of AuNPs. The last type of method is the biosynthesis of gold nanoparticles. Biosynthesis techniques provide a toxin-free and safer way to create biocompatible nanoparticles that can be used in a variety of therapeutic applications. Biopolymers, fungi, bacteria, plants, and enzymes are examples of natural materials that can be used as eco-friendly reduction substances [32]. To enhance their functionalities, AuNPs can be combined with a variety of nano-biological structures, such as proteins, oligonucleotides, and antibodies.

#### 9.2.1.2 Toxicity of Gold Nanoparticles

Gold nanoparticles are known for their resistance to oxidation, excellent biocompatibility, and low toxic properties. Because of their low cytotoxicity, they find widespread use in various biomedical research applications, including biosensors and medical imaging. Most of the time the toxicity comes from stabilizing capping agents and can be reduced by using more bio-friendly capping agents [33]. Li et al. showed that toxicity depends on concentration and size of the AuNPs; that is, toxicity is high for high-concentration and smaller-sized nanoparticles [34]. According to Lee et al., an oxaliplatin-loaded gold nanoparticle combines with cancer cells' nucleic acids, potentially causing an effect on lung cancer cells [35]. Additionally, it has been demonstrated that the 3-mercaptopropionic acid-coated gold nanoparticle interacts with nucleic acid, resulting in a therapeutic effect against leukemic K562/ADM cells (human leukemia cells) [36]. The biocompatibility of AuNPs can be enhanced by using polyethylene glycol (PEG) and ionic coatings [37]. Generally, the synthesis of gold nanoparticles through the biosynthesis method shows less toxicity or nontoxicity, which makes them suitable for biomedical use.

#### 9.2.1.3 Gold Nanoparticles in Magnetic Resonance Imaging

Debouttière et al. showed that incorporating gadolinium chelates onto gold nanoparticles renders them highly appealing for use as MRI contrast agents, a multilayered ligand called dithiolated ligand (DTDTPA) that can hold more gadolinium than the monolayered one [38]. Guan et al. reported the $Fe_3O_4$-functionalized Au nanoparticle is an excellent contrast agent due to its ultra-refined size, and they can increase the $T_1$–$T_2$ (relaxation time) weighed MRI signal. Following functionalization, the current multimodal contrasting agent nanoparticles demonstrated excellent biocompatibility, minimal accumulation, and a small average size [39]. The work by Hoskins et al. shows the potential of iron oxide core, poly(ethylenimine) intermediate, and gold-coated hybrid nanoparticles as an MRI contrast agent and for regulated, highly localized tissue heating [40]. Wang et al. showed that folate-conjugated $Fe_3O_4$@$SiO_2$@gold nanorods@mesoporous $SiO_2$ hybrid nanomaterials can be used as an MRI contrast agent due to their superparamagnetic iron oxide core, which is able to sped up the transverse relaxation of water protons. Also, through magnetic resonance imaging, the labeling of KB cells (human epithelial carcinoma cells) by hybrid nanomaterial is confirmed [41]. Lee et al. used the hepatitis B virus core protein as a framework to create ultrasmall gold nanoparticles with a particle dimension of 1.4 nm. Subcutaneous and deep-tissue tumors in living mice were significantly affected by these gold nanoparticles in terms of MRI and magnetic hyperthermia therapy [42]. Rajat et al. showed a significant increase in the uptake of MRI imaging contrast agent into malignant MDA-MB-231 cells (human breast cancer cells) by using functionalized gold nanoparticles with targeted oligonucleotide AS1411 [43].

#### 9.2.1.4 Gold Nanoparticles in Photoacoustic Imaging

Many forms of AuNPs, including nanostars, nanorods, nanospheres, nanocages, and nanoshells, are widely used as contrast agents in photoacoustic imaging because of their surface plasmon resonance effect [44]. Xu et al. reported that hyaluronic acid functionalized gold nanocages showed high local surface plasmon resonance, which increased the PA imaging property by acting as a contrasting agent. In order to determine the dimension, spot, and limit of a tumor, Au nanocages–hyaluronic acid revealed considerably intense photoacoustic signals, which made imaging-guided treatment easier [45]. Gold nanoparticles coated with glycol-chitosan have been employed as photoacoustic contrast agents for visualizing cancer cells. When comparing glycol-chitosan-coated AuNPs (GC-AuNPs) to polyethylene glycol-coated AuNPs (PEG-AuNPs), it was observed that GC-AuNPs exhibited higher accumulation within cancer cells. This heightened accumulation resulted in more pronounced photoacoustic signals within cell models mimicking tissue. These signals emanate from the surface plasmon resonance effect of GC-AuNPs following their uptake by cancer cells [46]. Since cancer medications like silicon phthalocyanine and recombinant human tumor necrosis factor (TNF) have been linked to gold nanospheres, PA imaging using gold nanospheres as contrast agents can be utilized to monitor drug delivery agents [47]. Three-dimensional photoacoustic images of tumor neovessels with excellent spatial detail and deep imaging depth can be achieved with the targeting properties of cyclic arginine-glycine-aspartic acid (cRGD) peptide-functionalized gold nanostars [48]. Panchanathan et al. reported that high-amplitude PA imaging signals were identified when MDA-MB-231 cells labeled with doxorubicin-loaded fucoidan-capped gold nanoparticles (DOX-Fu AuNPs) were added at elevated concentrations. He also suggested that with DOX-Fu AuNPs acting as cancer imaging contrast agents, it is possible to use PA imaging to image AuNPs inside cancer cells, which raises the possibility of image-guided cancer therapy [49].

#### 9.2.1.5 Gold Nanoparticles in Computed Tomography

The use of gold nanoparticles as a CT contrast agent to identify tumors in mice has been shown by Hainfeld et al. [50]. Kattumuri et al. showed gold nanoparticles stabilized by gum Arabic as a possible biocompatible X-ray CT contrast agent [51]. Wang et al. reported that CT imaging can be used to image the xenograft tumor model and that X-ray attenuation can be used to detect dendrimer-functionalized gold nanoparticles. Dendrimer-functionalized gold nanoparticles are primarily absorbed by the cells' lysosomes and do not exhibit cytotoxicity with a concentration between 0 and 3000 nM [52]. Boote et al. observed that the levels of Au in the tissues raised the liver's CT values by injecting swine with gum Arabic–functionalized AuNPs [53]. Mouse monoclonal (UM–A9) antibody–conjugated gold nanorods are used as target-specific agents for squamous cell carcinoma and show elevated X-ray attenuation in targeted carcinoma cells for the development of CT imaging agents [54]. These types of gold nanoparticles can also be used to identify neck and head cancer in vitro with CT imaging.

#### 9.2.1.6 Other Imaging Techniques

Ke et al. introduced gold nanoshell microcapsules as a means to improve ultrasound contrast imaging. When contrasted with empty microcapsules, the gold nanoshell microcapsules demonstrated enhanced ultrasound imaging outcomes for rabbit kidney after injection [55]. Lim et al. disclosed the development of highly consistent SERS-active nanostructures featuring a core-shell configuration with an exceedingly precise and uniform intra-nanogap of about 1.2 nanometers. They utilized these structures in high-resolution live-cell Raman imaging of cancer cells, employing a custom-made high-speed confocal Raman microscope [56]. Utilizing integrin-targeted radiotracers for tumor imaging can involve single photon emission computed tomography or positron emission tomography with a positron-emitting radionuclide [57].

### 9.2.2 Silver Nanoparticles

#### 9.2.2.1 Properties and Synthesis

The distinct physical and chemical characteristics of silver nanoparticles (AgNPs) have garnered growing attention within the scientific community, primarily because of their exceptional thermal conductivity, plasmonic attributes, chemical durability, and antimicrobial capabilities [58]. Silver nanoparticles show significant promise as effective tools for cancer-related diagnostics and probing. They exhibit potent effects against various cancer cell lines, providing numerous benefits [59]. They possess distinctive characteristics, including readily modifiable surfaces, optical attributes, consistently replicable synthesis methods, and a substantial surface-to-volume ratio, which render them well suited for applications in cancer therapy.

AgNPs can be synthesized through different methods such as physical, chemical, and green synthesis methods. Physical synthesis methods include evaporation, condensation, laser ablation, thermal decomposition, ultrasonic spray pyrolysis, and the arch discharge method. Chemical synthesis methods include chemical reduction, microemulsion, the sonochemical method, the sol-gel process, and the electrochemical synthetic method. The properties of AgNPs synthesized by different chemical process differ from each other. The sol-gel process can be used to synthesize AgNPs at low temperatures. Controllable-size NPs can be obtained from sonochemical and electrochemical methods. In the microemulsion method, the precursors used are a polar phase, a non-polar phase, and a surfactant, whereas in the chemical reduction process, metal precursors, reducing agents, and stabilizing/capping agents are used as precursors. The drawbacks of these methods include the use of hazardous and toxic materials in synthesis and purification challenges. Therefore, we can eliminate these drawbacks through green synthesis methods. AgNPs can be green-synthesized by means of plant extracts, which act as both a reducing agent and capping agent; microbes (yeasts, bacteria, fungi, actinomycetes, etc.); viral DNA; membranes; and diatoms. The chemical method is determined to be the most effective, practical, and manageable of all the approaches. This process is widely used to produce AgNPs through chemical reduction, and it can be highly productive at very low cost.

#### 9.2.2.2 Toxicity of Silver Nanoparticles

Numerous in vitro and in vivo studies have demonstrated toxicity occurring by means of cellular pathway perturbations. The level of cytotoxicity was found to be influenced by the size of internalized nanoparticles, as observed in the case of AgNPs. In a study by Liu and colleagues, the size-dependent cytotoxicity of AgNPs (5–50 nm) was demonstrated in four human cell lines, affecting cell morphology, viability, membrane integrity, and cell cycle progression. The results confirmed a stronger cytotoxic response associated with smaller AgNPs [60]. Several researchers have reported the cytotoxicity of AgNPs in liver cells, human lung cells, brain cancer cells, and human glioblastoma cells [61–63]. The extensive use of AgNPs in cancer diagnosis and treatment gives rise to concerns regarding their potential short- and long-term toxicity, particularly when considering their integration into clinical practice.

#### 9.2.2.3 Silver Nanoparticles in Optical Imaging

Fang et al. showed the optical imaging properties of silver nanoparticle–embedded microbubbles by dark-field microscopy (DFM) and surface-enhanced Raman scattering (SERS). The DFM results shown that AgNPs within a microbubble (MB) shell can vary in size and structure, and DFM imaging can be used to study how these AgNPs are organized within the polymer shell of the MBs [64]. Nancy et al. used peptide-functionalized silver nanoparticles to study the charge-dependent transport of embryos with the help of DFM [65]. Qingyan et al. synthesized silver nanoparticles embedded with a ligand-containing cyano group which acted as a SERS probe and showed Raman imaging membrane proteins on cells with the help of the SERS probe [66]. Cell imaging using Raman spectroscopy was successfully achieved with AgCu@Graphene (ACGs), leading to a notably reduced imaging time due to the robust Raman signals of ACGs enhanced by SERS. Researchers created alkyne polyethylene glycol molecules and modified the surface of ACGs to enable cell imaging within the Raman-silent range. Following the functionalization process, ACGs were precisely co-located within the cells alongside the Raman modes associated with D (disordered graphite band), G (graphite band), and alkyne signals [67]. Photoacoustic imaging successfully enabled the visualization of silver nanoplates conjugated with an epidermal growth factor receptor (EGFR) in a mouse model of orthotopic pancreatic cancer. This technology also distinguished the photoacoustic signal produced by the Ag nanoplates from the signal originating from the natural blood in the body [68].

#### 9.2.2.4 Silver Nanoparticles in Magnetic Resonance Imaging and Computed Tomography

Sandeep et al. synthesized chitosan-coated bimetallic nanoparticles (BNPs) consisting of silver and gadolinium (Ag-Gd) with uniform size and distribution. These nanoparticles were created for the purpose of "cancer theranosis" using an eco-friendly method involving microwave-assisted polyol synthesis. The Ag-Gd BNPs exhibit distinctive properties suitable for multimodal imaging purposes. Their paramagnetic qualities are guaranteed by the presence of gadolinium, resulting in effective contrast enhancement for both $T_1$ and $T_2$ weighted MRI scans. At 0.50 mg/mL, the BNPs' CT signal intensity was approximately 45 Hounsfield unit (HU), which is much higher than the previously reported HU values for standard contrast agents like barium sulphate, iodinated compounds, or other metal nanoparticles [69]. Silver-neodymium bimetallic nanoparticles (Ag-Nd BNPs) were prepared using microwave-assisted polyol synthesis method. Due to the synergistic effects of paclitaxel and their photothermal properties, these nanoparticles demonstrated a strong ability to destroy cancer cells in vitro at much lower doses than single therapies. Consequently, Ag-Nd BNPs offer a dual MRI/CT imaging and treatment approach [70]. According to Agnoli and colleagues, their in vivo MRI findings indicate that PEG-Ag-Fe NPs have an extended circulation time in comparison to endorem. Moreover, these nanoparticles exhibit minimal accumulation in the liver and do not show significant accumulation in the spleen and kidneys in the initial days following administration [71]. Alhasan et al. devised a novel approach by creating silver-doped manganese oxide nanoparticles linked with a conventional CT contrast agent, iohexol, forming I@Si-Ag: $Mn_3O_4$. This strategy aimed to enhance CT contrast imaging. The outcomes of this study confirmed that the I@Si-Ag: $Mn_3O_4$ preparation method was effective in overcoming the limitations of iohexol, particularly in terms of contrast value, brightness, and radiation dosage, ultimately resulting in improved CT contrast efficiency [72].

#### 9.2.2.5 Silver Nanoparticles in Ultrasound Imaging

Fang et al. synthesized silver nanoparticles embedded with microbubbles. When nanoparticles are found on the surface of microbubbles, it can alter the MBs' viscoelastic characteristics and enhance their nonlinear echogenic properties when they are subjected to ultrasound. This increased contrast is a result of the greater ultrasound echoes generated by silver nanoparticles in the MB shell [64]. By merging ultrasound and photoacoustic imaging, it becomes possible to detect variations in the number of blood vessels or oxygen saturation levels within tumors when compared to the neighboring healthy tissue. This is achievable because ultrasound imaging provides visual cues for anatomical references such as boundaries, size, and tumor location, which can be identified and segmented. Homan et al. used silver nanoplates conjugated with a-EGFR in ultrasound photoacoustic imaging, and tumor histology, stained with silver and nuclear fast red, confirmed heterogeneous nanoplate dispensation [68]. David et al. used European black elderberry extracts to synthesize silver nanoparticles. They also demonstrated that the ultrasound features of psoriasis lesions treated with an AgNP cream verified the anti-inflammatory effects of the nanomaterials under investigation, establishing the superior efficacy of AgNPs in comparison to hydrocortisone cream [73].

### 9.2.3 Platinum Nanoparticles

#### 9.2.3.1 Properties and Synthesis

Platinum nanoparticles (Pt NPs) are primarily recognized for their effective catalytic properties and their ability to combat oxidative stress, attributes that have been extensively investigated through the synthesis of nanoparticles via chemical methods. Platinum is characterized by its remarkable resistance to corrosion when inside the human body, and it exhibits specific forms that can impede cell division. These characteristics have spurred the development of nanomaterials based on platinum for various biomedical purposes. Other than this, platinum nanoparticles are also known for their anti-microbial, anti-parasitic, antioxidant, and photo-catalytic properties.

The manufacture of Pt NPs for use in biomedical applications relies heavily on their physicochemical characteristics and their distribution and stability within a biological setting. These factors are pivotal in determining the level of toxicity associated with Pt NPs. There are several methods are used to synthesize Pt NPs, such as physical methods, chemical methods, and green synthesis methods. Certain chemical methods such as galvanic displacement, chemical vapor deposition, wet chemical reduction (WCR), and electrochemical reduction have recently garnered increased attention [74–76]. These methods are employed to effectively control the physicochemical characteristics of Pt NPs during chemical processes. Presently, physical techniques like flame synthesis, electron beam–induced reduction, laser ablation, and aerosol-assisted deposition have gained significant popularity. They are being extensively explored to address certain drawbacks associated with chemical methods, such as the use of organic solvents and hazardous reagents [77–80]. Biological synthesis has been suggested as an alternative to chemical and physical approaches, with a focus on utilizing biological processes involving both single-cell and multicellular organisms. Effective biosynthesis of platinum nanoparticles is possible from a variety of sources, including bacteria, cyanobacteria, fungi, plants, seaweeds, and bio-derived materials like egg yolk and aqueous honey solutions [81–86].

#### 9.2.3.2 Toxicity in Platinum Nanoparticles

Pt NPs are instrumental in producing chemotherapy drugs like carboplatin, cisplatin, and oxaliplatin used for treating various cancers, including rectal and ovarian cancer. However, these drugs often lead to side effects like nephrotoxicity, neurotoxicity, and cytotoxicity. Employing a green synthesis approach for Pt NP formation can help mitigate these side effects [87]. Research that investigates the size of Pt NPs and their potential cytotoxicity is a critical factor influencing cellular pathways. Notably, Pt NPs with an 8-nm size exhibited no adverse effects, while Pt NPs measuring 1 nm induced dose-dependent cytotoxicity in renal cells within the same concentration range. In a neural cell line, Pt NPs ranging from 1 to 21 nm were assessed, revealing that Pt NPs of 5–6 nm was compatible with the cells, while other sizes caused cellular harm. Therefore, these studies have underscored the ability to mitigate or manage the toxicity associated with Pt NPs by carefully regulating their size during the preparation process. This, in turn, impacts their potential applications in the field of biomedicine [88–90].

#### 9.2.3.3 Platinum Nanoparticles in Magnetic Resonance Imaging

Lei et al. synthesized a double mesoporous nanosystem based on platinum nanoparticles functionalized with gadolinium, which is used for treating and imaging cancerous cells. The final product, mPt@m$SiO_2$-GdDTPA, displayed greater $R_1$ (relaxation rate) values compared to the medical contrast agent magnevist and has been effectively utilized for in vivo MRI experiments with Kunming mice [91]. Wang et al. synthesized superparamagnetic iron oxide NP (SPION)–based platinum composites which show diagnostic potential and specific accumulation of the composite in tumors, and it was confirmed through time-dependent $T_2$-weighted MRI scans conducted with MCF-7 cells and mice bearing tumors. Core-shell-structured upconversion nanoparticles (UCNPs) functionalized with platinum nanoparticles were used for drug delivery. They were also applied for both in vitro and in vivo T1-weighted magnetic resonance imaging and computed tomography (CT) imaging [92]. Chen et al. synthesized multifunctional FePt@MMT (montmorillonite) material with the capability to perform multiple tasks, including the enhancement of MRI signals for liver cancer (HCC-hepatocellular carcinoma) visualization, treatment of different diseases, and triggering magnetic fluid hyperthermia (MFH) [93]. Xu et al. designed yolk-shell magnetic nanoparticles composed of $Fe_3O_4$, carbon, platinum, and chlorin e6 (MCPtC$e_6$), which have the potential to be applied in magnetic resonance imaging as well as for tumor therapy [94]. Zhu et al. reported an in vivo and in vitro MRI analysis of the $Pt^{IV}$ prodrug (HSPt), which is created by modifying cisplatin with hydroxyl and succinate groups, which is linked to poly (ethylene glycol)–modified superparamagnetic iron oxide nanoparticles. Notably, in both in vivo and in vitro analysis, the nanoparticles accumulated in cancerous cells showed a negative contrast [95].

#### 9.2.3.4 Platinum Nanoparticles in Photoacoustic Imaging

Phan et al. synthesized iron–platinum nanoparticles (FePt NPs) with a polypyrrole (PPy) coating, which showed the PA imaging system successfully imaged FePt@PPy NPs within a phantom, demonstrating strong potential for image-guided cancer therapy [96]. Zhang et al. demonstrated that PEG-C$e_6$ functionalized platinum nanoparticles (Pt@PEG-C$e_6$) in PA imaging by injecting it into mice with cancer. The injection of Pt@PEG-C$e_6$ led to an increase in the signal from tumor blood vessels, signifying that Pt@ PEG-C$e_6$ holds promise for photoacoustic imaging [97]. For in vivo PA imaging, copper iron sulfide with chitosan-functionalized platinum nanoparticles (Cu$FeS_2$-CS-Pt) was employed. These nanoparticles generated potent PA signals when injected into the tumor area. However, no such signals were seen in the nanoparticle-free control group. The effectiveness of Cu$FeS_2$-CS-Pt nanoparticles as contrast agents for PA dual-modal imaging is thus confirmed [98].

Zhou et al. synthesized dendritic platinum–copper alloy nanoparticles (DPCNs) for cancer therapy. They observed that increasing DPCN concentrations in deionized water led to higher photoacoustic intensities with a linear relationship. After injecting DPCN, there was a notable increase in photoacoustic signals within the tumor, enhancing the distinction between the tumor and nearby normal tissues [99]. Liu et al. created safe Fe-Pt nanoparticles for photoacoustic imaging of breast tumors, even at depths of a few centimeters. Initially, they confirmed the targeting of Fe-Pt nanoparticles linked to anti-vascular endothelial growth factor (VEGFR) multiphoton microscopy and in vivo PA imaging. Additionally, they confirmed the capability of the anti-VEGFR-conjugated Fe-Pt nanoparticles to act as contrast agents for detecting breast cancer in vivo using photoacoustic imaging [100].

### 9.2.4 Iron Oxide Nanoparticles

Magnetic nanoparticles (MNPs) have long been employed in the production of nanosensors and in nanomedicine. The position of an MNP could be utilized to pinpoint the precise location of a medical issue because it is widely recognized that MNPs have regulated interaction with an external magnetic field created by utilizing permanent magnets. MNPs undergo a thermal heating process when exposed to an alternating magnetic field; this makes them excellent for usage as a medicine delivery agent for various illnesses as well as a contrast agent in nuclear magnetic resonance imaging. Because of their low toxicity and large spin magnetic moments, magnetic iron oxide nanoparticles are primarily studied and exploited in biological activities. Since iron is widely distributed throughout the crust, it is referred to as the foundation of the earth's chemical makeup [101]. It is inexpensive and important to many biological and geological processes. Iron oxide is created when iron (Fe) interacts with oxygen (O) molecules in its native state. Magnetite ($Fe_3O_4$), maghemite ($\gamma$-$Fe_2O_3$), and hematite ($\alpha$-$Fe_2O_3$) are the three most prevalent IONP forms that can be found in nature. Hematite is created when magnetite undergoes further oxidation. Compared to other types of IONPs, magnetite has attracted a lot of study interest recently.

#### 9.2.4.1 Properties

Due to the motions of the external magnetic field, they have several notable advantages, including superparamagnetism, a greater surface area, a substantial ratio of surface to volume, and a quick and simple technique of separating the target from the solution. As a result, IONPs, or magnetite, have been thoroughly researched and used in the biomedical domains [102].

MNPs have certain drawbacks when it comes to their application in biological activities, such as toxicity, which can be reduced by surface functionalization with biocompatible substances such as polymers, aptamers, antibodies, and fluorophores that cover the entire surface of the MNP and so offer biocompatibility.

#### 9.2.4.2 Synthesis of Magnetic Iron Oxide Nanoparticles

Three layers constitute the structural makeup of multifunction magnetic IONPs: the surface layer shows functionalized small molecules, doped metals, polymers, antibodies, and proteins; the shell layer shows chemical components distinct from the prominent component; and the core layer contains the primary chemical entity, which is typically a nanoparticle. The design and synthesis of multifunctional magnetic IONPs for biomedical applications have been described by numerous groups. Researchers have published many techniques for creating magnetic iron oxide nanoparticles. Their size, distribution, form, surface chemistry, and uses are all directly impacted by the techniques used. The physical techniques that are frequently employed to create magnetic iron oxide nanoparticles are mechanical grinding, gas phase deposition, and electron beam lithography. On the other hand, chemical synthesis includes thermal decomposition, sol-gel synthesis, co-precipitation, hydrothermal/solvothermal synthesis, micro-emulsion synthesis, and sol-chemical synthesis. Due to the convenience of tunability and size control, chemical synthesis methods are more suited for the synthesis of iron oxide nanoparticles in biomedical applications than physical synthesis methods. In fact, these methods have received the greatest attention in the field.

##### *9.2.4.2.1 Co-Precipitation Method*

Both $Fe^{2+}$ and $Fe^{3+}$ (1:2 ratio) salt solutions in aqueous alkaline media (such as ammonia or sodium hydroxide solution) are crucial to the synthesis of this process and lead to the final product, which has IONPs with regulated features like shape, size, size distribution, and crystallinity. The precipitation of iron salts mediates synthesis in the presence of sodium hydroxide or nitrate, occasionally using photochemical techniques. Additionally, MNPs are altered with diluted sodium citrate, nitric acid, sodium alginate, or organic stabilizing agents such as polyvinyl pyrrolidone. A blackish-brown precipitate is obtained at the end of a reaction, which is followed by three rinses with de-ionized water. The precipitate is suitable to keep for drying at room temperature.

##### *9.2.4.2.2 Hydrothermal Synthesis*

Researchers suggested that the appropriate parameters of hydrothermal techniques are used for specific size IONP synthesis, and the required reaction temperature is 180–220°C. This method is advantageous for scale-up processes. Hydrothermal methods are based on the principle of high temperature and pressure. Extreme conditions are provided in Teflon-coated stainless-steel autoclaves. The autoclave system includes three solution phases – solid, liquid, and gas – on which the reaction depends. Many benefits come with the hydrothermal technique: powder can be recovered straight from the solution, different precursor chemicals can be used to control the shape and size of the particle, and these highly reactive powders cause temperature sintering. However, the technique has drawbacks as well: prior knowledge of precursor solubility is necessary, and hydrothermal slurry is corrosive; thus, there is always a risk of an inadvertent explosion due to high pressure.

##### *9.2.4.2.3 Thermal Decomposition Method*

Specific size and shape manipulation is made possible in this technique by the IONP synthesis using organometallic compounds such as metal fatty acid salts, carbonyls, and metal acetylacetonates. It does, however, need a high temperature and a lengthy response time. It is conducted in the presence of stabilizing surfactants in organic solvents and surfactants, such as oleic acid, fatty acids, and hexadecylamine. Appropriate criteria, such as the ratios of the starting reagents, which include organometallic compounds, surfactants, and solvents, control the size and shape of the synthesized MNP. The ageing period, temperature, and reaction time are critical for controlling the shape and size of IONPs.

The most often used ingredient in the preparation of functionalized iron oxide nanoparticles is silica. The benefits of silica-coated NPs include the following: (1) it stabilizes iron oxide nanoparticles in solution; (2) it prevents interparticle interactions and agglomeration; (3) the preparation technology for size-tunable composite nanoparticles is already well developed, and controlling the shell thickness variation is not too difficult. Moreover, silica coating aids in the binding of different biological or other ligands at the surface of NPs for a variety of uses.

For example, Ashtari et al. have shown an efficient technique based on amino-modified silica-coated $Fe_3O_4$ NPs for the recovery of target ssDNA [103]. Silica coating will cause the particle size to rise and the composite NPs magnetic characteristics to alter. Notably, silica thickness can be adjusted between 5 and 200 nm by adjusting the ammonia concentration and the ratio of tetraethoxysilane (TEOS) to water. Furthermore, silica coating was easily applied to the surface of iron oxide NPs in an aqueous environment via hydroxyl groups, particularly when using the Steinber method and sol–gel process. Better dispersion and reduced agglomeration of the magnetic particles were the outcomes of this. The present emphasis of the study extends the function of silica-functionalized iron oxide NPs to produce magnetic composites that have good dimensional stability and a programmable structure.

Common materials coated on iron oxide NP surfaces have drawn increased interest recently because they often import the unique features of metal oxides. A chemical displacement process that produces air-stable $Fe_3O_4@Al_2O_3$ core-shell NPs was described by Li et al. [104]. The NPs are rather homogeneous, and because of their strong magnetic response, they say that they were able to deposit the collected phosphopeptides straight onto the matrix-assisted laser desorption/ionization (MALDI) target for analysis. According to Hong et al. [105], zinc acetate and ammonium carbonate were used to coat the $Fe_3O_4$ NPs directly, resulting in ZnO-coated magnetite core-shell NPs. They assert that the antioxidation capacity of the functionalized NPs is significantly higher than that of the naked $Fe_3O_4$ NPs. Nevertheless, the produced particles exhibit nonuniformity and a reduction in their saturation magnetization (*Ms*).

#### 9.2.4.3 Iron Oxide Nanoparticles in MRI and Biomarker Tracking

Without the use of radioactive radiation, MRI is an effective bio-imaging technique that provides both 3D and cross-sectional images of sensitive tissues. In addition to serving as an MRI contrasting agent, magnetic IONPs can be used to follow a particular biomarker, cell, or organ in conditions like cancer, stroke, or aortic or cerebral dysfunction. For this reason, the synergistic bioimaging and biomarker-tracking properties of magnetic IONPs make them multifunctional. The contrasting agents now in use, such as gadolinium, dysprosium, and manganese, are comparatively less sensitive, and their high dosage can cause organ poisoning.

Magnetic IONPs have been investigated as magnetic contrast agents during the past 10 years due to their unique physico-chemical characteristics, biocompatibility, simple surface functionalization, and strong magnetism. As contrast agents, IONPs have effectively contributed contrast properties that are essential for MRI. It is recommended that IONPs be employed to mitigate the drawbacks of the contrasts that are currently in use. MRI is a type of tomographic imaging that is non-invasive. The interaction among protons and nearby tissue molecules serves as its foundation. It is a promising method for imaging the heart, detecting tumors, and imaging the central nervous system. MRI contrast chemicals alter the surrounding tissues' relaxation times, which are then picked up by MRI. Several contrast chemicals are used in the MRI method. Nonetheless, in-vivo magnetic labels are used, such as hyperpolarized components, IONPs, and gadolinium contrasting agents. The most suitable one is Gd-DTPA, or gadolinium chelate. It is thought that superparamagnetic IONPs could be a viable option for MRI application. Many biomarkers from bodily tissues can be detected by quantum dots and gold nanoparticles. Unlike IONPs, these nanoparticles are not appropriate for numerous in-vivo MRI imaging procedures. The biomedical sector currently uses ferromagnetic and superparamagnetic IONPs, as they reduce $T_1$ spin lattice relaxation time and $T_2$ spin-spin relaxation time. The image is sharper and brighter as a result. By causing local field inhomogeneity, they can alter the spin-spin relaxation effects and reduce the $T_1$ and $T_2$ relaxation periods. PEG polymers and other functional entities are added to the surface of magnetic IONPs to enhance their biocompatibility, hydrophilicity, and degradability [106].

### 9.2.5 Bismuth Oxide Nanoparticles

#### 9.2.5.1 Properties

The efficiency of radiation in clinical settings is weakened by intrinsic radio resistance and imprecise tumor location. Hyaluronic acid-functionalized bismuth oxide nanoparticles (HA-$Bi_2O_3$ NPs) can be produced using a one-pot hydrothermal process to get beyond these restrictions and can be used for radio sensitization of tumors and target-specific computed tomography imaging. Following hyaluronic acid functionalization, $Bi_2O_3$ NPs had advantageous exceptional biocompatibility, water solubility, and selective absorption by cancer cells [107].

#### 9.2.5.2 Synthesis of HA-$Bi_2O_3$ NPs

Thirty mL of diethylene glycol (DEG) is mixed with 5.7 mmol $BiCl_3$ and continuously swirled to create a clear, thick solution. For 1 hour, the solution is heated to between 140°C and 160°C in a silicon oil bath. Next, 30 mL of DEG is added, along with 7.5 mmol of dissolved NaOH. Following the total dissolution of the reactants, the mixture is vigorously stirred while refluxed at 180°C for 30 minutes. Room temperature is maintained for the prepared NPs. Then 0.5 mmol hyaluronic acid and 0.5 mmol NaOH can be added. After vigorously swirling the solution for 6 hours at room temperature, a white precipitate will be obtained. Large-sized agglomerates can be eliminated from the transparent solution by filtering it through a 0.22-µm membrane once the system has naturally cooled to ambient temperature. Following preparation, the solution is dialyzed against water for 3 days [108].

#### 9.2.5.3 In Vivo and In Vitro CT Imaging Study and Biodistribution of HA-$Bi_2O_3$ NPs

A clinical 64-slice multidetector CT scanner can be utilized to get both in vitro and in vivo CT pictures. Using the commercial contrast agent Omnipaque (iohexol injection) as a reference, solutions of HA-$Bi_2O_3$ NPs varying in concentration from 0 to 10 mg/mL were applied to a 96-well culture plate in order to investigate the CT images in vitro. Using the average grey density data, appropriate software (ImageJ) and the medical imaging analysis system assessed the X-ray attenuation intensity. We used normal ICR mice as our model for in vivo CT imaging. Before doing a CT scan on the animals, HA-$Bi_2O_3$ NPs (40 mg/kg) were given intravenously and filtered through membrane filters with a pore size of 0.22 µm. Microwave digestion was applied to the matching dissected organs.

#### 9.2.5.4 HA-$Bi_2O_3$ NP Intracellular Tracking and Cytotoxicity

HA-$Bi_2O_3$ NPs are anticipated to have acceptable biocompatibility for their intended use as biomedical agents. All of the HA-$Bi_2O_3$ NPs at various doses have very little cytotoxicity. The cell viability was approximately 90% even at a dose of 400 µg/mL and a 24-hour exposure duration. The absorbance of the supernatant at 541 can be used to quantitatively assess the proportion of hemolysis at various HA-$Bi_2O_3$ NP concentrations. It has excellent hemobiocompatibility and extremely low cytotoxicity.

#### 9.2.5.5 Application in Computed Tomography

Using cancer research mice (ICR mice) as a model, the potential application of HA-$Bi_2O_3$ NPs as CAs for CT imaging in vivo can be investigated. Following a 20-mg/kg intravenous injection of HA-$Bi_2O_3$ NPs, various scan periods can be used to get CT images. Within 10 minutes, the tumors' CT signal intensities will gradually increase due to the presence of HA-$Bi_2O_3$ NPs. A significant contrast enhancement in the tumor is clearly visible as compared to preinjection, meaning that at 10 minutes after injection, HA-$Bi_2O_3$ NPs produced a time-dependent rise in signals and demonstrated good CT imaging capacity.

### 9.2.6 Copper Oxide Nanoparticles

#### 9.2.6.1 Properties

Due to its high copper concentration, copper oxide (CuO) has grown into a prevalent source of this metal. Nevertheless, CuO bulk particles remain undissolved in water, and the gastrointestinal tract is the primary site of solubilization for the particles' bioavailability [109]. CuO NPs on a nanoscale, such as those measuring 7 nm, may enter cells and organelles and are readily dissolved in water with good colloidal stability. Their large surface area might improve binding capacity and bioreactivity.

#### 9.2.6.2 Synthesis

Glacial acetic acid (0.009 mol) and copper acetate (0.003 mol) are combined after being dissolved in double-distilled water [110]. On the hot plate set, the solution is heated to reflux while being agitated. Next, a quantity of NaOH pellets (0.015 mol) is carefully added. The mixture will instantly transform from blue to dark brown, signifying the creation of CuO NPs. The suspension is then allowed to cool to room temperature. Centrifugation (10 min, 4000 rpm) can be used to separate the CuO NP precipitate, with the excess ionic species remaining in the supernatant. A stable colloidal solution is produced by redistributing the CuO NP sediment in water.

#### 9.2.6.3 Copper Oxide Nanoparticle MRI and Ultrasound Imaging

CuO NPs exhibit desirable acoustic and magnetic characteristics that make them appropriate for dual-modal MRI-ultrasound imaging. Dual-modal imaging and detection of CuO NPs, followed by improved destruction of the cancerous tissue surrounding them, may result in a potentially improved diagnostic and treatment procedure, as MRI and ultrasound are the common modes of choice for image-guided tumor ablation [111].

### 9.2.7 Manganese Oxide Nanoparticles

#### 9.2.7.1 Synthesis

Park et al. synthesized highly crystalline, monodisperse, and size-controllable $Mn_3O_4$ nanoparticles by thermally breaking down a single precursor, Mn(acac)2, in oleylamine in an inert atmosphere. Acacia stands for acetylacetonate [112]. Recently, at low temperature in an air atmosphere, Hyeon and colleagues generated homogeneous $Mn_3O_4$ nanocrystals with different forms [113]. Oleylamine-capped $Mn_3O_4$

NPs were created in a study using the technique that Park and colleagues previously published [112].

#### 9.2.7.2 Manganese Oxide in MRI

Hyeon et al. developed the first biocompatible $T_1$ MRI contrast agent based on MnO NPs. Their findings demonstrate the potential of conjugating a tumor-specific antibody to create functionalized MnO NPs that are specifically useful for imaging breast cancer cells. A new avenue for targeted therapy and biomedical imaging in the field of medicine is made possible by this novel class of MRI contrast agents [114]. Hollow manganese oxide nanoparticles were further developed by Lee and colleagues as an intracellular drug delivery carrier and as a dual contrast agent for $T_1$ and $T_2$-weighted MRI [115]. A biocompatible magneto fluorescent $T_1$ MRI contrast agent, silica-coated $Mn_3O_4$ NPs enable targeted fluorescence and magnetic resonance imaging of cancer cells. Studies using MRI, confocal microscopy, and in vitro flow cytometry demonstrated that the $Mn_3O_4$@$SiO_2$ (isothyocyanate-folate) NPs may precisely target cancer cells expressing folate receptors through the receptor-mediated delivery pathway in a brief incubation period [116]. $Mn_3O_4$ NPs coated in silica can be further coupled with other biomolecules, including peptides, sugars, and antibodies, to create a novel platform for bimodal imaging and targeting of other biological systems.

### 9.2.8 Zinc Oxide Nanoparticles

#### 9.2.8.1 Properties

Zinc is an important trace element that is widely distributed in all body tissues, such as the skin, muscles, bones, brain, and so forth. Zinc participates in the body's metabolism and is essential for the production of proteins and nucleic acids, hematopoiesis, and neurogenesis. It is the primary component of many enzyme systems [117–120].

In contrast to other metal oxide nanoparticles, zinc oxide nanoparticles (ZnO NPs) have a relatively low cost and less toxic properties and have good therapeutic uses, including anti-inflammation, wound healing, medication delivery, and antibacterial and diabetic treatment [121–124].

#### 9.2.8.2 Application in Bioimaging

Zinc oxide nanoparticles demonstrate effective blue and near-UV emissions, accompanied by green or yellow luminescence associated with oxygen vacancies. This expands the NPs' application in the bioimaging sector [125, 126].

The first stable aqueous ZnO@polymer core-shell nanoparticles (ZnO@ poly(ethylene glycol) methyl ether methacrylate (MAA-co-PEGMEMA)) were synthesized by Xiong et al. using a straightforward sol-gel technique [127]. These nanoparticles showed a high quantum yield and very stable wide photoluminescence in aqueous solutions. ZnO-1 (produced from LiOH) with an average size of 3 nm displayed green fluorescence within human hepatoma cells, but ZnO-2 (derived from NaOH) with an average size of 4 nm looked yellow.

It was noteworthy that at concentrations below 0.2 mg/mL, these nanoparticles did not exhibit any appreciable toxicity on human hepatoma cells. In addition, the cells were still alive after 45 minutes of exposure, and the luminescence was extremely stable during cell culture. Hence, ZnO@polymer core-shell nanoparticles can be employed as fluorescent probes for in vitro cell imaging as a kind of low-cost and safe luminous labeling [127].

ZnO nanosheets were created by Jiang et al. for the purpose of imaging cultured cells. They applied ZnO nanosheets to drug-sensitive leukemia line K562 (myelogenous leukemia cell line) cells, and at room temperature and UV irradiation (365 nm), a yellow-orange light emission was clearly visible surrounding or inside the cells [128]. The successful attachment and penetration of ZnO nanostructures into the cells raised the possibility that ZnO nanosheets with discernible yellow-orange emission could serve as a workable label for bioimaging.

Tang et al. used the chemical precipitation method to manufacture ZnO NPs. It displayed blue, green, yellow, and orange emission colors [129]. The precipitation solutions' pH can be adjusted to alter the emission color.

They formed ZnO@silica core-shell nanostructures by encasing ZnO NPs in silica, which helped stabilize the ZnO NPs in water. The resulting ZnO@silica core-shell nanoparticles showed good water stability and preserved the visible ZnO emissions [130]. It adhered to NIH/3T3 cells (also called Swiss 3T3), which are a type of fibroblast that was first generated from an embryo of an NIH/Swiss mouse. Additionally, it displayed a variety of fluorescent colors with various wavelengths of emission.

## 9.3 METAL NANOPARTICLES IN BIOLABELING

Wang et al. developed dual-mode (PET/CT) imaging agents by radio-labeling gold nanocages with 64Cu. Notably, the 30-nm gold nanocages displayed significantly improved biodistribution characteristics in contrast to the larger 55-nm gold nanocages [131]. Black et al. created dual-radiolabeled gold nanoparticles (labeled with $^{125}$I and $^{111}$In) as a versatile SPECT imaging contrast agent, incorporating a matrix metalloproteinase (MMP) cleavable peptide. They showcased distinct pharmacokinetic characteristics of this contrast agent among different tumors [132].

Gontero et al. reported highly luminescent gold core-shell nano labels were synthesized via metal-enhanced fluorescence (MEF) nanoimaging for the purpose of detecting and quantifying *E. coli* in flow cytometry, utilizing laser excitation and fluorescence detection. The effectiveness of this nanolabeling approach was verified using a conventional cell counter, and bacterial counts were validated through both bright field microscopy and laser fluorescence microscopy [133]. Cheng et al. used folic acid-coated gold nanoparticles functionalized with fluorophore FITC (Au-FA-FITC) and FITC-functionalized AuNPs (Au-FITC) as a biolabel, and they observed that the former primarily localized within the cytoplasm of primary human fibroblasts, and the latter entered the nuclei of cancer cells [134].

A highly sensitive method for detecting individual miRNA in lung cancer cells involves utilizing silver-enhanced fluorescence. This approach relies on a structure comprising a silver shell surrounding a silica core, labeled with oligonucleotides and containing Tris(bipyridine)ruthenium (II) complex $[Ru(bpy)_3]^{2+}$. It is chosen for its remarkable sensitivity and outstanding characteristics [135]. Xu et al. created an extremely sensitive fluorescence technique by introducing silver labels on plasmonic microarrays modified with AgNPs. This method is designed for the specific detection of platelet-derived growth factor-BB (PDGF-BB) and offers both a low limit of detection (LOD) and high specificity [136].

Chrastina et al. established a technique to radiolabel AgNPs with iodine-125 (I-125) for monitoring the in vivo tissue absorption of silver nanoparticles following systemic delivery using SPECT imaging. The analysis of biodistribution disclosed the accumulation of these nanoparticles in the liver and spleen 24 hours after administration [137]. Farrag et al. developed a quick method to radiolabel three types of AgNPs with I–125. Following intravenous injection into both normal and solid tumor-bearing mice, they observed long-lasting localization of I-125 AgNPs in the tumor site [138].

## 9.4 METAL OXIDE NANOPARTICLES IN BIOLABELING

Zhan et al. reported that in an in vitro experiment, neural stem cells were exposed to amino-superparamagnetic iron oxide nanoparticles linked to the Tat (FITC) peptide. The presence of green fluorescence in the labeled neural stem cells suggests that Tat (FITC) peptide-conjugated amino-SPIO nanoparticles can penetrate and enter these cells [139]. A novel method for labeling mesenchymal stem cells with newly developed anionic citrate-coated superparamagnetic iron oxide nanoparticles was reported by Andreas et al. This approach resulted in enhanced MRI sensitivity, enabling improved visualization of transplanted cells labelled with citrate SPIONs compared to conventional commercial SPIONs. As a result, this technique offers superior in vivo MRI tracking capabilities for transplanted cells [140].

Zang et al. prepared AET (amineothanethiol-HCl)-based water-soluble ZnO nanoparticles, which acted as biological labeling for HeLa cells, and they confirmed it through fluorescence imaging [141]. Matsuyama et al. showed that the fluorescent microscope clearly revealed the presence of a green emission in the nerve cells, signifying the attachment of ZnO-nanoparticle quantum dots through the binding of antibodies and avidin to the nerve cells of the spinal cord of the tested mouse [142]. Hong et al. prepared ZnO functionalized with $^{64}$Cu-NOTA-PEG-TRC105, which acted as a radio-labeled nanoparticle for cancer targeting purposes. The use of radio-labeled ZnO nanoparticles resulted in a notable and elevated radioactivity accumulation in the liver, tumor, spleen, and abdominal region. Importantly, there was a lack of accumulation in most other tissues, demonstrating effective tumor targeting and providing clear image contrast [143].

## 9.5 FUTURE SCOPE

Scientists frequently struggle to strike a balance between the beneficial medicinal effects of metal and metal oxide nanoparticles and their harmful side effects. The creation of novel nanomaterials with excellent biological performance and little toxicity is necessary to produce more effective treatments as well as accurate diagnostic instruments and apparatus.

## 9.6 CONCLUSION

The field of biomedical imaging is witnessing a continual and diverse evolution in the structural design of nanomaterials. Nanoparticle distribution, uptake, and elimination are predominantly influenced by their size and surface characteristics. Ideally, nanoparticles within the range of 10 to 100 nm, or slightly larger, are preferred for molecular imaging applications such as iron, gold, silver, zinc, and more. In this chapter, the role of different metal and metal oxide nanoparticles in bioimaging and biolabeling was discussed. The use of metal and metal oxide nanoparticles such as gold, silver, platinum, iron, and other metals as contrast agents has demonstrated their ability to enhance imaging modalities such as magnetic resonance imaging, computed tomography, and optical imaging. Their capacity for targeted delivery and controlled release of therapeutic agents further accentuates their potential in personalized medicine.

Challenges such as potential toxicity, long-term biocompatibility, and scalability must be overcome for successful clinical translation. Ongoing interdisciplinary research and collaboration between fields like chemistry, biology, and medicine are crucial to fully harness the potential of these nanoparticles, advancing diagnostic and therapeutic strategies in biomedicine.

### Keywords

Biocompatibility
Bioimaging
Biolabeling
Functionalization
Toxicity

## REFERENCES


1. Hahn, M. A.; Singh, A. K.; Sharma, P.; Nanoparticles as contrast agents for in-vivo bioimaging: current status and future perspectives. *Anal. Bioanal. Chem.*, 2011, 399, 3–27. https://doi.org/10.1007/s00216-010-4207-5
2. Lahoti, H. S.; Jogdand. S. D.; Bioimaging: evolution, significance, and deficit. *Cureus*, 2022, 14, 9, 28923. https://doi.org/10.7759/cureus.28923.
3. Friedman, A. D.; Claypool, S. E.; Liu, R.; The smart targeting of nanoparticles. *Curr. Pharm. Des.*, 2013, 19, 35, 6315–6329.
4. Nune, S. K.; Gunda, P.; Thallapally, P. K.; Lin, Y.-Y.; Laird Forrest, M.; Berkland, C. J.; Nanoparticles for biomedical imaging. *Expert Opin. Drug Deliv.*, 2009, 6, 11, 1175–1194. https://doi.org/10.1517/17425240903229031.
5. Aslan, N.; Ceylan, B.; Koç, M. M.; Findik, F.; Metallic nanoparticles as X-Ray computed tomography (CT) contrast agents: a review. *J. Mol. Struct.,* 2020, 1219, 128599. https://doi.org/10.1016/j.molstruc.2020.128599
6. Kim, D.; Park, S.; Lee, J. H.; Jeong, Y. Y.; Jon, S.; Antibiofouling polymer-coated gold nanoparticles as a contrast agent for in vivo X-ray computed tomography imaging. *J. Am. Chem. Soc.*, 2007, 129, 24, 7661–7665. https://doi.org/10.1021/ja071471p.
7. Willem, R.; 2D NMR applied to dynamic stereochemical problems. *J. Prog. Nuc. Mag. Res. Spect.*, 1988, 20, 1, 1–94. https://doi.org/10.1016/0079–6565(88)80008–6.
8. Plewes, D. B.; Kucharczyk, W.; Physics of MRI: a primer. *J. Mag. Res. Ima.*, 2012, 35, 5, 1038–1054. https://doi.org/10.1002/jmri.23642.
9. Nowogrodzki, A.; The world's strongest MRI machines are pushing human imaging to new limits. *Nature*, 2018, 563, 7729, 24–26. https://doi.org/10.1038/d41586-018-07182-7.
10. Sharma, P.; Brown, S.; Walter, G.; Santra, S.; Moudgil, B.; Nanoparticles for bioimaging. *Adv. Colloid Interface Sci.*, 2006, 123–126, 471–485. https://doi.org/10.1016/j.cis.2006.05.026.
11. Ahrén, M.; Selegård, L.; Klasson, A.; Söderlind, F.; Abrikossova, N.; Skoglund, C.; Bengtsson, T.; Engström, M.; Käll, P.; Uvdal, K.; Synthesis and characterization of PEGylated $Gd_2O_3$ nanoparticles for MRI contrast enhancement. *Langmuir*, 2010, 26, 8, 5753–5762. https://doi.org/10.1021/la903566y.
12. Buxton, R. B.; The physics of functional magnetic resonance imaging (fMRI). *Rep. Prog. Phys.*, 2013, 76, 9. https://doi.org/10.1088/0034–4885/76/9/096601
13. De León-Rodríguez, L. M.; Martins, A. F.; Pinho, M. C.; Rofsky, N. M.; Sherry, A. D.; Basic MR relaxation mechanisms and contrast agent design. *J. Magn. Reson. Imag.*, 2015, 42, 3, 545–565. https://doi.org/10.1002/jmri.24787.
14. Malhotra, N.; Lee, J.-S.; Liman, R. A. D.; Ruallo, J. M. S.; Villaflores, O. B.; Ger, T.-R.; Hsiao, C.-D.; Potential toxicity of iron oxide magnetic nanoparticles: a review. *Molecules*, 2020, 25, 14, 3159. https://doi.org/10.3390/molecules25143159.
15. Siddiqui, M. A.; Wahab, R.; Saquib, Q.; Ahmad, J.; Farshori, N. N.; Al-Sheddi, E. S.; Al-Oqail, M. M.; Al-Massarani, S. M.; Al-Khedhairy, A. A.; Iron oxide nanoparticles induced cytotoxicity, oxidative stress, cell cycle arrest, and DNA damage in human umbilical vein endothelial cells. *J. Trace Elem. Med. Biol.,* 2023, 80, 127302. https://doi.org/10.1016/j.jtemb.2023.127302.
16. Tearney, G. J.; Brezinski, M. E.; Bouma, B. E.; Boppart, S. A.; Pitris, C.; Southern, J. F.; Fujimoto, J. G.; In vivo endoscopic optical biopsy with optical coherence tomography. *Science*, 1997, 276, 5321, 2037–2039. https://doi.org/10.1126/science.276.5321.2037.
17. Sanderson, M. J.; Smith, I.; Parker, I.; Bootman, M. D.; Fluorescence microscopy. *Cold Spring Harbor Protocols*, 2014, 10. https://doi.org/10.1101/pdb.top071795
18. Wolfbeis, O. S.; An overview of nanoparticles commonly used in fluorescent bioimaging. *Chem. Soc. Rev.*, 2015, 44, 14, 4743–4768. https://doi.org/10.1039/C4CS00392F.
19. Alphandéry, E.; Ultrasound and nanomaterial: an efficient pair to fight cancer. *J. Nanobiotech.*, 2022, 20, 1, 139. https://doi.org/10.1186/s12951-022-01243-w.
20. Sun, I.-C.; Emelianov, S.; Gas-generating nanoparticles for contrast-enhanced ultrasound imaging. *Nanoscale*, 2019, 11, 35, 16235–16240. https://doi.org/10.1039/c9nr04471j.
21. Tarighatnia, A.; Reza Fouladi, M.; Nader, N. D.; Aghanejad, A.; Ghadiri, H.; Recent trends of contrast agents in ultrasound imaging: a review of the classifications and applications. *Mater. Adv.*, 2022, 3, 9, 3726–3741. https://doi.org/10.1039/D1MA00969A.
22. Han, X.; Xu, K.; Taratula, O.; Farsad, K.; Applications of nanoparticles in biomedical imaging. *Nanoscale*, 2019, 11, 3, 799–819. https://doi.org/10.1039/C8NR07769J.
23. Fu, Q.; Zhu, R.; Song, J.; Yang, H.; Chen, X.; Photoacoustic imaging: contrast agents and their biomedical applications. *Adv. Mater.*, 2019, 31, 6, 180587. https://doi.org/10.1002/adma.201805875.
24. Vigneshvar, S.; Senthilkumaran, B.; Current technological trends in biosensors, nanoparticle devices and biolabels: Hi-tech network sensing applications. *Med. Dev. Sens.*, 2018, e10011. https://doi.org/10.1002/mds3.10011
25. Zhang, Y.; Chu, W.; Foroushani, A. D.; Wang, H.; Li, D.; Liu, J.; Barrow, C. J.; Wang, X.; Yang, W.; New gold nanostructures for sensor applications: a review. *Materials*, 2014, 7, 5169–5201. https://doi.org/10.3390/ma7075169.
26. Zhou, Y.; Wang, C. Y.; Zhu, Y. R.; Chen, Z. Y.; A novel ultraviolet irradiation technique for shape-controlled synthesis of gold nanoparticles at room temperature. *Chem. Mater.*, 1999, 11, 2310–2312. https://doi.org/10.1021/cm990315h.
27. Zhang, Y.; Qian, J.; Wang, D.; Wang, Y.; He, S.; Multifunctional gold nanorods with ultrahigh stability and tunability for in vivo fluorescence imaging, SERS detection, and photodynamic therapy. *Angew. Chem. Int. Ed.*, 2013, 52, 1148–1151. https://doi.org/10.1002/anie.201207909.
28. Yang, M.; Liu, Y.; Hou, W.; Zhi, X.; Zhang, C.; Jiang, X.; Pan, F.; Yang, Y.; Ni, J.; Cui, D.; Mitomycin C-treated human-induced pluripotent stem cells as a safe delivery system of gold nanorods for targeted photothermal therapy of gastric cancer. *Nanoscale*, 2017, 9, 334–340. https://doi.org/10.1039/C6NR06851K.
29. Turkevich, J.; Stevenson, P. C.; Hillier, J.; A study of the nucleation and growth processes in the synthesis of colloidal gold. *Discuss. Faraday Soc.,* 1951, 11, 55–75. https://doi.org/10.1039/DF9511100055.
29. Herizchi, R.; Abbasi, E.; Milani, M.; Akbarzadeh, A.; Current methods for synthesis of gold nanoparticles. *Artificial Cells Nanomed. Biotechnol.*, 2016, 44, 2, 596–602. https://doi.org/10.3109/21691401.2014.971807.
30. Piella, J.; Bastus, N. G.; Puntes, V.; Size-controlled synthesis of sub-10-nanometer citrate-stabilized gold nanoparticles and related optical properties. *Chem. Mater.*, 2016, 28, 4, 1066–1075. https://doi.org/10.1021/acs.chemmater.5b04406.
31. Yang, Z.; Li, Z.; Lu, X.; He, F.; Zhu, X.; Ma, Y.; He, R.; Gao, F.; Ni, W.; Yi, Y.; Controllable biosynthesis and properties

of gold nanoplates using yeast extract. *Nano-Micro Lett.*, 2017, 9, 5. https://doi.org/10.1007/s40820-016-0102-8.
32. Vala, A. K.; Exploration on green synthesis of gold nanoparticles by a marine-derived fungus Aspergillus sydowii. *Environ. Prog. Sustain. Energy.*, 2015, 34 194–197. https://doi.org/10.1002/ep.11949.
33. Sasidharan, A.; Monteiro-Riviere, N.; Biomedical applications of gold nanomaterials: opportunities and challenges. *Wiley Interdiscip. Rev. Nanomed. Nanobiotechnol.*, 2015, 7 779–796. https://doi.org/10.1002/wnan.1341.
34. Li, X.; Hu, Z.; Ma, J.; Wang, X.; Zhang, Y.; Wang, W.; Yuan, Z.; The systematic evaluation of size-dependent toxicity and multi-time biodistribution of gold nanoparticles. *Colloids Surf. B Biointerfaces*, 2018, 167, 260–266. https://doi.org/10.1016/j.colsurfb.2018.04.005.
35. Lee, H. Y.; Mohammed, K. A.; Nasreen, N.; Nanoparticle-based targeted gene therapy for lung cancer. *Am. J. Cancer Res.*, 2016, 6, 5, 1118–1134.
36. Kapse-Mistry, S.; Govender, T.; Srivastava, R.; Yergeri, M.; Nanodrug delivery in reversing multidrug resistance in cancer cells. *Front. Pharmacol.*, 2014, 5, 159. https://doi.org/10.3389/fphar.2014.00159.
37. Piao, J.-G.; Gao, F.; Li, Y.; Yu, L.; Liu, D.; Tan, Z.-B.; Xiong, Y.; Yang, L.; You, Y.-Z.; pH-sensitive zwitterionic coating of gold nanocages improves tumor targeting and photothermal treatment efficacy. *Nano Res.*, 2018, 11, 6, 3193–3204. https://doi.org/10.1007/s12274-017-1736-7.
38. Debouttière, P.-J.; Roux, S.; Vocanson, F.; Billotey, C.; Beuf, O.; Favre-Réguillon, A.; Lin, Y.; Pellet-Rostaing, S.; Lamartine, R.; Perriat, P.; Tillement. O.; Design of gold nanoparticles for magnetic resonance imaging. *Adv. Funct. Mater.*, 2006. https://doi.org/10.1002/adfm.200600242
39. Wang, G.; Gao, W.; Zhang, X.; Au nanocage functionalized with ultra-small $Fe_3O_4$ nanoparticles for targeting $T_1$–$T_2$ dual MRI and CT imaging of tumor. *Sci. Rep.*, 2016, 6, 28258. https://doi.org/10.1038/srep28258.
40. Hoskins, C.; Min, Y.; Gueorguieva, M.; Hybrid gold-iron oxide nanoparticles as a multifunctional platform for biomedical application. *J. Nanobiotechnol.*, 2012, 10, 27. https://doi.org/10.1186/1477-3155-10-27.
41. Wang, D.-W.; Zhu, X.-M.; Lee, S.-F.; Chan, H.-M.; Li, H.-W.; Kong, S. K.; Yu, J. C.; Cheng, C. H. K.; Wang, Y.-X. J.; Leung, K. C.-F.; Folate-conjugated Fe3O4@SiO2@gold nanorods@mesoporous SiO2 hybrid nanomaterial: a theranostic agent for magnetic resonance imaging and photothermal therapy. *J. Mater. Chem.*, 2013, 1, 2934–2942. https://doi.org/10.1039/C3TB20090F
42. Kwon, K. C.; Jo, E.; Kwon, Y. W.; Lee, B.; Ryu, J. H.; Lee E. J.; Superparamagnetic gold nanoparticles synthesized on protein particle scaffolds for cancer theragnosis. *Adv. Mater.*, 2017, 29 1701146. https://doi.org/10.1002/adma.201701146.
43. Chauhan, R.; El-Baz, N.; Keynton, R. S.; James, K. T.; Malik, D. A.; Zhu, M.; El-Baz, A.; Ng, C. K.; Bates, P. J.; Malik, M. T.; O'Toole, M. G.; Targeted gold nanoparticle–oligonucleotide contrast agents in combination with a new local voxel-wise MRI analysis algorithm for in vitro imaging of triple-negative breast cancer. *J. Nanomater.*, 2019, 9, 709. https://doi.org/10.3390/nano9050709.
44. Li, W.; Chen, X.; Gold nanoparticles for photoacoustic imaging. *Nanomedicine*, 2015, 10, 2, 299–320. https://doi.org/10.2217/nnm.14.169.
45. Xu, X.; Chong, Y.; Liu, X.; Fu, H.; Yu, C.; Huang, J.; Zhang, Z.; Multifunctional nanotheranostic gold nanocages for photoacoustic imaging guided radio/photodynamic/photothermal synergistic therapy. *Acta Biomater.*, 2019, 84, 328–338. https://doi.org/10.1016/j.actbio.2018.11.043.
46. Sun, I.-C.; Ahn, C.-H.; Kim, K.; Emelianov, S.; Photoacoustic imaging of cancer cells with glycol-chitosan-coated gold nanoparticles as contrast agents. *J. Biomed. Opt.*, 2019, 24, 12, 121903. https://doi.org/10.1117/1.JBO.24.12.121903.
47. Cheng, Y.; Samia, A. C.; Meyers, J. D.; Panagopoulos I.; Fei, B.; Burda, C.; Highly efficient drug delivery with gold nanoparticle vectors for in vivo photodynamic therapy of cancer. *J. Am. Chem. Soc.*, 2008, 130, 32, 10643–10647. https://doi.org/10.1021/ja801631c.
48. Tian, Y.; Zhang, Y.; Teng, Z.; Tian, W.; Luo. S.; Kong, X.; pH-dependent transmembrane activity of peptide-functionalized gold nanostars for computed tomography/photoacoustic imaging and photothermal therapy. *ACS Appl. Mater. Interfaces.*, 2017, 9, 3, 2114–2122. https://doi.org/10.1021/acsami.6b13237.
49. Manivasagan, P.; Bharathiraja, S.; Bui, N. Q.; Jang, B.; Oh, Y.-O.; Lim, I. G.; Oh, J.; Doxorubicin-loaded fucoidan capped gold nanoparticles for drug delivery and photoacoustic imaging. *Int. J. Biol. Macromol.*, 2016, 91, 578–588, 0141–8130. https://doi.org/10.1016/j.ijbiomac.2016.06.007
50. Hainfeld, J. F.; Slatkin, D. N.; Focella, T. M.; Smilowitz HM. Gold nanoparticles: a new X-ray contrast agent. *Br. J. Radiol.*, 2006, 79, 939, 248–253. https://doi.org/10.1259/bjr/13169882.
51. Kattumuri, V.; Katti, K.; Bhaskaran, S.; Gum arabic as a phytochemical construct for the stabilization of gold nanoparticles: in vivo pharmacokinetics and X-ray-contrast-imaging studies. *Small*, 2007, 3, 2, 333–341. https://doi.org/10.1002/smll.200600427.
52. Wang, H.; Zheng, L. F.; Peng, C.; Guo, R.; Shen, M. W.; Shi X. Y.; Zhang, G. X.; Computed tomography imaging of cancer cells using acetylated dendrimer-entrapped gold nanoparticles. *Biomaterials*, 2011, 32, 2979–2988. https://doi.org/10.1016/j.biomaterials.2011.01.001.
53. Boote, E.; Fent, G.; Kattumuri, V.; Casteel, S.; Katti, K.; Chanda, N.; Kannan R.; Churchill, R.; Gold nanoparticle contrast in a phantom and juvenile swine: models for molecular imaging of human organs using X-ray computed tomography. *Acad. Radiol.*, 2010, 17, 410–417. https://doi.org/10.1016/j.acra.2010.01.006.
54. Popovtzer, R.; Agrawal, A.; Kotov, N. A.; Popovtzer, A.; Balter, J.; Carey, T. E.; Kopelman, R.; Targeted gold nanoparticles enable molecular CT imaging of cancer. *Nano Lett.*, 2008, 8, 12, 4593–4596. https://doi.org/10.1021/nl8029114.
55. Ke, H.; Wang, J.; Dai, Z.; Jin, Y.; Qu, E.; Xing, Z.; Guo, C.; Yue, X.; Liu, J.; Gold-nanoshelled microcapsules: a theranostic agent for ultrasound contrast imaging and photothermal therapy. *Chem. Int. Ed.*, 2011, 50, 3017–3021. https://doi.org/10.1002%2Fanie.201008286.
56. Lim, D.-K.; Jeon, K.-S.; Hwang, J.-H.; Kim, H.; Kwon, S.; Suh, Y. D.; Nam, J.-M.; Highly uniform and reproducible surface-enhanced Raman scattering from DNA-tailorable nanoparticles with 1-nm interior gap. *Nat. Nanotechnol.*, 2011, 6, 452–460. https://doi.org/10.1038/nnano.2011.79.
57. Zhou, Y.; Chakraborty, S.; Liu, S.; Radiolabeled cyclic rgd peptides as radiotracers for imaging tumor's and thrombosis by SPECT. *Theranostics*, 2011, 58–82. https://doi.org/10.7150%2Fthno%2Fv01p0058.
58. Chen, X.; Schluesener, H. J.; Nanosilver: a nanoproduct in medical application. *Toxicol. Lett.*, 2008, 176, 1–12. https://doi.org/10.1016/j.toxlet.2007.10.004.

59. Conde, J.; Doria, G.; Baptista, P.; Noble metal nanoparticles applications in cancer. *J. Drug Deliv.*, 2012, 751075. https://doi.org/10.1155/2012/751075.
60. Liu, W.; Wu, Y.; Wang, C.; Li, H. C.; Wang, T.; Liao, C. Y.; Cui, L.; Zhou, Q. F.; Yan, B.; Jiang, G. B.; Impact of silver nanoparticles on human cells: effect of particle size. *Nanotoxicology*, 2010, 4, 319–330. https://doi.org/10.3109/17435390.2010.483745.
61. Piao, M. J.; Kang, K. A.; Lee, I. K.; Kim, H. S.; Kim, S.; Choi, J. Y.; Choi, J.; Hyun, J. W.; Silver nanoparticles induce oxidative cell damage in human liver cells through inhibition of reduced glutathione and induction of mitochondria-involved apoptosis. *Toxicol. Lett.*, 2011, 201, 92–100. https://doi.org/10.1016/j.toxlet.2010.12.010.
62. AshaRani, P. V.; Sethu, S.; Lim, H. K.; Balaji, G.; Valiyaveettil, S.; Hande, M. P.; Differential regulation of intracellular factors mediating cell cycle, DNA repair and inflammation following exposure to silver nanoparticles in human cells. *Genome Integr.*, 2012, 3, 2. https://doi.org/10.1186/2041-9414-3-2.
63. Liang, P.; Shi, H.; Zhu, W.; Gui, Q.; Xu, Y.; Meng, J.; Guo, X.; Gong, Z.; Chen, H.; Silver nanoparticles enhance the sensitivity of temozolomide on human glioma cells. *Oncotarget*, 2017, 8, 7533–7539. https://doi.org/10.18632%2Foncotarget.13503.
64. Yang, F.; Wang, Q.; Gu, Z.; Fang, K.; Marriott, G.; Gu, N.; Silver nanoparticle-embedded microbubble as a dual-mode ultrasound and optical imaging probe. *ACS Appl. Mater. Interfaces*, 2013, 5, 18, 9217–9223. https://doi.org/10.1021/am4029747.
65. Lee, K. J.; Browning, L. M.; Nallathamby, P. D.; Xu, X.-H. N.; Study of charge-dependent transport and toxicity of peptide-functionalized silver nanoparticles using zebrafish embryos and single nanoparticle plasmonic spectroscopy. *Chem. Res. Toxicol.*, 2013, 26, 6, 904–917. https://doi.org/10.1021/tx400087d.
66. Hu, Q.; Tay, L.-L.; Noestheden, M.; Pezacki, J. P.; Mammalian cell surface imaging with nitrile-functionalized nanoprobes: biophysical characterization of aggregation and polarization anisotropy in SERS imaging. *J. Am. Chem. Soc.*, 2007, 14–15, 129. https://doi.org/10.1021/ja0670005.
67. Song, Z.-L.; Chen, Z.; Bian, X.; Zhou, L.-Y.; Ding, D.; Liang, H.; Zou, Y.-X.; Wang, S.-S.; Chen, L.; Yang, C.; Zhang, X.-B.; Tan, W.; Alkyne-functionalized superstable graphitic silver nanoparticles for raman imaging. *J. Am. Chem. Soc.*, 2014, 136, 39, 13558–13561. https://doi.org/10.1021/ja507368z.
68. Homan, K. A.; Souza, M.; Truby, R.; Luke, G. P.; Green, C.; Vreeland, E.; Emelianov, S.; Silver nanoplate contrast agents for in vivo molecular photoacoustic imaging. *ACS Nano* 6, 2012, 1, 641–650. https://doi.org/10.1021/nn204100n.
69. Mishra, S. K.; Kannan, S.; Doxorubicin-conjugated bimetallic silver–gadolinium nanoalloy for multimodal MRI-CT-optical imaging and pH-responsive drug release. *ACS Biomater. Sci. Eng.*, 2017, 3, 12, 3607–3619. https://doi.org/10.1021/acsbiomaterials.7b00498.
70. Mishra, S. K.; Kannan, S.; A bimetallic silver–neodymium theranostic nanoparticle with multimodal NIR/MRI/CT imaging and combined chemo-photothermal therapy. *Inorg. Chem.*, 2017, 56, 19, 12054–12066. https://doi.org/10.1021/acs.inorgchem.7b02103.
71. Amendola, V.; Guadagnini, A.; Agnoli, S.; Badocco, D.; Pastore, P.; Fracasso, G.; Gerosa, M.; Vurro, F.; Busato, A.; Marzola, P.; Polymer-coated silver-iron nanoparticles as efficient and biodegradable MRI contrast agents. *J. Colloid Interface Sci.*, 2021, 596, 332–341, 0021–9797. https://doi.org/10.1016/j.jcis.2021.03.096.
72. Alhasan, A.; Tajuddin, H. A.; Sani, S. A. F.; Ali, T. H.; Hisham, S.; Mokti, M. H.; Ung, N. M.; Ng, M. P.; Sim, K. S.; Iohexol functionalized Si-Ag: Mn3O4 hybrid nanoparticles-based contrast agent for computed tomography imaging. *Mater. Today Commun.*, 2022, 33, 104377. https://doi.org/10.1016/j.mtcomm.2022.104377
73. David, L.; Moldovan, B.; Vulcu, A.; Olenic, L.; Schrepler, M. P.; Fodor, E. F.; Florea, A.; Crisan, M.; Chiorean, I.; Clichici, S.; Filip, G. A.; Green synthesis, characterization and anti-inflammatory activity of silver nanoparticles using European black elderberry fruits extract. *J. Colloids Surfaces B Biointerfaces*, 2014, 122, 767–777, 0927–7765. https://doi.org/10.1016/j.colsurfb.2014.08.018.
74. Mahmoud, M.; Saira, F.; Sayed, M. E.; Experimental evidence for the nanocage effect in catalysis with hollow nanoparticles. *Nano Lett.*, 2010, 10, 9, 3764–3769. https://doi.org/10.1021/nl102497u.
75. Mahima, S.; Kannan, R.; Komath, I.; Aslam, M.; Pillai, V. K.; Synthesis of platinum Y-junction nanostructures using hierarchically designed alumina templates and their enhanced electrocatalytic activity for fuel cell applications. *Chem. Mater.*, 2008, 20, 3, 601–603. https://doi.org/10.1021/cm702102b.
76. Bönnemann, H.; Richards, R. M.; Nanoscopic metal particles synthetic methods and potential applications. *Eur. J. Inorg. Chem.*, 2001, 10, 2455–2480. https://doi.org/10.1002/1099-0682(200109)2001:10%3C2455::AID-EJIC2455%3E3.0.CO;2-Z\
77. Choi, I. D.; Lee, H.; Shim, Y.-B.; Lee, D.; A one-step continuous synthesis of carbon supported Pt catalysts using a flame for the preparation of the fuel electrode. *Langmuir,* 2010, 26, 13, 11212–11216. https://doi.org/10.1021/la1005264.
78. Ke, X.; Bittencourt, C.; Bals, S.; Tendeloo, G. V.; Low-dose patterning of platinum nanoclusters on carbon nanotubes by focused-electron-beam-induced deposition as studied by TEM. *Beilstein J. Nanotechnol.,* 2013, 4, 1, 77–86. https://doi.org/10.3762/bjnano.4.9.
79. Rakshit, R. K.; Bose, S. K.; Sharma, R.; Budhani, R. C.; Vijaykumar, T.; Neena, S. J.; Kulkarni, G. U.; Correlations between morphology, crystal structure, and magnetization of epitaxial cobalt-platinum films grown with pulsed laser ablation. *J. Appl. Phys.*, 2008, 103, 2, 023915. https://doi.org/10.1063/1.2832763.
80. Paschos, O.; Choi, P.; Efstathiadis, H.; Haldar, H.; Synthesis of platinum nanoparticles by aerosol assisted deposition method. *Thin Solid Films,* 2008, 516, 12, 3796–3801. https://doi.org/10.1016/j.tsf.2007.06.123.
81. Nellore, J.; Pauline, C.; Amarnath, K.; Bacopa monnieri phytochemicals mediated synthesis of platinum nanoparticles and its neuro rescue effect on 1-methyl 4- phenyl 1, 2, 3, 6 tetrahydropyridine-induced experimental parkinsonism in zebrafish. *J. Neurodegener. Dis.*, 2013. https://doi.org/10.1155/2013/972391.
82. Rashamuse, K.; Mutambanengwe, C.; Whiteley, C.; Enzymatic recovery of platinum (IV) from industrial wastewater using a biosulphidogenic hydrogenase. *Afr. J. Biotechnol.*, 2008, 7, 8.
83. Roberta, B.; Hélène, B.; Miryana, H.; Chakib, D.; Claude, Y.; Thibaud, C.; Jacques, L.; Fernand, F.; Alain, C.; Cyanobacteria as bioreactors for the synthesis of Au, Ag, Pd, and Pt nanoparticles via an enzyme-mediated route. *J. Nanosci. Nanotechnol.*, 2007, 7, 8, 2696–2708. https://doi.org/10.1166/jnn.2007.600.

84. Govender, Y.; Riddin, T.; Gericke, M.; Whiteley, C. G.; Bioreduction of platinum salts into nanoparticles: a mechanistic perspective. *Biotechnol. Lett.,* 2009, 31, 1, 95–100. https://doi.org/10.1007/s10529-008-9825-z.
85. P. Shiny, A. Mukherjee, N. Chandrasekaran. DNA damage and mitochondria-mediated apoptosis of A549 lung carcinoma cells induced by biosynthesised silver and platinum nanoparticles. *RSC Adv.,* 2016, 6, 33, 27775–27787. https://doi.org/10.1039/C5RA27185A.
86. Leo, A. J.; Oluwafemi, O. S.; Plant-mediated synthesis of platinum nanoparticles using water hyacinth as an efficient biomatrix source–An eco-friendly development. *Mater. Lett.,* 2017, 196, 141–144. https://doi.org/10.1016/j.matlet.2017.03.047.
87. Bendale, Y.; Bendale, V.; Paul, S.; Evaluation of cytotoxic activity of platinum nanoparticles against normal and cancer cells and its anticancer potential through induction of apoptosis. *Integr. Med. Res.,* 2017, 6, 2, 141–148. https://doi.org/10.1016/j.imr.2017.01.006.
88. Manikandan, M.; Hasan, N.; Wu, H.-F.; Platinum nanoparticles for the photothermal treatment of neuro 2A cancer cells. *Biomaterials,* 2013, 34, 23, 5833–5842. https://doi.org/10.1016/j.biomaterials.2013.03.077.
89. Buchtelova, H.; Dostalova, S.; Michalek, P.; Krizkova, S.; Strmiska, V.; Kopel, P.; Hynek, D.; Richtera, L.; Ridoskova, A.; Adam, P.; Kynicky, J.; Brtnicky, M.; Heger, Z.; Adam, V.; Size-related cytotoxicological aspects of polyvinylpyrrolidone-capped platinum nanoparticles. *Food Chem. Toxicol.,* 2017, 105 337–346. https://doi.org/10.1016/j.fct.2017.04.043.
90. Yamagishi, Y.; Watari, A.; Hayata, Y.; Li, X.; Kondoh, M.; Tsutsumi, Y.; Yagi, K.; Hepatotoxicity of sub-nanosized platinum particles in mice. *Die Pharm. Int. J. Pharm. Sci.,* 2013, 68, 3, 178–182. https://doi.org/10.1691/ph.2013.2141.
91. Zhao, L.; Ge, X.; Yan, G.; Wang, X.; Hu, P.; Shi, P.; Wolfbeis, O. S.; Zhang, H.; Sun, L.; Double-mesoporous core–shell nanosystems based on platinum nanoparticles functionalized with lanthanide complexes for in vivo magnetic resonance imaging and photothermal therapy. *Nanoscale,* 2017, 9, 16012–16023. https://doi.org/10.1039/C7NR04983H.
92. Wang, X.; Wang, X.; Guo, Z.; Functionalization of platinum complexes for biomedical applications. *Acc. Chem. Res.,* 2015, 48, 9, 2622–2631. https://doi.org/10.1021/acs.accounts.5b00203.
93. Chan, M. H.; Lu, C. N.; Chung, Y. L.; Magnetically guided theranostics: montmorillonite-based iron/platinum nanoparticles for enhancing in situ MRI contrast and hepatocellular carcinoma treatment. *J. Nanobiotech.,* 2021, 19, 308. https://doi.org/10.1186/s12951-021-01052-7.
94. Xu, Z.; Chen, J.; Li, Y.; Hu, T.; Fan, L.; Xi, J.; Han, J.; Guo, R.; Yolk-shell $Fe_3O_4$@Carbon@Platinum-Chlorin e6 nanozyme for MRI-assisted synergistic catalytic-photodynamic-photothermal tumor therapy. *J. Colloid Interface Sci.,* 2022, 628, 1033–1043. https://doi.org/10.1016/j.jcis.2022.08.006.
95. Zhu, Z.; Wang, Z.; Hao, Y.; Zhu, C.; Jiao, Y.; Chen, H.; Wang, Y.-M.; Yan, J.; Guo, Z.; Wang, X.; Glutathione boosting the cytotoxicity of a magnetic platinum (IV) nano-prodrug in tumor cells. *Chem. Sci.,* 2016, 7, 2864–2869. https://doi.org/10.1039/C5SC04049C
96. Phan, T. T. V.; Bui, N. Q.; Moorthy, M. S.; Synthesis and in vitro performance of polypyrrole-coated iron–platinum nanoparticles for photothermal therapy and photoacoustic imaging. *Nanoscale Res. Lett.,* 2017, 12, 570. https://doi.org/10.1186/s11671-017-2337-9
97. Zhang, L.; Li, M.; Zhou, Q.; Dang, M.; Tang, Y.; Wang, S.; Fu, J.; Teng, Z.; Lu, G.; Computed tomography and photoacoustic imaging guided photodynamic therapy against breast cancer based on mesoporous platinum with insitu oxygen generation ability. *J. Acta Pharm. Sin. B,* 2020, 10, 1719–1729. https://doi.org/10.1016/j.apsb.2020.05.003.
98. Ding, B.; Yu, C.; Li, C.; Deng, X.; Ding, J.; Cheng, Z.; Xing, B.; Ma, P.; Lin, J.; cis-Platinum pro-drug-attached $CuFeS_2$ nanoplates for in vivo photothermal/photoacoustic imaging and chemotherapy/photothermal therapy of cancer. *Nanoscale,* 2017, 9, 16937–16949. https://doi.org/10.1039/C7NR04166G
99. Zhou, Z.; Hu, K.; Ma, R.; Yan, Y.; Ni, B.; Zhang, Y.; Wen, L.; Zhang, Q.; Cheng, Y.; Dendritic platinum–copper alloy nanoparticles as theranostic agents for multimodal imaging and combined chemophotothermal therapy. *J. Adv. Funct. Mater.,* 2016, 26, 5971–5978. https://doi.org/10.1002/adfm.201601754.
100. Liu, Y.; Wu, P.-C.; Guo, S.; Chou, P.-T.; Deng, C.; Chou, S.-W.; Yuan, Z.; Liu, T.-M.; Low-toxicity FePt nanoparticles for the targeted and enhanced diagnosis of breast tumors using few centimeters deep whole-body photoacoustic imaging. *Photoacoustics,* 2020, 19, 100–179, 2213–5979. https://doi.org/10.1016/j.pacs.2020.100179.
101. Khan, I.; Saeed, K.; Khan, I.; Nanoparticles: properties, applications and toxicities. *Arab. J. Chem.,* 2019, 12, 7, 908–931. https://doi.org/10.1016/j.arabjc.2017.05.011.
102. Manescu, V.; Paltanea, G.; Antoniac, I.; Vasilescu, M.; Magnetic nanoparticles used in oncology. *Materials,* 2021, 14, 20, 5948. https://doi.org/10.3390/ma14205948.
103. Ashtari, P.; He, X.; Wang, K.; Gong, P.; An efficient method for recovery of target ssDNA based on amino-modified silica-coated magnetic nanoparticles. *Talanta,* 2005, 67, 548. https://doi.org/10.1016/j.talanta.2005.06.043.
104. Li, Y.; Liu, Y.; Tang, J.; Lin, H.; Yao, N.; Shen, X.; $Fe_3O_4$@$Al_2O_3$ magnetic core–shell microspheres for rapid and highly specific capture of phosphopeptides with mass spectrometry analysis. *J. Chromatogr. A,* 2007, 1172, 57. https://doi.org/10.1016/j.chroma.2007.09.062.
105. Hong, R. Y.; Zhang, S. Z.; Di, G. Q.; Li, H. Z.; Zheng, Y.; Ding, J.; Preparation, characterization and application of Fe3O4/ZnO core/shell magnetic nanoparticles. *Mater. Res. Bull.,* 2008, 43, 2457. https://doi.org/10.1016/j.materresbull.2007.07.035.
106. Gambhir, R. P.; Rohiwal, S. S.; Tiwari, A. P.; Multifunctional surface functionalized magnetic iron oxide nanoparticles for biomedical applications: a review. *Appl. Surf. Sci.,* 2022, 11. https://doi.org/10.1016/j.apsadv.2022.100303.
107. Du, F.; Lou, J.; Jiang, R.; Fang, Z.; Zhao, X.; Niu, Y.; Zou, S.; Zhang, M.; Gong, A.; Wu, C.; Hyaluronic acid-functionalized bismuth oxide nanoparticles for computed tomography imaging-guided radiotherapy of tumor. *Int. J. Nanomed.,* 2017, 12, 5973–5992. https://doi.org/10.2147/IJN.S130455.
108. Wang, Y.; Yang, T.; Ke, H.; Zhu, A.; Wang, Y.; Wang, J.; Shen, J.; Liu, G.; Chen, C.; Zhao, Y.; Chen, H.; Smart albumin-biomineralized nanocomposites for multimodal imaging and photothermal tumor ablation. *Adv. Mater.,* 2015, 27, 26, 3874–3882. https://doi.org/10.1002/adma.201500229.
109. Perlman, O.; Weitz, I. S.; Azhari, H.; Copper oxide nanoparticles as contrast agents for MRI and ultrasound dual-modality imaging ac. *Phys. Med. Biol.,* 2015, 60, 5767. https://doi.org/10.1088/0031–9155/60/15/5767.
110. Zhu, J.; Li, D.; Chen, H.; Yang, X.; Lu, L.; Wang, X.; Highly dispersed CuO nanoparticles prepared by a novel

quick-precipitation method. *Mater. Lett.*, 2004, 58, 3324–3327. https://doi.org/10.1016/j.matlet.2004.06.031
111. Ke, H.; Wang, J.; Tong, S.; Jin, Y.; Wang, S.; Qu, E.; Bao, G.; Dai, Z.; Gold nanoshelled liquid perfluorocarbon magnetic nanocapsules: a nanotheranostic platform for bimodal ultrasound/magnetic resonance imaging guided photothermal tumor ablation. *Theranostics*, 2014, 4, 1, 12–23. https://doi.org/10.7150%2Fthno.7275
112. Seo, W. S.; Jo, H. H.; Lee, K.; Kim, B.; Oh, S. J.; Park, J. T.; Size-dependent magnetic properties of colloidal Mn3O4 and MnO nanoparticles. *Angew. Chem. Int. Ed.*, 2004, 43, 1115. https://doi.org/10.1002/anie.200352400
113. Yu, T.; Moon, J.; Park, J.; Park, Y. I.; Na, H. B.; Kim, B. H.; Song, I. C.; Moon, W. K.; Hyeon, T.; Various-shaped uniform Mn3O4 nanocrystals synthesized at low temperature in air atmosphere. *Chem. Mater.*, 2009, 21, 2272. https://doi.org/10.1021/cm900431b.
114. Na, H. B.; Lee, J.-H.; An, K.; Park, Y. I.; Park, M.; Lee, I. S.; Nam, D.-H.; Kim, S. T.; Kim, S.-H.; Kim, S.-W.; Lim, K.-H.; Kim, K.-S.; Kim, S.-O.; Hyeon, T.; Development of a $T_1$ contrast agent for magnetic resonance imaging using MnO nanoparticles. *Angew. Chem. Int. Ed.*, 2007, 46, 5397. https://doi.org/10.1002/ange.200604775.
115. Shin, J.; Anisur, R. M.; Ko, M. K.; Im, G. H.; Lee, J. H.; Lee, I. S.; Hollow manganese oxide nanoparticles as multifunctional agents for magnetic resonance imaging and drug delivery. *Angew. Chem. Int. Ed.*, 2009, 48, 321. https://doi.org/10.1002/anie.200802323.
116. Yang, H.; Zhuang, Y.; Hu, H.; Du, X.; Zhang, C.; Shi, X.; Wu, H.; Yang, S.; Silica-coated manganese oxide nanoparticles as a platform for targeted magnetic resonance and fluorescence imaging of cancer cells. *Adv. Funct. Mater.*, 2010. https://doi.org/10.1002/adfm.200902445.
117. Smijs, T. G.; Pavel, S.; Titanium dioxide and zinc oxide nanoparticles in sunscreens: focus on their safety and effectiveness. *Nanotechnol. Sci. Appl.*, 2011, 4, 95–112. https://doi.org/10.2147/NSA.S19419.
118. Ruszkiewicz, J. A.; Pinkas, A.; Ferrer, B.; Peres, T. V.; Tsatsakis, A.; Aschner, M.; Neurotoxic effect of active ingredients in sunscreen products, a contemporary review. *Toxicol. Rep.*, 2017, 4, 245–259. https://doi.org/10.1016/j.toxrep.2017.05.006.
119. Kolodziejczak-Radzimska, A.; Jesionowski, T.; Zinc oxide–from synthesis to application: a review. *Materials*, 2014, 7, 4, 2833–2881. https://doi.org/10.3390/ma7042833.
120. Sahoo, S.; Maiti, M.; Ganguly, A.; George, J. J.; Bhowmick, A. K.; Effect of zinc oxide nanoparticles as cure activator on the properties of natural rubber and nitrile rubber. *J. Appl. Polym. Sci.*, 2007, 105, 4, 2407–2415. https://doi.org/10.1002/app.26296.
121. Mishra, P. K.; Mishra, H.; Ekielski, A.; Talegaonkar, S.; Vaidya, B.; Zinc oxide nanoparticles: a promising nanomaterial for biomedical applications. *Drug Discov. Today*, 2017, 22, 12, 1825–1834. https://doi.org/10.1016/j.drudis.2017.08.006.
122. Zhang, Z. Y.; Xiong, H. M.; Photoluminescent ZnO nanoparticles and their biological applications. *Materials*, 2015, 8, 6, 3101–3127. https://doi.org/10.3390/ma8063101.
123. Kim, S.; Lee, S. Y.; Cho, H. J.; Doxorubicin-wrapped zinc oxide nanoclusters for the therapy of colorectal adenocarcinoma. *Nanomaterials*, 2017, 7, 11, 354. https://doi.org/10.3390/nano7110354.
124. Xiong, H. M.; ZnO nanoparticles applied to bioimaging and drug delivery. *Adv. Mater.*, 2013, 25, 37, 5329–5335. https://doi.org/10.1002/adma.201301732.
125. Zhang, Y.; Nayak, T. R.; Hong, H.; Cai, W.; Biomedical applications of zinc oxide nanomaterials. *Curr. Mol. Med.*, 2013, 13, 10, 1633–1645. https://doi.org/10.2174/1566524013666131111130058.
126. Zhu, P.; Weng, Z.; Li, X.; Liu, X.; Wu, S.; Yeung, K. W. K.; Wang, X.; Cui, Z.; Yang, X.; Chu, P. K.; Biomedical applications of functionalized ZnO nanomaterials: from biosensors to bioimaging. *Adv. Mater. Interfaces*, 2016, 3, 1, 1500494. https://doi.org/10.1002/admi.201500494.
127. Xiong, H. M.; Xu, Y.; Ren, Q. G.; Xia, Y. Y.; Stable aqueous ZnO@polymer core-shell nanoparticles with tunable photoluminescence and their application in cell imaging. *J. Am. Chem. Soc.*, 2008, 130, 24, 7522–7523. https://doi.org/10.1021/ja800999u.
128. Jiang, H.; Wang, H.; Wang, X.; Facile and mild preparation of fluorescent ZnO nanosheets and their bioimaging applications. *Appl. Surf. Sci.*, 2011, 257, 15, 6991–6995. https://doi.org/10.1016/j.apsusc.2011.03.053
129. Tang, X.; Choo, E. S. G.; Li, L.; Ding, J.; Xue, J.; Synthesis of ZnO nanoparticles with tunable emission colors and their cell labeling applications. *Mater. Chem.*, 2010, 22, 11, 3383–3388. https://doi.org/10.1021/cm903869r.
130. Jiang, J.; Pi, J.; Cai, J.; The advancing of zinc oxide nanoparticles for biomedical applications. *Bioinorg. Chem. Appl.*, 2018, 1062562, 18. https://doi.org/10.1155/2018/1062562.
131. Wang, Y.; Liu, Y.; Luehmann, H.; Xia, X.; Brown, P.; Jarreau, C.; Welch, M.; Xia, Y.; Evaluating the pharmacokinetics and in vivo cancer targeting capability of au nanocages by positron emission tomography imaging. *ACS Nano.*, 2012, 6, 5880–5888. https://doi.org/10.1021/nn300464r.
132. Black, K. C.; Akers, W. J.; Sudlow, G.; Xu, B.; Laforest, R.; Achilefu, S.; Dual-radiolabeled nanoparticle SPECT probes for bioimaging. *Nanoscale*, 2015, 7, 440–444. https://doi.org/10.1039/C4NR05269B.
133. Gontero, D.; Veglia, A. V.; Bracamonte, A. G.; In flow metal-enhanced fluorescence for biolabelling and biodetection. *Photochem. Photobiol. Sci.*, 2020, 19, 1168–1188. https://doi.org/10.1039/d0pp00145g
134. Cheng, J.; Gu, Y.-J.; Cheng, S. H.; Wong, W.-T.; Surface functionalized gold nanoparticles for drug delivery. *J. Biomed. Nanotechnol.*, 2013, 9, 8, 1362–1369. https://doi.org/10.1166/jbn.2013.1536.
135. Zhang, J.; Fu, Y.; Mei, Y.; Jiang, F.; Lakowicz, J. R.; Fluorescent metal nanoshell probe to detect single miRNA in lung cancer cell. *Anal. Chem.*, 2010, 82, 4464–4471. https://doi.org/10.1021/ac100241f.
136. Li, H.; Wang, M.; Qiang, W.; Hu, H.; Li, W.; Xu, D.; Metal-enhanced fluorescent detection for protein microarrays based on a silver plasmonic substrate. *Analyst*, 2014, 139, 1653–1660. https://doi.org/10.1039/C3AN01875J.
137. Chrastina, A.; Schnitzer, J. E.; Iodine-125 radiolabeling of silver nanoparticles for in vivo SPECT imaging. *Int. J. Nanomed.*, 2010, 5, 653–659. https://doi.org/10.2147/IJN.S11677.
138. Farrag, N. S.; El-Sabagh, H. A.; Al-Mahallawi, A. M.; Amin, A. M.; AbdEl-Bary, A.; Mamdouh, W.; Comparative study on radiolabeling and biodistribution of core-shell silver/polymeric nanoparticles-based theranostics for tumor targeting. *Int. J. Pharm.*, 2017, 529, 1–2, 123–133. https://doi.org/10.1016/j.ijpharm.2017.06.044.
139. Ma, M.; Zhan, Y.; Shen, Y.; Synthesis of amino-group functionalized superparamagnetic iron oxide nanoparticles and applications as biomedical labeling probes. *J. Nanopart. Res.*, 2011, 13, 3249–3257. https://doi.org/10.1007/s11051-011-0239-9

140. Andreas, K.; Georgieva, R.; Ladwig, M.; Mueller, S.; Notter, M.; Sittinger, M.; Ringe, J.; Highly efficient magnetic stem cell labeling with citrate-coated superparamagnetic iron oxide nanoparticles for MRI tracking. *Biomaterials*, 2012, 33, 18, 4515–4525. https://doi.org/10.1016/j.biomaterials.2012.02.064.
141. Zang, Z.; Tang, X.; Enhanced fluorescence imaging performance of hydrophobic colloidal ZnO nanoparticles by a facile method. *J. Alloys Compd.*, 2015, 619, 98–101. https://doi.org/10.1016/j.jallcom.2014.09.072.
142. Matsuyama, K.; Ihsan, N.; Irie, K.; Mishima, K.; Okuyama, T.; Muto, H.; Bioimaging application of highly luminescent silica-coated ZnO-nanoparticle quantum dots with biotin. *J. Colloid Interface Sci.*, 2013, 399, 19–25. https://doi.org/10.1016/j.jcis.2013.02.047.
143. Hong, H.; Wang, F.; Zhang, Y.; Graves, S. A.; Eddine, S. B. Z.; Yang, Y.; Cai, W.; Red fluorescent zinc oxide nanoparticle: a novel platform for cancer targeting. *ACS Appl. Mater. Interfaces*, 2015, 7, 5, 3373–3381. https://doi.org/10.1021/am508440j.

## LIST OF ABBREVIATIONS

**3D** Three-Dimensional
**AuNPs** Gold Nanoparticles
**BC** Biocompatible
**$Bi_2S_3$** Bismuth Sulfide
**BL** Bioluminescent
**BMI** Biomedical Imaging
**BNPs** Bismuth Nanoparticles
**CA** Contrast Agent
**CFSE** Carboxyfluorescein Diacetate Succinimidyl Ester
**cRGD** Cyclic Arginine-Glycine-Aspartic Acid
**CT** Computed Tomography
**Cu-64** Copper-64
**DEN** Dendrimer
**DOX** Doxorubicin
**DTPA** Diethyl Triamine Penta Acetic Acid
**EGFR** Epidermal Growth Factor Receptor
**EPR** Extravasation Permeation Retention
**F-18** Fluorine-18
**FITC** Fluorescein Isothiocyanate
**FL** Fluorescent
**Fu** Fucoidan
**FZ** Functionalized
**GA** Gum Arabic
**Gd** Gadolinium
**Gd III** Gadolinium-III
**GNPs** Gold Nanoparticles
**HA** Hyaluronic Acid
**Hb** Hemoglobin
**HECC** Human Epithelial Carcinoma Cells
**HNC** Head and Neck Cancer
**In-111** Indium-111
**IO** Iron Oxide
**IONP** Iron Oxide Magnetic Nanoparticles
**IV** In Vivo
**LSPR** Localized Surface Plasmon Resonance
**MA** Malignant
**MHT** Magnetic Hyperthermia Therapy
**MM** Multimodal
**MRI** Magnetic Resonance Imaging
**NMR** Nuclear Magnetic Resonance
**NPs** Nanoparticles
**OL** Oligonucleotide
**PA** Photoacoustics
**PEG** Polyethylene Glycol
**PET** Positron Emission Tomography
**PFOB** Perfluorooctyl Bromide
**PVPh** Poly (4-Vinyl Phenol)
**rHTNF** Recombinant Human TNF
**ROS** Reactive Oxygen Species
**RNS** Reactive Nitrogen Species
**SCC** Squamous Cell Carcinoma
**SERS** Surface-Enhanced Raman Spectroscopy
**Si NPs** Silica Nanoparticles
**SPC** Silicon Phthalocyanine
**SPECT** Single Positron Emission Computed Tomography
**SPIO** Superparamagnetic Iron Oxide
**SPR** Surface Plasmon Resonance
**TH** Therapeutic
**TNF** Tumor Necrosis Factor
**TR** Transverse Relaxation
**UM-A9** Mouse Monoclonal Antibody
**USCAs** Ultrasound Contrast Agents
**USPIO** Ultra-Small Paramagnetic Iron Oxide
**XA** X-ray Attenuation
**XRA** X-ray Absorption
**ZP** Zeta Potential

# 10 Sensing Applications and Antimicrobial Properties of Metal and Metal Oxide Nanoparticles

*V. S. Kavitha*[1], *A. K. Visnuprasad*[2], *and Riju K. Thomas*[3]

[1]Laser and Plasma Technology Division, Bhabha Atomic Research Centre, Trombay, Mumbai, India

[2]Government Polytechnic College, Kollam, Kerala, India

[3]Department of Physics, Bharata Mata College, Thrikkakara, Edappally, Kochi, Kerala, India

## 10.1 INTRODUCTION

Nanotechnology is one of the fast-growing fields of science and technology, which is guiding different industries in every part of the world in the past decade. This multifaceted field of science and technology finds application in various industries like the chemical, food, medicine, mechanical, agricultural, cosmetics, healthcare, defense, electronics, and gas industries. According to estimates, the global market for nanotechnology will increase from $68.0 billion in 2023 to $183.7 billion in 2028, expanding at a compound annual rate (CAGR) of 22.0% throughout that time [1]. Although the concept of nanotechnology, which involves production and manipulation of nano-sized materials, has been around for a while, the introduction of this idea is associated with the famous lecture "There's Plenty of Room at the Bottom" by renowned physicist Richard P. Feynman at the meeting of the American Physical Society in December 1959 [2]. A material is nanomaterial if any of its external dimensions, internal structure, or surface structure is in the nanoscale regime (size range: 1-100 nm). They can be in the form of nanoparticles, nanotubes, nanofibers, and nanocomposites. Nanomaterials' properties are significantly different from their bulk counterparts, mainly due to their small size, high surface-to-volume ratio, and high reactivity, which affects properties like chemical reactivity and electrical, mechanical, optical, and magnetic properties [3]. For example, the typical colors of 20-nm gold (Au), silver (Ag), platinum (Pt), and palladium (Pd) nanoparticles are wine red, black, yellowish gray, and dark black, respectively. These nanoparticles display distinctive colors and properties with variation in their size and shape, making them useful for bio-imaging applications. The color of a nanoparticle solution is influenced by variations in the aspect ratio, nanoshell thickness, and particle concentration. When any of the previously mentioned factors are changed, the absorption properties of the nanoparticles are affected, and different absorption colors are observed [4].

The properties of metal and metal oxide nanoparticles are numerous and include antimicrobial, anti-cancerous, anti-insecticidal, anti-inflammatory, and non-toxic properties. As a result, they have numerous uses in the biomedical industry for the identification and management of serious illnesses. They find application in different fields such as medicine, catalysis, environmental protection, and molecular sensing [57]. Nanoparticles are prepared in a controlled manner to achieve targeted and repeatable functionalities. The nanoparticle synthesis route plays a vital role in the fine-tuning of their properties to achieve specific functionalities. Top-down and bottom-up approaches are used to synthesize metal and metal oxide nanoparticle with the desired properties. Lithography, milling, and machining are examples of top-down methods, whereas plasma-assisted deposition, the liquid phase method, vapor phase deposition, the sol-gel method, the epitaxial method, and electrodeposition are examples of bottom-up methods [8]. In the current chapter, we will only focus on the antimicrobial and sensing properties of metal and metal oxide nanoparticles.

## 10.2 METAL NANOPARTICLES

Metal nanoparticles (MNPs) are single-element nanomaterials made up of single atoms or collections of several atoms. They can be synthesized as solid material or colloidal fluids using different synthesis techniques. Generally, both top-down and bottom-up approaches, including physical, chemical, and biological methods, can be used to prepare and stabilize metal nanoparticles; however, chemical methods, such as chemical reduction, photochemical reduction, and electro-chemical techniques, are most commonly employed [9, 10]. In chemical methods, the interaction kinetics between metal ions and the reducing agent, the adsorption process of the stabilizing agent, and various experimental conditions all have a significant impact on the size, shape, and physico-chemical properties of the resulting nanoparticles, so choosing the right preparation method is crucial. Figure 10.1 presents the different synthesis methods for metal nanoparticles. Nanoparticles of Au, Ag, Cu,



DOI: 10.1201/9781032660639-10

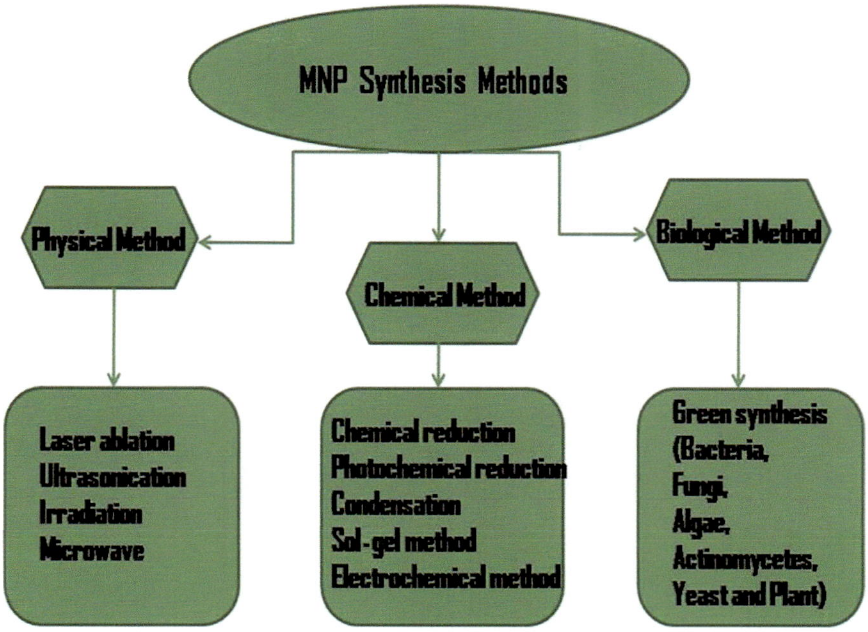


**FIGURE 10.1** Methods for the synthesis of metal nanoparticles.

Fe, Zn, Pt, Al, Pd, Ni, and Co are commonly prepared for application in numerous fields [11]. MNPs can be prepared from their respective metal precursors and can be altered with different chemical functional groups, enabling them to be coupled with desired antibodies, ligands, and drugs, opening up a variety of applications in the biomedical field [12]. Their broad absorption band in the electromagnetic spectrum, strong reactivity, and localized surface plasmon resonance (LSPR) behavior are among their noteworthy qualities. Owing to their enhanced optical, opto-electric, catalytic, anti-microbial, and cancer-fighting properties, MNPs are highly attractive materials for a wide range of practical applications [11]. The main thing that fascinates scientists about them is the relationship between their size and shape and physico-chemical reactivity. These parameters can also affect other properties via quantum confinement, which happens when the NP size forces delocalized electronic states to become localized [13].

MNPs behave completely differently from their bulk materials due to their large surface-to-volume ratio, small size, high surface energy, and spatial confinement; these properties are what make them significant in the field of catalysis, electronics, agriculture, ground water purification, and biomedical analysis. Due to their wide range of antimicrobial properties MNPs like Ag, Au, Cu, Zn, and Fe are used in food packaging, as drug delivery catheters, and for wound dressings. Considering their strong antimicrobial properties, silver nanoparticles are primarily used in the field of biomedical science. Furthermore, MNPs are also used in the development of biosensors for various substances associated with agriculture and the environment, in gene delivery, in cell labeling, and in medicines [14].

### 10.2.1 Metal Nanoparticles for Biosensing

Biosensors are electronic devices used to detect or quantify biological analytes, like microorganisms, biomolecules, and biological structures. Nanomaterials are perfect for creating new and improved biosensing devices because of their distinct chemical and physical properties [15]. Since metal nanoparticles can be synthesized in the same size range as proteins (1100 nm), they have found many applications in the biomedical field. Their special qualities, which include a high percentage of atoms and molecules on the surface and a large surface-to-volume ratio, can be used to enhance sensing and detection of a number of significant biomolecules in fields related to healthcare [16]. Metallic nanoparticles have the potential to improve biosensing by acting as reporting molecules or as a surface for the binding and immobilization of analytes [17]. MNP-based sensors offers a novel way to improve their sensitivity and selectivity. The main parts of conventional biosensors are a biological receptor, which can be an enzyme, nucleic acid, antibody, or entire cell; a transducer that converts the binding event into a detectable electrochemical or optical signal; and a signal readout that shows the analyte molecules' presence and concentration. The miniaturization of biosensor components is made possible by a variety of nanomaterials, which leads to the realization of portable biosensor kits [15].

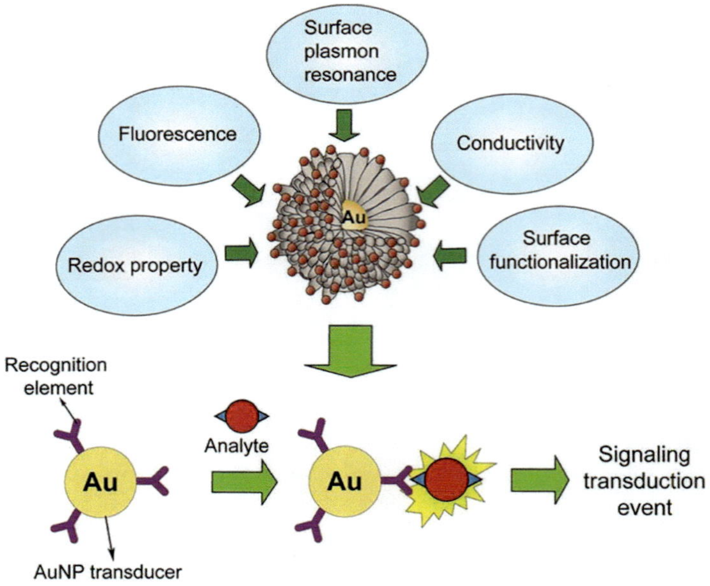


**FIGURE 10.2** Different physical characteristics of AuNPs and a schematic diagram of an AuNP-based sensing system.

**Source: Reproduced with permission from [20], Copyright © 2012, American Chemical Society.**

### 10.2.1.1 Gold Nanoparticle–Based Biosensors

Since gold nanoparticles (AuNPs) have size-dependent properties like chemical, optical, and electronic properties, they are currently employed in many biomedical applications. AuNPs have gained much interest because of their simple preparation, high surface-to-volume ratio, inertness, biocompatibility, and vibrant colors. Moreover, the metal nanoparticle dimensions, which fall between 2 and 20 nm, are similar to those of biomolecules like DNA and proteins, indicating a structural compatibility between them. Furthermore, the ease of surface modification of AuNPs with a variety of biomolecules enables the development of new biosensors [18]. AuNPs with a diameter ranging from 1 to 100 nm offer a stable immobilization of a significant number of biomolecules while preserving their bioactivity due to their high surface-to-volume ratio and surface energy. Fast and direct electron transfer between a variety of electro-active species and electrode materials is made possible by AuNPs. Additionally, AuNPs can be used as signal amplification tags in a variety of biosensors due to their enhanced light-scattering and the local electromagnetic field enhancement property [19]. Figure 10.2 shows the different physical properties of AuNPs and a schematic illustration of an AuNP-based detection system.

#### *10.2.1.1.1 AuNP-Based Plasmonic Biosensors*

The optical properties of AuNPs are utilized for optical biosensing; these applications ultimately stem from the collective oscillations of conduction electrons, or "plasmons", in response to external electromagnetic radiation called surface plasmon resonance. Of the various optical sensing modes available for AuNPs, the most extensive research has been focused on SPR. In SPR-based biosensors, by monitoring the change in the parameters of the resonance coupling of incoming light wave to the propagating surface plasmon wave, one can determine the amount of a targeted analyte that a surface bio-receptor has adsorbed. The electromagnetic properties of MNPs and a dielectric (sample medium) interface determine the existence of this surface plasmon wave. The resonance coupling manifests as a dip in the light spectrum's reflectivity, which is typically monitored by measuring the reflected light's wavelength, intensity, or incident angle (Figure 10.3). A high-index prism or a periodic grating surface is necessary for coupling light to the surface plasmon wave because of electromagnetic reasons [20, 21].

*10.2.1.1.1.1 Sensors Based on the Wavelength Shift in the LSPR Absorption of AuNPs* The shift in the LSPR peak resulting from local dielectric changes produced by analyte adsorption is the basic idea underlying LSPR-based sensors. LSPR analysis has been performed on surfaces coated with monolayers of nanoparticles as well as in the solution phase [23, 24]. The observed shift in the LSPR wavelength is found to be proportional to the quantity of ligands [25]. However, the majority of AuNP-based SPR sensors are created by immobilizing NPs on surfaces [26]. Because of their high dielectric constants and the electromagnetic coupling they have with the metal film on the surface, AuNPs can be introduced onto the sensing surface to effectively increase the sensitivity of SPR sensors [27]. Many substrates have been used to support AuNPs, including quartz, ITO glass, optical fibers, sol-gel matrix, and more. This has made it possible to detect a wide range of analytes, including human serum

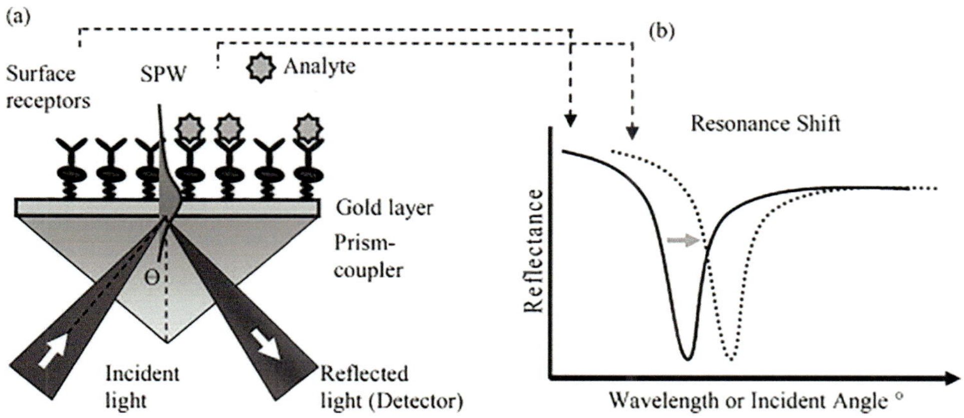


**FIGURE 10.3** (a) Surface plasmon resonance-based detection mechanism in prism-coupled configuration, and (b) the resonance shift in the detected reflectance spectrum.

**Source: Reproduced with permission from [22], Copyright © 2007, Elsevier B.V.**

albumin [28], BSA [29], streptavidin [30], human IgG [31], and propanethiol [32]. Accumulation of AuNPs changes the surface absorption band, leading to a visible color change. Using this principle, researchers developed colorimetric sensors with oligonucletide-functionalized AuNPs for use in DNA hybridization assays. It has been reported that other SPR-based sensors that use AuNP aggregation can detect lectin [33] and proteins [34].

*10.2.1.1.1.2 Sensos Based on LSPR Signal Amplification by AuNPs* AuNPs are thought to be capable of amplifying developed SPR signals. The electronic coupling interaction between the propagating surface plasmons and the localized surface plasmon of AuNPs produces the signal amplification, which is dependent on size, shape, and distance from the MNP producing SPR [35]. Using this AuNP-amplified SPR phenomenon, protein sensing can be done from antigen–antibody interaction detection. A number of competitive and sandwich immunoassays have been created based on this SPR amplification to detect human tissue inhibitor of metalloproteinases-2, TNT, allergens, testosterone, and human IgE [20]. Using LSPR-coupled fluorescence fiber optic sensors, the sensitivity of these assays can be improved by employing AuNP-decorated fluorescence-labeled antibodies [36].

Researchers demonstrated that by employing an AuNP-amplified SPR technique; oligonucleotide detection sensitivity can be increased. L. He [37] developed a novel method for the ultrasensitive detection of DNA hybridization based on SPR amplification of Au nanoparticles. When AuNP tags are used, the angle shift increases by more than ten times, which improves the target oligonucleotide's sensitivity by more than a thousand times in comparison to unamplified binding.

*10.2.1.1.1.3 Sensors Based on Plasmon Resonance Scattering of AuNPs* Apart from alterations in LSPR absorption, AuNP plasmon resonance scattering offers a useful technique for sensor design. Compared to quantum dots or dyes, the plasmon scattering of AuNPs with a diameter of 36 nm is 10,100 times stronger. Several groups have created immunoassays to identify kanamycin, human IgG, and lysozyme in human urine by taking advantage of this nanoscale phenomenon [20]. I. H. El-Sayed et al. described a simple method for the detection of oral epithelial cancer cells in vitro using SPR scattering images and SPR absorption spectra of anti-EGFR functionalized AuNPs. They reported a 600% stronger binding of the anti-EGFR functionalized AuNPs to malignant cells than to normal cells, leading to sharper SPR absorption [38].

### 10.2.1.1.2 AuNP-Based Colorimetric Biosensors

AuNPs are the perfect colorimetric reporters for biological analysis because of their distinct LSPR behavior and the color variation due to the aggregation and the dispersion of the nanoparticles. The design of colorimetric biosensors can be facilitated by the easy functionalization of AuNPs, which exhibit varying colors based on their shape, size, and level of aggregation. Over the past ten years, numerous biosensors have been designed to take advantage of the degree of color changes that occur when nanoparticles either disperse or aggregate in the presence of analytes [39]. AuNPs' ability to exploit the interparticle distance-dependent LSPR property is basic for colorimetric sensing. The dispersion and aggregation of the AuNPs can be easily controlled by monitoring biomolecular interactions and biological processes. Controlled particle aggregation can change the color of the AuNPs from pink to violet to pale blue, depending on their size [40]. Home pregnancy test, analyses for particular gene sequences, and colorimetric detection of a range of analytes have all made use of this phenomenon [41, 42]. Colorimetric detection using AuNPs as signal transducers is made possible by the expected color changes during aggregate dispersion (or aggregation). In addition to differing in color from scattered AuNPs, aggregated AuNPs also have different surface enhancement capacities for Raman scattering [43]. NP biosensors typically include a biological substrate or an immobilized ligand that changes when an analyte is present. The substrate composition is altered by the analyte, which modifies the NP environment and alters their aggregation status, which causes observable color changes [44, 45].

It is reported that when the distances between particles are less than 2.5 times the diameters of each individual particle, the electromagnetic coupling of NPs becomes effective [46]. The coupling of AuNP plasmon modes can be induced by aggregation, leading to a broadening and red shift of the SPR band, which is linked to the longitudinal resonance in the optical spectrum. Consequently, aggregation typically turns the AuNPs dispersion's original red color into purple or blue [47, 48]. Variations in the external environment have the potential to cause the AuNPs to disperse or aggregate, which will change the colloidal solution's color and cause a shift in the AuNPs absorption spectrum. Since the amount of aggregation or re-dispersion is correlated with the absorption peak shift, the degree of signal alteration can be measured, providing a clear indicator of the impact of the triggering agent, such as an active enzyme [49]. Figure 10.4 shows the color change of AuNP solution from red to purple or blue relative to the degree of aggregation of NPs by the addition of DPP4/CD26 enzymes to VP-EN-DC and GPDC substrate-functionalized AuNPs [50]. The systematic increase in the aggregation leads to a systematic decrease in the plasmon absorption peak of the sample around 520 nm and the emergence of new peak in the rage of 600–700 nm. On extreme aggregation, the solution becomes colorless, and correspondingly the position of the plasmon peak shifts to the infrared region. Particle size and composition of NPs have an impact on the molar extinction coefficients of the plasmon bands, which in turn affect the sensitivity of a colorimetric assay. Because the surface plasmons of larger NPs have higher molar extinction coefficients, they offer improved sensitivity [51]. The degree of aggregation can be measured using the aggregation parameter, which is the ratio of maximum absorption at the higher wavelength to the absorption at the shorter (original) wavelength [52].

#### *10.2.1.1.3 AuNP-Based Electrochemical Biosensors*

AuNP-based electrochemical biosensors are attractive because of their improved analytical performance compared to other designs. These bioelectro-analytical devices' remarkable potential stems from the special qualities of gold nanoparticles. One significant benefit of the creation of AuNP-based electrochemical biosensors is the capacity of gold nanoparticles to offer stable immobilization of biomolecules while maintaining their bioactivity. In the case of AuNPs, thiol derivatives (-SH) are used as stabilizers to functionalize the metal surface by creating a monolayer on the surface with enhanced co-ordination. Because of this functionalization, metal nanoparticles can now be modified while retaining their active properties by adding a variety of biomolecules and organic ligands [53]. Moreover, direct transfer of electrons between redox proteins and bulk electrodes is made possible by gold nanoparticles, eliminating the need for electron transfer mediators in electrochemical sensing. The ability of gold nanoparticles to reduce the distance between proteins and metal particles, their high surface area-to-volume ratio, their high surface energy, and their ability to act as an electron-conducting medium between the electrode surface and the prosthetic groups are cited as reasons for why they facilitate electron transfer between redox proteins and electrode material surfaces [54, 55]. Three methods are available for preparing AuNPs modified electrode surfaces: (1) directly depositing NPs onto the electrode surface, (2) binding AuNPs to the functional groups of self-assembled monolayers, and (3) combining AuNPs with other elements in the electrode matrix to incorporate colloidal Au into the electrode. The biomolecules can then be immobilized by adsorbing them onto NPs, cross-linking with bifunctional agents, or combining them with the other elements of composite electrodes to create biosensors [56].

*10.2.1.1.3.1 AuNP-Based Electrochemical Enzymatic Sensors* A major portion of the investigation into AuNP-based biosensors has focused on enzyme electrodes. The AuNP-based detection of enzyme activity can be achieved through a range of electrochemical assays. There are several methods for detecting charge and current responses to changes brought about by enzyme action, including chronocoulometry, square-wave voltammetry, differential pulse voltammetry, cyclic voltammetry, and impedance spectroscopy. Typically, AuNPs are used as signal amplifiers in sensor devices [57]. Studies show that majority of macromolecules adsorb to colloidal Au while retaining their bioactivity. These findings, coupled with the widely acknowledged conductivity of bulk gold, have led certain scientific groups to develop biosensing electrodes utilizing enzymes adsorbed to colloidal Au [53].

One practical and simple method is the direct deposition of NPs onto the electrode surface. For instance, a biosensor with high activity towards different phenolic compounds was created by immobilizing tyrosinase by cross-linking onto a GCE modified with electrodeposited AuNPs [58]. Researchers reported the fabrication of xanthine oxidase (XOD) biosensors for the amperometric assessment of hypoxanthine (Hx) using carbon paste electrodes functionalized with electrodeposited AuNPs. XOD-nAu-CPE, where the enzyme is immobilized by cross-linking with bovine serum albumin and glutaraldehyde, shows a higher amperometric signal for Hx. Usually, Hx detection is carried at a potential of ±600 mV versus Ag/AgCl, while the GA-BSA-XOD-nAu-CPE facilitates this around 0.00 V, thereby reducing the potential interferences from ascorbic acid like electrochemically oxidizable substances. At this lowest potential (0.00 V), the detection limit for Hx is ~$2.2 \times 10^{-7}$mol $l^{-1}$, which is the best value reported with biosensors operating at the highest potential [59].

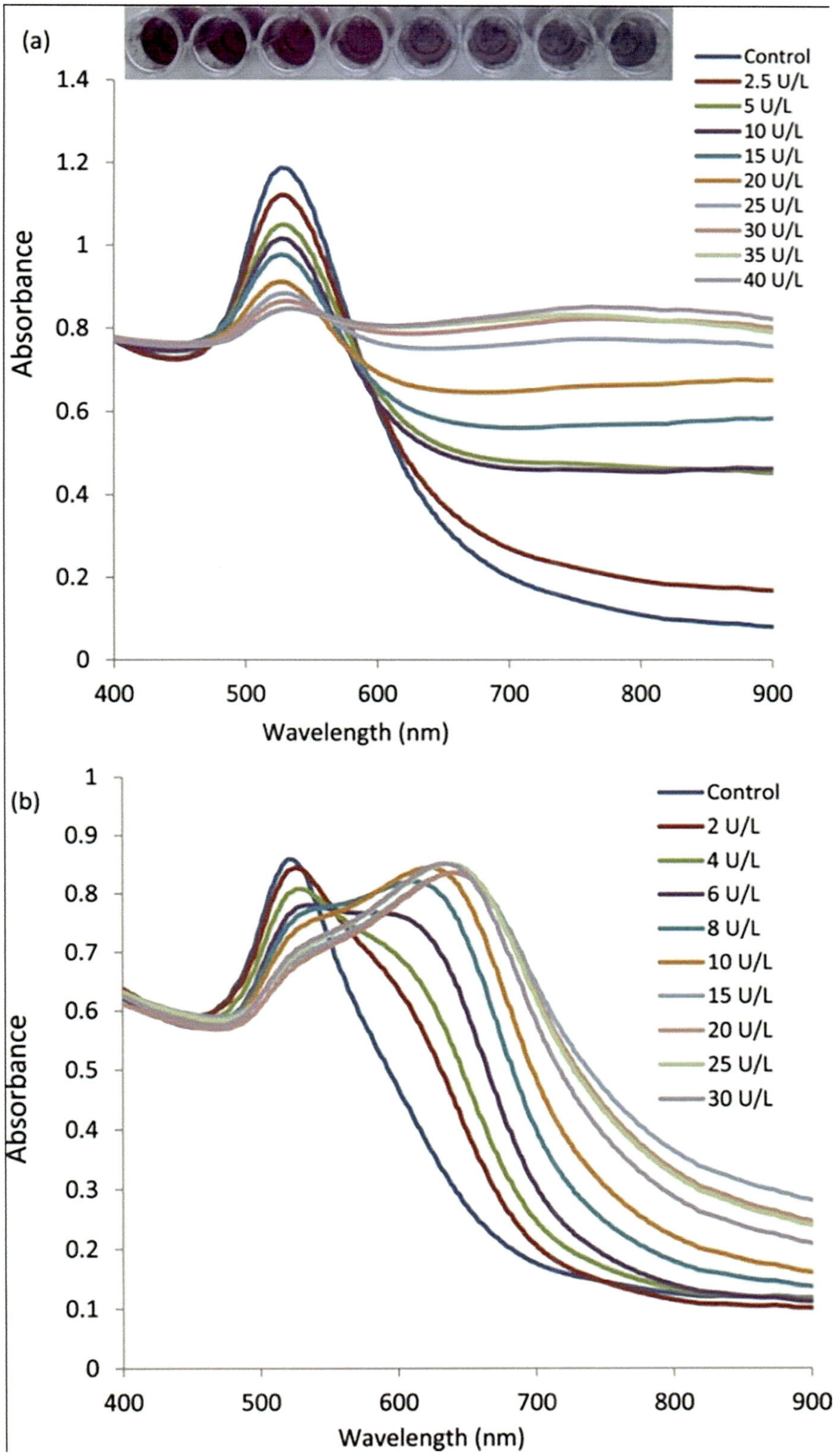


**FIGURE 10.4** (a) Absorption spectra of VP-EN-DC–functionalized AuNPs, and (b) GPDC-functionalized AuNPs after incubation with DPP4. Inset of figure (a) shows the systematic color variation due to the increased activity of DPP4 enzyme in the VP-EN-DC AuNP solution due to the enhanced aggregation of NPs.

**Source: Reproduced with permission from [50], Copyright © 2017, Elsevier B.V.**

Another method for creating enzymatic biosensors is to integrate nanomaterials into composite electrode matrices. This method offers the privilege of composite electrodes, like low background currents, simple surface generation, and the flexibility to add different materials to the bulk electrode matrix. In this regard, numerous studies have been conducted with the goal of fusing the benefits of carbon paste electrodes with colloidal gold. An HRP [60] and Tyr [61] biosensor and glucose biosensor using direct transfer of electrons of Gox [62] have been designed by immobilizing the respective enzymes onto the electrode containing a combination of colloidal gold and carbon paste. In order to increase AuNP binding efficiency with electrode matrices, they are frequently conjugated to other NPs. An efficient electrocatalytic property is shown by electrodes in which AuNPs are conjugated with carbon nanotubes (CNTs), facilitating efficient detection of glucose, hydroxylamine, cytochrome c, bisphenol A, and tryptophan [35]. H. Feng et al. fabricated an electrochemical biosensor using electropolymerization by incorporating CNT and AuNPs onto poly(thionine) film. Because of their synergic effect and the presence of efficient redox polymer, the design led to quick electron transfer and improved enzyme immobilization for the sensing of $H_2O_2$ compared to the design with either CNT or AuNPs [63].

In the presence of AuNPs, electro-polymerization offers a method for binding biomolecules on electrode surfaces. Using bioelectrodes like AuNP-(3-mercaptopropyl)-trimethoxysilane, Au-polypyrrole, and Au-polyaniline, researchers developed glucose sensors [20]. An amperometric glucose biosensor based on GOx was created by assembling AuNPs onto a modified GCE that contained an AgCl@polyaniline nanocomposite. Detecting glucose at 4 pM, the hybrid electrode system demonstrated improved electrocatalytic activity and reproducibility [64]. A novel biosensor for the amperometric detection of $H_2O_2$ was fabricated using anchored HRP on a platinum disk containing AuNPs, gelatin, polythionine (PTn), and nafion as matrixes. The enzyme was immobilized due to the combined binding of AuNPs and gelatin. The test results showed that the biosensor was highly stable and exhibited a high response (<20 s). It had a working range of 0.05 to 30.6 mM with a detection limit of 0.02 mM for $H_2O_2$ [65]. A biopolymer, chitosan is used by several researchers to immobilize enzymes due to their considerable mechanical strength and biocompatibility. Figure 10.5 shows AuNPs interfacing with chitosan hydrogel using the electrochemical process of tetrachloroauric (III) acid and chitosan on Au electrodes. Fast amperometric sensing of the pesticides monocrotophos and malathion was accomplished with a sensing limit of 1 ng/mL using AChE as a model enzyme [66].

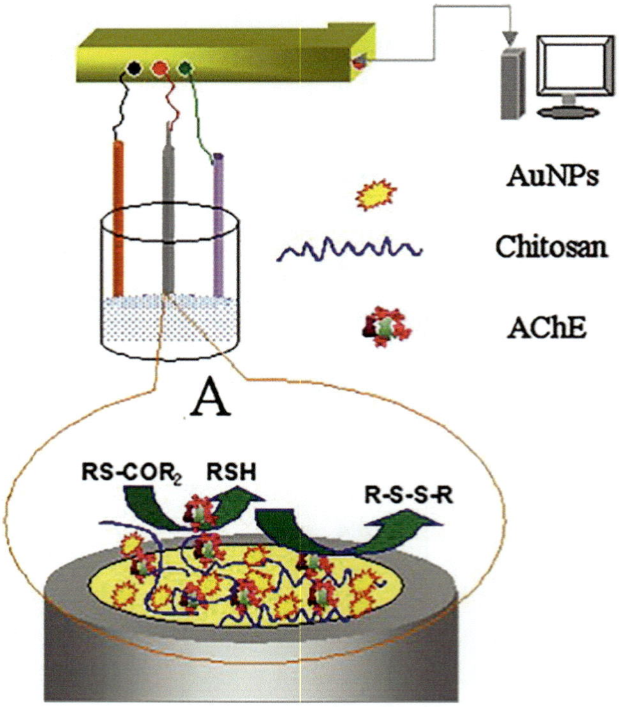


**FIGURE 10.5** Schematic illustration showing the mechanism of the developed biosensor using one-step electrodeposition: The letter A in the figure shows the megascopic interface of AChE/Chitosan–Au nanoparticles modified Au electrode.

**Source: Reproduced with permission from [66], Copyright © 2007, Elsevier B.V.**

*10.2.1.1.3.2 AuNP-Based Electrochemical Immune Sensors* Immunoassay is an important analytical technique employed in clinical examination, biochemical analysis, and environmental analysis since it measures the antigen or antibody concentrations on the basis of biospecific recognition interactions. AuNPs are essential to electrochemical immunosensors because they can increase the quantity of attached immune reagents in a fixed mode and improve the electrochemical signal that transduces the binding of antigens to antibody immobilized surfaces [54]. An immunosensor's ability to function depends on antibodies that have a specific reaction with the corresponding antigens. The quantification of an immunosensor is typically accomplished by measuring a particular activity of the label, such as surface resonance plasmon, bioluminescence, fluorescence, electrochemical signals, radioactivity, or enzyme or chemical amplification [67].

Hepatitis B antigen (HBsAg) detection using AuNP-based electrochemical immunoassays has been the subject of much research. Researchers developed an amperometric immunosensor that uses HRP and gold nanoparticles to modify a gold electrode and immobilize hepatitis B antibodies. In order to prevent non-specific adsorption and enhance the antigen-antibody reaction, HRP was added to the immunosensor in place of BSA in order to block any potential active sites that might still be present in the monolayer of gold nanoparticles [68]. A potentiometric immunosensor for the quick and direct sensing of diphtherotoxin was fabricated based on a PtE. By using a combination of gold and silica nanoparticles as well as polyvinyl butyral as matrices, monoclonal diphtheria antibody was used to self-assemble onto the PtE. The nanoparticle surface adsorbs the anti-Diph, which was subsequently entrapped within a polyvinyl butyral sol-gel on the PtE. In response to DiphAg, the immobilized DiphAb showed a direct potentiometric reaction. Compared to immunosensors built with silica or gold nanoparticles alone, the one using the nanoparticle mixture showed significantly improved long-term stability, reproducibility, and sensitivity [69]. The direct measurement of human IgG was made possible by the development of a capacitive sensing technique that uses gold nanoparticles that self-assemble to an electrode modified with sol-gel. The ultrathin nature of the formed mercapto propyltriethoxysilane film resulted in a high antibody immobilization density due to the assembled gold nanoparticle's high surface-to-volume ratio. There was no evidence of cross-reactivity between the capacitive immunosensor and other proteins, and it offered high sensitivity [70].

##### *10.2.1.1.3.3 AuNP-Based Electrochemical DNA Sensors*

For the detection of DNA, electrochemical DNA sensors offer a straightforward, reliable, and affordable platform. Direct and indirect DNA electrochemistry, electron-mediated charge transport, DNA-specific redox indicator detection, and nanoparticle-based electrochemical signal amplification are some of the detection techniques used in electrochemical DNA sensing [71]. The methods for electrochemical analysis that are most frequently employed include cyclic voltammetry, differential pulse voltammetry, electrochemical impedance spectroscopy, and chronoamperometry. Labeling molecules or identifying the peaks of NPs in the cyclic voltammogram allows for the quantitative detection of DNA hybridization in cyclic voltammetry. Differential pulse voltammetry is an improved form of cyclic voltammetry that provides distinct peaks. Impedimetry detects the system's impedance across a range of frequencies, and chrono-amperometry provides values at various time intervals [72].

In electrochemical sensors, AuNPs work well as labels and can be used to alter electrode surfaces, raising the sensor's detection limit. The use of electrochemical DNA sensing techniques has significantly advanced DNA diagnostics because they generate direct electronic signals without the need for costly signal transduction instrument and offer a straightforward, accurate, and affordable platform [71]. The electrode surface-immobilization method to create stable single-stranded DNA (ssDNA) monolayers is the key concept in the design of DNA sensors. AuNPs films give a proper method of ssDNA immobilization. Gold nanoparticles can be modified with thiolated oligonucleotides and immobilized in aqueous buffers, leading to the development of DNA-nanoparticle probes with different properties for a wide range of applications [72].

Functionalizing AuNPs with probe DNA provides improved stability, robust binding, and electrochemical activity that leads to the development of DNA sensors with improved performance [73]. The detection of DNA hybridization relies on the identification of AuNPs through the use of any of their characteristics or by identifying the electrochemical tags that are affixed to Au or Au-DNA complexes [74]. In electrochemical DNA sensors, AuNPs can be used in the following ways: (1) signal strength can be enhanced using oligonucleotide functionalized AuNPs, (2) AuNP labels can be used for DNA hybridization detection, (3) electrodes modified with AuNPs can be used to increase the adsorption of DNA, and (4) AuNPs can be used as carriers for other electroactive labels [72]. AuNPs are used as labels in breast cancer gene detection. Here the electrons from the redox mediator can be transferred efficiently to electrodes using AuNPs by binding AuNPs to the ssDNA on the Au electrode. Target DNA induces hybridization, and the double-stranded hybridized DNA separates the AuNPs from the DNA, reducing the transfer of electrons to the electrode. Target DNA was detected with a limit of 1 pM by observing the signal difference.

Utilizing AuNPs' distinct conducting features, they are also employed as signal amplifiers in a variety of electrochemical DNA sensors to boost sensitivity. Several reports describe the use of AuNPs to enhance hybridization event transduction in DNA sensors. In this instance, amplification was achieved by combining redox labels into a single hybridization process. AuNPs have a large surface area and more binding sites, so they are also used as carriers for electrochemical tags in DNA sensors. This results in the immobilization of electrochemical tags on AuNPs. Analyte signal enhancement in a single recognition process makes this method appealing. AuNPs' strong catalytic activity makes them useful as an electrocatalyst as well. The NPs' high surface area and surface energy give them a lot of catalytically active sites on their surfaces. When AuNPs are employed as catalysts, signal production from the NP active sites may enable greater amplification. Moreover, AuNPs are used as an immobilization platform in DNA sensors because of their superior electrical conductivity [72].

### 10.2.2 Metal Nanoparticles as Antimicrobial Agents

The emergence of pathogens resistant to antibiotics has become a major health concern. Many studies have been published to enhance the available antimicrobial therapies. The antibacterial uniqueness of noble metals and their NPs has been verified throughout history through their use as antimicrobial agents in food and food packaging, initially to guarantee food quality. Later, their applications were expanded to include medical uses, waste water treatment, and electronic gadgets. High antimicrobial activity of MNPs was demonstrated against a range of bacteria. For centuries, people have been aware of and used metals with antimicrobial properties such as Ag, Au, Zn, Ti, and Cu, each of which has unique properties, potential, and spectra of activity [75]. Two important factors that influenced the antimicrobial efficacy of the final product were the type of materials used to prepare the NPs and their size [76]. An increase in the fraction of surface molecules occurs when the size is reduced to the nanoscale, improving various particle properties such as heat treatment, dissolution rate, mass transfer, and catalytic activity [77]. Although the precise mechanisms by which nanometals have an antimicrobial effect are still being studied, two more widely accepted theories have been put forth in this regard: (1) free metal ion toxicity, which results from the metals dissolving from the surface of the NPs, and (2) oxidative stress caused by the production of reactive oxygen species (ROS) on the surfaces of the NPs [78]. Additionally, it has been demonstrated that the physicochemical and morphological properties of the nanometals influence their antimicrobial activity. The shape of the NPs also determines their antimicrobial properties. The most potent bactericidal effect of nanoparticles is known to be observed in small sizes. The MNPs' positively charged surface helps them to bind to the negatively charged surface of the bacteria, potentially enhancing the bactericidal effect [75]. Figure 10.6 gives a view of the different mechanisms leading to the antimicrobial activity of MNPs.

#### 10.2.2.1 Silver Nanoparticles as Antimicrobial Agents

Because of their extensive antimicrobial properties and strong antimicrobial efficacy against a wide range of bacteria, fungi, and viruses, silver nanoparticles (AgNPs) are the most extensively studied antibacterial nanoagent among the many engineered nanoparticles [79]. Extensive research has been conducted to determine the source of AgNP antibacterial activity. AgNP antibacterial activity is thought to be significantly influenced by a number of physicochemical characteristics, including size, shape, stability, and surface chemistry [80]. Understanding the antibacterial system's mode of action is essential for designing AgNPs that are more effective. There are two unique steps in this, and each will affect its efficiency. The first is how the system will act in the relevant environment, where changes may take place on a physical or chemical level. The second phase focuses on how the species that contain silver interact with the bacterial cell to cause cellular death [81]. AgNPs show particle-specific toxicity, meaning that the physicochemical characteristics of the particle have a significant influence on the pathways leading to the antibacterial activity. The current consensus is that the antibacterial properties of AgNPs involve their attachment to cell walls and membranes, the damage they cause to intracellular biomolecules and structures, and the oxidative stress they induce. These mechanisms contribute to the antibacterial actions either independently or simultaneously [82].

##### *10.2.2.1.1 Importance of Silver Nanoparticles in Antimicrobial Action*

Since the antimicrobial effect of AgNPs was first noticed, most of the reports regarding it naturally assumed that the antimicrobial effect is due to the $Ag^0$ core. Silver is usually regarded as a Lewis acid that reacts with Lewis bases, such as biomolecules containing phosphorus and sulfur, which are important building blocks of DNA bases, proteins, and cell membranes. AgNPs have a tendency to build up at the bacterial membrane and form aggregates when they come into contact with bacteria. Therefore, transmission electron microscopy detects morphological changes due to AgNP accumulation on the cell wall, such as cytoplasm shrinkage, membrane detachment, multiple electron dense pits, and ultimately a disrupted membrane, leading to cellular death [81, 82]. Studies showed that the surface charge of the bacteria changes towards neutrality upon treatment with AgNPs. This neutralization changes the permeability of the cell membrane, which in turn leads to cell death [83]. Additionally, AgNPs have the ability to bind to membrane proteins, such as transport and respiratory chain proteins, which can significantly impair ion transport, cell division, membrane permeability, and the respiration chain [84]. Also, AgNPs can enter bacteria after adhering to their cell

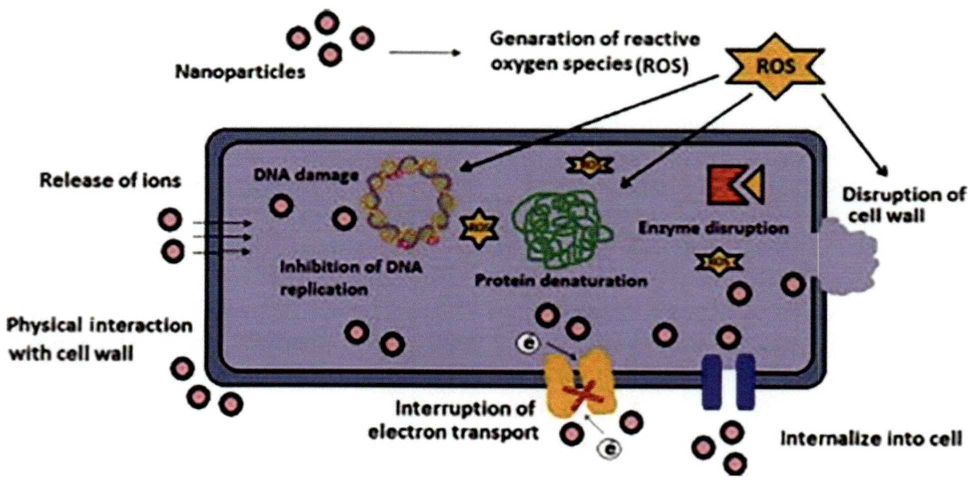


FIGURE 10.6 Different methods leading to the antimicrobial activity of metal nanoparticles.

Source: Reproduced with permission from [75], Copyright © 2014, Elsevier B.V.

membranes and interacting with their biomolecules like DNA, lipids, and proteins and other cell structures, leading to the blockage of cellular processes like protein synthesis, translation, and transcription [85]. Meanwhile, AgNP antibacterial activity has also been linked to DNA denaturation or shearing as well as the inhibition of sugar metabolism [86]. Reactive oxygen specie production is another mode of AgNP cytotoxicity effect. Numerous investigations revealed that cells treated with AgNPs had elevated ROS levels. Under these circumstances, the cells experience a high level of oxidative stress, which causes the cells to become inactive.

##### 10.2.2.1.2 Importance of Silver Ions in Antimicrobial Action

Several studies have demonstrated the connection between AgNPs' antibacterial properties and silver ion ($Ag^+$) release. Certain systems, like salts, ionomers, or zeolites, are initially composed of $Ag^+$ species, which can be liberated through ion exchange or simple dissolution. In the case of metallic silver ($Ag^0$) nanoparticles, it is not true; here, $Ag^+$ species release is dependent on the presence of an oxidizer for the oxidative dissolution of $Ag^0$ NPs. In this situation, the antibacterial agent is the $Ag^+$ species, and the $Ag^0$ NPs serves as a reservoir. Additionally, $Ag^+$ species have a strong affinity towards electron-donating groups found in membranes and proteins, such as phosphates, organic amines, and thiols. Silver ions bind to thiol groups to form a quasi-covalent AS-Ag bond (binding energy ~ 65 kcal/mol), which can change a protein's three-dimensional structure and obstruct active binding sites [87, 88]. Therefore, silver ions ($Ag^+$) can hinder the release and transport of potassium ions ($K^+$) from the target cell, thereby blocking the production of adenosine triphosphate (ATP) [89]. Furthermore, a variety of other biomolecules, including RNA, DNA, and peptides, have been found to be $Ag^+$ ions' targets, forming insoluble compounds, which prevent cell division and reproduction [90].

##### 10.2.2.1.3 Importance of Oxidative Stress

Several studies explained the cytotoxic effect of AgNPs by considering their active oxygen species generation as the mode of action, contrary to the belief that AgNPs hinder a wide range of metabolic pathways. ROS is the common name of some transient reactive oxidants like hydroxyl radicals (-OH), superoxide radicals, hydrogen peroxide ($H_2O_2$), and singlet oxygen. These are normal byproducts of oxygenic respiration, and scavengers such as glutathione (GSH) typically maintain their intracellular concentrations at a low level. AgNPs have been shown to raise the rate of respiration, which causes intracellular ROS to be produced. Mitochondria are the primary location of ROS production inside the cell. It is widely accepted that mitochondrial damage is the main destruction during AgNP-induced early apoptosis. AgNP deposition in the mitochondria can change their normal function by interfering with the electron transport mechanism, which eventually leads to the generation of ROS and a low ATP production. The catalytic activity of AgNPs in the presence of dissolved oxygen generates free radicals, and it prevents the scavenging mechanisms by linking with the thiol groups of the particular enzyme and GSH, which leads to the enhancement of ROS and free radical concentration [91]. Although results show that ROS production is the primary effect of AgNP ingestion, it can also induce many secondary effects like DNA damage, protein damage, and lipid peroxidation [92]. Studies on the toxic effect of AgNPs on *D. melanogaster* showed that AgNPs induce oxidative stress, DNA damage, heat shock stress, and apoptosis [93]. At a high ROS level, all the aforementioned mechanisms are triggered in conjunction with increased damage to the cell constituent parts, which can result in apoptosis or sometimes necrosis [94].

##### 10.2.2.1.4 Structural Effects

Silver nanoparticles' bactericidal action is dependent on a number of structural factors, like particle size, shape, surface chemistry, and surface charge. Size is an important factor that affects the antimicrobial activity of AgNPs. Usually, AgNPs with reduced size exhibit higher antimicrobial activity. A general trend regarding the particle size of AgNP and their antibacterial properties is that smaller Ag particles, especially with sizes below 10 nm, show an improved antibacterial effect compared to their larger counterparts. Lok et al. [95] reported that AgNPs around 9 nm show approximately ninefold antimicrobial activity compared to nanoparticles with size around 60 nm. Since smaller AgNPs are easily oxidized, they postulated that AgNP oxidation ($Ag^+$) is primarily responsible for the observed antibacterial properties of silver nanoparticles. To confirm this hypothesis, they compare the antibacterial activity of partially oxidized AgNPs with zerovalent particles and observed that oxidized nanoparticles are more active than zerovalent against *E. coli* cells. They also found that for effective antimicrobial activity, chemisorbed $Ag^+$ is needed on the surface of oxidized AgNPs. Since smaller AgNPs possess a larger value for the surface area-to-mass ratio, they can adsorb a large number of $Ag^+$ ions on their surface, leading to increased antimicrobial activities.

In addition to size, the shape or morphology of the nanoparticles also plays a vital role in the antimicrobial efficiency of AgNPs. Antibacterial activity based on shape is dependent on how well the nanomaterial makes contact with the bacterial cell membrane. Goyal et al. [96] compared the antimicrobial activity of plate like hexagonal- and triangular-shaped AgNPs having same size but with different edge shape (sharp or round). They observed enhanced activity for those with sharp edges due to the higher charge density at the sharp edges in comparison to round edges. This higher charge density will lead to a better change in the permeability of the bacterial membrane, resulting in easier rupture of the membrane. Cheon et al. [97] studied the antibacterial activity of disk-like, spherical, and triangular plate AgNPs against *S. aureus, E. coli,* and *P. aeruginosa*. Here, spherical particles imparted more damage to cells compared

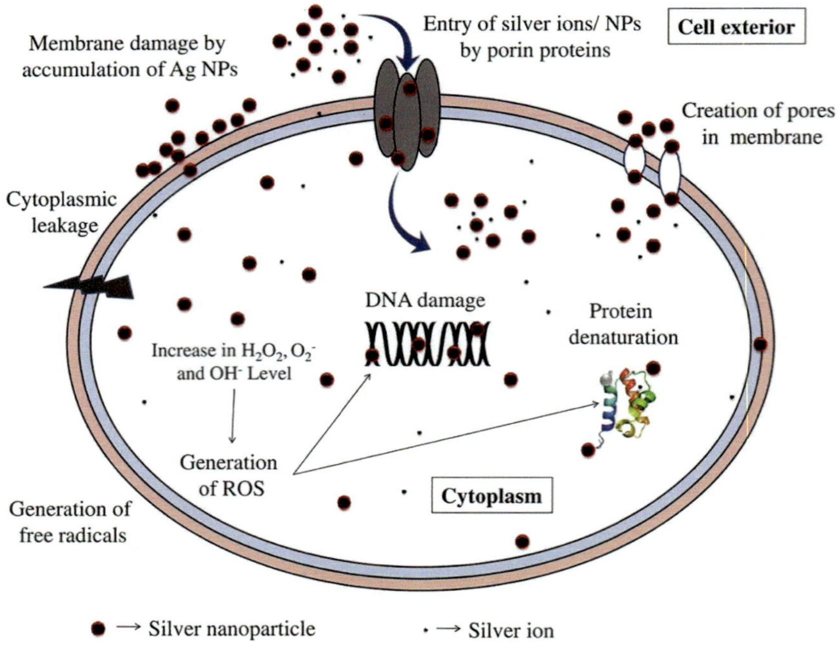


**FIGURE 10.7** Different AgNP cell destruction mechanism, leading to the apoptosis of bacterial cells.

**Source: Reproduced with permission from [101], Copyright © 2018, Elsevier B.V.**

to other shapes due to their higher surface area, leading to the enhanced release of $Ag^+$. In brief, crystal facets, surface area, edge sharpness, and geometry of interaction are the main morphological parameters determining the antibacterial properties of AgNPs. Each shape has particular crystal facets and corresponding reactivity. Also, sharp edges facilitate improved activity due to the enhanced interaction with the bacterial membrane. Geometry that leads to closeness of the nanoparticle and bacterial surface shows improved activity. Moreover, geometries with large surface area possess better antibacterial properties [98].

Surface chemistry also plays an important role in the antibacterial activity of AgNPs. Usually, the AgNP is exposed to different liquid environments like blood, cultural media, and body fluids to study their antibacterial properties in vitro and in vivo, so the stability of AgNPs in these media is an important issue because it affects the nanoparticle size, morphology, surface-to-volume ratio, and $Ag^+$ ion release kinetics, which affects their antimicrobial properties [99]. The stability concern of AgNPs in these experimental environments is mainly the chemical stability against the release of silver ions ($Ag^+$) due to the oxidative dissolution of AgNPs and also the dispersion stability in opposition to aggregation of AgNPs. Additionally, the surface charge also has been demonstrated to be one of the most important factors that govern the antibacterial activity of AgNPs. Abbaszadegan et al. [100] studied the role of surface charge of AgNPs on the antibacterial effect of a set of Gram-positive and Gram-negative bacteria. They synthesized AgNPs with positive, negative, and neutral surface charges and found that AgNPs with a positive charge were the most efficient antimicrobial agents against the entire tested bacterial group. Figure 10.7 shows the different bacterial cell destruction mechanism of AgNPs [101].

## 10.3 METAL OXIDE NANOPARTICLES

Metal oxide nanoparticles (MONPs) belong to the field of material chemistry and are gaining much attention due to their potential use in different fields like sensors, biomedical, catalysis, optoelectronic materials, energy, and agriculture [102]. The distinct physical and chemical properties of MNOPs are due to their high density and limited size of edges and corners [103]. MONPs are typically ionic and can be arranged in crystal morphologies with different reactive sites and corners. In particular, MONPs have a number of benefits that make them a versatile tool for biomedical applications. These benefits include high stability, easy preparation, better control over shape, size, and porosity, ease of incorporation into hydrophilic and hydrophobic systems, and a simple functionalization mechanism by different molecules because of the presence of a negative charge on the surface [104].

Physical and chemical methods are the commonly employed synthesis routes of MONPs. The synthesis method adopted can alter their properties; therefore, the reaction mechanism and the functionality of the MONPs depends on their, size, shape, composition, surface stoichiometry, and crystallographic structure. Chemical vaporization, spray pyrolysis, and ultrasonication are some of the physical methods used, and chemical methods constitute the oxidation-reduction reaction or solution synthesis

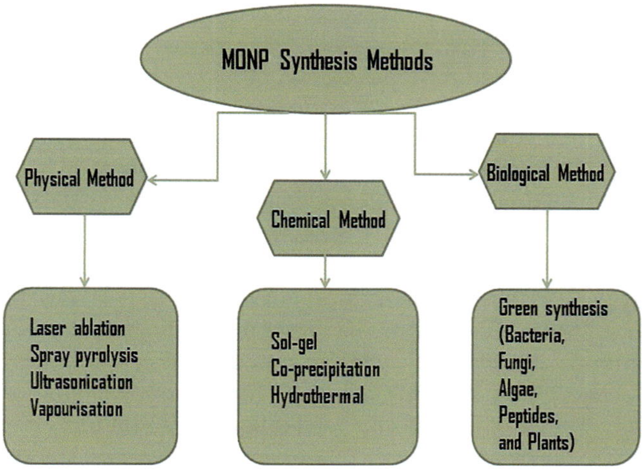


**FIGURE 10.8** Different synthesis routes for metal oxide nanoparticles.

process. In these approaches, physical methods have numerous drawbacks, including difference in particle size and the need for expensive equipment. Conventional chemical procedures, although successful in producing a wide range of morphologies for inorganic nanoparticles, typically involve the use of hazardous chemicals that pose a risk to both living things and the environment. As a result, new, ecologically friendly green chemistry-based synthetic techniques have been developed. Green synthesis methods include the sol-gel method and biomolecules like proteins, microorganism, amino acids, carbohydrates, plants, and living cells [105]. Figure 10.8 shows the different synthesis routes for MONPs.

### 10.3.1 Metal Oxide Nanoparticles for Biosensing

Of all the nanomaterials that have been developed, nanostructured metal oxides have gained a lot of attention recently as an immobilizing platform for the design of biosensors because of their special morphologies and excellent chemical and physical properties. With their specific orientation and high biological activity, MONPs offer an effective surface for immobilizing biomolecules that results in improved sensing properties. Electrical, optical, and molecular properties of nanostructured metal oxides, combined with desired functionalities and surface charge behavior, offer integration of bio-recognition elements with transducers for better signal amplification [106, 107]. Nanostructured oxides of metals like Zn, Sn, Fe, Ti, Mg, Zr, and Ce exhibit astonishing non-toxic and biocompatible properties. Together with their high surface area, strong adsorption capacity, better biocompatibility, high chemical stability, and improved electron-transfer kinetics, these materials offer ideal microenvironments for the binding of biomolecules, which improves electron transfer and yields better biosensing properties. In biosensors, MONPs are used as signal transduction and/or a target identification element. They provide different means for bridging biological recognition process and electronic signal transduction to create novel electronic devices [107].

#### 10.3.1.1 Zinc Oxide Nanoparticles as Biosensors

Zinc oxide (ZnO) nanostructured materials, including films and nanoparticles, have interesting material properties that make them a promising candidate for the development of high-performance biosensors. A wide energy gap, high exciton binding energy, improved electrochemical activity, biocompatibility, inexpensive synthesis methods, non-toxicity, and chemical and photochemical stability are some of the important properties of ZnO nanoparticles [108]. As an oxide semiconducting material, ZnO nanoparticles are highly advisable for biosensors and gas sensors. Biosensing applications are highly promising due to their particular properties like high adsorption ability and strong catalytic efficiency. Recent strong interest in ZnO biosensors is due to their large isoelectric value (IEP) of ~9.5, which provides a better platform for protein (with low IEP) adsorption through electrostatic interaction [109]. ZnO nanostructures possess different morphologies, which is advantageous for the design of biosensors. To date, ZnO nanorods, nanotubes, and nanocombs are used in enzyme biosensors, immunosensors, and other types of sensors [110]. In particular, ZnO quantum dots (QDs), having comparable size to biomolecules, offer a favorable environment for biomarker interaction [111]. Moreover, their high electron communication capability helps in the binding of biomolecules without the need for an electron mediator, giving a better way for the development of implantable biosensors [112].

##### *10.3.1.1.1 ZnO-Based Electrochemical Biosensors*

ZnO nanostructures can be used to develop efficient electrochemical biosensors to detect a large number of analytes such as DNA, cholesterol, glucose, uric acid, and dopamine. ZnO-based electrochemical sensors have fast response time and allow early detection of biological defects. ZnO nanostructures can be classified in to 0D, 1D, 2D, and 3D structures. Electrochemical sensors based on 0D ZnO nanostructures have shown high potential in detecting bioelement targets and have produced favorable results in sensing tests. 0D ZnO structures are giving new opportunities for the miniaturization of sensors due to their size range of 10–100 nm. Moreover, 0D nanoparticles offer improved surface binding and reactivity due to their increased surface area. For the detection of carbohydrate antigen 19–9, a crucial marker of pancreatic cancer, Gu et al. [113] developed a sandwich-style immunoassay using quantum dots of ZnO as the electrochemical and fluorescent labels.

However, 0D ZnO nanoparticles possess low mobility because of their large grain boundaries, which led to the development of 1D ZnO nanoparticles for sensing application. In addition to a large surface area, 1D ZnO NPs offer a stable and direct pathway for fast electron transfer. Liu et al. [114] fabricated a glucose biosensor using carbon-decorated

ZnO nanowire electrodes with a detection limit of ~1 μM and improved sensitivity. The low cost and biocompatibility of ZnO nanowires, combined with their high electron transfer capacity and high carbon conductivity, make the C-ZnO nanowire array a proper choice for the mediator-free enzymatic biosensors. Studies show that 1D ZnO nanostructures also impose some drawbacks in the complementary metal oxide semiconductor (CMOS) process for biosensors, like reduced signal strength, problems regarding electrical contacts, inconsistency with CMOS functioning, and the usual synthesis methods with high temperatures [115]. As a solution to this problem, Vabinna et al. [116] developed a label-free electrochemical sensor for cortisol detection using antibody-decorated 2D ZnO structures (nanoflakes) with improved selectivity and sensitivity. Here antibody loading and sensing activity were improved due to the higher surface charge density and catalytic activity of (0001) plane of ZnO. ZnO nanostructures have also been fabricated as 3D structures with enhanced bio-recognition properties resulting from increased surface area and high porosity. Fang et al. [117] developed an AuNP-functionalized 3D ZnO nanostructure-based electrochemical glucose sensor with a detection limit of 0.02 mM. The Au-ZnO structure helps in the immobilization of glucose oxidase (GOD) and the perforation of glucose molecules, which catalyzes the oxidation of glucose and improves electron transfer for GOD.

#### *10.3.1.1.2 ZnO-Based Optical Biosensors*

The optical characteristics of ZnO provide a better platform for biosensor applications. Optical biosensors are very sensitive and selective tools for identifying extremely low concentrations of chemicals and biological substances. They allow real-time analysis of samples without the need for substantial sample volumes or pretreatment [118]. The primary benefit of an optical sensor is its ability to measure material properties contactlessly, without the need for electrical fields. As a result, it is unaffected by external electric and electromagnetic fields. For ZnO biosensors, the most developed techniques are SPR, SERS and fluorescence. Nevertheless, these techniques are highly costly and require labels for analyte detection, which makes the experiment more difficult and time consuming. Sometime photoluminescence is also employed as the optical technique for ZnO-based biosensors; this method does not show high sensitivity. The importance of optical biosensors compared to other types includes their specificity, selectivity, remote sensing, quick and real-time detection, isolation from electromagnetic fields, compact design, and multi-parameter detection [119].

Usually, MNPs like Au and Ag are used for the design of surface plasmon–based devices. Mixing these MNPs with ZnO alters their structural, optical, and electrical properties. In ZnO nanostructure-based SPR biosensors, ZnO can be used as both active and passive component. Kaur et al. [120] developed an SPR biosensor using a ZnO/Au matrix for the detection of *Neisseria meningitidis* DNA. The biosensor shows good linearity, a lower detection limit, and improved specificity over a range of DNA concentrations. Using this sensor, researchers demonstrated a label-free, efficient method for the detection of *N. meningitides* with high stability.

SERS technique–based biosensors were also studied for the effective detection of biomolecules. Usually ZnO and ZnO/metal nanostructures are employed as a template for SERS. The SERS activity of ZnO-based sensors is due to the charge transfer process between the adsorbed organic species and the ZnO structures. Picciolini et al. [121] developed SERS substrate of zinc oxide tetrapods incorporated with Au nanoparticles, which shows a SERS enhancement factor around $10^7$. This substrate shows activity against a broad range of biomolecules and drugs. Moreover, this substrate can be used to study single cancer cells since it shows enhancement in the Raman signal from the part of the cell interacting with the 3D ZnO substrate.

Photoluminescence emission spectroscopy is a simple method to study the surface defects of nanomaterials. Using this technique, one can identify what type of defect is responsible for a particular PL emission band. By detecting the variation in the PL emission band arising from the biological impact, one can identify the defect responsible for the variation. PL measurements can be used to design two types of biosensors; one uses the variation in the PL signal from the nanostructure surface, and the other uses PL quantum dots as the label. Gua et al. [113] used ZnO quantum dots as fluorescent and electrochemical labels for the immunoassay of carbohydrate antigen 19–9 (CA 19–9), a usual label for cancer in the pancreas, using a sandwich-type biosensor. The immunoassay using PL of ZnO quantum dots shows a detection limit of around 0.25 U/ml and a linear response of 1–180 U/ml (Figure 10.9). The electrochemical assay also shows improved sensitivity of 0.47 A/U/ml and

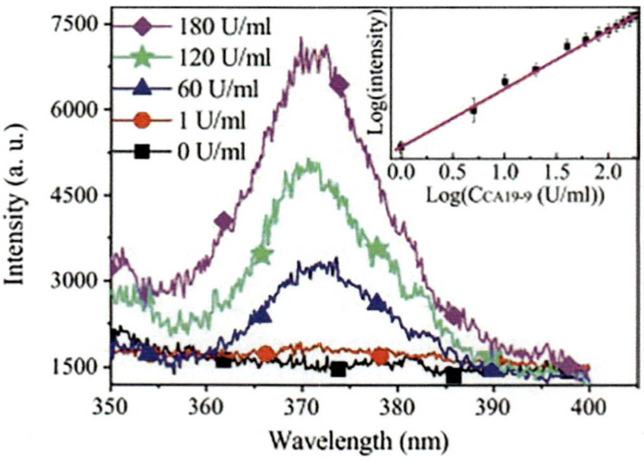


**FIGURE 10.9** Variation in the PL spectra of the immunosensor for different concentrations of CA19–9 (excitation 325 nm); inset of figure shows the logarithmic scale variation in the spectral intensity as a function of CA 19–9 concentration.

**Source: Reproduced with permission from [113], Copyright © 2010, Elsevier B.V.**

lower detection limit of 0.04 U/ml in a wider linear range of 0.1–180 U/ml. The better performance of the sensor may be due to the stable immobilization arising from the high IEP of ZnO quantum dots and also due to their uniform smaller size. ZnO quantum dots are also anticipated to develop additional amperometric immunosensors to identify additional significant antigens based on the effective assembly of antibodies, using similar techniques. In the second scenario, the change in the PL spectra of the nanostructured material is utilized to detect and quantify the analyte. The basic principle of this type of biosensor is the variation in the PL spectrum before and after immobilization of the biological selective points as well as after interaction with the analyte molecule. Sodzel et al. [122] constructed an optical glucose biosensor by immobilizing glucose oxidase (GOx) on ZnO nanoparticles. Glucose reacts with GOx and breaks down into gluconic acid and $H_2O_2$. The presence of $H_2O_2$ near the surface of ZnO nanoparticles causes the PL intensity of the nanoparticles (both NBE and DLE) to decrease. They found that the PL quenching was proportionate to the glucose concentration. This idea can be applied to a wide range of other enzymatic tests that rely on oxidase production of $H_2O_2$ or peroxidase consumption of $H_2O_2$.

### 10.3.2 Metal Oxide Nanoparticles as Antimicrobial Agents

Metal oxide nanoparticles with unique physical, chemical, optical, electrical, magnetic, and biological properties are promising antimicrobial agents against a wide range of microorganisms. It has been shown that MONPs engage in electrostatic interactions with bacterial cells to modify the prokaryotic cell wall and enzyme or DNA pathways by producing ROS. MONPs of Ti, Zn, and Mg are widely used as antimicrobial agents, mainly because they can be synthesized in the desired size and shape with cost-effective methods. In MONP-mediated antimicrobial activity, ROS production is the main mechanism of cell destruction. Other mechanisms of action include genotoxicity, photokilling, disruption of metal/metal ion homeostasis, damage to cell membranes through electrostatic interaction, and malfunction of proteins and enzymes (Figure 10.10) [123, 124].

The main sources of MONPs' antimicrobial activity are their unique physical and chemical nature. For example, semiconductor photocatalysts nanoparticles like$TiO_2$ have the ability to stop the growth of even bacteria that are resistant to ultraviolet (UV) radiation and desiccation [125].

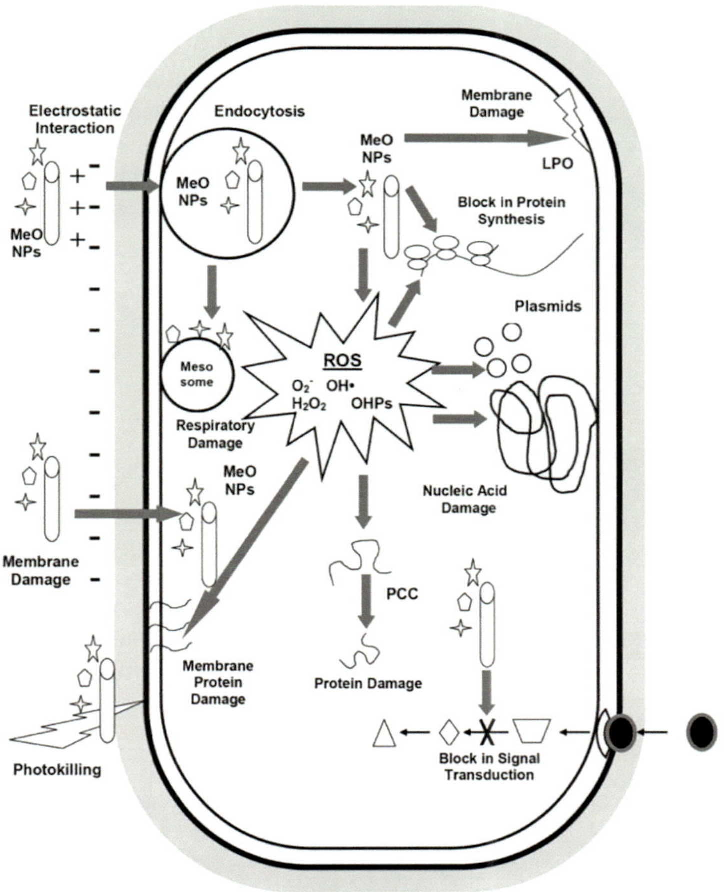


FIGURE 10.10 Different antimicrobial pathways of metal oxide nanoparticles.

**Source: Reproduced with permission from [123], Copyright © 2017, Elsevier B.V.**

Also, the alkalinity of the MgO and CaO NPs surfaces provides better antimicrobial activity. These alkaline MONPs are more soluble, giving an alkaline nature to the medium [126]. In the case of positively charged $CeO_2$ nanoparticles, the electrostatic nature of the NPs determines their antimicrobial properties [123].

#### 10.3.2.1 Titanium Dioxide Nanoparticles as Antimicrobial Agent

Titanium dioxide is one of the most often used metal oxide nanoparticles in the fields of microbiology and medicine. It is nontoxic and biologically inert at the micro scale and is well explored for efficient use of photocatalytic activity under UV irradiation. The importance of $TiO_2$ includes high oxidizing power, better stability, and low toxic effect on humans. $TiO_2$ NPs can be efficiently used as antimicrobial agents for a broad range of bacteria, including Gram-positive and Gram-negative. The antimicrobial properties of $TiO_2$ mainly depend on their morphology and crystal structure. ROS production below 385 nm and photo-induced superhydrophilicity are proposed to be important mechanisms of $TiO_2$ NP (anatase structure) antimicrobial activity [124]. It is clear that UV-irradiated $TiO_2$ causes bacteria and spores to become non-viable. Studies show that ROS production is the main cause of bacterial inactivation in the case of $TiO_2$ NPs. The $TiO_2$ nanoparticles undergo a photocatalytic reaction and produce free radicals. During metabolism, aerobic cells produce ROSs that include hydroxyl anions ($OH^-$), super oxide anions ($.O_2^-$), hydroxyl radicals (.OH), and hydrogen peroxide. Electrostatically adsorbed $TiO_2$ NPs (positive charge) on the bacterial surface are necessary for ROS generation, because they can improve cell membrane permeability, attenuate fluidity, and assist in lipid peroxidation of the membrane. As shown in Figure 10.11, ROSs directly smash the cell membrane constituents, destroy its integrity, and cross the cell membrane [127]. The primary ROSs responsible for the antibacterial and antiviral effects of $TiO_2$ NPs are free, surface-bound OH radicals. .OH has an important role in the obstruction of microbial growth. These free radicals have an impact on the phospholipid, peptidoglycan, and lipopolysaccharide bilayers because of peroxidation [128, 129]. Lipid peroxidation is a sign of oxidative stress due to $TiO_2$ NPs since it is the outcome of membrane lipid oxidation. Increased lipid peroxidation is an indication of high ROS levels. $TiO_2$ NPs enter the cell after breaking the cell wall and membrane, then attach to organelles like mitochondria or bind to macromolecules, which causes denaturation of proteins and cytoplasm flow-out.

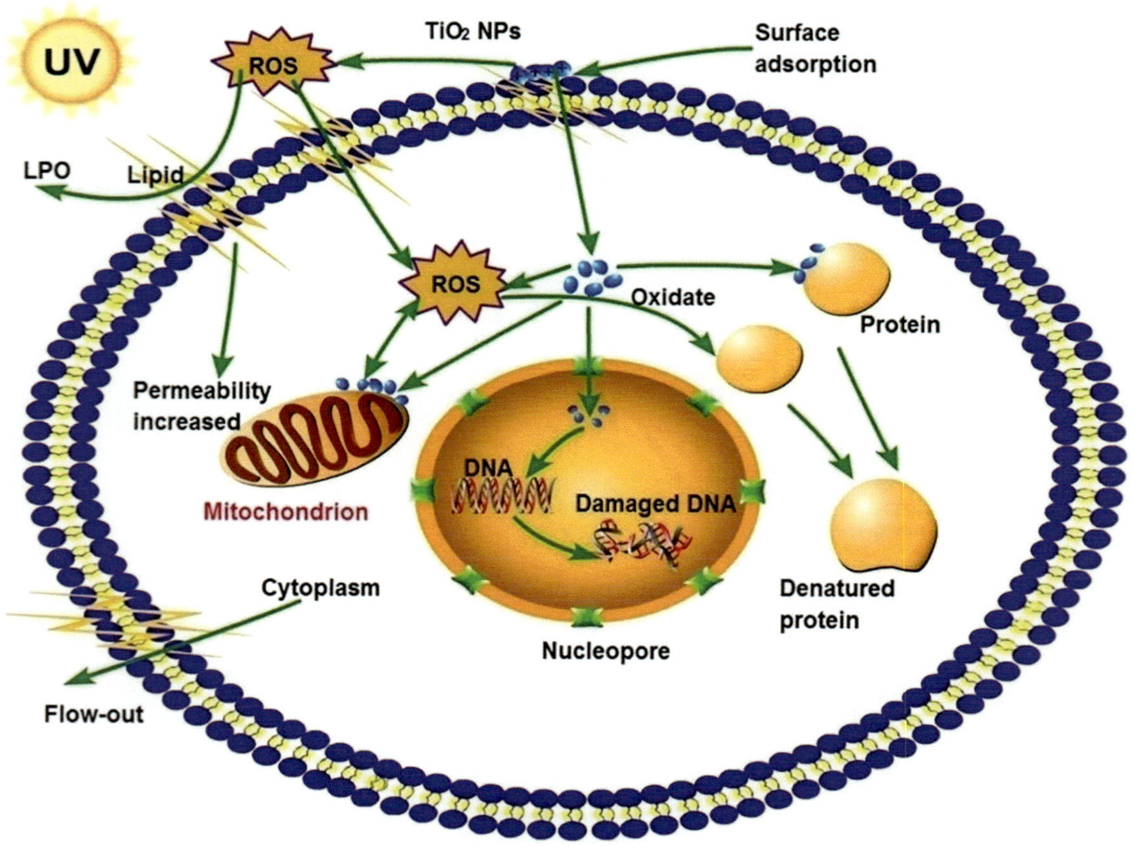


FIGURE 10.11 Different toxicity pathways of $TiO_2$ nanoparticles.

Source: Reproduced with permission from [127], Copyright © 2018, Elsevier B.V

In a cell, first proteins undergo oxidation, then lipid membranes, and then DNA in the nuclei of bacterial cells. In dark conditions, $TiO_2$ NP-treated bacteria produces remarkably fewer ROSs than bacteria exposed to light. Due to light exposure, electron–hole pairs are produced, and by the transfer of electrons, the generated holes attract electrons from $OH^-$ or from the absorbed water molecules, thereby producing .OH or react with molecular oxygen to generate $.O_2^-$. Moreover, $H_2O_2$ and singlet oxygen ($^1O_2$) can be produced by the oxidation of $.O_2^-$ due to the interdependence of hydroxyl radicals. As a result, under light, $TiO_2$ NPs can cause ROS production, which puts bacteria under oxidative stress [127].

The first report on the antimicrobial properties of UVA-activated $TiO_2$ was given by Matsunaga et al. [130]; since then, there have been numerous studies on the photocatalytic properties of $TiO_2$ for the inactivation of bacteria, viruses, fungi, protozoa, and algae. Tsuang et al. [129] studied the photokilling effects of $TiO_2$ NPs on an orthopedic implant. They investigated the antibacterial effect of $TiO_2$ NP-coated metal plates on *E. coli*. In the absence of $TiO_2$ nanoparticles, all bacteria flourished, and the dosage of ultraviolet light employed in this experiment had no effect on their viability. The microorganism survival curve in the presence of $TiO_2$ nanoparticles demonstrated that after 50 minutes of UV light, almost total death was attained. Significant reductions were also observed in the formation of bacterial colonies above the metal plates coated with $TiO_2$ nanoparticles.

$TiO_2$ is commonly used as a surface coating due to its effective surface area and smaller size of particles, which leads to efficient surface reactions. The enhanced surface area produces better interaction with the incoming light and thus increases the production of ROSs [131]. Ohko et al. [132] reported the efficient performance of $TiO_2$-coated silicone catheters or medical tubes on the bleaching effect of methylene blue and the photocatalytic antibacterial effect on *E. coli*. They found that this kind of catheter can be cleaned and sterilized with a low-intensity UV light exposure, which makes it helpful in preventing bacterial infections linked to catheter use. Yeung et al. [133] studied the photosensitive reactivity of $TiO_2$ nanoparticles, nanotextured $TiO_2$ film, and titania-silica aerogel for the photocatalytic oxidation of trichloroethylene (TCE) and inactivation of *E. coli* and *B. subtilis* microbial cells. This study suggests that the photo-generated reactive species that cause TCE to be photo-oxidized on $TiO_2$ may have a direct or indirect role in the inactivation of microbes. Also, titania-aerogels and nanotextured $TiO_2$ film exhibit improved bactericidal activities, demonstrating the significant influence of the nanoscale environment on these materials' antimicrobial activity.

However, $TiO_2$ NPs under UV light induce genetic damage in human cells and tissues; therefore, their use is restricted. As a solution to this problem, scientists use metal ion doping in a $TiO_2$ matrix, which enhances the photocatalytic and antimicrobial properties of $TiO_2$ NPs. Here the doping of metal ions changes the light absorption range of $TiO_2$ NPs to the visible region, and therefore the need for UV light can be eliminated [75]. Transition metal ion doping of $TiO_2$ NPs suppresses the rate of electron–hole recombination and causes the metal ions to produce photo-generated electrons. In brief, doping $TiO_2$ NPs with metal ions pushes their light absorption spectrum toward visible light and boosts their photocatalytic activity by blocking charge exchanges [134]. Moreover, $TiO_2$ NPs are more stable and chemically inert; hence they produce persistent antimicrobial activity following exposure to visible light. Akhavan [135] investigated the photodegradation of *E. coli* bacteria by Ag-$TiO_2$/Ag/a-$TiO_2$ film under solar irradiation. The nanocomposite film was developed by a sol-gel coating of a 30-nm-thick Ag-$TiO_2$ layer on a silver-doped anatase $TiO_2$ (a-$TiO_2$) film. The nanocomposite film shows an antibacterial effect ~5.1 times higher than activity of the anatase $TiO_2$ in dark conditions. The bactericidal effect under solar irradiation was measured to be 6.90 and 1.35 times higher than the effect of a-$TiO_2$ and Ag/a-$TiO_2$ films, respectively. Compared to the Ag/a-$TiO_2$ film, the nanocomposite film's durability was at least 11 times greater. Hence, they suggested an Ag-$TiO_2$/Ag/a-$TiO_2$ photocatalyst as one of the most potent and durable antibacterial nanocomposite materials. The $TiO_2$ nanoparticle toxicity issues can also be addressed by conjugating the particles with nontoxic polymers.

## 10.4 CONCLUSION

Metal and metal oxide nanoparticles are now playing a vital role in the diagnostics and therapeutics of some life-threatening diseases. During the current pandemic situation, these bio-inspired nanoparticles are not only helping in detecting microorganisms but are also used to develop different means to inactivate them. This chapter reviews the importance of some metal and metal oxide (Au, Ag, ZnO, and $TiO_2$) nanoparticles for sensing and antimicrobial properties. Gold nanoparticles are a widely used noble metal nanoparticle in the field of biosensing because they offer stable immobilization of a large number of biomolecules while retaining their bioactivity. They can be used as signal amplification tags in a variety of biosensors. Au nanoparticles are used in different types of sensors like plasmonic, electrochemical, and calorimetric sensors. Biosensing applications of ZnO nanoparticles are highly promising due to their particular properties like high adsorption ability and strong catalytic efficiency. They can be used in optical and electrochemical sensors. Optical biosensors based on ZnO nanoparticles use properties such as SERS, SPR, and fluorescence. Silver and $TiO_2$ nanoparticles are efficient antimicrobial agents against a broad spectrum of microbes such as bacteria, fungi, and viruses. The antimicrobial activity of Ag nanoparticles is mainly due to the action of $Ag^0$ particles or $Ag^+$ ions, oxidative stress, and structural effects. In the case of $TiO_2$ nanoparticles, ROS production is the main cause of bacterial inactivation under UV irradiation. The doping of metal ions in the $TiO_2$ lattice shifts their absorption edge to the visible region, which eliminates the need for UV light for their activation, hence reducing the UV-induced toxicity of $TiO_2$ nanoparticles.

## REFERENCES


1. *Global Nanotechnology Market*, BCC Publishing, 2023.
2. Feynman, R. P. http://www.zyvex.com/nanotech/feynman.html.
3. Roduner, E.; Size matters: why nanomaterials are different. *Chem. Soc. Rev.,* 2006, 35, 583–592.
4. Dreaden, E. C.; Alkilany, A. M.; Huang, X.; Murphy, C. J.; El-Sayed, M. A.; The golden age: gold nanoparticles for biomedicine. *Chem. Soc. Rev.,* 2012, 41, 2740–2779.
5. Livage, J.; Henry, M.; Sanchez, C.; Sol-gel chemistry of transition metal oxides. *Prog. Solid State Chem.,* 1988, 18, 259–341.
6. Hoffmann, M.R.; Martin, S.T.; Choi, W.; Bahnemann, D. W.; Environmental applications of semiconductor photocatalysis. *Chem. Rev.* 1995, 95, 69–96.
7. Laurent, S.; Forge, D.; Port, M.; Roch, A.; Robic, C.; Elst, L. V.; Muller, R. N.; Magnetic iron oxide nanoparticles: synthesis, stabilization, vectorization, physicochemical characterizations, and biological applications. *Chem. Rev.,* 2008, 108, 2064–2110.
8. Cao, G.; *Nanostructures and Nanomaterials: Synthesis, Properties and Applications*, Imperial College Press, London, 2004.
9. Chen, W.; Cai, W.; Zhang, L.; Wang, G.; Zhang, L.; Sonochemical processes and formation of gold nanoparticles within pores of mesoporous silica. *J. Colloid Interface Sci.,* 2001, 238, 291–295.
10. Frattini, A.; Pellegri, N.; Nicastro, D.; de Sanctis, O.; Effect of amine groups in the synthesis of Ag nanoparticles using aminosilanes. *Mater. Chem. Phys.,* 2005, 94, 148–152.
11. Saleh, T.A.; Properties of nanoadsorbents and adsorption mechanisms. *Interface Sci. Technol.,* 2022, 34, 233–263.
12. Tiwari, A. P.; Rohiwal, S. S.; Synthesis and bioconjugation of hybrid nanostructures for biomedical applications. *Hybrid Nanostructures for Cancer Theranostics*, 2019, 17–41.
13. El-Sayed, M. A.; Some interesting properties of metals confined in time and nanometer space of different shapes. *Acc. Chem. Res.,* 2001, 34, 257–264.
14. Khandel, P.; Yadaw, R. K.; Soni, D.K.; Kanwar, L.; Shahi, S.K.; Biogenesis of metal nanoparticles and their pharmacological applications: present status and application prospects. *J. Nanostruct. Chem.,* 2018, 8, 217–254.
15. Mostafavi, E.; Medina-Cruz, D.; Truong, L.B.; Kaushikand, A.; Iravani, S.; Selenium-based nanomaterials for biosensing applications. *Mater. Adv.,* 2022, 3, 7742–7756.
16. Fritea, L.; Banica, F.; Costea, T.O.; Moldovan, L.; Dobjanschi, L.; Muresan, M.; Cavalu, S.; Metal nanoparticles and carbon-based nanomaterials for improved performances of electrochemical (bio)sensors with biomedical applications. *Materials*, 2021, 14, 6319.
17. Mostafavi, E.; Zarepour, A.; Barabadi, H.; Zarrabi, A.; Truong, L.B.; Medina-Cruz, D.; Antineoplastic activity of biogenic silver and gold nanoparticles to combat leukemia: beginning a new era in cancer theragnostic. *Biotechnol. Rep.,* 2022, 34, e00714.
18. Aldewachi, H.; Chalati, T.; Woodroofe, M. N.; Bricklebank, N.; Sharrack, B.; Gardiner, P.; Gold nanoparticle-based colorimetric biosensors. *Nanoscale,* 2018, 10, 18–33.
19. Li, Y.; Schluesener, H. J.; Xu, S.; Gold nanoparticle-based biosensors. *Gold Bull.,* 2010 43, 29–41.
20. Saha, K.; Agasti, S. S.; Kim, C.; Li, X.; Rotello, V. M.; Gold nanoparticles in chemical and biological sensing. *Chem. Rev.,* 2012, 112, 2739–2779.
21. Raether, H.; *Surface Plasmons on Smooth and Rough Surfaces and on Gratings*, Springer-Verlag, Berlin, 1988.
22. Hoa, X. D.; Kirk, A. G.; Tabrizian, M.; Towards integrated and sensitive surface plasmon resonance biosensors: a review of recent progress. *Biosens. Bioelectron.,* 2007, 23, 151–160.
23. Fahnestock, K.J.; Manesse, M.; McIlwee, H.A.; Schauer, C.L.; Boukherroub, R.; Szunerits, S.; Selective detection of hexachromium ions by localized surface plasmon resonance measurements using gold nanoparticles/chitosan composite interfaces. *Analyst,* 2009, 134, 881–886.
24. Hone, D.C.; Haines, A.H.; Russell, D.A.; Rapid, quantitative colorimetric detection of a lectin using mannose-stabilized gold nanoparticles. *Langmuir,* 2003, 19, 7141–7144.
25. Englebienne, P.; Hoonacker, A.V.; Verhas, M.; High-throughput screening using the surface plasmon resonance effect of colloidal. *Analyst,* 2001, 126, 1645–1651.
26. Zhu, S.L.; Zhang, J.B.; Yue, L.Y.L.; Hartono, D.; Liu, A.Q.; Label-free protein detection via gold nanoparticles and localized surface plasmon resonance. *Adv. Mater. Res.,* 2009, 74, 95–98.
27. Wang, J.; Nanomaterial-based amplified transduction of biomolecular interactions. *Small*, 2005, 1, 1036–1043.
28. Frederix, F.; Friedt, J.M.; Choi, K.H.; Laureyn, W.; Campitelli, A.; Mondelaers, D.; Maes, G.; Borghs, G.; Biosensing based on light absorption of nanoscaled gold and silver particles. *Anal. Chem.,* 2003, 75, 6894–6900.
29. Azzam, E. M. S.; Bashir, A.; Shekhah, O.; Alawady, A. R. E.; Birkner, A.; Grunwald, C.; Woll, C.; Fabrication of a surface plasmon resonance biosensor based on gold nanoparticles chemisorbed onto a 1,10-decanedithiol self-assembled monolayer. *Thin Solid Films,* 2009, 518, 387–391.
30. Lin, Y.B.; Zou, Y.; Mo, Y.Y.; Guo, J.P.; Lindquist, R.G.; E-beam patterned gold nanodot arrays on optical fiber tips for localized surface plasmon resonance biochemical sensing. *Sensors,* 2010, 10, 9397–9406.
31. Wang, J.; Wang, L. Y.; Sun, Y.; Zhu, X. N.; Cao, Y. B.; Wang, X. H.; Zhang, H. Q.; Song, D.; Surface plasmon resonance biosensor based on Au nanoparticle in titania sol–gel membrane. *Colloids Surf. B.,* 2010, 75, 520–525.
32. Briglin, S. M.; Gao, T.; Lewis, N. S.; Detection of organic mercaptan vapors using thin films of alkylamine-passivated gold nanocrystals. *Langmuir,* 2004, 20, 299–305.
33. Otsuka, H.; Akiyama, Y.; Nagasaki, Y.; Kataoka, K.; Quantitative and reversible lectin-induced association of gold nanoparticles modified with R-Lactosyl-ω-mercapto-poly (ethylene glycol). *J. Am. Chem. Soc.,* 2001, 123, 8226–8230.
34. Thanh, N. T. K.; Rosenzweig, Z.; Development of an aggregation-based immunoassay for anti-protein A using gold nanoparticles. *Anal. Chem.,* 2002, 74, 1624–1628.
35. Chah, S.; Hutter, E.; Roy, D.; Fendler, J. H.; Yi, J.; The effect of substrate metal on 2-aminoethanethiol and nanoparticle enhanced surface plasmon resonance imaging. *Chem. Phys.,* 2001, 272, 127–136.
36. Huang, J. C.; Chang, Y. F.; Chen, K. H.; Su, L. C.; Lee, C. W.; Chen, C. C.; Chen, Y. M. A.; Chou, C.; Detection of severe acute respiratory syndrome (SARS) coronavirus nucleocapsid protein in human serum using a localized surface plasmon coupled fluorescence fiber-optic biosensor. *Biosens. Bioelectron.,* 2009, 25, 320–325.
37. He, L.; Musick, M. D.; Nicewarner, S. R.; Salinas, F. G.; Benkovic, S. J.; Natan, M. J.; Keating, C. D.; Colloidal Au-enhanced surface plasmon resonance for ultrasensitive detection of DNA hybridization. *J. Am. Chem. Soc.,* 2000, 122, 9071–9077.

38. El-Sayed, I. H.; Huang, X.; El-Sayed, M. A.; Surface plasmon resonance scattering and absorption of anti-EGFR antibody conjugated gold nanoparticles in cancer diagnostics: applications in oral cancer. *NanoLett.*, 2005, 5, 829–834.
39. Aldewachi, H.; Chalati, T.; Woodroofe, M. N.; Bricklebank, N.; Sharrack, B.; Gardiner, P.; Gold nanoparticle-based colorimetric biosensors. *Nanoscale,* 2017, 10, 18–33.
40. Von Raben, K. U.; Chang, R. K.; Laube, B. L.; Surface enhanced raman scattering of $Au(CN)_2^-$ ions adsorbed on gold colloids. *Chem. Phys. Lett.*, 1981, 79, 465–469.
41. Rojanathanes, R.; Sereemaspun, A.; Pimpha, N.; Buasorn, V.; Ekawong, P.; Wiwanitkit, V.; Gold nanoparticle as an alternative tool for a urine pregnancy test. *Taiwan J. Obstet. Gynecol.*, 2008, 47, 296–299.
42. Ghosh, S. K.; Pal, T.; Interparticle coupling effect on the surface plasmon resonance of gold nanoparticles: from theory to applications. *Chem. Rev.*, 2007, 107, 4797–4862.
43. Joseph, V.; Matschulat, A.; Polte, J.; Rolf, S.; Emmerling, F.; Kneipp, J.; SERS enhancement of gold nanospheres of defined size. *J. Raman Spectrosc.*, 2011, 42, 1736–1742.
44. Link, S.; El-Sayed, M. A.; Spectral properties and relaxation dynamics of surface plasmon electronic oscillations in gold and silver nanodots and nanorods. *J. Phys. Chem. B,* 1999, 103, 8410–8426.
45. Dykman, N.; Khlebtsov, L.; Gold nanoparticles in biomedical applications: recent advances and perspectives. *Chem. Soc. Rev.*, 2012, 41, 2256–2282.
46. Sardar, R.; Funston, A. M.; Mulvaney, P.; Murray, R. W.; Gold nanoparticles: past, present and future. *Langmuir,* 2009, 25, 13840–13851.
47. Link, S.; El-sayed, M. A.; Shape and size dependence of radiative, non-radiative and photothermal properties of gold nanocrystals. *Int. Rev. Phys. Chem.*, 2000, 19, 409–453.
48. Link, S.; El-Sayed, M. A.; Size and temperature dependence of the plasmon absorption of colloidal gold nanoparticles. *J. Phys. Chem. B,* 1999, 103, 4212–4217.
49. Lim, S.; Koo, O. K.; You, Y. S.; Lee, Y. E.; Kim, M.; Chang, P.; Kang, D. H.; Yu, J.; Choi, Y. J.; Gunasekaran, S.; Enhancing nanoparticle-based visible detection by controlling the extent of aggregation. *Sci. Rep.*, 2012, 2, 456–461.
50. Aldewachi, H. S.; Woodroofe, N.; Turega, S.; Gardiner, P. H. E.; Optimization of gold nanoparticle-based real-time colorimetric assay of dipeptidyl peptidase IV activity. *Talanta,* 2017, 169, 13–19.
51. Reynolds, R. A.; Mirkin, C. A.; Letsinger, R. L.; Homogeneous, nanoparticle-based quantitative colorimetric detection of oligonucleotides. *J. Am. Chem. Soc.,* 2000, 122, 3795–3796.
52. Laromaine, A.; Koh, L.; Murugesan, M.; Ulijn, R. V.; Stevens, M. M.; Protease-triggered dispersion of nanoparticle assemblies. *J. Am. Chem. Soc.,* 2007 129, 4156–4157.
53. Hernandez-Santos, D.; Gonzalez-Garcỉa, M. B.; Garcỉa, A. C.; Metal-nanoparticles based electroanalysis. *Electroanalysis,* 2002, 14, 1225–1235.
54. Pingarron, J. M.; Yanez-Sedeno, P.; Gonzalez-Cortes, A.; Gold nanoparticle-based electrochemical biosensors. *Electrochim. Acta,* 2008, 53, 5848–5866.
55. Liu, S.; Leech, D.; Ju, H.; Application of colloidal gold in protein immobilization, electron transfer and biosensing. *Anal. Lett.*, 2003, 36, 1–19.
56. Yanez-Sedeno, P.; Pingarron, J. M.; Gold nanoparticle-based electrochemical biosensors. *Anal. BioanalChem.*, 2005, 382, 884–886.
57. Hutter, E.; Maysinger, D.; Gold-nanoparticle-based biosensors for detection of enzyme activity. *Trends in Pharmacol. Sci.*, 2013, 34, 497–507.
58. Sanz, V. C.; Mena, M. L.; Gonzalez-Cortes, A.; Yanez-Sedeno, P.; Pingarron, J. M.; Development of a tyrosinase biosensor based on gold nanoparticles-modified glassy carbon electrodes: application to the measurement of a bioelectrochemical polyphenols index in wines. *Anal. Chim. Acta,* 2005, 528, 1–8.
59. Agui, L.; Manso, J.; Yanez-Sedeno, P.; Pingarron, J. M.; Amperometric biosensor for hypoxanthine based on immobilized xanthine oxidase on nanocrystal gold-carbon paste electrodes. *Sens. Actuators B Chem.*, 2006, 113, 272–280.
60. Liu, S. Q.; Ju, H. X.; Renewable reagentless hydrogen peroxide sensor based on direct electron transfer of horseradish peroxidase immobilized on colloidal gold-modified electrode. *Anal. Biochem.*, 2002, 307, 110–116.
61. Liu, S.; Yu, J.; Ju, H.; Renewable phenol biosensor based on a tyrosinase-colloidal gold modified carbon paste electrode. *J. Electroanal. Chem.*, 2003, 540, 61–67.
62. Liu, S. Q.; Ju, H. X.; Reagentless glucose biosensor based on direct electron transfer of glucose oxidase immobilized on colloidal gold modified carbon paste electrode. *Biosens. Bioelectron.*, 2003, 19, 177–183.
63. Feng, H.; Wang, H.; Zhang, Y.; Yan, B.; Shen, G.; Yu, R.; A direct electrochemical biosensing platform constructed by incorporating carbon nanotubes and gold nanoparticles onto redox poly(thionine) film. *Anal. Sci.*, 2007, 23, 235–239.
64. Yan, W.; Feng, X. M.; Chen, X. J.; Hou, W. H.; Zhu, J. J.; A super highly sensitive glucose biosensor based on Au nanoparticles-AgCl@polyaniline hybrid material. *Biosens. Bioelectron.*, 2008, 23, 925–931.
65. Zhu, Q.; Yuan, R.; Chai, Y. Q.; Zhuo, Y.; Zhang, Y.; Li, X.; Wang, N.; A novel amperometric biosensor for determination of hydrogen peroxide based on nafion and polythionine as well as gold nanoparticles and gelatin as matrixes. *Anal. Lett.*, 2006, 39, 483–494.
66. Du, D.; Ding, J.; Cai, J.; A. Zhang, One-step electrochemically deposited interface of chitosan–gold nanoparticles for acetylcholinesterase biosensor design. *J. Electroanal. Chem.*, 2007, 605, 53–60.
67. Ho, J. A.; Chang, H.; Shih, N.; Wu, L.; Chang, Y.; Chen, C.; Chou, C.; Diagnostic detection of human lung cancer-associated antigen using a gold nanoparticle-based electrochemical immunosensor. *Anal. Chem.*, 2010, 82, 5944–5950.
68. Zhuo, Y.; Yuan, R.; Chai, Y.; Zhang, Y.; Wang, N.; Li, X.; Zhu, Q.; Wang, N.; An amperometricimmunosensor based on immobilization of hepatitis B surface antibody on gold electrode modified gold nanoparticles and horseradish peroxidase. *Anal. Chim. Acta,* 2005, 548, 205–210.
69. Tang, D.; Ren, J. J.; Direct and rapid detection of diphtherotoxin via potentiometric immunosensor based on nanoparticles mixture and polyvinyl butyral as matrixes. *Electroanalysis,* 2005, 17, 220822–16.
70. Wu, Z. S.; Li, J. S.; Luo, M. H.; Shen, G. L.; Yu, R. Q.; A novel capacitive immunosensor based on gold colloid monolayers associated with a sol-gel matrix. *Anal. Chim. Acta,* 2005, 528, 235–242.
71. Drummond, T. G.; Hill, M. G.; Barton, J. K.; Electrochemical DNA sensors. *Nat. Biotechnol.*, 2003, 21, 1192–1199.
72. Rasheed, P. A.; Sandhyarani, N.; Electrochemical DNA sensors based on the use of gold nanoparticles: a review on recent developments. *MicrochimActa,* 2017, 184, 981–1000.
73. Wang, Z.; Zhang, J.; Zhu, C.; Wu, S.; Mandler, D.; Marks, R. S.; Zhang, H.; Amplified detection of femtomolar DNA based on a one-to-few recognition reaction between DNA-Au conjugate and target DNA. *Nanoscale*, 2014, 6, 3110–3115.

74. Castaneda, M. T.; Alegret, S.; Merkoai, A.; Electrochemical sensing of DNA using gold nanoparticles. *Electroanal,* 2007, 19(7), 743–753.
75. Dizaj, S. M.; Lotfipour, F.; Barzegar-Jalali, M.; Zarrintan, M. H.; Adibkia, K.; Antimicrobial activity of the metals and metal oxide nanoparticles. *Mater. Sci. Eng. C.,* 2014, 44, 278–284.
76. Seil, J. T.; Webster, T. J.; Antimicrobial applications of nanotechnology: methods and literature. *Int. J. Nanomedicine,* 2012, 7, 2767–2781.
77. Rai, V. R.; Bai, A. J.; Nanoparticles and their potential application as antimicrobials. *Science Against Microbial Pathogens: Communicating Current Research and Technological Advances,* 2011, 197–209.
78. Besinis, A.; De Peralta, T.; Handy, R. D.; The antibacterial effects of silver, titanium dioxide and silica dioxide nanoparticles compared to the dental disinfectant chlorhexidine on streptococcus mutans using a suite of bioassays. *Nanotoxicology,* 2014, 8, 1–16.
79. Kong, H.; Jang, J.; Antibacterial properties of novel poly (methyl methacrylate) nanofiber containing silver nanoparticles. *Langmuir,* 2008, 24, 2051–2056.
80. Dakal, T. C.; Kumar, A.; Majumdar, R. S.; Yadav, V.; Mechanistic basis of antimicrobial actions of silver nanoparticles. *Front. Microbiol.,* 2016, 7, 1831.
81. Ouay, B. L.; Stellacci, F.; Antibacterial activity of silver nanoparticles: a surface science insight. *Nanotoday,* 2015, 10, 339–354.
82. Tang, S.; Zheng, J.; Antibacterial activity of silver nanoparticles: structural effects. *Adv. Healthcare Mater.,* 2018, 17(3), 1701503.
83. Ramalingam, B.; Parandhaman, T.; Das, S. K.; Antibacterial effects of biosynthesized silver nanoparticles on surface ultrastructure and nanomechanical properties of gram-negative bacteria viz. *Escherichia coli* and *Pseudomonas aeruginosa. ACS Appl. Mater. Interfaces,* 2016, 8, 4963–4976.
84. Ahearn, D. G.; May, L. L.; Gabriel, M. M., Adherence of organisms to silver-coated surfaces. *J. Ind. Microbiol.,* 1995, 15, 372–376.
85. Mijakovic, I.; Petranovic, D.; Macek, B.; Cepo, T.; Mann, M.; Davies, J.; Jensen, P. R.; Vujaklija, D.; Bacterial single-stranded DNA-binding proteins are phosphorylated on tyrosine. *Nucleic Acids Res.,* 2006, 34, 1588.
86. Li, W.; Xie, X.; Shi, Q.; Duan, S.; Ouyang, Y.; Chen, Y.; Antibacterial effect of silver nanoparticles on Staphylococcus aureus. *Biometals,* 2011, 24, 135–141.
87. Bragg, P. D.; Rainnie, D. J.; The effect of silver ions on the respiratory chain of *Escherichia coli. Can. J. Microbiol.,* 1974, 20, 883.
88. Klueh, U.; Wagner, V.; Kelly, S.; Johnson, A.; Bryers, J. D.; Efficacy of silver-coated fabric to prevent bacterial colonization and subsequent device-based biofilm formation. *J. Biomed. Mater. Res.,* 2000, 53, 621.
89. Yamanaka, M.; Hara, K.; Kudo, J.; Bactericidal actions of a silver ion solution on *Escherichia coli*, studied by energy-filtering transmission electron microscopy and proteomic analysis. *Appl. Environ. Microbiol.,* 2005, 71, 7589.
90. Jung, W. K.; Koo, H. C.; Kim, K. W.; Shin, S.; Kim, S. H.; Park, Y. H.; Antibacterial activity and mechanism of action of the silver ion in *Staphylococcus aureus* and *Escherichia coli. Appl. Environ. Microbiol.,* 2008, 74, 2171.
91. Fenoglio, I.; Corazzari, I.; Francia, C.; Bodoardo, S.; Fubini, B.; The oxidation of glutathione by cobalt/tungsten carbide contributes to hard metal-induced oxidative stress. *Free Radical Res.,* 2008, 42, 437.
92. You, C.; Han, C.; Wang, X.; Zheng, Y.; Li, Q.; Hu, X.; Sun, H.; The progress of silver nanoparticles in the antibacterial mechanism, clinical application and cytotoxicity. *MolBiol Rep,* 2012, 39, 9193–9201.
93. Ahamed, M.; Posgai, R.; Gorey, T. J.; Nielsen, M.; Hussain, S. M.; Rowe, J. J.; Silver nanoparticles induced heat shock protein 70, oxidative stress and apoptosis in Drosophila melanogaster. *ToxicolApplPharmacol,* 2010, 242, 263–269.
94. Boonstra, J.; Post, J. A.; Molecular events associated with reactive oxygen species and cell cycle progression in mammalian cells. *Gene,* 2004, 337, 1–13.
95. Lok, C.; Ho, C.; Chen, R.; He, Q.; Yu, W.; Sun, H.; Tam, P. K.; Chiu, J.; Che, C.; Silver nanoparticles: partial oxidation and antibacterial activities. *J. Biol. Inorg. Chem.,* 2007, 12, 527–534.
96. Goyal, D.; Kaur, G.; Tewari, R.; Kumar, R.; Correlation of edge truncation with antibacterial activity of plate-like anisotropic silver nanoparticles. *Environ. Sci. Pollut. Res.,* 2017, 24, 20429–20437.
97. Cheon, J. Y.; Kim, S. J.; Rhee, Y. H.; Kwon, O. H.; Park, W. H.; Shape-dependent antimicrobial activities of silver nanoparticles. *Int. J. Nanomed.,* 2019, 14, 2773–2780.
98. Menichetti, A.; Mavridi-Printezi, A.; Mordini, D.; Montalti, M.; Effect of size, shape and surface functionalization on the antibacterial activity of silver nanoparticles. *J. Funct. Biomater.,* 2023, 14, 244.
99. Yang, X.; Gondikas, A. P.; Marinakos, S. M.; Auffan, M.; Liu, J.; Hsu-Kim, H.; Meyer, J. N.; Mechanism of silver nanoparticle toxicity is dependent on dissolved silver and surface coating in caenorhabditiselegans. *Environ. Sci. Technol.,* 2012, 46, 1119–1127.
100. Abbaszadegan, A.; Ghahramani, Y.; Gholami, A.; Hemmateenejad, B.; Dorostkar, S.; Nabavizadeh, M.; Sharghi, H.; The effect of charge at the surface of silver nanoparticles on antimicrobial activity against gram-positive and gram-negative bacteria: a preliminary study. *J. Nanomater.,* 2015, 16, 53.
101. Pareek, V.; Gupta, R.; Panwar, J.; Do physico-chemical properties of silver nanoparticles decide their interaction with biological media and bactericidal action? A review. *Mater. Sci. Eng. C,* 2018, 90, 739–749.
102. Kamat, P. V.; Photochemistry on nonreactive and reactive (semiconductor) surfaces. *Chem. Rev.,* 1993, 93, 267–300.
103. Chavali, M. S.; Nikolova, M. P.; Metal oxide nanoparticles and their applications in nanotechnology. *SN Appl. Sci.,* 2019, 1, 607.
104. Sánchez-Moreno, P.; Ortega-Vinuesa, J. L.; Peula-García, J. M., Marchal, J. A., Boulaiz, H.; Smart drug-delivery systems for cancer nanotherapy. *Curr. Drug Targets,* 2018, 19, 339–359.
105. Zikalala, N.; Matshetshe, K.; Parani, S.; Oluwafemi, O. S.; Biosynthesis protocols for colloidal metal oxide nanoparticles. *Nano-Struct. Nano-Objects,* 2018, 16, 288–299.
106. Ansari, A. A.; Solanki, P. R.; Malhotra, B. D.; Sol-gel derived nanostructured cerium oxide film for glucose sensor. *Appl. Phys. Lett.,* 2008, 92, 263901.
107. Solanki, P. R.; Kaushik, A.; Agrawal, V. V.; Malhotra, B. D.; Nanostructured metal oxide-based biosensors. *NPG Asia Mater.,* 2011, 3, 17–24.
108. Mehta, S. K.; Singh, K.; Umar, A.; Chaudhary, G. R.; Singh, S.; Ultra-high sensitive hydrazine chemical sensor based on low-temperature grown ZnO nanoparticles. *Electrochim Acta,* 2012, 69, 128–133.
109. Degen, A.; Kosec, M.; Effect of pH and impurities on the surface charge of zinc oxide in aqueous solution. *J Eur. Ceram. Soc.,* 2000, 20, 667–673.

110. Rahman, M. M.; Umar, A.; Sawada, K.; Ultra-sensitive cholesterol biosensor based on low-temperature grown ZnO nanoparticles. *Electrochem. Commun.*, 2009, 11, 118–121.
111. Somers, R. C.; Bawendi, M. G.; Nocera, D. G.; CdSe nanocrystal based chem-/bio-sensors. *Chem. Soc. Rev.*, 2007, 36, 579–591.
112. Gizeli, E.; Design considerations for the acoustic waveguide biosensor. *Smart Mater. Struct.*, 1997, 6, 700–706.
113. Gu, B.; Xu, C.; Yang, C.; Liu, S.; Wang, M.; ZnO quantum dot labeled immunosensor for carbohydrate antigen 19–9. *Biosens. Bioelectron.*, 2011, 26, 2720–2723.
114. Liu, J.; Guo, C.; Li, C. M.; Li, Y.; Chi, Q.; Huang, X.; Liao, L.; Yu, T.; Carbon-decorated ZnO nanowire array: a novel platform for direct electrochemistry of enzymes and biosensing applications. *Electrochem. Comm.*, 2009, 11, 202–205.
115. Napi, M. L. M.; Sultan, S. M.; Ismail, R.; How, K. W.; Ahmad, M. K.; Electrochemical-based biosensors on different zinc oxide nanostructures: a review. *Materials*, 2019, 12, 2985.
116. Vabbina, P. K.; Kaushik, A.; Pokhrel, N.; Bhansali, S.; Pala, N.; Electrochemical cortisol immunosensors based on sonochemically synthesized zinc oxide 1D nanorods and 2D nanoflakes. *Biosens. Bioelectron.*, 2015, 63, 124–130.
117. Alenezi, M. R.; Alzanki, T. H.; Almeshal, A. M.; Alshammari, A. S.; Beliatis, M. J.; Henley, S. J.; Silva, S. R. P.; Hierarchically designed ZnO nanostructure based high performance gas sensors. *RSC Adv.*, 2014, 4, 49521.
118. Bayan, S.; Chakraborty, P.; Secondary ion mass spectrometry and photoluminescence study on microstructural characteristics of chemicallysynthesized ZnOnanowalls. *Appl. Surf. Sci.*, 2014, 303, 233–240.
119. Tereshchenko, A.; Bechelany, M.; Viter, R.; Khranovskyy, V.; Smyntyna, V.; Starodub, N.; Yakimova, R.; Optical biosensors based on ZnO nanostructures: advantages andperspectives. A review. *Sens. Actuators B: Chem.*, 2016, 229, 664–677.
120. Kaur, G.; Paliwal, A.; Tomar, M.; Gupta, V.; Detection of Neisseria meningitidis using surface plasmon resonance-based DNA biosensor. *Biosens. Bioelectron.*, 2016, 78, 106–110.
121. Picciolini, S.; Castagnetti, N.; Vanna, R.; Mehn, D.; Bedoni, M.; Gramatica, F.; Villani, M.; Calestani, D.; Pavesi, M.; Lazzarini, L.; Zappettini, A.; Morasso, C.; Branched gold nanoparticles on ZnO 3D architecture as biomedical SERS sensors. *RSC Adv.*, 2015, 5, 93644–93651.
122. Sodzel, D., Khranovskyy, V.; Beni, V.; Turner, A. P. F.; Viter, R.; Eriksson, M. O.; Holtz, P.; Janot, J.; Bechelany, M.; Balme, S.; Smyntyna, V.; Kolesneva, E.; Dubovskaya, L.; Volotovski, I.; Ubelis, A.; Yakimova, R.; Continuous sensing of hydrogen peroxide and glucose via quenching of the UV and visible luminescence of ZnO nanoparticles. *MicrochimActa*, 2015, 182, 1819–1826.
123. Raghunath, A.; Perumal, E.; Metal oxide nanoparticles as antimicrobial agents: a promise for the future. *Int. J. Antimicrob. Agents*, 2017, 49, 137–152.
124. Gold, K.; Slay, B.; Knackstedt, M.; Gaharwar, A. K.; Antimicrobial activity of metal and metal-oxide based nanoparticles. *Adv. Therap.*, 2018, 1700033.
125. Sadiq, I. M.; Chandrasekaran, N.; Mukherjee, A.; Studies of effect of $TiO_2$ nanoparticles on growth and membrane permeability of *Escherichia coli*, *Pseudomonas aeruginosa* and *Bacillus subtilis*. *Curr. Nanosci.*, 2010, 6, 381–387.
126. Sawai, J.; Himizu, K.; Yamamoto, O.; Kinetics of bacterial death by heated dolomite powder slurry. *Soil Biol. Biochem.*, 2005, 37, 1484–1489.
127. Hou, J.; Wang, L.; Wang, C.; Zhang, S.; Liu, H.; Li, S.; Wang, X.; Toxicity and mechanisms of action of titanium dioxide nanoparticles in living organisms. *J. Environ. Sci.*, 2019, 75, 40–53.
128. Kong, H.; Song, J.; Jang, J.; Photocatalytic antibacterial capabilities of $TiO_2$-biocidal polymer nanocomposites synthesized by a surface-initiated photopolymerization. *Environ. Sci. Technol.*, 2010, 44, 5672–5676.
129. Tsuang, Y.; Sun, J.; Huang, Y.; Lu, C.; Chang, W. H.; Wang, C.; Studies of photokilling of bacteria using titanium dioxide nanoparticles. *Artif. Organs*, 2008, 32, 167–174.
130. Matsunaga, T.; Sterilization with particulate photosemiconductor. *J Antibact. Antifung. Agents*, 1985, 13, 211–220.
131. Peng, T.; Zhao, D.; Dai, K.; Shi, W.; Hirao, K.; Synthesis of titanium dioxide nanoparticles with mesoporous anatase wall and high photocatalytic activity. *J. Phys. Chem. B*, 2005, 109, 4947–4952.
132. Ohko, Y.; Utsumi, Y.; Niwa, C.; Tatsuma, T.; Kobayakawa, K.; Satoh, Y.; Y. Kubota, Fujishima, A.; Self-sterilizing and self-cleaning of silicone catheters coated with $TiO_2$ photocatalyst thin films: a preclinical work. *J. Biomed. Mater. Res.*, 2001, 58, 97–101.
133. Yeung, K. L.; Leung, W. K.; Yao, N.; Cao, S.; Reactivity and antimicrobial properties of nanostructured titanium dioxide. *Catal. Today*, 2009, 143, 218–224.
134. Stengl, V.; Houskova, V.; Bakardjieva, S.; Murafa, N.; Photocatalytic activity of boron-modified titania under UV and visible-light illumination. *ACS Appl. Mater. Interfaces*, 2010, 2, 575–580.
135. Akhavan, O.; Lasting antibacterial activities of Ag-TiO2/Ag/a-TiO2 nanocomposite thin film photocatalysts under solar light irradiation. *J. Colloid Interface Sci.*, 2009, 336, 117–124.

## LIST OF ABBREVIATIONS

| | |
|---|---|
| **0D** | Zero dimensional |
| **1D** | One dimensional |
| **2D** | Two dimensional |
| **3D** | Three dimensional |
| **BSA** | Bovine serum albumin |
| **CNT** | Carbon nanotube |
| **DNA** | Deoxyribonucleic acid |
| **DPP4** | Dipeptidyl peptidase 4 |
| **GCE** | Glassy carbon electrode |
| **GOx** | Glucose oxidase |
| **GPDC** | Gly-Pro-Asp-Cys |
| **$H_2O2$** | Hydrogen peroxide |
| **HRP** | Horseradish peroxidase |
| **ITO** | Indium tin oxide |
| **PtE** | Platinum electrode |
| **QDs** | Quantum dots |
| **TNT** | Trinitrotoluene |
| **Tyr** | Tyrosinase |
| **VP-EN-DC** | Val-Pro-NH-$(CH2)_2$-NH-Asp-Cys |

# 11 LPSR-Based Biodetection of Metal and Metal Oxide Nanoparticles

*Beena Mol*
Department of Physics, Mar Thoma College, Chungathara, Kerala, India

## 11.1 INTRODUCTION

Surface plasmon (SP) resonance is the most remarkable optical characteristic of metallic nanostructures. It consists of a collective oscillation of conduction electrons excited by the suitable electromagnetic field of light. SP corresponds to an interaction of the electromagnetic field of light with matter (Garcia M. A., 2011; Loiseau A. et al., 2019). When a metallic particle's size is reduced to the nanoscale range, its optical properties are significantly modified by the appearance of SP, and its behaviour differs completely from that of the bulk metal. Bulk gold, for instance, is yellow in colour. However, when gold particles are small enough, their colour becomes ruby red, due to their significant absorption of green light at around 520 nm, which corresponds to the frequency at which gold has a plasmon resonance (Loiseau A. et al., 2019). Similarly in the nano range, silver also appears yellow in colour (Mol B. et al., 2016). Noble metal particles have a distinctive high absorption band in the visible spectrum. This is a small particle effect, as the individual atom and the bulk will not contain any of these characteristics (Mayer K. M. et al., 2008). The coherent oscillation of conduction band electrons caused by the interacting electromagnetic field is the physical source of the light absorption by metal nanoparticles (Garcia M. A., 2011).

Localized surface plasmons resonance (LSPR) is an optical phenomenon that arises due to the collective oscillation of the valence electrons and subsequent absorption within the ultraviolet-visible (UV-Vis) band when an electromagnetic radiation interacts with the conduction band of a noble metal nanostructures (Agrawal A. et al., 2018; Hammond J. L. et al., 2014). When a very small, subwavelength conductive nanoparticle scatters in an oscillating electromagnetic field, these modes naturally occur. The propelled electrons are subject to an effective restoring force from the particle's curved surface, and resonance can occur, which induces field amplification within and outside the particle in the near-field zone. This resonance is known as localized plasmon resonance or localized surface plasmon resonance (Figure 11.1) (Mayer K. M. and Hafner J. H., 2011; Hammond J. L. et al., 2014).

Both LSPR and surface plasmon resonance (SPR) waves decay exponentially, but they differ from each other in their propagation and decay length. Surface plasmon polaritons (SPPs) are propagating, dispersive electromagnetic waves coupled to the electron plasma of a conductor at a dielectric interface, while localized surface plasmons are *non-propagating* excitations of the conduction electrons of metallic nanostructures coupled to the electromagnetic field. The decay length of the electromagnetic field observed in the surface plasmon is on the order of 200 nm, whereas the decay length of the electromagnetic field in localized surface plasmons is on the order of 6 nm (Hammond J. L. et al., 2014). The reduced field-decay length of LSPR minimizes sensitivity to refractive-index variations in the bulk solution, while enhancing sensitivity to refractive-index changes at the surface. This property of LPSR can be exploited as the foundation for biosensing applications (Hammond J. L. et al., 2014).

SPR is one of many examples of an intriguing physical phenomenon leading to applications that are important to both science and society. SP allows us to amplify, focus, and alter light at the nanoscale, breaking the classical optics diffraction limit and enhancing the resolution and sensitivity of optical probes (Garcia M. A., 2011). The discovery of the phenomena of localized surface plasmon resonance has led to the basis for many research areas, from materials science to biosensing. SP has wide range of applications in various fields, including biomedicine (Du Y. et al., 2022; Sasidevi S. et al., 2024), energy, environmental protection (Aljbar N. A. et al., 2024; Long F et al., 2013), sensing (Loiseau A. et al., 2019), and information technology (Sopan N. et al., 2024; Chen T. et al., 2023). LSPR has provided a transduction platform for the production of affordable and portable devices for a wide range of applications (Hammondet J. L. et al., 2014). Nowadays, there are well-established uses of SP, which are rapidly expanding as our ability to create and manipulate nanoparticle advances. From a technological perspective, we will be able to fully utilize the potential of SP phenomena to customise the properties of nanoparticles. This chapter discusses localized surface plasmon resonance phenomena and the potential applications of LSPR-based biodetection of metal and metal oxide nanoparticles.


DOI: 10.1201/9781032660639-11

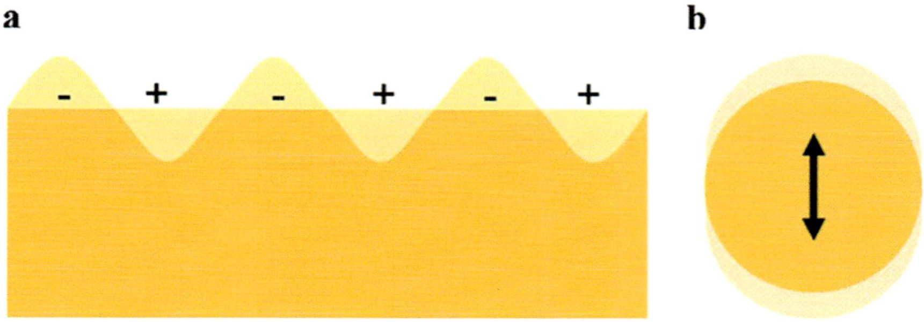


FIGURE 11.1 Schematic representation of the formation (a) surface plasmons and (b) localized surface plasmons on the surface of a metallic nanoparticle.

Source: Adapted from Mayer K. M. and Hafner J. H., 2011, Copyright ©2011, American Chemical Society. All rights reserved.

## 11.2 LIGHT MATTER INTERACTION IN NANOMATERIALS

LSPR can occur in noble metallic structures due to the restriction of the movement of electrons through their internal lattice when their size is reduced to the nano level (<100 nm). This phenomena can be explained using Mie's solution to Maxwell's equations (Hammond J. L. et al., 2014). Light matter interaction of an arbitrary-shaped particle and its surrounding medium with the dielectric properties $\varepsilon_p$ and $\varepsilon_m$ is governed by the fundamental laws of electromagnetism (Hu M. et al., 2006; Giannini V. et al., 2011). The incoming oscillatory electromagnetic waves set the oscillatory charges in motion inside a material and result in the reradiation of light in all directions. This reradiation of light is referred to as scattering, and the associated thermal losses during interaction are referred to as absorption of the material. Knowing the size and shape of the particles in a given electromagnetic field, the resultant electric and magnetic fields can be obtained by solving the macroscopic Maxwell equations (Garcia M. A., 2011). Thus, optical properties such as absorption, extinction, and scattering of the particles can be evaluated using simplified Maxwell equations (Equation 11.1), assuming the material's magnetic permeability is unity inside the closed system (Agrawal A. et al., 2018).

$$\partial\nabla \times \nabla \times \mathrm{E} = -\mu_0 \frac{\partial D}{\partial t} \tag{11.1}$$

where $E$ is the electric field, $D$ is the displacement field and $\mu_0$ is the magnetic permeability in free space.

When light interacts with matter, the electric field around the materials has two components; one is the incident field ($\overrightarrow{E_{in}}$), and the other is the field produced due to the scattered radiation ($\overrightarrow{E_{sca}}$). The light extinction can be calculated as the sum of $\overrightarrow{E_{in}}$ and $\overrightarrow{E_{Sca}}$, and it represents the total electric field outflowing from the material surface (Figure 11.2) (Giannini V. et al., 2011; Agrawal A. et al., 2018). The absorption of light by particles is determined by the difference between extinction and scattering losses, whereas the power of the scattered light or the energy outflow can be determined by $\overrightarrow{E_{sca}}$ (Garcia M. A., 2011).

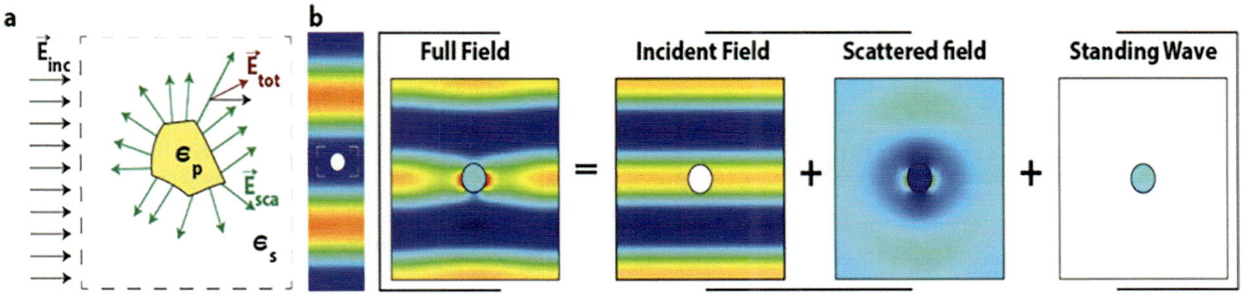


FIGURE 11.2 (a) Schematic diagram illustrating the interaction of light with matter. Light interacts with an object of a specific shape, size, and dielectric function, producing a net electric field equal to the sum of scattered and incident fields. (b) The total, incident, dispersed, and standing electric fields for a semiconductor nanocrystal. Each figure is normalized, and the colour scale spans +1 (blue) to –1 (red).

Source: Adapted from Agrawal A. et al., 2018.Copyright © 2018, American Chemical Society. Adapted from Giannini V. et al., 2011, Copyright © 2011, American Chemical Society.

### 11.2.1 Mie Theory

The electric field inside and outside a spherical particle can be solved using Mie theory and is given in Equations 11.2 and 11.3 (Agrawal A. et al., 2018).

$$E_{in}(x,y,z)=\left[\frac{3\varepsilon_p}{\varepsilon_p+3\varepsilon_m}\right]E_0\hat{z} \quad (11.2)$$

$$E_{out}(x,y,z)=E_0\hat{z}-\left[\frac{\varepsilon_p-\varepsilon_m}{\varepsilon_p+2\varepsilon_m}\right]a^3$$

$$E_0\left[\frac{\hat{z}}{r^3}-\frac{3z}{r^5}(x\hat{x}+y\hat{y}+z\hat{z})\right] \quad (11.3)$$

where $E_{in}$ and $E_{out}$ are the net electric fields inside and outside the particle, respectively; $E_0$ is the magnitude of incident electric field ($\overline{E_{in}}$); $x$, $y$, and $z$ are the spatial coordinates with respect to the centre of the particle; $\hat{x},\hat{y},\hat{z}$ are the unit vectors along the $x$, $y$, and $z$ directions, respectively; and $r$ is the distance from the centre of the particle.

Mie theory is applicable only to spherical particles, and for all other geometries Maxwell's equations cannot be solved numerically. To calculate the electric field inside and outside the particle, for all other geometries, several computational methods, such as discrete dipole approximation (Draine B. T. and Flatau P. J., 1994), finite difference time domain (Ahmed I. et al., 2011.), the boundary value problem (Waxenegger J. et al., 2015), and the finite-element method (Burger S. et al., 2010; Khoury C. G. et al., 2009), are used. Using Mie theory approximations, in the case of spherical particles whose size is much smaller than the wavelength of the incident light (quasistatic approximation), the absorption and scattering cross-sections can be found using Equations 11.4 and 11.5 (Agrawal A. et al., 2018).

$$C_{ext}(\omega)=4\pi R^2\left[k(\varepsilon_H)^{1/2}R\right]Im\left\{\frac{\varepsilon_p(\omega)-\varepsilon_H}{\varepsilon_p(\omega)+2\varepsilon_H}\right\} \quad (11.4)$$

$$C_{sca}(\omega)=4\pi R^2\left[k(\varepsilon_H)^{1/2}R\right]^4\left|\frac{\varepsilon_p(\omega)-\varepsilon_H}{\varepsilon_p(\omega)+2\varepsilon_H}\right|^2 \quad (11.5)$$

where $k$ is the wave vector of the incident light, $R$ is the radius of the nanoparticle, and $\varepsilon_H$ is the dielectric constant of the surrounding environment of the nanoparticle.

The expressions of cross-sections (Equations 11.4 and 11.5) also reveal that while the extinction of small materials or nanoparticles can lead primarily to the absorption of light above a diameter threshold of approximately $d = \lambda_{LSPR}/20$, nanoparticles will begin to scatter light with increasing efficiency (Agrawal A. et al., 2018). From Equations 11.4 and 11.5, it is clear that the dielectric function ($\varepsilon_p$) directly affects the scattering cross-sections. The form of the dielectric function, whether it is real with a positive component or complex with a negative real component, decides the type of optical interaction. Optical interactions will consist of only pure scattering if the dielectric function is a positive quantity. In this case, there will be no absorption. Optical interactions will include Rayleigh scattering if the dielectric function is complex with a positive real component.

When the dielectric function is complex with a negative real component, then from the extinction cross-section expression given in Equation 11.4, it is clear that, if the real part of the dielectric function goes negative, or when $\varepsilon_p(\omega) = -2\varepsilon_H$, extinction will be maximized and optical resonance will occur. This resonant optical phenomenon is the physical basis of LSPR. This type of behaviour, one that is complex but has a negative real component of the dielectric function, is exhibited by conductive materials such as Ag, Au, or degenerately doped semiconductors with high free carrier density (Agrawal A. et al., 2018).

The LSPR response of nanomaterial depends on its free carrier density, size, and shape and the surrounding dielectric environment. Depending on these parameters, when an oscillating electromagnetic field interacts with a nanomaterial, there can arise a coherent collective oscillation of free carriers inside the material, which leads to the phenomenon of LSPR (Agrawal A. et al., 2018; Garcia M. A., 2011).

LSPR quality factor is defined as $$Q_{LPSR}=\frac{\omega_{LPSR}}{fwhm_{LPSR}} \quad (11.6)$$

where $\omega_{LPSR}$ is the LSPR frequency. It is the central resonance frequency which depends primarily on the free carrier concentration and the carrier effective mass. The FWHM (full width at half maximum) of LSPR represents the resonance linewidth, which can be determined from the scattering or damping effects associated with the motion of free carriers. At the LSPR frequency ($\omega_{LPSR}$), both the extinction cross-section and electric field around the nanomaterial (Equation 11.3) are maximized, and this concentration of electric field at the surface of the nanomaterial is termed the near field enhancement (NFE). A high value of NFE is usually responsible for a high LSPR quality factor (Agrawal A. et al., 2018).

### 11.2.2 Gans Theory

Mie theory as formulated is strictly applicable only to spherical particles. For spheroidal particles of any aspect ratio in the small particle approximation, Richard Gans generalized Mie theory and obtained an expression for the absorption cross-section for a prolate spheroid (Mayer K. M. and Hafner J. H., 2011; Gans R., 1912).

$$\sigma_{abs}=\frac{\omega}{3c}\varepsilon_m^{3/2}V\sum_j\frac{\left(\frac{1}{P_j^2}\right)\varepsilon_2}{\left\{\varepsilon_1+\left[\left(1-P_j\right)/P_j\right]\varepsilon_m\right\}^2+\varepsilon_2^2} \quad (11.7)(A)$$

Considering the three dimensions of the particle, here we take the sum over $j$. For a prolate spheroid, $A > B = C$. $P_j$ includes the depolarization factors $P_A$, $P_B$, and $P_C$ along the three axes of the particle.

The depolarization factors anisotropically modify the values of $\varepsilon_1$ and $\varepsilon_2$ and bring about the LSPR peak frequencies. $P_A$, $P_B$, and $P_C$ can be obtained as follows.

$$P_A = \frac{1-e^2}{e^2}\left[\frac{1}{2e}\ln\left(\frac{1+e}{1-e}\right)-1\right] \tag{11.8}$$

$$P_B = P_C = \frac{1-P_A}{2} \tag{11.9}$$

where $e$ is a factor that includes the aspect ratio $R$ of the particle and can be obtained as follows:

$$e = \left[1-\left(\frac{B}{A}\right)^2\right]^{1/2} = \left(1-\frac{1}{R^2}\right)^{1/2} \tag{11.10}$$

From these relations, it is clear that Equation 11.7 provides a clear understanding of the effect of aspect ratio on the LSPR peak wavelength of the nanoparticle (Link S. et al., 1999; Link, S. et al., 2005). For nanoparticles other than these spheres and spheroids, particle shape plays a significant role in determining the LSPR spectrum and cannot be found analytically as previously. Such particles can be studied by using numerical methods for plasmonic nanoparticles, including discrete dipole approximation (DDA), finite difference time domain (FDTD), and the finite-element method (FEM) (Mayer K. M. and Hafner J. H., 2011; Khoury C. G. et al., 2009).

### 11.2.3 Sensing Modes of LPSR

LSPR spectroscopy offers sensing through transduction of refractive index changes in close proximity to the surface of the noble metal nanostructure. The formation of adlayers and biorecognition events on the surface causes quantifiable shifts in the LSPR extinction wavelength maximum ($\lambda_{max}$), due to the dependence of the adlayer refractive index ($n$) (Hammond J. L. et al., 2014). Two main methods can generally be adopted for LPSR-based sensing. One is wavelength-shift sensing, and the other is surface-enhanced Raman spectroscopic techniques (Loiseau A. et al., 2019). In the first method, UV-Vis spectroscopy can be used to monitor the change in wavelengths of light, which occurs due to the collective oscillation of conduction band electrons on the surface. Here, the change in wavelength of the extinction curves can be monitored as a function of changes in the adlayer refractive index caused by target analyte adsorption (Hammond J. L. et al., 2014).

The surface-enhanced Raman spectroscopic techniques, such as surface-enhanced Raman scattering (SERS) and surface-enhanced Raman resonance scattering (SERRS), employ the formation of enhanced electromagnetic fields that spread out from the nanostructure surface as the basis of LPSR-based sensing. These techniques offer detection of a target analyte by monitoring changes in the enhanced electromagnetic fields (Hammond J. L. et al., 2014; Zhao J. et al., 2008).

## 11.3 REFRACTIVE INDEX DEPENDENCE

The dependence of the LSPR peak wavelength on the dielectric function of the surrounding medium can be quantitatively evaluated using the Drude model of the electronic structure of metals (Jensen T. R. et al., 2000; Mayer K. M. and Hafner J. H., 2011). The Drude model is a classical model which provides a very good approximation of the conductivity of noble metals. It describes the electronic transport in conductors in terms of the collisions between freely moving electrons and a lattice of heavy, stationary ionic cores. From the Drude model, one can get the analytical, frequency-dependent form for the real part of the complex metal dielectric function ($\varepsilon_1$) (Mayer K. M. and Hafner J. H., 2011).

$$\varepsilon_1 = 1-\frac{\omega_p^2}{\omega^2+\gamma^2} \tag{11.11}$$

where $\omega_p$ is the plasma frequency, and $\gamma$ is the damping parameter of the bulk metal.

For visible and near infrared frequencies, $\gamma \ll \omega_p$, so the previous equation can be simplified to

$$\varepsilon_1 = 1-\frac{\omega_p^2}{\omega^2} \tag{11.12}$$

Applying the resonance condition and using this expression $\varepsilon_1 = 2\varepsilon_m$, the previous equation becomes

$$\omega_{max} = \frac{\omega_p}{\sqrt{2\varepsilon_m+1}} \tag{11.13}$$

where $\omega_{max}$ is the LSPR peak frequency (resonant frequency). We know the relationship of frequency and wavelength, $\lambda = 2\pi c/\omega$, and then from the dielectric constant to index of refraction, $\varepsilon_m = n^2$, the previous equation reduces to

$$\lambda_{max} = \lambda_p\sqrt{2n_m^2+1} \tag{11.14}$$

where $\lambda_{max}$ is the LSPR peak wavelength, and $\lambda_p$ is the wavelength corresponding to the plasma frequency of the bulk metal. From experimental observations using the previous expression, it is clear that the dependence of LSPR peak wavelength on the refractive index ought to be approximately linear at optical frequencies over small ranges of the refractive index ($n$). Figure 11.3 illustrates this result for a plot for $\lambda_p$ = 300 nm over a reasonable range of $n$ (Mayer K. M. and Hafner J. H., 2011).

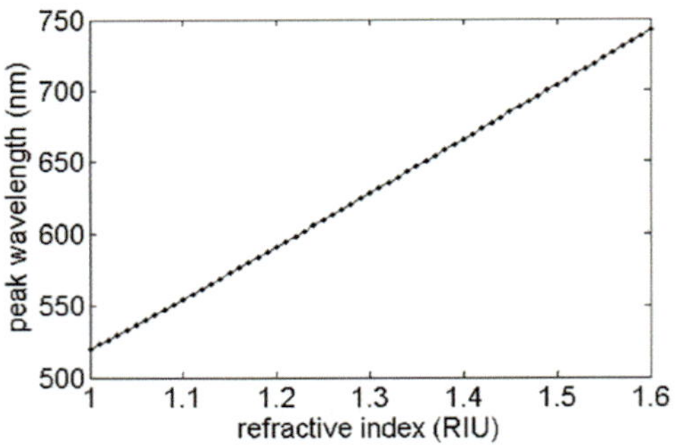


**FIGURE 11.3** The prediction of the Drude model, which gives a nearly linear relationship between LSPR peak wavelength and refractive index.

**Source: Adapted from Mayer K. M. and Hafner J. H., 2011. Copyright © 2011, American Chemical Society.**

## 11.4 REFRACTIVE INDEX SENSING USING LPSR

The LSPR peak wavelength shift is almost linear with variations in the refractive index of the surrounding medium. Consequently, the refractive index sensitivity ($S$) of a certain nanoparticle type is typically reported in nanometres of peak shift per refractive index unit (nm/RIU) (Mayer K. M. and Hafner J. H., 2011).

$$S = \frac{d\lambda_p}{dn} \tag{11.15}$$

Figure 11.4 shows an example of this type of sensing for a gold nanorod film in different types of solvents of different refractive indices such as air ($n$ = 1.000), water ($n$ = 1.333), ethanol ($n$ = 1.361), and formamide ($n$ = 1.428). From this figure, it can be understood that, even though the plasmon resonance wavelength is not strictly linear with index of refraction, it is linear over small ranges of $n$ (Mayer K. M. et al., 2008).

LSPR sensing is based on spectral peak shifts, and the precision that can be attained with respect to the changes in the refractive index depends on the sensitivity $S$ and the peak line width. The figure of merit (FOM) is obtained by dividing the sensitivity by the resonance line width and is widely used to measure the sensing capabilities of nanoparticles (Figure 11.5) (Mayer K. M. and Hafner J. H., 2011; Sherry L. J. et al., 2005). Nanoparticles of large size have relatively high sensitivities, but due to radiative damping and multipolar excitations, their spectral peaks are broadened.

$$FOM = \frac{S}{\Delta\lambda} \tag{11.16}$$

Complex plasmonic nanostructures (like metamaterials, etc.) lack a simple, single line shape, and it is difficult to obtain a consistent LSPR line width $\Delta\lambda$. Becker et al. suggested a general and alternative expression for figure of merit (FOM*) which does not include the line width by comparison between different nanostructures with different sensing abilities (Equation 11.17). FOM* is defined in terms of the relative intensity change (dI/I) that arises at a given wavelength ($\lambda_0$) upon a small change to the local refractive index (d$n$) (Becker J. et al., 2010). In this manner, it is possible to directly compare sensors based on peak shifts with those based on intensity. From the equation, it is clear that FOM* can be defined in terms of both the change in wavelength that gives the maximum intensity with change in refractive index and also in terms of the sensitivity in (nm/RIU).

$$FOM^* = \left(\frac{\frac{dI}{dn}}{I}\right)_{max} = \left(\frac{\frac{dI}{d\lambda}\cdot\frac{d\lambda}{dn}}{I}\right)_{max} = \left(\frac{S\frac{dI}{d\lambda}}{I}\right)_{max} \tag{11.17}$$

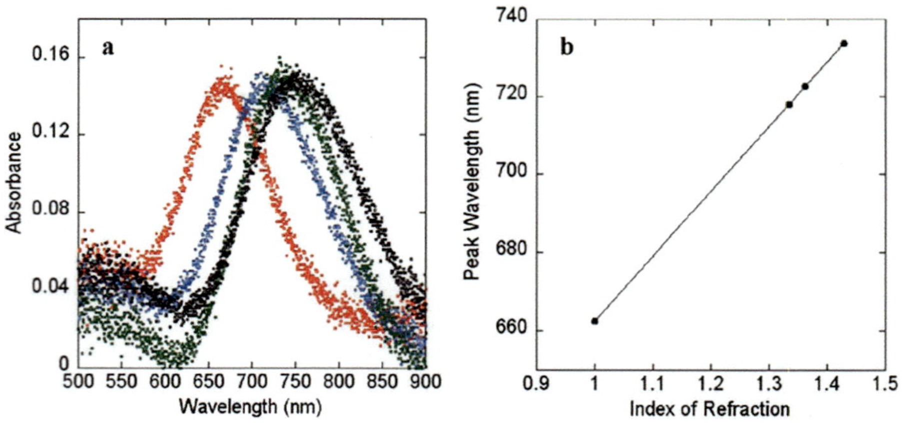


**FIGURE 11.4** (a) Optical absorption spectra of gold nanorod film in air (red), water (blue), ethanol (green), and formamide (black). (b) Plot of the peak wavelength versus index of refraction yields a sensitivity (slope) of 170 nm/RIU.

**Source: Adapted from Mayer K. M. et al., 2008. Copyright © 2008, American Chemical Society.**

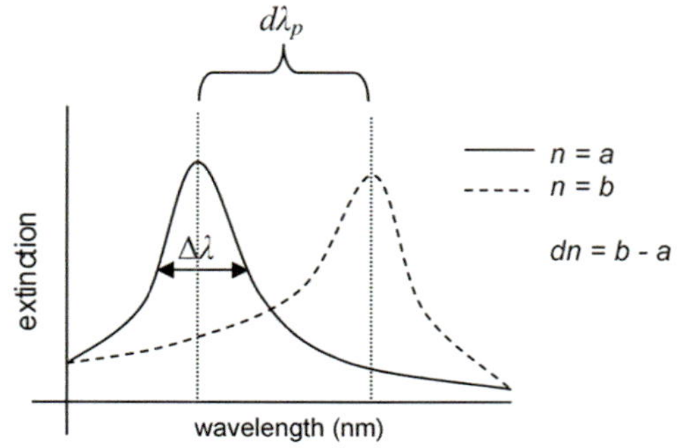


FIGURE 11.5 Using Equations 11.15 and 11.16, defining the LPSR figure of merit (FOM).

**Source: Adapted from Mayer K. M. and Hafner J. H., 2011 Copyright © 2011, American Chemical Society.**

## 11.5 NANOBIOSENSORS

In clinical and biological research, the collection and analysis of molecular data from diseased patients is essential to determine the best possible treatment plans. Precise diagnosis and reliable prognoses of disease depend on the sensitive detection and analysis of disease-associated biomolecules in tissues and biological fluids, such as proteins and nucleic acids. Numerous diagnostic techniques, such as polymerase chain reaction (PCR), western blot, and enzyme-linked immunosorbent assays (ELISAs) (Yates A. M. et al., 1999), have certain drawbacks. One significant drawback is the requirement for an antibody labelling procedure, which may or may not involve an enzyme or nucleic acid (such as a PCR primer) that can bind to the target biomolecule. Also, an amplification process is needed for PCR to detect nucleic acid targets, which results a delay in detection and increases the risk of contamination during analysis due to the presence of dead bacteria (Hong Y. et al., 2012; Olsen J. E. et al., 1995). As a result, numerous nanotechnology-based diagnostic biosensors are now being developed for detecting low biomolecule concentrations and identifying biomolecular interactions (Naresh V. and Lee N., 2021).

A biosensor is a device that detects chemical compounds by sensing optical, thermal, or electrical signals through a precise biochemical reaction mediated by the immune system, isolated enzymes, organelles, or tissues. A nanobiosensor is characterized as a tool that has a high affinity and specificity for identifying biological processes at the molecular level and can convert signals into information that can be measured (Naresh V. and Lee N., 2021; Hong Y. et al., 2012). Within the range of nanobiosensors that are accessible today, LSPR-based nanobiosensors are thought to be one of the most effective tools in the biotechnology and biosensor domains. The unique properties of metallic nanostructured materials, including noble metal nanoparticles, are resonant with the collective oscillations of conduction electrons at a particular wavelength when excited by incident photon radiation. The peaks found in the LSPR spectra are sensitive to the dielectric medium on the surface of the metal nanoparticles and can be employed for identifying biomolecules (Zhao J. et al., 2006; Garcia M. A., 2011).

For the detection of biomolecules, nanobiosensors based on LSPR offer advantages such as (1) high sensitivity through the detection of changes in refractive index, (2) no labelling is required due to the sensing of spectral shifts, (3) real-time assay accessibility by the use of microfluidic systems, (4) good reproducibility through the use of nanoparticle substrates, and (5) low cost and simple instrumental setup (Hong Y. et al., 2012; Naresh V. and Lee N., 2021). These advantages indicate that nanobiosensors can be used in a variety of fields, including drug testing, environmental monitoring, food safety, and medicine (Naresh V. and Lee N., 2021; Hong Y. et al., 2012).

## 11.6 ENGINEERING SILVER AND GOLD NANOPARTICLES FOR BIOSENSING

Silver (Ag) and gold (Au) nanoparticles have the most intriguing physical characteristics for biosensing applications among all other metallic NPs (Mayer K. M. et al., 2008). Gold nanoparticles (AuNPs) continue to be the most researched material for this application due to their excellent chemical stability and biocompatibility (Daniel M. C. and Astruc D., 2004; Dreaden E. C. et al., 2012), while silver nanoparticles (AgNPs) provide greater sensitivity rates (Rycenga M. et al., 2011). The phenomenon of localized surface plasmon resonance, which gives the colloidal suspensions of nanoparticles their vibrant colour, is one of the most distinctive physical characteristics of metallic nanoparticles (NPs) (Garcia M. A., 2011).

The size, uniformity, shape, dispersion, composition (ratio Au:Ag), and dielectric constant of the surrounding medium all affect the position of the LSPR band of AuNPs and AgNPs (Garcia M. A., 2011; Anker J. N. et al., 2008; Lee K. S. and El-Sayed M. A., 2006). Consequently, wavelength shifts in the absorption spectrum can be produced by altering any of these parameters. In particular, the LSPR band is red-shifted, or moved to a higher wavelength, when the size of the NP increases (Garcia M. A., 2011; Mol B. et al., 2016). Likewise, NP aggregation can also cause a red-shift in the absorption spectrum. Anisotropic NPs can exhibit multiple absorption bands, whereas isotropic (spherical) NPs only have one (Garcia M. A., 2011). The refractive index sensitivity (RIS), which is measured as a shift in the LSPR peak's wavelength and is expressed in nm/refractive index unit (RIU) (nanometre per refractive index unit), indicates the sensitivity of the biosensor to slight variations in refractive index (RI).

### 11.6.1 Shape-Properties Correlation

Metallic NPs can be synthesized using physical, chemical, or biological methods. The most widely used synthetic method for preparing AgNPs is chemical reduction using a bottom-up strategy. Either an organic or aqueous solvent is used for the process. The size, shape, and colour of NPs are influenced by various parameters, including temperature and solution acidity (Alqadi M. et al., 2014; Dong X. et al., 2009; Garcia M. A., 2011). The AgNP form is very strongly influenced by the solution's acidity. Dong et al. used citrate as a stabilizing and reducing agent and demonstrated that pH had a substantial impact on the morphology of AgNPs (Alqadi M. et al., 2014; Dong X. et al., 2009). They discovered that the majority AgNP morphologies at low pH were polygonal and triangular, while the shape of AgNPs at high pH was a mixture of spherical and rod-like particles. NP size changed in accordance with pH. AgNPs were in fact smaller as a result of the enhanced reducing activity of ascorbic acid (AA) or citrate when pH increased, which in turn caused a reduction in nuclei (Alqadi M. et al., 2014; Qin Y. et al., 2010). It is observed that the type of reductant influences the form of AgNPs by influencing the precursor's (Ag+) pH-dependent reduction rate and pH-dependent redox potential. AgNP shape can be controlled more effectively by fine-tuning the nucleation and growth equilibrium more precisely. In contrast to citrate, when ascorbic acid is used as a reducing agent, it typically produces spherical particles within the pH range of 6 to 11. Studies indicate that acidity affects the LSPR band's position (Alqadi M. et al., 2014). The LSPR band also is blue-shifted and become more intense as the pH increases, and it has also been observed that bands get sharper when the size of the nanoparticles decreases (Qin Y. et al., 2010).

### 11.6.2 Anisotropic Shapes

Anisotropic nanoparticles have gained significant interest due to their improved structural, optical, electrical, magnetic, and catalytic capabilities compared to spherical nanoparticles of comparable size (Li N. et al., 2014). Anisotropic and hollow NPs exhibit a plasmon band at longer wavelengths (near-infrared area) compared to spherical NPs (Hirsch L. R. et al., 2006; Jakab A. et al., 2011). By using a wet chemical reduction technique, Zaheer and Rafiuddin were able to synthesize flower-like silver nanoparticles at room temperature (2011). Modulating the experimental conditions, such as temperature, Ag precursor concentration, and pH of the solution, and using different reducing agents (citric acid, AA, and $NaBH_4$) and methods (chemical, physical, or biological) helps us to achieve different morphologies such as the transition from spherical to nanowire particles through nanorods, nanobars, nanowires, nanocubes, pyramids, nanoprisms, nanorice, and nanoflower-shaped silver nanoparticles (Figure 11.6) (Murphy C. J. and Jana N. R., 2002; Cobley C. M. et al., 2009; Khodashenas B. and Ghorbani H. R., 2019).

Silver nanoplates (AgNPls) have a distinctive anisotropy due to their lateral dimensions being greater than their height, making them a promising material for LSPR biosensors. The systems employed for LSPR biosensing have primarily been composed of ordered arrays of triangular NPls (TNPls) produced by nanosphere lithography (NSL) (Rycenga M. et al., 2011; Lu L. et al., 2006; Pastoriza-Santos I. and Liz-Marzán L. M., 2008). AgNPl optical resonance can be precisely tuned between 500 and 1100 nm by varying the thickness and diameter of the plate (Liz-Marzán L. M., 2004). They exhibit extraordinarily wide absorption and scattering cross-sections in the visible and near-infrared (NIR) ranges (Tan T. et al., 2013). AgNPls are useful in SERS, photovoltaics, molecular detection, and photothermal-based therapeutics.

Another distinctive morphology is nanoprisms (NPrs). They appear to have a traditional triangle shape, but closer examination reveals that the triangles' apexes are flat. The antibacterial properties of NPrs with flat apexes and (111) crystal planes were higher than those of spherical nanoparticles (Millstone J. E. et al., 2009). By modifying the size, shape, and short-range sensing length scale based on the characteristic decay length of the local electromagnetic field, Haes and van Duyne developed very good refractive index sensitivity (2002).

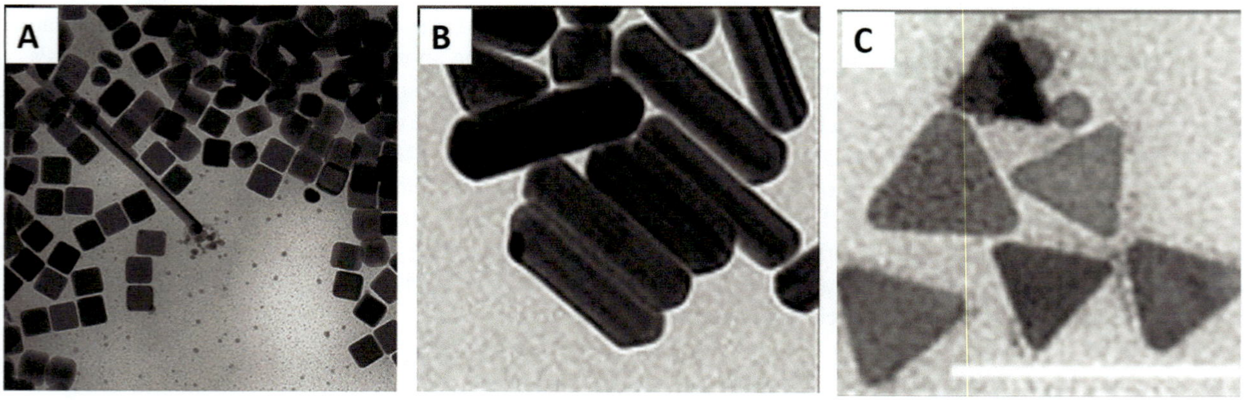


FIGURE 11.6 TEM images of silver nanoparticles of anisotropic shapes: (A) silver nanocubes. Adapted from Zhou S. et al., 2016, Copyright © 2016, American Chemical Society. (B) Silver nanorods. Adapted from Jakab A. et al., 2011, Copyright © 2011, American Chemical Society, (C) Silver nanoprism. Adapted from Lee G. P. et al., 2013, Copyright © 2013, American Chemical Society.

### 11.6.3 Plasmonic Nanoparticles Based on Silver and Gold

The theoretical refractive index sensitivity of silver nanoparticles is higher than that of the two other plasmonic nanometals, copper and gold (Navas M. P. and Soni R. K., 2015; Lismont M. and Dreesen L., 2012). For AgNPs, AuNPs, and CuNPs with diameters ranging from 5 to 50 nm, it has been observed that RIS increased from 153 to 265 nm/refractive index unit, 128 to 233 nm/RIU, and 117 to 212 nm/RIU, respectively. The LSPR absorption of spherical AgNPs is stronger than that of gold, having an absorption peak at 400 nm, and is five to ten times more intense (Navas M. P. and Soni R. K., 2015; Lismont M. and Dreesen L., 2012). However, AgNPs have some disadvantages for biosensing despite their higher sensitivity, which has been demonstrated both experimentally and theoretically. They are less stable and biocompatible, and their reversibility in sensing is reduced when exposed to light, making their use in repeated cycles less dependable than that of AuNPs (Jakab A. et al., 2011). Combining AgNPs with other metals, especially gold, has resulted in numerous advancements, and a combination of these two plasmonic metals also allowed for increased efficiency. Ag@Au core@shell structures (also known as Ag@AuNPs or Au@AgNPs) or AgAu alloys were primarily formed to achieve this combination.

#### 11.6.3.1 Silver-Gold Alloy Nanoparticles

Silver-gold alloy nanoparticles (AgAuNPs) are a mixture of Ag and Au atoms with no spatial differentiation between gold and silver. Spherical AgAuNPs can be produced by the chemical synthesis methods that involve co-reducing $AgNO_3$ and $HAuCl_4$ with sodium citrate (Link S. et al., 1999a,b; Singh A. V. et al., 2005). The concentration of $AgNO_3$ and $HAuCl_4$ determines the mole fraction of each metal in the alloy. Compared to gold-silver core-shell nanoparticles of the same size and shape, AgAu alloys exhibit greater stability (Qi W. and Lee S., 2010). For spherical AgNPs and AuNPs, the plasmon peak is approximately 400 and 520 nm, respectively; however, the AgAuNP absorption peak can be tuned continuously from 400 to 520 nm by varying the alloy composition (Garcia M. A., 2011; Mallin M. P. and Murphy C. J., 2002; Murphy C. J. and Jana N. R., 2002; Liz-Marzán L. M., 2006).

In contrast to core@shell NPs, Qi et al. demonstrated that alloy NPs became less stable as the Ag molar ratio increased (2010). Thus, the ratio of one metal to the other, as well as the size and form of the metals, determines the optical properties of the alloys. AgAuNPs only exhibit one peak, whose wavelength and intensity are determined by the molar proportion of Au or xAu (Liz-Marzán L. M., 2006; Link S. et al., 1999). Link et al. reported that for spherical NPs with an average size of 18 nm, the single LSPR peak of the AgAuNPs changed from 400 nm (pure AgNPs) to 520 nm (pure AuNPs) with an increase in gold molar content (Link, S. et al., 1999). In fact, the LSPR band shift from 400 to 520 nm and has been shown to be linear and proportional to the mole percent of gold in xAu. This result is both theoretically anticipated using Mie theory and is physically observed (Figure 11.7).

#### 11.6.3.2 Silver and Gold Core@Shell Nanostructures

Silver and gold core@shell nanostructures consist of two spatially separated layers, each containing a different element, a shell composed of the second metal (Au or Ag, respectively) and a core composed of the first metal (Ag or Au). Ag@AuNP denotes a silver core coated by a gold shell and vice versa. In the core@shell notation, the core is placed first. Whether isotropic or anisotropic, shell creation maintains the original shape of the NPs as enforced by the metallic core (Pérez-Juste J. et al., 2005). In the case of Ag@AuNPs, in order to preserve the chemical stability of the previously synthesized AgNPs, the gold shell, which is frequently quite thin, effectively serves as a protective layer on the Ag core.

The relationship between shell thickness and LSPR peak position is established by Mie theory, and it can be understood from UV-Vis spectroscopy measurements at various ablation times. For biosensing applications, the LSPR band position can be tuned at the required wavelength by adjusting the Ag content in AuNS. Very thin Au shells can also be produced by progressively adding gold solution after chemical reduction of $HAuCl_4$ at the AgNPls surface (Yang X. et al., 2014). Au@AgNPs can be synthesized by the chemical reduction of $AgNO_3$ at the surface of AuNPs. To synthesize spherical and rod-shaped Ag coated Au core–shell structures through chemical deposition, a reducing agent such as citrate or ascorbate is needed (Ma Y. et al., 2013; Zhu J. et al., 2013; Sun L. et al., 2014; Liz-Marzán L. M., 2006; Pérez-Juste J. et al., 2005).

Unique structures known as multilayer bimetallic nanostructures such as Au@Ag@AuNPs can also be produced by electrodepositing a thin coating of Au on already synthesized Au@AgNPs (Dong P. et al., 2013; Sui M. et al., 2019). Nanoparticles composed of various metallic elements have been attracting significant attentions in the field of catalysis, sensing, biology, energy, and electronics because of their superior characteristics compared to their monometallic counterparts. Multi-metallic alloy NPs can provide proficient advantages over monometallic NPs, such as multi-functionality, site-specific response, dynamic surface plasmon resonance, and electronic heterogeneity induced by the elemental variation. Additionally, $AgNO_3$ and $HAuCl_4$ were successively reduced in the presence of AA and cetyltrimethylammonium bromide (CTAB) on prepared Au seeds, resulting in multilayer bimetallic nanoparticles (Au@Ag, Au@Ag@Au, and Au@Ag@Au@AgNPs) (Liz-Marzán L. M., 2006; Rodríguez-González B. et al., 2005).

Mie theory states that the LSPR spectrum for isotropic nanoparticles resulting from the hybridization of two distinct plasmonic nanoparticles should show two

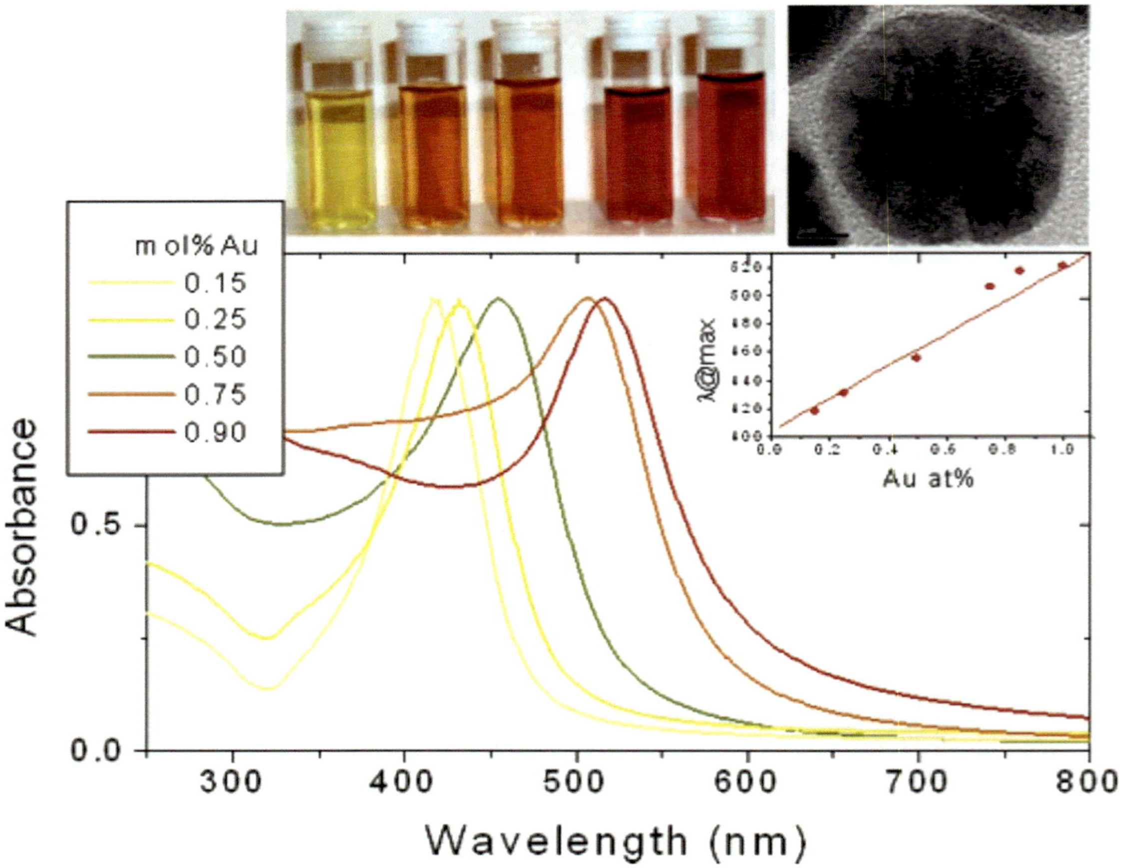


**FIGURE 11.7** UV-Vis absorption spectra and colloidal suspensions of AuAgNPs with increasing Au concentration. The inset figure shows the position of the experimental absorption band (dots) plotted as a function of increasing Au mole fraction composition, and the solid line is a linear fit to values obtained using Mie theory. The HRTEM image represents the homogeneous distribution of Au and Ag atoms within the particles.

**Source: Adapted from Liz-Marzán L. M., 2004, Copyright © 2006, American Chemical Society.**

peaks, one originating from the surface of the shell and the other from the core@shell interface between the two metals (Navas M. P. and Soni R. K., 2015; Chen Y. et al., 2012). The former peak's position is mostly determined by the core metal, whereas the latter peak's position is primarily determined by the shell metal and its thickness. Regarding Ag@AuNPs, it is envisaged that the thinnest Au shells will exhibit the best LSPR characteristics; however, for Au@AgNPs, the Ag shell may be thicker. Zhu and colleagues have shown that an increase in Au shell thickness results in a red-shift in the peak of Ag@Au nanowires and a decrease in intensity. Moreover, at low Au shell thicknesses, the shell peak is almost absent. When the Ag shell's thickness was raised, the peak blue-shifted, and the Au@Ag nanowires' shell peak was more intense.

### 11.6.4 Surface Modification of Nanoparticles

Studies using bare AgNPs/AuNPs as plasmonic biosensors are quite rare. The majority of biosensors function ex vivo. The application of bare AgNPs in biosensing is limited due to their poor stability, toxicity, and complex surface chemistry The coating method has a significant impact on the stability and toxicity of NPs in a particular environment (Zhang Q., 2019; Li X. et al., 2010; Badawy A. M. E. et al., 2010). The coating on NPs with organic or inorganic shells offers electrostatic, steric, and electrosteric stabilization and also prevents aggregation (Badawy A. M. E. et al., 2010).

Organic chemicals such as citrate and polyvinyl pyrrolidone (PVP) are commonly used as capping agents of NPs to stabilize and inhibit particle aggregation through adsorption or covalent attachment to their surface. They have been shown to affect the size and shape control of NPs (Wiley B. et al., 2005). The optical characteristics of NPs are strongly influenced by the coating reactant and layer thickness. NPs come in a variety of morphologies, including quasi-spheres, nanotubes, rods, and triangular nanoplates (TNPls), requiring distinct coating techniques and chemical agents. Citrate-and PVP-capped AgNPs were stabilized against aggregation using natural organic matter (NOM)

(Gunsolus I. L. et al., 2015). AgNPs are stabilized and protected through the application of thiol-capping agents as anchoring groups (Battocchio C. et al., 2012). Thiol-capping agents are grafted onto the surface of AgNPs via Ag-S bonds, creating an exterior layer with an Ag core and an $Ag_2S$-like shell (Battocchio C. et al., 2012; Cheng X. et al., 2015).

#### 11.6.4.1 Polymer Coatings

Steric interactions are the primary source of interest for polymer coatings. Polymers, whether grafted or adsorbed on NPs, enhance dispersion in the solution (Papa A. L. et al., 2015). High molecular weight and high grafting density of the polymers also enhance colloidal stability. Poly(ethylene) glycol (PEG) is hydrophilic, biocompatible, and neutral and has been extensively explored as a stabilizing or coating agent for nanoparticles (Papa A. L. et al., 2015; Maurizi L. et al., 2015; Karakoti A. S. et al., 2011). Because of the excellent biocompatibility of chitosan, it has often been employed to encapsulate metallic nanoparticles. It provides NPs strong stability and dispersibility and exhibits good affinity for the metallic surface (Sallem F. et al., 2017; Jiang H. et al., 2012).

#### 11.6.4.2 Silica Coating

While considering plasmonic nanoparticle capping, inorganic silica is one of the most preferred coating materials for choice due to its biocompatibility, chemical inertness, good mechanical strength, and large 3D porous network (Guerrero-Martínez A. et al., 2010; Liu S. and Han M. Y., 2010). Also, the biocompatible protective shell is tunable in thickness, preventing aggregation as a result of electrostatic repulsion, and is stable in a variety of solvents. Mesoporous $SiO_2$ has an interconnected pore structure that is open to the atmosphere and also has a wide range of transparency in the UV–visible region (Mol B. et al., 2016; Bi H. et al., 2002). Surface passivation of silica over metallic nanoparticles is mainly achieved through the use of the sol-gel and/or Stöber method to accomplish nanometric control of thickness, porosity, and homogeneity. The use of such a mesoporous matrix gives an interacting nanostructure, and the presence of silanol groups on the surface of silica makes it easier to modify chemically in order to introduce different surface functionalities (such as COOH, CHO, $NH_2$, or NCO) with readily available coupling agents (Aissaoui N. et al., 2012; Hanske C. et al., 2018).

$SiO_2$-coated core shell nanostructures of AgNPs were synthesized by the sol-gel method using tetraethyl orthosilicate (TEOS) as the main precursor, and researchers observed a red-shift of the LSPR band for thicknesses in the range of 28 to 48 nm and found that the TEOS concentration regulated the shell thickness (Mol B. et al., 2016; Kobayashi Y. et al., 2005). Increasing silica shell thicknesses of 57–76 nm cause the plasmon band to blue-shift and reduce intensity. This indicates that thicker silica shells significantly increase scattering at shorter wavelengths. Their results also agree well with the theoretical stimulations of Mie theory. The protective layer of silica has no impact on biosensing capacity as long as it is no thicker than a few nanometres and also exhibits diverse optical properties (Kobayashi Y. et al., 2005).

## 11.7 APPLICATIONS OF LPSR

LSPR technology has wide range of the applications of in agriculture, medicine, electronic industries, health care, and environmental monitoring.

### 11.7.1 Medical Applications

A biosensor is an analytical tool that combines a physical transducer, such as an optical, mass, or electrochemical one, with a biological sensing element, such as an enzyme or an antibody, so that the interaction between the target and the bio-recognition molecules is converted into a quantifiable electrical signal (Figure 11.8). Optical biosensors are effective substitutes for traditional analytical techniques because they utilize the properties of light such as absorption, fluorescence, luminescence, reflectance, Raman scattering, and refractive index in order to overcome the diffraction limit of traditional optical devices (Long F. et al., 2013).

The identification of tiny molecules is one of the primary uses of LSPR in biosensing.

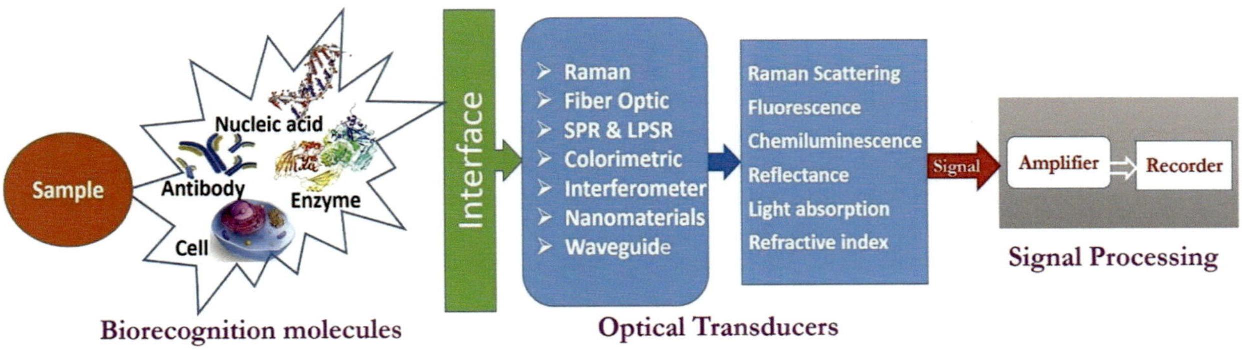


FIGURE 11.8 Schematic representation of an optical biosensor.

Since the sizes of bioreceptors (such as enzymes, antigens, and antibodies) are comparable to those of nanostructures, they can be regarded as structurally compatible. Because of this dimensional compatibility, a wide variety of accessible bioreceptors, fast surface biofunctionalization techniques, and nanostructure properties can all be combined to create extremely miniaturized signal transducers (Doria G. et al., 2012; Zhu W. et al., 2016; Craciun A. et al., 2017).

Noble metal nanostructures possess specific catalytic, electrical, and photonic characteristics. To design an LSPR biosensor, conjugates of nanostructure and biomolecule must first be formed. Numerous biomolecules like proteins, DNA, enzymes, peptides, antibodies, antigens, biotin, streptavidin, oligonucleotides, and aptamers, can be functionalized into these nanostructures to create novel substrates (Naresh V. and Lee N., 2021; Long F. et al., 2013). Covalent binding particular affinity interaction, chemisorption, and electrostatic adsorption are the primary synthetic methods used to create these conjugates. Covalent binding by the use of bifunctional linkers, in which an anchor group binds to the substrate and a functional group is free to covalently connect to the target biomolecule, is one of the most promising methods. These kinds of novel substrates that obtained using these techniques reduce the loss of bioactivity/biorecognition and can be used in a variety of biomedical applications, like sensing, imaging, diagnosis, and therapy (Garcia M. A., 2011; Loiseau A. et al., 2019).

#### 11.7.1.1 LSPR-Based Optical Diagnostics Using Nanoparticles

LSPRs can be used in biosensing applications because of their sensitivity to the NP surroundings and interparticle interaction (Wang X. H. et al., 2010; Loiseau A. et al., 2019). Also, local electromagnetic fields can be amplified by LSPRs, which makes AuNPs perfect for usage in MEF (Loiseau A. et al., 2019) and SERS (Song C. Y. et al., 2016). Modulating the decay rate and quantum efficiency of emitters around AuNPs is possible due to the increased photonic local density of states caused by LSPR in tiny mode volume (Bharadwaj P. et al., 2009). Spectral modulation is the basis for the use of AuNPs in molecular diagnostics (Ou J. et al., 2019).

Au and AgNPs have an extinction coefficient that is significantly higher than that of traditional dyes, which enables them to be used as agents for optical sensing applications with lower limits of detection (LOD) and higher sensitivity (Link S. and El-Sayed M. A., 2003; Ou J. et al., 2019). The colour change for the sample solution, which is visible to the naked eye, in particular makes it possible to determine the spectral shift (LSPR). For example, the "red-to-blue" colour change (spectral red shift) that AuNPs exhibit in their "dispersion-to-aggregation" state enables numerous sensing applications to be read with the naked eye. The dielectric environment close to the surface of AuNPs changes as they bind to target molecules, such as proteins or DNA strands. As a result, the change in LSPR peaks can be used to track the binding process (Garcia M. A., 2011).

The conventional illness screening technique known as enzyme-linked immunosorbent assay has limitations in precision molecular diagnostics because of its low sensitivity and accuracy. The restrictions of the conventional ELISA test are overcome by AuNP-enhanced ELISA due to their large surface area-to-volume ratio, which allows for significantly more attachment of antibody-enzyme complexes. Enzyme-mediated growth or aggregation of AuNPs has led to significant advancements in the development of AuNP-based plasmonic sensors for ELISA (Nie X. M. et al., 2014; Xianyu Y. L. et al., 2014).

De la Rica et al. introduced plasmonic ELISA for the ultralow-concentration detection of disease biomarkers (de la Rica R. and Stevens M. M., 2012). ELISA's enzyme label controlled the production of AuNPs and produced coloured solutions with a recognizable, unique tone. Prostate-specific antigen (PSA) and HIV-1 capsid antigen (p24) were detected at ultralow concentrations of $1 \times 10^{-18}$ g/mL in total serum. The identification of p24 in the serum of HIV-positive individuals was carried out successfully by this plasmonic ELISA. The POC diagnosis for p24 in pseudo-serum was recently exploited by Ruiz-Sanchez et al. through the development of a tunable plasmonic gold dendrimer nanochain immunoassay (2017).

#### 11.7.1.2 Electrochemical Biosensors for Virus Detection

Because of the excellent properties of electrochemical sensing technology such as simplicity, speed, and high sensitivity, it has been widely used for biosensing (Khan M. Z. H. et al., 2020). Electrochemical activities such as the reduction and oxidation of biomolecules or chemicals can be analysed using electrochemical technologies, such as cyclic voltammetry (CV), electrochemical impedance spectroscopy (EIS), and amperometry. The incorporation of noble metal nanoparticles in electrochemical biosensors has received increasing attention because of their attractive properties, such as high electrical conductivity, high surface area, excellent optical properties, and the effectiveness in binding to biological molecules. Due to these properties, they can enhance the electrical response signal and enable target virus detection either directly or indirectly with rapid response (Cho J. W. et al., 2020; Noori J. S. et al., 2020).

Immobilizing biomolecules on noble metal nanoparticles is of great importance for electrochemical biosensors, as they act as transducers, signal inducers, and capture agents for target molecules (Zhang L. Y. et al., 2020). AgNPs can interact with viruses, such as influenza, and the virus–metal nanoparticle has inherent electrochemical activity in solution. Due to the excellent antiviral activity of AgNPs, they can be used for low-level detection of viruses (Galdiero S. et al., 2011). Using the interaction between AgNPs and the virus, Sepunaru et al. developed an influenza electrochemical biosensor (2016). The Ag NP-modified virus displayed a stronger electrochemical signal than the original virus due to the oxidation of AgNPs. In their study, Sepunaru et al. observed that the AgNP-tagged virus allows the electrochemical biosensor to quickly detect the influenza viruses at a single viral level (2016).

Antibodies can be immobilized on the electrode by AuNPs due to their ability to physisorb or interact electrostatically with proteins (Choi K. H. et al., 2021). A fluorine-doped tin oxide electrode containing AuNPs and a SARS-CoV antibody was developed using these characteristics (Mahari S. et al., 2020). The peak current of the electrode increased when the antibody-antigen reaction immobilized the SARS-CoV spike protein on the electrode. The biosensor effectively identified the SARS-CoV spike protein using DPV, with a limit of detection of $7.8 \times 10^{-1}$ pg/mL (Mahari S. et al., 2020).

Noble metal nanoparticles have vast development potential for a variety of uses, including hybridizing with other materials and modifying their structural properties. Advancements in biosensors including noble metal nanoparticles may facilitate more effective early detection of serious new diseases, such as respiratory conditions that can trigger an epidemic or pandemic. This can lead to better health outcomes and global wellness (Choi K. H. et al., 2021).

### 11.7.2 Metal-Enhanced Fluorescence–Based Biosensors

Fluorescence is a very sensitive technique that makes it easy to identify single molecules. But the quantum yield, auto-fluorescence of the materials, and photo-stability of the fluorophores often limit the amount of information that can be detected. For metal-enhanced fluorescence (MEF), metallic nanostructures like silver can be used to alter the spectrum characteristics of fluorophores and reduce some of these shortcomings (Aslan K. et al., 2004, 2005).

For MEF-based biosensors, AgNPs functionalized with different thiolates have also been investigated (Aslan K. et al., 2005). Thiolated oligonucleotides coupled to AgNPs on a glass substrate were employed to study DNA hybridization experiments with metal-enhanced fluorescence. The use of AgNPs increased the number of detected photons per fluorophore molecule by a factor of ten or more and also indicates improving the sensitivity of DNA detection (Malicka J. et al., 2003).

Ag@$SiO_2$ NPs have also been used to develop MEF-based biosensors and as transducers for DNA hybridization (Aslan K. et al., 2005, 2007). Cyanine-labelled $SiO_2$-coated AgNPs colloids and their aggregation in suspension can yield three to five times stronger fluorescence signals. This inert coating allows a wide range of chemistries for biomolecule attachment while also reducing the near proximity quenching caused by noble metals. AgNPs@$SiO_2$ may therefore develop into a solution-based platform for increased fluorescence sensing (Aslan K. et al., 2005, 2007).

Silver island films (SIFs), formed by depositing AgNPs onto glass slides, were used to develop a sensor that provides an improved detection limit for a sandwich-format immunoassay of the cardiac marker myoglobin (Aslan K. et al., 2005). Anti-myoglobin antibodies were coated on silver island films and glass surfaces before being treated with antibodies labelled with fluorescence. With this metal-enhanced planar immunoassay method, fluorescence emission on the SIFs was observed to be 10–15 times higher than on the naked substrate. The results also showed that myoglobin concentrations were identified in the range of 10–1000 ng/mL.

### 11.7.3 Agriculture and Environmental Monitoring

The growing population, climate change, rising temperature, and depletion of natural resources have all contributed to a significant decline in agricultural production. Also, water pollution and contamination are both caused by agriculture as a result of improper use of chemicals as fertilizers for farming practices. In order to enable the detection of pollutants and contaminants and the identification of the most likely source, the production, processing, and distribution chain must be carefully screened (Ukhurebor K. E., 2021). Through continuous environmental monitoring, agricultural output might be enhanced and can be made more sustainable through improved use of soil, water, and fertilizers, as well as higher crop resilience to disease. With the development of inexpensive biosensing technology using nanomaterials, polymers, and microbe built-biosensors, it will be possible to improve farming practices and sustainable environment (Ukhurebor K. E, 2021; Hammond J. L. et al., 2014).

Crop quality and output can be increased by the monitoring of plant nutrients. Many techniques are currently available for analysing plant nutrients, such as atomic absorption spectrometry (AAS), X-ray fluorescence spectrometry (XRF), inductively-coupled plasma mass spectrometry (ICP-MS), and inductively-coupled plasma atomic emission spectroscopy (ICP/AES), and all these devices have to be placed in a laboratory. The use of these devices has various limitations, including their size, continuous cooling systems, and supplies of necessary gases needed for proper working. The effectiveness of an LSPR-based sensor technology demonstrates a quick and easy method for monitoring of plant nutrients (Hammond J. L. et al., 2014; Long F. et al., 2013).

Adverse health effects, including the development of cancerous tumours, can result from the use of organophosphate (OP) pesticides, industrial chemicals such as polychlorinated biphenyls (PCBs), pharmaceuticals, dioxins, and agrochemicals like atrazine. These substances can enter our bodies and build up, causing abnormalities throughout the body and causing serious illness. Acetylcholinesterase (AChE) covalently coupled to nanostructures has been used to demonstrate LSPR detection of OP compounds at 0.234 ppb LOD (Lin T. J. et al., 2006). LSPR in conjunction with SERS has been used to quantify PCBs down to the ppb level (50 pM). Compatible and sandwich immunoassays have been used to demonstrate the detection of dioxins, PCBs, and atrazine (Shimomura M. et al., 2001; Bantz K. C. and Haynes C. L., 2009). The detection of pesticides like malathion and chlorpyrifos, which are usually common in surface waters, can be done by using a basic UV-Vis spectrometer. The use of $Na_2SO_4$ enhanced gold nanostructures' ability to aggregate, enabling the observation of amplified LSPR spectrum changes at low concentrations (Lisha K. P. and Pradeep T., 2009). Most metal concentrations are hazardous to live cells regardless of whether they are necessary. By measuring the affinity difference between Hg and Au nanostructures with DNA, mercury can be detected; higher Hg concentrations alter the aggregation state of Au nanostructures, which results in higher LSPR intensities (Liu Z. D. et al., 2009).

### 11.7.4 LSPR Optical Fibre Sensors

In order to increase the sensitivity of the LSPR optical sensor, a novel method is adopted by depositing NPs on an optical fibre. Since plasmon resonance is the basis of LSPR optical fibre sensors, their surfaces are coated, making them more sensitive to the variations in the surrounding medium (Ortega-Mendoza G. J. et al., 2014; Chen J. et al., 2015). When compared to traditional methods, this technology is label free, and real-time detection has a significant advantage. The development of AgNP-based LSPR optical fibre sensors has not received much attention, despite the fact that nanosilver films have been shown to be far more sensitive to changes in the surrounding medium than other metal films (Sagle L. B. et al., 2011). Chen et al. suggested an AgNP-based reflecting LSPR optical fibre sensor is more sensitive and stable (2015).

A robust and sensitive reflecting LSPR optical fibre sensor based on AgNPs was proposed by Chen et al. among the investigations in order to optimize the fabrication process, taking into account the coating duration and the sensing length (2015). Once an antibody had functionalized the surface of AgNPs placed on the optical fibres, the antigen-antibody binding process was optically seen by monitoring the wavelength shift in real time. A substantially higher RIS of 387 nm/RIU was obtained with this approach than was previously reported for AgNPs suspended in colloidal form. By photo-depositing AgNPs on the optical fibre end, another study discovered an RIS of 67.6 nm/RIU. A higher increase of RI causes the LSPR peak wavelength to shift linearly to a longer wavelength (Ortega-Mendoza G. J. et al., 2014).

## 11.8 CONCLUSIONS

Localized surface plasmon resonance is a powerful phenomenon that can be utilized in biosensing applications due to its sensitivity to changes in the local refractive index near metallic nanoparticle surfaces. This property enables the detection and quantification of biomolecular interactions in real time without the need for labels or extensive sample preparation. By analysing shifts in the LSPR wavelength, one can deduce binding events on the nanoparticle surface, making LSPR a versatile tool in various biomedical and environmental applications. LSPR sensing is based on a simple optical extinction measurement and can be used for the development of robust and portable optical sensors, biological sensors, and chemical sensors for point-of care diagnostics deployed in global outreach programs. The low cost of LSPR sensors makes them appealing for commercial development. However, detecting analytes in complex solutions like blood, serum, and urine poses a challenge due to the sensor's specificity and the need to design for non-optically transparent solutions. Bimetallic sensors use the composition and spatial distribution of various metals for sensing purposes. Controlling bimetallic nanoprobes and precisely modifying their LSPR signals can improve sensor sensitivity and reproducibility. Furthermore, investigating the LSPR tunability of Cu and other low-cost metals, as well as increasing their chemical stability and reducing toxicity, may significantly reduce the cost of bimetallic plasmonic sensors in the future.

## REFERENCES


Agrawal, A.; Cho, S. H., Zandi, O.; Ghosh, S.; Johns, R. W.; Milliron, D. J.; Localized surface plasmon resonance in semiconductor nanocrystals. *Chem. Rev.*, 2018, 118, 3121–3207.

Ahmed, I.; Khoo, E. H.; Kurniawan, O.; Li, E. P.; Modeling and simulation of active plasmonics with the FDTD method by using solid state and Lorentz-Drude dispersive model. *J. Opt. Soc. Am. B.*, 2011, 28, 352–359.

Aissaoui, N.; Bergaoui, L.; Landoulsi, J.; Lambert, J. F.; Boujday, S.; Silane layers on silicon surfaces: Mechanism of interaction, stability, and influence on protein adsorption. *Langmuir.*, 2012, 28, 656–665.

Aljbar, N. A.; Khalid A. H, Abdulwahid, F. S.; Attallah, A. H.; Haider, A. J.; Mahdi, B. R.; Aljbar, N. A.; Enhanced surface plasmon resonance (SPR) fiber optic sensor for environmental monitoring: A coreless fiber–based design. *Plasmonics,* 2024, 1–10.

Alqadi, M.; Noqtah, O. A.; Alzoubi, F.; Alzouby, J.; Aljarrah, K.; pH effect on the aggregation of silver nanoparticles synthesized by chemical reduction. *Mater. Sci. Pol.*, 2014, 32, 107–111.

Anker, J. N.; Hall, W. P.; Lyandres, O.; Shah, N. C.; Zhao, J.; van Duyne, R. P.; Biosensing with plasmonic nanosensors. *Nat. Mater.*, 2008, 7, 442.

Aslan, K.; Gryczynski, I.; Malicka, J.; Matveeva, E.; Lakowicz, J. R.; Geddes, C. D.; Metal-enhanced fluorescence: An emerging tool in biotechnology. *Curr. Opin. Biotechnol.*, 2005, 16, 55–62.

Aslan, K.; Lakowicz, J. R.; Szmacinski, H.; Geddes, C. D.; Metal-enhanced fluorescence solution-based sensing platform. *J. Fluoresc.*, 2004, 14, 677–679.

Aslan, K.; Wu, M.; Lakowicz, J. R.; Geddes, C. D.; Fluorescent core-;shell Ag@SiO2 nanocomposites for metal-enhanced fluorescence and single nanoparticle sensing platforms. *J. Am. Chem. Soc.*, 2007, 129, 1524–1525.

Badawy, A. M. E.; Luxton, T. P.; Silva, R. G.; Scheckel, K. G.; Suidan, M. T.; Tolaymat, T. M.; Impact of environmental conditions (pH, ionic strength, and electrolyte type) on the surface charge and aggregation of silver nanoparticles suspensions. *Environ. Sci. Technol.*, 2010, 44, 1260–1266.

Bantz, K. C.; Haynes, C. L.; Surface-enhanced Raman scattering detection and discrimination of polychlorinated biphenyls. *Vib. Spectrosc.*, 2009, 50, 29–35.

Battocchio, C.; Meneghini, C.; Fratoddi, I.; Venditti, I.; Russo, M. V.; Aquilanti, G.; Maurizio, C.; Bondino, F.; Matassa, R.; Rossi, M.; Mobilio, S.; Polzonetti, G.; Silver nanoparticles stabilized with thiols: A close look at the local chemistry and chemical structure. *J. Phys. Chem. C.,* 2012, 116, 19571–19578.

Becker, J.; Trügler, A.; Jakab, A.; Hohenester, U.; Sönnichsen, C.; The optimal aspect ratio of gold nanorods for plasmonic bio-sensing. *Plasmonics,* 2010, 5, 161–167.

Bharadwaj, P.; Deutsch, B.; Novotny, L.; Optical antennas. *Adv. Opt. Photonics.,* 2009, 1, 438.

Bi, H.; Cai, W.; Kan, C.; Zhang, L.; Martin, D.; Trager, F.; Optical study of redox process of Ag nanoparticles at high temperature. *J. Appl. Phys.,* 2002, 92, 7491.

Burger, S.; Zschiedrich, L.; Pomplun, J.; Schmidt, F.; Finite element method for accurate 3D simulation of plasmonic waveguides. *Proc. SPIE,* 2010, 7604, 76040.

Chen, J.; Shi, S.; Su, R.; Qi, W.; Huang, R.; Wang, M.; Wang, L.; He, Z.; Optimization and application of reflective lspr optical fiber biosensors based on silver nanoparticles. *Sensors,* 2015, 15, 12205–12217.

Chen, T.; Xin, J.; Chang, S. J.; Chen, J.; Liu, T.; Surface plasmon resonance (SPR) combined technology: A powerful tool for investigating interface phenomena. *Adv. Mater. Interfaces,* 2023, 10, 2202202.

Chen, Y.; Wu, H.; Li, Z.; Wang, P.; Yang, L.; Fang, Y.; The study of surface plasmon in Au/Ag core/shell compound nanoparticles. *Plasmonics,* 2012, 7, 509–513.

Cheng, X.; Liu, M.; Zhang, A.; Hu, S.; Song, C.; Zhang, G.; Guo, X.; Size-controlled silver nanoparticles stabilized on thiol-functionalized MIL-53(Al) frameworks. *Nanoscale,* 2015, 7, 9738–9745.

Cho, I. H.; Kim, D. H.; Park, S.; Electrochemical biosensors: Perspective on functional nanomaterials for on-site analysis. *Biomater. Res.,* 2020, 24, 6.

Choi, H. K.; Lee, M. J.; Lee, S. N.; Kim, T. H.; Oh, B. K.; Noble metal nanomaterial-based biosensors for electrochemical and optical detection of viruses causing respiratory illnesses. *Front. Chem.,* 2021, 9, 672739.

Cobley, C. M.; Skrabalak, S. E.; Campbell, D. J.; Xia, Y.; Shape-controlled synthesis of silver nanoparticles for plasmonic and sensing applications. *Plasmonics,* 2009, 4, 171–179.

Craciun, A. M.; Focsan, M.; Magyari, K.; Vulpoi, A.; Pap, Z.; Surface plasmon resonance or biocompatibility – key properties for determining the applicability of noble metal nanoparticles. *Materials,* 2017, 10, 836.

Daniel, M. C.; Astruc, D.; Gold nanoparticles: Assembly, supramolecular chemistry, quantum-size-related properties, and applications toward biology, catalysis, and nanotechnology. *Chem. Rev.,* 2004, 104, 293–346.

de la Rica, R.; Stevens, M. M.; Plasmonic ELISA for the ultrasensitive detection of disease biomarkers with the naked eye. *Nat. Nanotechnol.;* 2012, 7, 821–824.

Dong, P.; Wu, Y.; Guo, W.; Di, J., Plasmonic biosensor based on triangular Au/Ag and Au/Ag/Au core/shell nanoprisms onto indium tin oxide glass. *Plasmonics.,* 2013, 8, 1577–1583.

Dong, X.; Ji, X.; Wu, H.; Zhao, L.; Li, J.; Yang, W.; Shape control of silver nanoparticles by stepwise citrate reduction. *J. Phys. Chem. C.,* 2009, 113, 6573–6576.

Doria, G.; Conde, J.; Veigas, B.; Giestas, L.; Almeida, C.; Assunção, M.; Rosa, J.; Baptista, P. V.; Noble metal nanoparticles for biosensing applications. *Sensors,* 2012, 12, 1657–1687.

Draine, B. T.; Flatau, P. J.; Discrete-dipole approximation for scattering calculations. *J. Opt. Soc. Am. A.,* 1994, 11, 1491–1499.

Dreaden, E. C.; Alkilany, A. M.; Huang, X.; Murphy, C. J.; El-Sayed, M. A.; The golden age: Gold nanoparticles for biomedicine. *Chem. Soc. Rev.,* 2012, 41, 2740–2779.

Du, Y.; Qu, X.; Wang, G.; Applications of surface plasmon resonance in biomedicine. *HSET,* 2022, 3, 137–143.

Galdiero, S.; Falanga, A.; Vitiello, M.; Cantisani, M.; Marra, V.; Galdiero, M.; Silver nanoparticles as potential antiviral agents. *Molecules,* 2011, 16, 8894–8918.

Gans, R.; über die Form ultramikroskopischer Goldteilchen. *Ann. Phys.,* 1912, 342, 881–900.

Garcia, M. A., Surface plasmons in metallic nanoparticles: Fundamentals and applications. *J. Phys. D: Appl. Phys.,* 2011, 44 (28), 283001.

Giannini, V.; Fernandez-Domínguez, A. I.; Heck, S. C.; Maier, S. A.; Plasmonic nanoantennas: Fundamentals and their use in controlling the radiative properties of nanoemitters. *Chem. Rev.,* 2011, 111, 3888–3912.

Guerrero-Martínez, A.; Perez-Juste, J.; Liz-Marzan, L. M.; Recent progress on silica coating of nanoparticles and related nanomaterials. *Adv. Mater.,* 2010, 22, 1182–1195.

Gunsolus, I. L.; Mousavi, M. P.; Hussein, K.; Bühlmann, P.; Haynes, C. L.; Effects of humic and fulvic acids on silver nanoparticle stability, dissolution, and toxicity. *Environ. Sci. Technol.,* 2015, 49, 8078–8086.

Haes, A. J.; van Duyne, R. P.; A nanoscale optical biosensor: Sensitivity and selectivity of an approach based on the localized surface plasmon resonance spectroscopy of triangular silver nanoparticles. *J. Am. Chem. Soc.,* 2002, 124, 10596–10604.

Hammond, J. L.; Bhalla, N.; Rafiee, S. D.; Estrela, P.; Localized surface plasmon resonance as a biosensing platform for developing countries. *Biosensors,* 2014, 4, 172–188.

Hanske, C.; Sanz-Ortiz, M. N.; Liz-Marzán, L. M.; Silica-coated plasmonic metal nanoparticles in action. *Adv. Mater.,* 2018, 30, 1707003.

Hirsch, L. R.; Gobin, A. M.; Lowery, A. R.; Tam, F.; Drezek, R. A.; Halas, N. J.; West, J. L.; Metal nanoshells. *Ann Biomed. Eng.,* 2006, 34, 15–22.

Hong, Y.; Huh, M.; Yoon, D. S.; Yang, J.; Nanobiosensors based on localized surface plasmon resonance for biomarker detection. *J. Nanomater.,* 2012, 759830, 1–14.

Hu, M.; Chen, J.; Li, Z.-Y.; Au, L.; Hartland, G. V.; Li, X.; Marquez, M.; Xia, Y.; Gold nanostructures: Engineering their plasmonic properties for biomedical applications. *Chem. Soc. Rev.,* 2006, 35, 1084.

Jakab, A.; Rosman, C.; Khalavka, Y.; Becker, J.; Trügler, A.; Hohenester, U.; Sönnichsen, C.; Highly sensitive plasmonic silver nanorods. *ACS Nano.,* 2011, 5, 6880–6885.

Jensen, T. R.; Duval, M. L.; Kelly, K. L.; Lazarides, A. A.; Schatz, G. C.; Van Duyne, R. P. Nanosphere lithography: Tunable localized surface plasmon resonance spectra of silver nanoparticles. *J. Phys. Chem. B.,* 2000, 103, 9846.

Jiang, H.; Chen, Z.; Cao, H.; Huang, Y.; Peroxidase-like activity of chitosan stabilized silver nanoparticles for visual and colorimetric detection of glucose. *Analyst,* 2012, 137, 5560–5564.

Karakoti, A. S.; Das, S.; Thevuthasan, S.; Seal, S.; PEGylated inorganic nanoparticles. *Angew. Chem. Int. Ed.,* 2011, 50, 1980–1994.

Khan, M. Z. H.; Hasan, M. R.; Hossain, S. I.; Ahommed, M. S.; Daizy, M.; Ultrasensitive detection of pathogenic viruses with electrochemical biosensor: State of the art. *Biosens. Bioelectron.,* 2020, 166, 112431.

Khodashenas, B.; Ghorbani, H. R.; Synthesis of silver nanoparticles with different shapes. *Arab. J. Chem.,* 2019, 12, 1823–1838.

Khoury, C. G.; Norton, S. J.; Vo-Dinh, T.; Plasmonics of 3-D nanoshell dimers using multipole expansion and finite element method. *ACS Nano,* 2009, 3, 2776–2788.

Kobayashi, Y.; Katakami, H.; Mine, E.; Nagao, D.; Konno, M.; Liz-Marzan, L. M.; Silica coating of silver nanoparticles using a modified Stöber method. *J. Colloid Int. Sci.,* 2005, 283, 392–396.

Lee, G. P.; Shi, Y.; Lavoie, E.; Daeneke, T.; Reineck, P.; Cappel, U. B.; Huang, D. M.; Bach, U.; Light-driven transformation processes of anisotropic silver nanoparticles. *ACS Nano,* 2013, 7, 5911–5921.

Lee, K. S.; El-Sayed, M. A.; Gold and silver nanoparticles in sensing and imaging: Sensitivity of plasmon response to size, shape, and metal composition. *J. Phys. Chem. B.,* 2006, 110, 19220–19225.

Li, N.; Zhao, P.; Astruc, D.; Anisotropic gold nanoparticles: Synthesis, properties, applications, and toxicity. *Angew. Chem. Int. Ed.,* 2014, 53, 1756–1789.

Li, X.; Lenhart, J. J.; Walker, H. W.; Dissolution-accompanied aggregation kinetics of silver nanoparticles. *Langmuir,* 2010, 26, 16690–16698.

Lin, T. J.; Huang, K. T.; Liu, C. Y.; Determination of organophosphorous pesticides by a novel biosensor based on localized surface plasmon resonance. *Biosens. Bioelectron.,* 2006, 22, 513–518.

Link, S.; El-Sayed, M. A.; Optical properties and ultrafast dynamics of metallic nanocrystals. *Annu. Rev. Phys. Chem.,* 2003, 54, 331–366.

Link, S.; El-Sayed, M. A.; Erratum: Simulation of the optical absorption spectra of gold nanorods as a function of their aspect ratio and the effect of the medium dielectric constant. *J. Phys. Chem. B.,* 2005, 109, 10531–10532.

Lisha, K. P.; Pradeep, T.; Enhanced visual detection of pesticides using gold anoparticles. *J. Environ. Sci. Health B.,* 2009, 44, 697–705.

Lismont, M.; Dreesen, L.; Comparative study of Ag and Au nanoparticles biosensors based on surface plasmon resonance phenomenon. *Mater. Sci. Eng. C.,* 2012, 32, 1437–1442.

Liu, S.; Han, M. Y.; Silica-coated metal nanoparticles. *Chem. Asian J.,* 2010, 5, 36–45.

Liu, Z. D.; Li, Y. F.; Ling, J.; Huang, C. Z.; A localized surface plasmon resonance light-scattering assay of mercury (II) on the basis of Hg2+-DNA complex induced aggregation of gold nanoparticles. *Environ. Sci. Technol.,* 2009, 43, 5022–5027.

Liz-Marzán, L. M.; Nanometals: Formation and color. *Mater. Today,* 2004, 7, 26–31.

Liz-Marzán, L. M.; Tailoring surface plasmons through the morphology and assembly of metal nanoparticles. *Langmuir,* 2006, 22, 32–41.

Loiseau, A.; Asila, V.; Boitel-Aullen, G.; Lam, M., Salmain, M., Boujday, S.; Silver-Based Plasmonic Nanoparticles for and Their Use in Biosensing. *Biosens.,* 2019, 9, 1–39.

Long, F.; Zhu, A.; Shi, H.; Recent advances in optical biosensors for environmental monitoring and early warning. *Sensors,* 2013, 13, 13928–13948.

Lu, L.; Kobayashi, A.; Tawa, K.; Ozaki, Y.; Silver nanoplates with special shapes: Controlled synthesis and their surface plasmon resonance and surface-enhanced Raman scattering properties. *Chem. Mater.,* 2006, 18, 4894–4901.

Ma, Y.; Zhou, J.; Zou, W.; Jia, Z.; Petti, L.; Mormile, P.; Localized surface plasmon resonance and surface enhanced Raman scattering responses of Au@Ag core-shell nanorods with different thickness of Ag shell. *Nanosci. Nanotechnol.,* 2013, 13, 1–6.

Mahari, S.; Roberts, A.; Shahdeo, D.; Gandhi, S.; eCovSen-sultrasensitive novel in-house built printed circuit board based electrochemical device for rapid detection of nCovid-19 antigen, a spike protein domain 1 of SARS-CoV−2. *BioRxiv.,* 2020, 059204.

Malicka, J.; Gryczynski, I.; Lakowicz, J. R.; DNA hybridization assays using metal-enhanced fluorescence. *Biochem. Biophys. Res. Commun.,* 2003, 306, 213–218.

Mallin, M. P.; Murphy, C. J.; Solution-phase synthesis of Sub-10 nm Au-Ag alloy nanoparticles. *Nano Lett.,* 2002, 2, 1235–1237.

Maurizi, L.; Papa, A.-L.; Dumont, L.; Bouyer, F.; Walker, P.; Vandroux, D.; Millot, N.; Influence of surface charge and polymer coating on internalization and biodistribution of polyethylene glycol-modified iron oxide nanoparticles. *J. Biomed. Nanotechnol.,* 2015, 11, 126–136.

Mayer, K. M.; Hafner, J. H.; Localized surface plasmon resonance sensors. *Chem. Rev.,* 2011, 111, 3828–3857.

Mayer, K. M.; Lee, S.; Liao, H.; Rostro, B. C.; Fuentes, A.; Scully, P. T.; Nehl, C. L.; Hafner, J. H.; A label-free immunoassay based upon localized surface plasmon resonance of gold nanorods. *ACS Nano,* 2008, 2 (4), 687–692.

Millstone, J. E.; Hurst, S. J.; Metraux, G. S.; Cutler, J. I.; Mirkin, C. A.; Colloidal gold and silver triangular nanoprisms. *Small,* 2009, 5, 646–664.

Mol, B.; Joy, L. K.; Thomas, H.; Thomas, V.; Joseph, C.; Narayanan, T. N.; Harthi, S. A.; Unnikrishnan, N. V.; Anantharaman, M. R.; Evidence for enhanced optical properties through plasmon resonance energy transfer in silver silica nanocomposites. *Nanotechnology,* 2016, 27, 085701.

Murphy, C. J.; Jana, N. R.; Controlling the aspect ratio of inorganic nanorods and nanowires. *Adv. Mater.,* 2002, 14, 80–82.

Naresh, V.; Lee, N.; A review on biosensors and recent development of nanostructured materials-enabled biosensors. *Sensors,* 2021, 21, 1109.

Navas, M. P.; Soni, R. K.; Laser-generated bimetallic Ag-Au and Ag-Cu core-shell nanoparticles for refractive index sensing. *Plasmonics,* 2015, 10, 681–690.

Nie, X. M.; Hong, R.; Dong, C. X.; Tang, L. J.; Gui, R.; Jiang, J. H.; Plasmonic ELISAfor the ultrasensitive detection of Treponema pallidum. *Biosen. Bioelectron.,* 2014, 58, 314–319.

Noori, J. S.; Mortensen, J.; Geto, A.; Recent development on the electrochemical detection of selected pesticides: A focused review. *J. Power Sources,* 2020, 20, 2221.

Olsen, J.; Aabo, S.; Hill, W.; Notermans, S.; Wernars, K.; Granum, P.; Popovic, T.; Rasmussen, H.; Olsvik, Ø.; Probes and polymerase chain reaction for detection of food-borne bacterial pathogens. *Int. J. Food Microbiol.,* 1995, 28, 1–78.

Ortega-Mendoza, G. J.; Padilla-Vivanco, A.; Toxqui-Quitl, C.; Zaca-Morán, P.; Villegas-Hernández, D.; Chávez, F.; Optical fiber sensor based on localized surface plasmon resonance using silver nanoparticles photodeposited on the optical fiber end. *Sensors,* 2014, 14, 18701–18710.

Ou, J.; Zhou, Z.; Chen, Z.; Tan, H.; Optical diagnostic based on functionalized gold nanoparticles. *Int. J. Mol. Sci.,* 2019, 20, 4346.

Papa, A. L.; Boudon, J.; Bellat, V.; Loiseau, A.; Bisht, H.; Sallem, F.; Chassagnon, R.; Berard, V.; Millot, N.; Dispersion of titanate nanotubes for nanomedicine: Comparison of PEI and PEG nanohybrids. *Dalton Trans.,* 2015, 44, 739–746.

Pastoriza-Santos, I.; Liz-Marzán, L. M.; Colloidal silver nanoplates. State of the art and future challenges. *J. Mater. Chem.,* 2008, 18, 1724–1737.

Pérez-Juste, J.; Pastoriza-Santos, I.; Liz-Marzán, L. M.; Mulvaney, P.; Gold nanorods: Synthesis, characterization and applications. *Coord. Chem. Rev.,* 2005, 249, 1870–1901.

Qi, W.; Lee, S.; Phase stability, melting, and alloy formation of Au-Ag bimetallic nanoparticles. *J. Phys. Chem. C.,* 2010, 114, 9580–9587.

Qin, Y.; Ji, X.; Jing, J.; Liu, H.; Wu, H.; Yang, W.; Size control over spherical silver nanoparticles by ascorbic acid reduction. *Colloids Surf. A Physicochem. Eng. Asp.,* 2010, 372, 172–176.

Rodríguez-González, B.; Burrows, A.; Watanabe, M.; Kiely, C. J.; Liz Marzán, L. M.; Multishell bimetallic AuAg nanoparticles: Synthesis, structure and optical properties. *J. Mater. Chem.,* 2005, 15, 1755–1759.

Ruiz-Sanchez, A. J.; Parolo, C.; Miller, B. S.; Gray, E. R.; Schlegel, K.; McKendry, R. A.; Tuneable plasmonic gold dendrimer nanochains for sensitive disease detection. *J. Mater. Chem. B.,* 2017, 5, 7262–7266.

Rycenga, M.; Cobley, C. M.; Zeng, J.; Li, W.; Moran, C. H.; Zhang, Q.; Qin, D.; Xia, Y.; Controlling the synthesis and assembly of silver nanostructures for plasmonic applications. *Chem. Rev.,* 2011, 111, 3669–3712.

Sagle, L. B.; Ruvuna, L. K.; Ruemmele, J. A.; van Duyne, R. P.; Advances in localized surface plasmon resonance spectroscopy biosensing. *Nanomedicine,* 2011, 6, 1447–1462.

Sallem, F.; Boudon, J.; Heintz, O.; Severin, I.; Megriche, A.; Millot, N.; Synthesis and characterization of chitosan-coated titanate nanotubes: Towards a new safe nanocarrier. *Dalton Trans.,* 2017, 46, 15386–15398.

Sasidevi, S.;, Kumarganesh S.; Saranya, S.; Thiyaneswaran, B.; Shree, K. V. M.; Martin, S. K.; Pandey, B. K.; Pandey, D.; Design of Surface Plasmon Resonance (SPR) Sensors for Highly Sensitive Biomolecular Detection in Cancer Diagnostics. *Plasmonics.,* 2024, 20, 677–689.

Sepunaru, L.; Plowman, B. J.; Sokolov, S. V.; Young, N. P.; Compton, R. G.; Rapid electrochemical detection of single influenza viruses tagged with silver nanoparticles. *Chem. Sci.,* 2016, 7, 3892–3899.

Sherry, L. J.; Chang, S. H.; Schatz, G. C.; van Duyne, R. P.; Localized surface plasmon resonance spectroscopy of single silver nanocubes. *Nano Lett.,* 2005, 5, 2034–2038.

Shimomura, M.; Nomura, Y.; Zhang, W.; Sakino, M; Lee, K. H.; Ikebukuro, K.; Karube, I.; Simple and rapid detection method using surface plasmon resonance for dioxins, polychlorinated biphenylx and atrazine. *Anal. Chim. Acta.,* 2001, 434, 223–230.

Singh, A. V.; Bandgar, B. M.; Kasture, M.; Prasad, B. L. V.; Sastry, M.; Synthesis of gold, silver and their alloy nanoparticles using bovine serum albumin as foaming and stabilizing agent. *J. Mater. Chem.,* 2005, 15, 5115–5121.

Song, C. Y.; Yang, Y. J.; Yang, B. Y.; Min, L. H.; Wang, L. H.; Combination assay of lung cancer associated serum markers using surface-enhanced Raman spectroscopy. *J. Mater. Chem. B.,* 2016, 4, 1811–1817.

Sopan, N.; Mahajan, M.; Patil, P.; Chapter 3: Surface plasmon-based diagnostic technology. *Nanoscience and Nanotechnology for Smart Prevention, Diagnostics and Therapeutics: Fundamentals to Applications*, 2024, 35–59. https://doi.org/10.1002/9781394175222.ch3

Sui, M.; Kunwar, S.; Pandey, P.; Lee, J.; Strongly confined localized surface plasmon resonance (LSPR) bands of Pt, AgPt, AgAuPt nanoparticles. *Sci. Rep.,* 2019, 9, 16582.

Sun, L.; Li, Q.; Tang, W.; Di, J.; Wu, Y.; The use of gold-silver core-shell nanorods self-assembled on a glass substrate can substantially improve the performance of plasmonic affinity biosensors. *Microchim. Acta.,* 2014, 181, 1991–1997.

Tan, T.; Tian, C.; Ren, Z.; Yang, J.; Chen, Y.; Sun, L.; Li, Z.; Wu, A.; Yin, J.; Fu, H.; LSPR-dependent SERS performance of silver nanoplates with highly stable and broad tunable LSPRs prepared through an improved seed-mediated strategy. *Phys. Chem. Chem. Phys.,* 2013, 15, 21034–21042.

Ukhurebor, E. K.; The role of biosensor in climate smart organic agriculture toward agricultural and environmental sustainability. *IntechOpen,* 2021, 8.

Wang, X. H.; Li, Y.; Wang, H. F.; Fu, Q. X.; Peng, J. C.; Wang, Y. L.; Du, J.; Zhou, Y.; Zhan, L. S.; Gold nanorod-based localized surface plasmon resonance biosensor forsensitive detection of hepatitis B virus in buffer, blood serum and plasma. *Biosens. Bioelectron.,* 2010, 26, 404–410.

Waxenegger, J.; Trügler, A.; Hohenester, U.; Plasmonics simulations with the MNPBEM toolbox: Consideration of substrates and layer structures. *Comput. Phys. Commun.,* 2015, 193, 138–150.

Wiley, B.; Sun, Y.; Mayers, B.; Xia, Y.; Shape-controlled synthesis of metal nanostructures: The case of silver. *Chem. Eur. J.,* 2005, 11, 454–463.

Xianyu, Y. L.; Wang, Z.; Jiang, X. Y.; A plasmonic nanosensor for immunoassay via enzyme-triggered click chemistry. *ACS Nano,* 2014, 8, 12741–12747.

Yang, X.; Ren, Y.; Gao, Z.; Silver/gold core-shell nanoprism-based plasmonic nanoprobes for highly sensitive and selective detection of hydrogen sulfide. *Chem. Eur. J.,* 2014, 21, 988–992.

Yates, A. M.; Elvin, S. J.; Williamson, D. E.; The optimisation of a murine TNF-α ELISA and the application of the method to other murine cytokines. *J. Immunoassay,* 1999, 20, 31–44.

Zaheer, Z.; Rafiuddin; Multi-branched flower-like silver nanoparticles: Preparation and characterization. *Colloids Surf. A Physicochem. Eng. Asp.,* 2011, 384, 427–431.

Zhang, L. Y.; Ouyang, Y.; Wang, S.; Gong, Y.; Jiang, M.; Yuan, W.; Li, C. M.; Ultrafast synthesis of uniform 4–5 atoms-thin layered tremella-like Pd nanostructure with extremely large electrochemically active surface area for formic acid oxidation. *J. Power Sources,* 2020, 447, 227248.

Zhang, Q. (Ed.); *Nanotoxicity: Methods and Protocols*, 1st ed., vol XII. New York, NY, USA: Humana Press, Springer, 2019, p. 371.

Zhao, J.; Dieringer, J. A.; Zhang, X.; Schatz, G. C.; van Duyne, R. P.; Wavelength-scanned surface-enhanced resonance Raman excitation spectroscopy. *J. Phys. Chem. C.,* 2008, 112, 19302–19310.

Zhao, J.; Zhang, X; Yonzon, C. R.; Haes, A. J.; Van Duyne, R. P.; Localized surface plasmon resonance biosensors. *Nanomedicine,* 2006, 1, 219–228.

Zhou, S.; Li, J; Gilroy, K. D.; Tao, J., Zhu, C.; Yang, X.; Sun, X.; Xia, Y.; Facile synthesis of silver nanocubes with sharp corners and edges in an aqueous solution. *ACS Nano,* 2016, 10, 9861–9870.

Zhu, J.; Zhang, F.; Li, J.-J.; Zhao, J.-W.; Optimization of the refractive index plasmonic sensing of gold nanorods by non-uniform silver coating. *Sens. Actuators B Chem.;* 2013, 183, 556–564.

Zhu, W.; Esteban, R.; Borisov, A. G.; Baumberg, J. J.; Nordlander, P.; Lezec, H. J.; Aizpurua, J.; Crozier, K. B.; Quantum mechanical effects in plasmonic structures with subnanometre gaps. *Nat. Commun.,* 2016, 7, 11495.

## LIST OF ABBREVIATIONS

| | |
|---|---|
| **AA** | Ascorbic acid |
| **Ag** | Silver |
| **$AgNO_3$** | Silver nitrate |
| **AgNPls** | Silver nanoplates |
| **AgNPs** | Silver nanoparticles |
| **Au** | Gold |
| **AuNPs** | Gold nanoparticles |
| **Cu** | Copper |
| **ELISA** | Enzyme-linked immunosorbent assay |
| **FOM** | Figure of merit |
| **fwhm** | Full width at half maximum |
| **LOD** | Limit of detection |
| **LSPR** | Localized surface plasmon resonance |
| **MEF** | Metal-enhanced fluorescence |
| **NIR** | Near-infrared |
| **NPls** | Nanoplates |
| **NPs** | Nanoparticles |
| **OP** | Organophosphate |
| **PCBs** | Polychlorinated biphenyls |
| **POC** | Point-of-care |
| **PVP** | Polyvinyl pyrrolidone |
| **RI** | Refractive index |
| **RIS** | Refractive index sensitivity |
| **RIU** | Refractive index unit |
| **SERS** | Surface-enhanced Raman spectroscopy |
| **SPP** | Surface plasmon polariton |
| **SPR** | Surface plasmon resonance |
| **TEM** | Transmission electron microscopy |
| **TEOS** | Tetraethyl orthosilicate |
| **TNPls** | Triangular NPls |

# 12 Magnetic Properties and Biomedical Applications of Iron Oxide Nanoparticles

*Vidhya Sivan*[1,2], *Namitha Rajeev*[2], *Navya S. L.*[2], *Pius Augustine*[1,2], and *Ann Rose Abraham*[1,2]

[1]Materials Research Laboratory, Sacred Heart College (Autonomous), Thevara, Kochi, India

[2]Department of Physics, Sacred Heart College (Autonomous), Thevara, Kochi, India

## 12.1 INTRODUCTION TO IRON OXIDE NANOPARTICLES

Nanoparticles (NPs) are driving the rapid advancement of nanotechnology. These extraordinary materials are significant in different domains due to their specific size-dependent characteristics.[1] In recent years, substantial study was undertaken on various aspects of iron oxide nanoparticles.[2] In the biomedical realm, iron oxide nanoparticles (IONPs) serve as contrast agents (CAs) in magnetic resonance imaging (MRI), targeted drug delivery, and magnetic fluid hyperthermia (MFH). They enhance image clarity, which is beneficial for accurate diagnosis of medical conditions. Beyond this, IONPs find utility in environmental remediation, focusing on the removal of contaminants from soil and water. In addition, catalytic properties have also been shown to facilitate a wide range of chemical reactions, indicating applicability in industrial processes.[2, 3]

Iron oxides, in their natural form, iron (III) oxide, are widely used in various biological and geological processes, as catalysts, long-lasting pigments (used in paints, coatings, and coloured concrete), and iron ores in thermite.[4] The three primary forms of iron oxide occurring naturally are hematite ($\alpha$-$Fe_2O_3$), maghemite ($\gamma$-$Fe_2O_3$), and magnetite ($Fe_3O_4$), which are explored in various applications, highlighting their essential role in scientific technology.[5, 6]

Among numerous applications,[7] IONPs hold significant promise in magnetic resonance imaging,[8] which uses strong magnetic fields and radio waves for imaging internal structures of the body. The superparamagnetic properties of IONPs at the nanoscale dimension are mainly exploited for MRI. IONPs[8] are used as contrast agents to enhance the image quality by enhancing tissue contrast between different organs. The introduction of IONPs into the body alter the relaxation times of water protons thus helps in highlighting different regions in MRI scans. In view of this, the introduction of IONPs into MRI technology is considered a leap forward not only in diagnostic imaging but also in developing strategies for treatment, underscoring the enormous transformative prospects that nanotechnology introduces into medicine.[9]

## 12.2 MAGNETIC PROPERTIES OF IRON OXIDE NANOPARTICLES

The magnetic properties of iron oxide nanoparticles are imparted to the presence of unpaired electrons in the 3D shell of iron atoms which impart their characteristic magnetic behaviour. The four unpaired electrons in the 3D shell give it a significant magnetic moment. Moreover, the magnetic properties of IONPs depend on the crystal structure. In paramagnetic materials, atomic magnetic moments are randomly oriented, and net magnetic moment is zero in the absence of a magnetic field. These materials possess good relative magnetic permeability. But in ferromagnetic substances, in the absence of an external field, all the atomic moments are aligned. Hence, a ferromagnetic material has a magnetic moment, while antiferromagnetic crystals have no net magnetic moment.[5] In a ferromagnetic material, magnetization ($M$) increases with applied magnetic field ($H$) and reaches saturation. Residual magnetization will be present in the absence of an external field.

Superparamagnetic materials are single-domain magnetic materials that do not have a hysteresis loop. The magnetic properties of IONPs also depend on temperature. As the temperature increases, the order of magnetic moments in antiferromagnetic, and ferromagnetic materials weakens. Above the Curie temperature, ferromagnetic materials lose their magnetism and become paramagnetic.

Magnetite and maghemite nanoparticles are extensively used Superparamagnetic Iron Oxide Nanoparticles (SPIONs) for biomedical applications. At ambient temperature, magnetite and maghemite nanoparticles are both superparamagnetic. The superparamagnetic attributes of iron oxide nanoparticles are size dependent, and when the size drops below 10 nm, thermal fluctuations can change the

DOI: 10.1201/9781032660639-12

magnetization direction of the whole crystal. When a material contains a large number of these crystals, it acts in the same way as a paramagnet, except that instead of individual atoms, the momentum of the entire crystal is different. Magnetization studies in both constant and time-varying magnetic fields are often used to measure the essential properties of iron oxide nanoparticles. These unique magnetic properties and their sensitivity to external magnetic fields make IONPs auspicious for various biomedical applications.

## 12.3 BIOMEDICAL APPLICATIONS OF IRON OXIDE NANOPARTICLES

In the field of nanobiotechnology, nanomaterials play a vital role, particularly in fields such as prosthetics, implants, drug delivery systems,[10] and diagnostic tools. The compatibility with living tissue, the ability to degrade over time, high surface area, durability, unique chemical behaviours, structural modulation, and low toxicity level of INOPs have sparked considerable research interests for medical purposes. These nanoparticles, known for their magnetism and semiconducting properties, have exciting possibilities in biomedical fields for marking cells, combating cancer, resisting bacteria, cancer therapy, targeted drug delivery, and magnetic resonance imaging.

Radiation-assisted cancer treatment, surgery, and anticancer drugs from IONPs have gained much research interest.[11, 12] For the past few decades, magnetic nanoparticles with reduced side effects have been used as a better choice in biomedicine[13] and are available as a therapeutic alternative for magnetic resonance imaging and cancer treatment.[14] They easily interact with biomolecules internally and externally on cells due to their small size.[15] In addition, the FDA has approved iron oxide nanoparticles in cancer hyperthermia and cancer diagnostics.

Here, we also enumerate the clinical experience of administering iron oxide nanoparticles with systemic liposomal drug delivery as a parenteral treatment for iron deficiency anaemia (IDA). Notably, the creators of liposomal drug delivery aim to avoid the innate immune system's natural interaction with nanoparticles, an interaction that injectable iron uses to its advantage in clinical trials.[16] Recent studies have shown that iron oxide nanoparticles and immune cells can interact to produce anticancer immune responses. The clinical utility and practice of iron oxide nanoparticles provide compelling case studies of the promise and challenges of cancer nanomedicine. They also demonstrate the need for interdisciplinary approaches in the development and use of nanomedicines in clinical cancer care. Iron oxides are more sensitive to magnetic fields because of their high magnetic permeability compared to other material systems.[17] Hence, they have been recognized to be utilized in micelles as magnetic nanoformulations, which produce superior stability and biocompatibility and increased cellular uptake by cancerous cells in vitro.[17] The various biomedical applications of IONPs are shown in Figure 12.1.

### 12.3.1 Iron Oxide Nanoparticles in MRI

Numerous studies have been done for the past few years on various biomedical nanomaterials that could act as agents to fulfil cell labelling, sensing, and diagnosis. The US Food and Drug Administration has authorized MRI contrast agents based on iron oxide nanoparticles that can minimize proton transverse relaxation T2. However, MRI CA based on IONPs exhibited low contrast ability because of poor

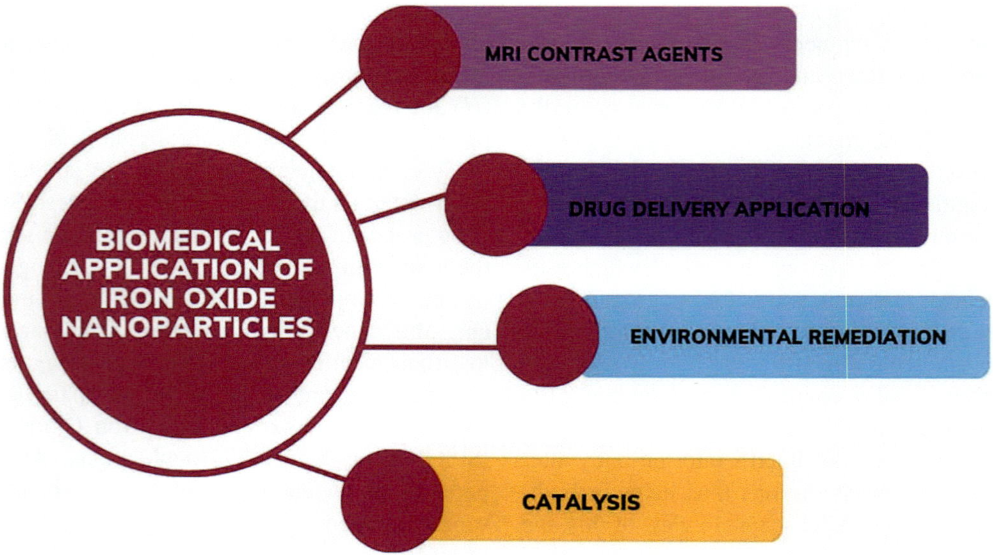


FIGURE 12.1 Biomedical applications of iron oxide nanoparticles.

crystallinity and surface modulation, which can lead to low accuracy and sensitivity, which hindered its application to an extent.[17–20]

Many studies were then focused on modulating the relaxivity of IONPs by crystal structure engineering. Depending on relaxivity–structure analysis, IONPs with large T2 relaxivity have exhibited excellent features such as saturation magnetization, large size, and high non-uniform magnetic field. Numerous strategies were also used to develop IONP-based multi-modal imaging CAs, including positron emission tomography and computed tomography.[21]

In recent studies, receptive MRI CAs featuring modal switchable capacity have shown a significant signal difference between normal tissue and lesions. Similar to conventional MRI, there are two modals of switchable tumour microenvironment (TEM)–responsive MRI contrast imaging: T1 (longitudinal relaxation time) contrast switch to T2 and T2 contrast switch to T1 – a detailed overview of MRI and how T1 and T2 switchable modes work is elucidated.[18]

MRI is one of the most influential diagnostic techniques that offer excellent temporal response and spatial resolution compared to X-ray imaging techniques. The fundamental principle of MRI involves capturing relaxation signals from proton spins within the human body. These proton spins are excited by radiofrequency waves while being subjected to a strong magnetic field.

MRI used in the biomedical field images the water-containing soft tissues of the human body. The images are obtained by aligning the magnetic poles of the water protons along the direction of the applied magnetic field. A small radiofrequency pulse is applied at a frequency matching the resonance frequency of hydrogen atoms. When the pulse stops, the water protons release RF signals, which are then detected and produce images. The relaxation time is the time taken by the protons to realign to the initial configuration.[19, 22, 23]

The two important types of relaxation times in MRI are T1, known as the longitudinal relaxation time, and T2, referred to as the transverse relaxation time. In the MRI, the shades of tissues and fluids are examined using the intensity which correlates with relaxation times. High-intensity signals are used to form images, such as white on grayscale, which correspond to the relaxation time T1. The relaxation time T2 corresponds to low-intensity signals, which appear as black. The intermediate signal strength can be observed in greyscale, indicating a balance between T1 and T2. Brighter parts suggest greater intensity scales, whereas darker sections imply lower intensity signals, which are depicted in greyscale in MRI scans.[22]

The longitudinal relaxation time, T1, spin-lattice relaxation, defines the process of returning the magnetic vector into equilibrium in the direction of the applied external magnetic field. Tissues with small T1 relaxation time appear brighter on T1-weighted images, such as bright/dark. Meanwhile, the T2 or spin relaxation vector runs transverse to the external magnetic field, and tissues along the T2 look bright on T2-weighted imaging. These are found useful in pathology identification, such as inflammation and edema.[19]

The T1 and T2 relaxation times are the basic concepts in MRI and have a significant role in image contrast. They are specific to different tissues and depend on the water content, which helps differentiate various tissues.[19, 22, 23]

The imaging ability of MRI can be further enhanced using contrast agents that can alter the relaxation time. Magnetic nanoparticles are identified as good contrast agents that can overcome current hindrances and revolutionize medical imaging. Magnetic nanoparticles have emerged as good MRI contrast agents capable of altering T1 and T2 relaxation times. Superparamagnetic iron oxide nanoparticles can diminish the intensity of signals in T2-weighted sequences by inducing tissue inhomogeneity while increasing T1-weighted sequences. T2 relaxation time rises as core diameter increases, whereas T1 increases as core diameter decreases. Commercially available SPIONs are usually T2 contrast agents with diameters ranging from 16 to 150 nm. In addition to T1 and T2, various MR sequences, such as gradient echo sequences, are available and used for imaging SPIONs. Other breakthroughs in tissue contrast are susceptibility-weighted imaging (SWI).[24–26]

### 12.3.2 Iron Oxide Nanoparticles for Drug Delivery Applications

Iron oxide nanoparticles, featuring solubility, stability, drug release control, customized dosage, and efficient delivery of therapeutics to desired sites with low toxicity and side effects, have been widely used in therapeutic agent remediation. With the advantage of low toxicity, biodegradability, and ease of surface functionalization and production, magnetic nanoparticles (MNPs), particularly iron, were the most widely used drug delivery system (DDS).[27, 28] However, the exceptional superparamagnetism of IONP drug delivery systems distinguishes them from other DDSs in terms of therapeutic (e.g., hyperthermia, drug delivery, and magnetic targeting) and diagnostic (e.g., use as contrast agents for magnetic resonance imaging) uses. Also, being superparamagnetic, IONPs in their magnetite ($Fe_3O_4$) or maghemite ($Fe_2O_3$) display excellent biocompatibility.[27–29]

Moreover, even at large dosages, IONPs do not generate oxidative stress due to their biodegradability, which allows for effective removal from the body.[29] Because of these exceptional attributes, IONPs have been able to show promise for bench-to-clinic translation. Because of their variable surface chemistry, IONPs can be loaded with a variety of pharmaceuticals and bioactive molecules to generate drug-conjugated systems.[29] Depending on the desired therapeutic applications, drug molecules are loaded using a range of surface functionalization strategies, including non-covalent conjugation and covalent conjugation (chemical surface modification).[30, 31]

### 12.3.3 Environmental Remediation

Rising consumer demands and population growth have shown a drastic increase in production and industrial installations. However, pollution of the environment by toxic compounds released from various human sources is a severe problem that requires immediate attention. Nanotechnology offers many useful and reasonably priced ways to prevent the discharge of harmful substances into the environment. One of the latest advancements in this field is the application of nanoparticles for environmental remediation. Surface modification, removal of heavy metals, and organic pollutants can be beneficial. Moreover, the catalytic properties of IONPs can break down pollutants and degrade organic contaminants. The application of nanotechnology is a controversial subject due to the limited understanding of how nanoparticles interact with the environment.[32] Despite the advantages, the toxicological effects of employing nanoparticles in the environment are being carefully considered. As it is, study places a strong emphasis on using nanoparticles to remove environmental pollutant materials at the nanoscale and nanoparticle sizes that have been effectively applied to environmental remediation. To reduce the impact of nanoparticles, it has been proposed that nanoparticles be manufactured via biological or green synthesis methods. The employment of artificial nanoparticles as a medicinal treatment has also been contemplated.

### 12.3.4 Catalysis

IONPs are used as effective catalysts for various chemical processes due to their large surface area and distinctive electrical characteristics.[32] INOPs are superparamagnetic in nature, have a size in micro diameters, are recyclable, and environmentally safe making them useful in numerous sectors.[32] IONPs are frequently employed as efficient nanocatalysts for organic processes. It has been determined that IONPs act as catalysts for organic reactions. These include arylation of C-H; hydroboration of styrene; oxidation of m-cresol; couplings between C–N, C–C, C–S, C–O, and S–S; and the production of bioactive chemicals.[33] Iron nanocomposite catalysts will ultimately unlock opportunities for the development of sustainable and environmentally friendly technologies.

### 12.3.5 Toxicity

The relationship between NPs and biological systems still needs to be explored. The intricacy of the particles arises from their capacity to interact and form bonds with biological matter, as well as from their capacity to change their surface properties in response to their environment. There has been recent studies on the mechanisms controlling NP interactions with cells. It is more challenging to understand the processes and pathways that exist inside cells. Even the same particle can exhibit vastly different behaviours depending on factors like particle size and surface coating. This makes it challenging to categorize the behaviour of NPs upon interaction with biological systems, hence complicating the identification of nanoparticle dangers.[34, 35]

In vivo toxicity studies have found that the use of PEI-coated IONPs like SEI-10[36] exhibits significant toxicity, and the mice in the experiments have shown high mortality rates with the maximum rate of tolerance dose. Haematological and biochemical analyses have shown an increase in alanine aminotransferase (ALT) levels, creating liver stress. Similarly, in cytotoxicity cell studies, PEI-coated IONPs like SEI-10 caused severe toxic effects in both macrophages and cancer cells through mechanisms like reactive oxygen species (ROS) production and apoptosis.[36] In summary, the toxicity of IONPs is dependent on their particle size and surface coating; that is, PEI-coated IONPs have shown more toxicity than PEGylated ones.

## 12.4 CONCLUSION

Iron oxide nanoparticles with unique characteristics and a multitude of applications are outlined in this chapter. The chapter begins with an introduction to the fundamental characteristics of IONPs, specifically the magnetic characteristics that make these nanoparticles particularly valuable in diverse applications. The utility of IONPs in biomedical applications, notably in MRI and as contrast agents, is discussed in detail. Further, the potential of IONPs for drug delivery systems, with targeted therapy ensuring low side effects, is also discussed. Beyond the medical applications and the efficiency of IONPs, the factors influencing their performance are also examined. Additionally, the chapter addresses concerns regarding toxicity and biocompatibility associated with IONPs, evaluating critical parameters to ensure the safe integration of these materials in various applications. This chapter thus provides insights into the advancing technology using IONPs in various domains of nanoscience and nanotechnology.

**Keywords:**

Catalysis
Drug delivery
Iron oxide nanoparticles
Magnetic resonance imaging
Superparamagnetism

## REFERENCES

1. Salata, O. V. Journal of Nanobiotechnology. *J. Nanobiotechnology* 6, 1–6 (2004).
2. Schneider, M. G. M. et al. Biomedical applications of iron oxide nanoparticles: Current insights progress and perspectives. *Pharmaceutics* 14, (2022).
3. Liu, G., Gao, J., Ai, H. & Chen, X. Applications and potential toxicity of magnetic iron oxide nanoparticles. *Small* 9, 1533–1545 (2013).

4. Laurent, S. et al. Magnetic iron oxide nanoparticles: Synthesis, stabilization, vectorization, physicochemical characterizations and biological applications. *Chem. Rev.* 108, 2064–2110 (2008).
5. Teja, A. S. & Koh, P. Y. Synthesis, properties, and applications of magnetic iron oxide nanoparticles. *Prog. Cryst. Growth Charact. Mater.* 55, 22–45 (2009).
6. Cornell, R. M., Schwertmann, U. & Bodenkunde, L. Introduction I. Book reviews: Announcements. *Corros. Rev.* 15, 533–559 (2011), Universität Tübingen.
7. Laurent, S. et al. Erratum: Magnetic iron oxide nanoparticles: Synthesis, stabilization, vectorization, physicochemical characterizations, and biological applications. *Chem. Rev.* 110, 2574 (2010). (*Chemical Reviews* 108, 2064 (2008)).
8. Sangaiya, P. & Jayaprakash, R. A review on iron oxide nanoparticles and their biomedical applications. *J. Supercond. Nov. Magn.* 31, 3397–3413 (2018).
9. Ali, A. et al. Synthesis, characterization, applications, and challenges of iron oxide nanoparticles. *Nanotechnol. Sci. Appl.* 9, 49–67 (2016).
10. Faraji, A. H. & Wipf, P. Nanoparticles in cellular drug delivery. *Bioorganic Med. Chem.* 17, 2950–2962 (2009).
11. Pucci, C., Martinelli, C. & Ciofani, G. Innovative approaches for cancer treatment: Current perspectives and new challenges. *Ecancermedicalscience* 13, 1–26 (2019).
12. Obaidat, I. M., Narayanaswamy, V., Alaabed, S., Sambasivam, S. & Muralee Gopi, C. V. V. Principles of magnetic hyperthermia: A focus on using multifunctional hybrid magnetic nanoparticles. *Magnetochemistry* 5, (2019).
13. Cole, A. J., Yang, V. C. & David, A. E. Cancer theranostics: The rise of targeted magnetic nanoparticles. *Trends Biotechnol.* 29, 323–332 (2011).
14. Sharma, A., Goyal, A. K. & Rath, G. Recent advances in metal nanoparticles in cancer therapy. *J. Drug Target.* 26, 617–632 (2018).
15. Conde, J., Doria, G. & Baptista, P. Noble metal nanoparticles applications in cancer. *J. Drug Deliv.* 2012, 1–12 (2012).
16. Soetaert, F., Korangath, P., Serantes, D., Fiering, S. & Ivkov, R. Cancer therapy with iron oxide nanoparticles: Agents of thermal and immune therapies. *Adv. Drug Deliv. Rev.* 163–164, 65–83 (2020).
17. Faria, M. R. et al. Synthesis and characterization of magnetoliposomes for MRI contrast enhancement. *Int. J. Pharm.* 446, 183–190 (2013).
18. Jeon, M., Halbert, M. V, Stephen, Z. R. & Zhang, M. *Prospectives* 33, 1–41 (2022).
19. Jeon, M., Halbert, M. V., Stephen, Z. R. & Zhang, M. Iron oxide nanoparticles as T1 contrast agents for magnetic resonance imaging: Fundamentals, challenges, applications, and prospectives. *Adv. Mater.* 33, 1–18 (2021).
20. Tombácz, E., Turcu, R., Socoliuc, V. & Vékás, L. Magnetic iron oxide nanoparticles: Recent trends in design and synthesis of magnetoresponsive nanosystems. *Biochem. Biophys. Res. Commun.* 468, 442–453 (2015).
21. Zhao, Z. et al. Bioactive materials recent advances in engineering iron oxide nanoparticles for effective magnetic resonance imaging. *Bioact. Mater.* 12, 214–245 (2022).
22. Lapusan, R., Borlan, R. & Focsan, M. Advancing MRI with magnetic nanoparticles: A comprehensive review of translational research and clinical trials. *Nanoscale Adv.* 6, 2234–2259 (2024).
23. Girardet, T. et al. SPIONs magnetic nanoparticles for MRI applications: Microwave synthesis and physicochemical, magnetic and biological characterizations. *Mater. Today Commun.* 36, (2023).
24. German, S. V. et al. Liposomes loaded with hydrophilic magnetite nanoparticles: Preparation and application as contrast agents for magnetic resonance imaging. *Colloids Surfaces B Biointerfaces* 135, 109–115 (2015).
25. Yang, H. M., Park, C. W., Park, S. & Kim, J. D. Cross-linked magnetic nanoparticles with a biocompatible amide bond for cancer-targeted dual optical/magnetic resonance imaging. *Colloids Surfaces B Biointerfaces* 161, 183–191 (2018).
26. Naha, P. C. et al. Dextran-coated cerium oxide nanoparticles: A computed tomography contrast agent for imaging the gastrointestinal tract and inflammatory bowel disease. *ACS Nano* 14, 10187–10197 (2020).
27. Na, H. Bin, Song, I. C. & Hyeon, T. Inorganic nanoparticles for MRI contrast agents. *Adv. Mater.* 21, 2133–2148 (2009).
28. Herrmann, I. K., Wood, M. J. A. & Fuhrmann, G. Extracellular vesicles as a next-generation drug delivery platform. *Nat. Nanotechnol.* 16, 748–759 (2021).
29. Skorjanc, T., Benyettou, F., Olsen, J. C. & Trabolsi, A. Design of organic macrocycle-modified iron oxide nanoparticles for drug delivery. *Chem. – A Eur. J.* 23, 8333–8347 (2017).
30. El-Boubbou, K. Magnetic iron oxide nanoparticles as drug carriers: Preparation, conjugation and delivery. *Nanomedicine* 13, 929–952 (2018).
31. Sun, C. et al. Tumor-targeted drug delivery and MRI contrast enhancement by chlorotoxin-conjugated iron oxide nanoparticles. *Nanomedicine* 3, 495–505 (2008).
32. Jung, H. et al. Preparation of biotic and abiotic iron oxide nanoparticles (IOnPs) and their properties and applications in heterogeneous catalytic oxidation. *Environ. Sci. Technol.* 41, 4741–4747 (2007).
33. Mohapatra, M. & Anand, S. Synthesis and applications of nano-structured iron oxides/hydroxides – a review. *Int. J. Eng. Sci. Technol.* 2, 127–146 (2011).
34. Elsaesser, A. & Howard, C. V. Toxicology of nanoparticles *Adv. Drug Deliv. Rev.* 64, 129–137 (2012).
35. Sa, E. et al. Homogeneous conjugation of peptides onto gold nanoparticles enhances macrophage response. *ACS Nano* 3, 1335–1344 (2009).
36. Feng, Q. et al. Uptake, distribution, clearance, and toxicity of iron oxide nanoparticles with different sizes and coatings. *Sci. Rep.* 8, 1–13 (2018).

# 13 Recent Developments in Metal and Metal Oxide Nanoparticles in the Field of Diagnosis

*Neson Varghese and Devika S.*
Bharata Mata College, Thrikkakkara, Edapally, Cochin, Kerala, India

## 13.1 INTRODUCTION

Metal nanoparticles have shown immense potential in the field of diagnosis due to their unique optical, magnetic, and electronic properties. Gold and silver nanoparticles are commonly used in biosensors for detecting biomolecules such as proteins, DNA, and small molecules. Their high surface area-to-volume ratio allows for efficient binding of target molecules, leading to sensitive detection. Metal nanoparticles can enhance various imaging techniques such as magnetic resonance imaging (MRI), computed tomography (CT), and photoacoustic imaging. They can be functionalized with targeting ligands to specifically accumulate at disease sites, providing contrast enhancement and improving diagnostic accuracy. Gold nanoparticles are frequently used in lateral flow assays, such as pregnancy tests and infectious disease diagnostics. Their ability to produce a visible color change upon binding to target analytes makes them suitable for rapid and easy-to-interpret diagnostic tests. They are utilized in surface-enhanced Raman scattering (SERS) for highly sensitive detection of analytes. They enhance the Raman signal of molecules adsorbed on their surface, enabling the detection of low concentrations of biomolecules for early disease diagnosis. They can be engineered to have dual functions for both diagnosis and therapy, known as theranostics. They can serve as contrast agents for imaging while also delivering therapeutic agents or photothermal therapy for targeted treatment of diseases. Recent advancements in nanoparticle-based biosensors have indeed opened up new avenues for accelerating diagnosis across various health conditions and diseases. The combination of high surface area, exceptional sensitivity, specificity, and optical characteristics of metal and metal oxide nanoparticles has revolutionized diagnostic approaches.

Metal and metal oxide nanoparticles offer unparalleled sensitivity, allowing for the detection of biomarkers and pathogens at extremely low concentrations. This heightened sensitivity enables early detection of diseases like cancer and viral infections, facilitating prompt intervention and treatment. Functionalization of metal nanoparticles with specific ligands or biomolecules enhances their specificity for target analytes. This specificity reduces false-positive results and ensures accurate diagnosis, crucial for reliable healthcare interventions. They can be tailored to detect a wide range of chemical and biological targets, including cancer biomarkers, viral RNA/DNA, and various disease-related molecules. This versatility enables their application in diverse diagnostic scenarios, addressing multiple healthcare needs. Advances in nanoparticle synthesis techniques have led to the development of innovative tools for chemical and biological sensing. Techniques such as wet chemical synthesis, sol-gel methods, and microfluidic-assisted synthesis offer precise control over nanoparticle size, shape, and surface properties, enhancing their performance in diagnostic applications. Metal nanoparticles hold promise for in vivo imaging applications, allowing real-time visualization of physiological processes and disease progression. Their optical characteristics, combined with targeting moieties, enable non-invasive imaging modalities for monitoring treatment efficacy and disease dynamics.[1]

Overall, the recent progress in nanoparticle-based biosensors is reshaping the landscape of diagnostic technologies, offering rapid, sensitive, and specific tools for early disease detection and personalized healthcare management. These advancements hold great potential for improving patient outcomes and reducing healthcare burdens globally. Among various nanomaterials, metals and metal oxides have attracted a lot of attention due to their special structure and considerable mechanical, thermal, optical, and electrical properties. The emerging use of metal or metal oxide nanoparticles in virus targeting formulations shows the promise of improved diagnostic or therapeutic ability of the agents while uniquely enhancing the prospects of targeted drug delivery. Although a number of nanoparticles varying in composition, size, shape, and surface properties have been approved for human use, the candidates being tested or approved for clinical diagnosis and treatment of viral infections are relatively less in number.[2]

Metal nanoparticles also have many applications in medicine for tumor visualization, drug delivery, and early diagnosis, such as X-ray imaging, computed tomography,


DOI: 10.1201/9781032660639-13

magnetic resonance imaging, positron emission tomography (PET), and treatment with radiation.[3]

This chapter aims to discuss the properties of different types of metal and metal oxide nanoparticles in the field of diagnosis, how we can cure diseases using those nanoparticles, and recent developments happening in the field. Among the different varieties of metal and metal oxide nanoparticles, this chapter focuses on gold and silver as metal nanoparticles and iron oxide and zinc oxide as metal oxide nanoparticles.

## 13.2 GOLD NANOPARTICLES IN THE FIELD OF DIAGNOSIS

The use of gold nanoparticles (GNPs) in the diagnosis and treatment of gastrointestinal cancer is indeed a promising avenue of research (Figure 13.1). Their unique optical properties and biocompatibility make them valuable tools in cancer therapy.[1–3] Their ability to serve as visualization agents helps in tracking drugs and identifying the location and boundaries of cancer, which can improve treatment precision.[4] Moreover, GNPs can be employed in photodynamic and photothermal therapies, further enhancing their potential in cancer treatment.[5, 6]

Michael Faraday's discovery of the Faraday-Tyndall effect and its relevance to GNPs add a historical perspective to their use in modern medicine.[7] The historical context of gold nanoparticles and their diverse applications, like scattering properties of suspended gold particles, could be used for treatment of solid tumors.[8] The explanation of how GNPs' size and morphology can be tailored for specific applications is informative. Various GNP shapes such as nanorods, nanocages, and nanoclusters have shown great relevance in gastrointestinal cancer research.[9]

Gold nanorods can be selectively absorbed by tumor tissue while maintaining optimal light absorption properties, making them ideal for photothermal treatment. The use of gold nanorods to enhance the delivery efficiency of photodynamic drugs and their effects on inducing cell apoptosis in gastric cancer treatment are noteworthy advancements.[5, 6, 9] The application of gold nanorods in targeting pancreatic ductal adenocarcinoma (PDA) cells and oral squamous cell carcinoma (OSCCs) is also promising.[10] These developments show the potential of gold nanorods to be utilized effectively in various cancer therapies. It would be beneficial to elaborate on the specific mechanisms of action and experimental findings in each of these applications for a more comprehensive understanding of their potential in cancer treatment.

The description of gold nanocages' structural characteristics, including their hollow cage-like structure, high specific surface area, surface modifiability, and high drug loading rate, emphasizes their advantageous properties for drug delivery.[11] The use of gold nanocages in combination with various agents to inhibit pancreatic cancer and colon cancer is a promising approach. Their multi-mode imaging ability, including near-infrared fluorescence and magnetic resonance imaging, adds a valuable diagnostic component to cancer detection and monitoring.[12, 13] The construction of PPHAuNCsTNCs with miR-26a loading for hepatocellular carcinoma (HCC) therapy, along with the ability to monitor treatment progress through fluorescence and photoacoustic tomography imaging, showcases the potential for personalized and precise cancer treatment.[10] Further elaboration on the mechanisms and specific experimental results in each of these applications would enhance the understanding of gold nanocages' role in cancer therapy.

Gold nanoclusters and their applications in cancer imaging and treatment are quite informative. The description of gold nanoclusters as ultra-small particles with low cytotoxicity and excellent red fluorescence characteristics, especially in avoiding an autofluorescence background in vivo, highlights their advantages in medical imaging.[14–16] The use of gold nanoclusters coated with folic acid-conjugated

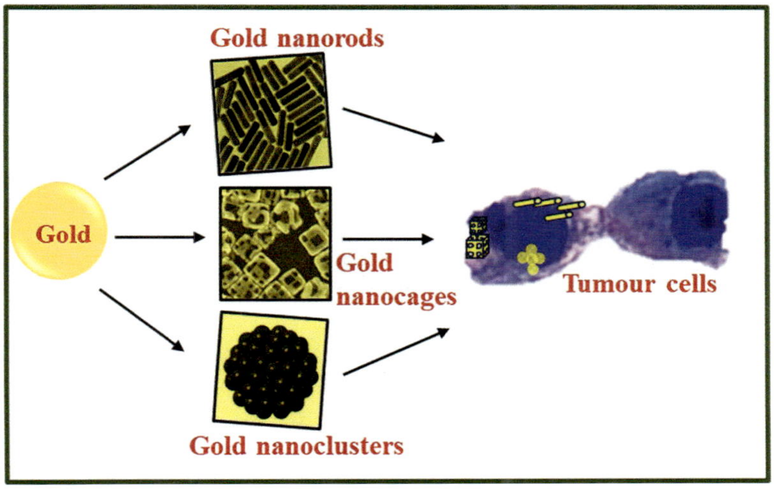


FIGURE 13.1 Gold nanoparticles with different structures.

silica for targeting gastric cancer cells and their application in optical and CT dual-mode imaging of gastric cancer is a promising development. This approach has the potential to aid in the early detection of gastric cancer, which is essential for improving patient outcomes.[17] The modification of gold nanoclusters with near-infrared fluorescent dye and albumin nanoparticles to create a complex with a photothermal effect for colorectal cancer treatment demonstrates their versatility in therapeutic applications.[18] The development of AuNCs@GSH-FA for fluorescence/CT dual-mode imaging in gastric cancer is another significant contribution to cancer diagnosis. It's essential to continue exploring how gold nanoclusters can be used in various diagnostic and treatment approaches.[10]

## 13.3 VERSATILE ROLE IN DRUG DELIVERY

Gold nanoparticles have versatile applications in cancer treatment. They can passively accumulate at tumor sites due to their small particle size. Additionally, when combined with specific active molecules, they can actively target and influence tumor cells (Figure 13.2). For example, the use of the ramucirumab antibody linked to gold nanorods targets gastric tumors and enhances the anti-cancer effect of the drug.[19, 20] GNPs serve as effective carriers for delivering chemotherapeutics and genes to the tumor site, significantly improving the efficacy of active molecules. They can also adapt to the changing tumor microenvironment and inhibit tumor growth. Various types of gold nanodelivery systems have been designed to release active molecules in response to the conditions of the tumor microenvironment, such as acidity or redox potential. These properties make GNPs promising tools in cancer therapy, so GNPs were used for targeted delivery of chemotherapy drugs to achieve precision of therapy for gastrointestinal cancer. The targeting drug trastuzumab (Tmab) was combined with AuNCs (T-AuNPs) to create new nanocomposites that can target human epidermal growth factor receptor-2 (HER2) to induce autophagy on both Tmab-sensitive and Tmab-resistant gastric cancer cells. The anti-tumor drug epigallocatechin gallate (EGCG) is delivered by GNPs to gastric cancer cells and tissues in a time-dependent manner. This significantly inhibits proliferation compared to direct injection of the EGCG drug for gastric cancer and has no toxic effect on normal epithelial tissue. By encapsulating GNPs with cisplatin and glucose, cisplatin can be effectively delivered into head and neck squamous cancer (HNSC). Compared to cisplatin used as a free drug, GNPs with cisplatin effectively inhibit tumor cell proliferation and enhance radiotherapy sensitivity in HNSC. Furthermore, as a CT contrast agent, they can be used as an adjuvant cancer therapy in CT imaging for HNSC diagnosis and treatment. Plectin-1-targeted multifunctional peptides were used to modify GNPs to target pancreatic ductal adenocarcinoma (PDAC). Furthermore, the anti-cancer drug Gemcitabine (GEM) was injected into the surface of GNPs to form GNP-Gem, which can selectively deliver GEM into cancer cells and exert a significant anti-proliferative effect on PDAC cell lines.[21, 22]

## 13.4 APPLICATIONS AND PROGRESS IN MEDICAL DIAGNOSIS

The use of gold nanoparticles in the treatment of gastrointestinal cancers, particularly through photothermal therapy (PTT) and photodynamic therapy (PDT), represents a promising approach for cancer treatment. These therapies selectively target cancerous tissues while minimizing damage to healthy surrounding tissues. These nanoparticles have several advantages, such as biocompatibility and the ability to accumulate in tumor sites, making them attractive for various imaging modalities like computed tomography and magnetic resonance imaging.

**CT Imaging:** GNPs can serve as effective CT contrast agents due to their high x-ray attenuation coefficient and biocompatibility. They offer advantages over traditional iodine-based contrast agents, as they provide longer circulation times and fewer side effects. Conjugating GNPs with specific targeting agents, like monoclonal antibodies, allows for precise imaging of cancer cells, even at the single-cell level.[23, 24]

**MRI:** GNPs can also be used in MRI by conjugating them with T1 or T2 contrast agents. This enables the enhancement of tissue contrast and the ability to visualize individual cancer cell positions. These hybrid nanoparticles, such as Fe3O4@Au@β-CD,

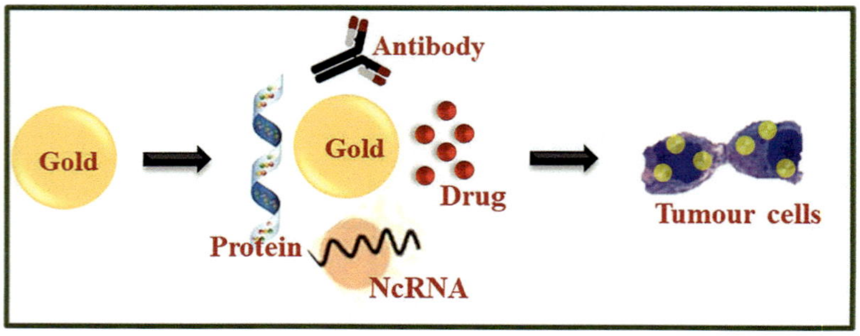


FIGURE 13.2 Delivery using GNPs to regulate tumor cell fate.

have the potential for both diagnosis and treatment of cancer, like gastric cancer.[25]

**Optical Coherence Tomography (OCT):** Gold nanoclusters can enhance the sensitivity of OCT when combined with microneedles and ultrasound. This improves the detection of early neoplasia, allowing for more accurate diagnosis and monitoring.[26, 27]

These advancements in nanoparticle-based imaging are valuable for early cancer detection and treatment planning. Innovative approaches have been developed, such as the construction of a PTT/PDT (Figure 13.3) composite nanosystem that couples ICG on the surface of hollow gold nanospheres (HAuNS) and targets the endoplasmic reticulum (ER). When irradiated with NIR, this system increases temperature and generates reactive oxygen species (ROS), inducing ER stress to enhance the immune response.

Another novel approach involves using branched polyethylene glycol (PEG) as a linker to conjugate ICG molecules to gold nanospheres. This nanosystem can inhibit tumor growth and metastasis through combined PTT and PDT therapy mediated by NIR. Additionally, it is amenable to NIR fluorescence imaging, making it a promising approach for cancer therapy.[28–30]

The advantages of using photoacoustic imaging (PAI) as a cost-effective and portable method for cancer diagnosis, especially when compared to more expensive and less accessible imaging modalities like MRI and CT. GNPs play a significant role in enhancing the capabilities of PAI as contrast agents. Their unique properties, such as high absorption cross-section due to surface plasmon resonance (SPR), stability, and biocompatibility, make them promising candidates for photoacoustic imaging. These advantages allow GNPs to produce strong photoacoustic signals, leading to high-resolution imaging in deep tissues.[31] Several innovative approaches have been developed using GNPs for cancer diagnosis with PAI. For example, the use of a 5-nm molecularly activated plasmonic nanosensor (MAP) targeting the epidermal growth factor receptor (EGFR) enhances sensitivity and specificity in detecting head and neck squamous cell carcinoma (HNSCC). Additionally, combining paclitaxel (PTX), gold nanorods, perfluorohexane (PFH), and folic acid-bovine serum protein (FA-BSA) forms a complex that acts as both an ultrasound contrast agent and a drug delivery system. This approach allows for efficient photoacoustic imaging and on-demand drug release. The ability of GNPs to serve as both imaging agents and potential therapeutic tools in cancer diagnosis and treatment is indeed a promising development in the field of medical imaging.[32, 33]

The field of nanomedicine, particularly the use of gold nanoparticles, has significantly advanced cancer treatment and diagnosis in recent years, particularly in the context of gastrointestinal tumors. GNPs, with their remarkable biocompatibility and versatility, offer a range of benefits for these applications. While there are challenges to overcome in terms of drug metabolism, safety concerns, in vivo efficacy, biocompatibility, stability, preparation costs, and immunogenicity, GNPs remain valuable in the realm of gastrointestinal cancer therapy and diagnosis. The extensive and successful research in biological imaging and cancer treatment using GNPs holds great promise for overcoming these challenges and improving the clinical management of gastrointestinal cancer.

## 13.5 SILVER NANOPARTICLES IN THE FIELD OF DIAGNOSIS

Silver nanoparticles indeed show promising applications, particularly in cancer diagnosis and treatment. The potential of silver-based nanoparticles for cancer treatment and diagnosis is due to their bioactivity and biocompatibility. The antimicrobial properties of silver seem promising in addressing complications from antibiotic resistance. Quick and low-adverse-reaction remedies for cancer are indeed crucial. With unique physical and chemical features, including conductance, durability, catalysis, and antimicrobial capabilities, antimicrobial properties of Ag and Ag-based compounds have been demonstrated against that range of microbes, including fungi, bacteria, and parasites. The versatility of silver nanoparticles makes them valuable for targeted therapy compositions, showcasing exciting advancements in biomedical imaging and pharmacology. These innovations shed light on the potential impact on cancer treatments.[34–36]

## 13.6 SYNTHESIS METHODS AND PROPERTIES OF SILVER NANOPARTICLES

It's fascinating how silver nanoparticles can be synthesized through both "top to bottom" and "bottom to top" approaches. The efficiency of the bottom to top method in producing a large quantity of nanoparticles quickly is impressive, especially with its high yield in chemical

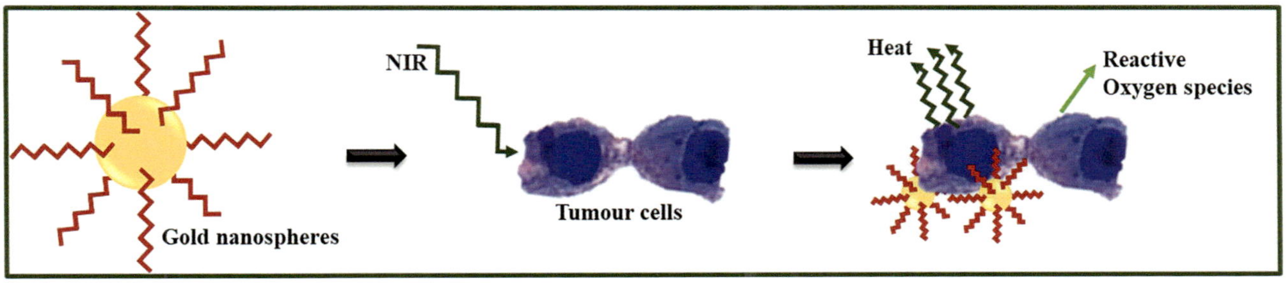


FIGURE 13.3 The use of PTT and PDT combination therapy to treat tumors.

methods. The challenges in the top to bottom approach highlight the need for more cost-effective and environmentally friendly synthesis methods, leading to the development of green syntheses. In the top to bottom approach, suitable bulk material is broken down into smaller fine particles by size reduction using various techniques like grinding, milling, sputtering, and thermal/laser ablation. Thus, the top-down method consists of mechanical grinding of bulk metals with subsequent stabilization using colloidal protecting agents, while in the bottom to top approach, nanoparticles are synthesized using chemical and biological methods by self-assembly of atoms to new nuclei, which grow into nanosized particles. Bottom-up methods include chemical reduction, electrochemical methods, and decomposition. In the case of top to bottom approaches, nanoparticles are generally synthesized by the evaporation condensation technique with the help of a tube furnace at atmospheric pressure. In this method, the primary material placed in the center of the furnace is vaporized into a carrier gas within a boat. One of the biggest limitations of this method is imperfections in the surface structure of the product, and the other physical properties of nanoparticles are highly dependent on the surface structure in reference to surface chemistry. Hence, a cost-effective and environment friendly alternative synthetic route was inevitable and culminated in green syntheses.[37, 38] Green synthesis of nanoparticles is progressively emerging as a key branch of nanotechnology where nanoparticles are produced with the help of biological entities like microorganisms, plant extracts, or plant biomass and could be a substitute for chemical and physical methods in an eco-friendly manner. Green synthesis is preferred over physical and chemical methods; as the former is environmentally friendly; cost effective; easily scalable to large-scale syntheses; and has no need to use high temperatures, energy, or toxic chemicals. Among biological routes, the use of plant extracts is particularly advantageous, as it is easier to optimize, poses fewer biohazard risks, and avoids the laborious maintenance of cell cultures required for microbial synthesis.[39]

AgNPs are used as a superior antibacterial agent and have been found to be effective against a broad spectrum of vancomycin-resistant strains in Gram-positive and Gram-negative bacteria. Silver nanoparticles seem to have a multi-pronged approach against bacteria. The disruption of the electron transport chain and interactions with DNA and proteins showcase their effectiveness. The antimicrobial mechanism of AgNPs is explained in Figure 13.4. AgNPs exert antifungal properties by interrupting the cell membrane and inhibit the normal asexual reproduction process by destroying the membrane integrity. Silver-based NPs are powerful in preventing bacterial infections and inflammation.[40–42]

## 13.7 APPLICATION OF SILVER NANOPARTICLES

Collagen is a natural biopolymer, and the usage of biopolymers in wound healing showed a better positive output in clinical trials. Collagen sponges and glycosaminoglycans act as a double-layered artificial skin and heal the wound. These act as a good carrier for AgNP to release drugs in a sustained manner. They are used in wound dressing, in the case of toxic epidermal and necrolysis, and for severe burns. AgNPs take about 3.35 days on average to heal a wound, and the bacterial load decreases in the wound area with no further effects.[43, 44] They also have wide application in artificial joint replacement in arthritic disease. Nanosilver-coated bone cement resulted in antibacterial activity against broad range of bacteria, including methicillin-resistant *Staphylococcus aureus* (MRSA). Taken further, the nanoparticles did not show any cytotoxicity. Ultra-high molecular weight polyethylene (UHMWPE) was used for

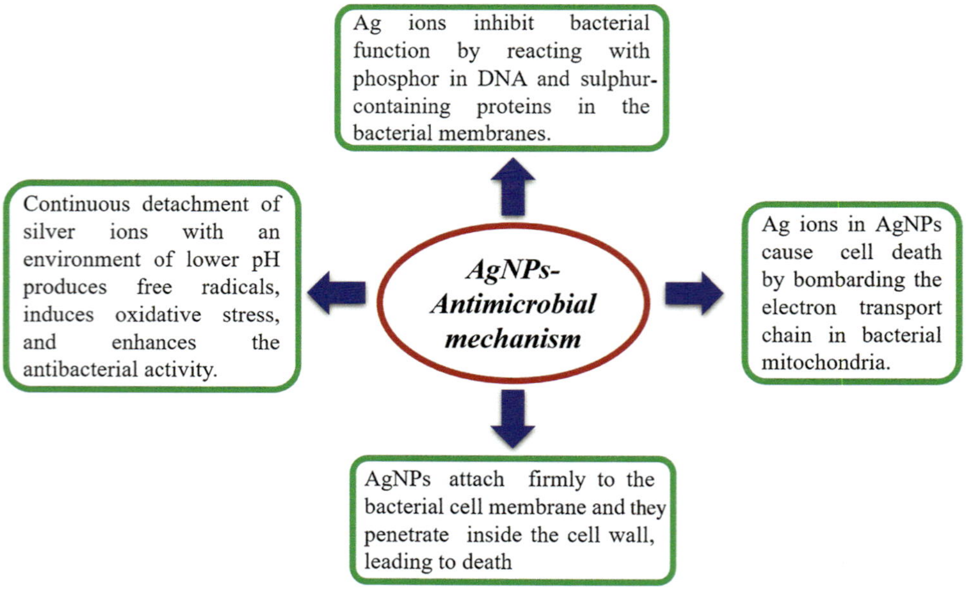


FIGURE 13.4 Antimicrobial mechanism of AgNPs.

artificial joint replacement, but the only disadvantages were wear and tear and associated debris generation, which led to inflammation and failure of joints in the body. This major problem was overcome by the integration of AgNPs with bone cement. This protocol consequently reduced polymer debris formation. The rate of infection was reduced by adding AgNPs to the outer layer of the implants.[45–47] AgNP is a promising agent used in dentistry. This property is because of the integration of antimicrobial properties in dental biomaterials. Incorporation of AgNPs into polymers used as a denture base and tissue conditioners in stomatitis have shown superior antimicrobial activity and the capacity to fight against oral infections. The nanobacteriostatic agent silver nanofluoride has been shown to inhibit the growth of *Streptococcus mutans*. It can be applied once a year, is easy to use, and offers a favorable cost -benefit ratio.[48] Many studies showed that AgNPs are localized inside the perinuclear space of cytoplasm and endolysosomal compartment cells by entering cells through endocytosis. AgNPs affect the respiration of cells and produce reactive oxygen species.[49] AgNPs can induce oxidative stress, leading to DNA damage, mitochondrial dysfunction, and apoptosis in cancer cells. Studies showed that AgNPs can affect the activity of vascular endothelial growth factors, which are involved in angiogenesis.

Theranostics is the mixture of diagnostics and therapy. Biosynthesized AgNPs possess theranostic applications as they have anti-cancer properties and are used in targeted drug delivery and bioimaging vehicles. This is a cost-effective, safe, simple, and eco-friendly approach.[50]

The uses of silver nanoparticles are varied and many, but the most utilized and desired aspect is their antimicrobial and anti-inflammatory capacity. The varying degrees of toxicity induced by silver nanoparticles remain a major limitation to their broad biomedical application. It is suggested that higher concentrations of silver nanoparticles are toxic and can cause various health problems and induce various ecological problems if released into the environment. Silver nanoparticles are widely used in wound dressings, coatings for medical devices, and nanoparticle-impregnated textiles. The continuous release of silver ions from both the inner and outer coatings enhances their long-term antimicrobial efficacy. On the whole, silver nanoparticles, due to the unique properties of silver and nanosize, appear to be promising in pharmaceutical, biomedical, and allied fields, provided safety data is generated to prove their safety and simultaneously rule out their toxicity. Future research is necessary to examine the safety aspects of nanosilver.

## 13.8 METAL OXIDE NANOPARTICLES

### 13.8.1 Iron Oxide Nanoparticles in Medical Diagnosis

Iron oxide nanoparticles have promising applications, from targeted drug delivery to magnetic resonance imaging. The combination of their magnetic properties and biocompatibility opens up exciting possibilities in various fields. Indeed, the versatility of iron oxide nanoparticles makes them invaluable in both diagnostic and therapeutic realms. Their role in targeted imaging for liver, lymph nodes, inflammation, and vascular imaging showcases their diagnostic potential. Meanwhile, their involvement in iron supplementation, macrophage modulation, hyperthermia, and drug targeting highlight their therapeutic applications. The intersection of diagnosis and therapy in theranostics adds another layer of significance to their use. Iron oxide nanoparticles, which belong to the ferrimagnetic class of magnetic materials, are used for many different biomedical and bioengineering applications. The main types of iron oxide based nanoparticles include magnetite, maghemite, and mixed ferrites ($MFe_2O_4$), where M represents metals such as Co, Mn, Ni or Zn. Upon surface-modification, the resulting superparamagnetic iron oxide nanoparticles (SPIONs) can be employed for magnetic resonance imaging; magnetic particle imaging (MPI); targeted delivery of drugs, proteins, antibodies, and nucleic acids; hyperthermia; biosensing; tissue repair; and separation of biomolecules. Iron oxide nanoparticles (IONPs) in the diagnosis and treatment of cancer have attracted more and more attention. Furthermore, iron-based nanoparticles have a high transverse relaxation rate, good biocompatibility, and long circulation in vivo and have been widely used as a magnetic resonance imaging contrast agent for tumor diagnosis. Iron oxide nanoparticles have unique magnetic properties, and they can be used as magnetic thermotherapy, photothermal therapy, and magnetic targeted drug delivery. Based on their characteristics, they have been used more and more in hematological tumors in recent years.[51–54]

## 13.9 SYNTHESIS OF IRON OXIDE NANOPARTICLES

There are many methods for preparing magnetic iron oxide nanoparticles, such as physical methods, chemical methods, and microbial technology. The physical properties of the prepared magnetic ferric oxide nanoparticles are closely related to the preparation methods. Therefore, it is very important to choose a suitable method to synthesize magnetic nanoparticles. Among the many methods, the most commonly used are chemical methods, which have lower synthesis costs and higher yields. Chemical methods of preparation include the thermal decomposition method, hydrothermal method, co-precipitation method, and micro-emulsion method. The most common method for preparation is co-precipitation. Although the preparation is mature, there are still several problems in production and application, such as the formation of aggregation due to the instability of nanoparticles. Therefore, it is necessary to modify the surface. The appropriate surface modification has ability to improve the stability, targeting ability, and biocompatibility of nanoparticles.[55, 56]

### 13.9.1 Different Methods of Synthesis of Iron Oxide Nanoparticles[33]

| Method Name | Principle | Advantages | Disadvantages | Impact Factors |
|---|---|---|---|---|
| Thermal decomposition method | The organic metal precursor is decomposed into metal oxide at high temperatures. | Monodisperse and uniform in diameter. | The product surface is hydrophobic and should be modified. | Temperature, heating rate, reaction time. |
| Hydrothermal method | The hot solvent is converted into nanoparticles under high temperature and pressure. | Simple operation, high crystallinity, good hydrophilicity, and no need for surface modification. | Reaction equipment is expensive, so the temperature, pressure, and device use is high. | Temperature, reaction time, type of solvent. |
| Coprecipitation method | Ferric and ferric precipitates in alkaline solutions. | Good hydrophilicity, simple procedure, and high yield. | Product size polydispersity. | Temperature, pH concentration, type of iron salt. |
| Microemulsion method | Surfactants and bases are added to form a microemulsion system in which the coprecipitation reaction takes place. | Size controlled, and the product is monodispersing. | Toxic surfactant. | Emulsifier, surfactant concentration ratio, reaction temperature, and time. |

## 13.10 APPLICATIONS OF IRON OXIDE NANOPARTICLES

In addition to diagnostic purposes, iron oxide nanoparticles have also been used for therapeutic application, for instance, to supplement iron in individuals with iron deficiency.[57] Ferumoxytol is clinically used to treat anemia in patients with chronic kidney disease (CKD). This SPION formulation was initially developed for sentinel lymph node and atherosclerotic plaque imaging but did not manage to outperform alternative diagnostic probes and protocols. The term theranostics refers to the combination of diagnosis and therapy. From a clinical and translational point of view, it relates to an intimate combination of diagnostic and therapeutic interventions, such as for staging and treatment of patients with thyroid cancer with radioactive iodine or to perform patient selection and treatment of somatostatin receptor-expressing neuroendocrine neoplasia with DOTATOC/TATE-based theranostic pairs (e.g., gallium and yttrium).[58]

Indeed, the integration of iron oxide nanoparticles in medical applications, including drug delivery and treatment of leukemia, showcases the potential of nanotechnology in advancing healthcare. The versatility of IONPs, from tumor imaging to magnetic hyperthermia and biomacromolecule magnetic separation, highlights their multifaceted role in medical science. As research progresses, exploring synergies between IONPs and various therapeutic methods may open new avenues for enhanced treatment outcomes. Continued innovation in this field holds promise for further improving clinical strategies. There are many challenges in the clinical translation of iron oxide nanoparticles for leukemia diagnosis and treatment. Addressing issues such as standardized preparation methods, thorough toxicity testing, understanding long-term effects, and elucidating the detailed mechanisms involved in regulating leukemic cell metabolism are crucial steps. Additionally, the underexplored potential of IONPs' magnetic properties, such as magnetic hyperthermia and separation, presents an opportunity for further investigation. Overcoming these hurdles and accelerating clinical trials could pave the way for a more comprehensive understanding and effective application of IONPs in leukemia therapy, especially with a focus on multi-treatment approaches integrating imaging, diagnosis, and therapy. Iron oxide nanoparticles possess many unique and attractive properties, explaining their extensive use in biomedical research.

### 13.10.1 Characteristics of Different Modifying Materials for IONPs[33]

| Classification | Representative | Advantage | Application |
|---|---|---|---|
| Small molecules | Carboxylic acid, oleic acid | High hydrophilicity | Targeted drug, vacuum sealing |
| Biomolecules | Nucleic acid, polypeptide, protein | High targeting | Targeted drugs, immunotherapy, gene therapy |
| Inorganic Materials | Au, Ag, MOS2 | High stability, simple preparation | Targeted drug delivery, multimodality imaging and photothermal therapy |
| Mesoporous materials | Mesoporous silicon dioxide, organic metal frame MOFs | High hydrophilicity, high drug loading rate | Fluorescence imaging |
| Polymer | PEG, PVP, PEI | Long circulation, high biocompatibility | Bioimaging |

## 13.11 ZINC OXIDE NANOPARTICLES IN MEDICAL DIAGNOSIS

Zinc oxide nanoparticles (ZONPs) are extensively used across various fields due to their unique physical and chemical properties.[59, 60] ZONPs exhibit a wide array of biomedical applications, leveraging their antimicrobial, regenerative, and mechanical properties. From enhancing antibacterial properties in restorative materials to serving as anti-sensitivity agents in toothpaste, ZONPs contribute significantly to oral healthcare. Their versatility extends to applications such as dental implant coatings, combating pathogenic oral microflora, improving denture base antifungal effects, remineralizing dentinal lesions, and stabilizing local drug delivery agents. ZONP are increasingly used in diverse industrial fields such as concrete production, photocatalysis, electronics, and more. Apart from the applications mentioned, zinc acts as a vital trace element present in all body organs. As the main component of various enzyme systems, zinc takes part in the body's metabolism and plays crucial roles in proteins and nucleic acid synthesis, hematopoiesis, and neurogenesis. Nano-ZnO helps with easier absorption of zinc by the body, making it a commonly used food additive. Compared with other MONPs, ZONP, having relatively less toxic properties, exhibit excellent biomedical applications, such as anticancer, drug delivery, antibacterial, anti-inflammation, and bioimaging.[61–65] This section aims to discuss the various research works that address synthesis and the potential use of ZONPs for antibacterial activity.

## 13.12 SYNTHESIS OF ZINC OXIDE NANOPARTICLES

Recently, a screen printed des-carboxy-prothrombin (DCP) was developed using ZnO nanoparticles in detection of liver cancer. An enhanced interdigitated electrode (IDE) for detecting viral oncogenes was also developed by coating gold-doped ZnO nanorods on top of IDE chips. ZnO nanosheets were also prepared using zinc nitrate and triethanolamine at neutral pH and low temperature. Similarly, stable ZnO polymers were prepared using the sol-gel method which exhibits the property of photoluminescence. They are very safe and cheap, so they can be used in cell imaging.

## 13.13 APPLICATIONS

Zinc oxide nanoparticles exhibit distinctive properties that contribute to their broad applications in dentistry. Their white, odorless powder form with a molecular weight of 81.38 g/mol aligns with FDA's recognition as a generally recognized as safe (GRAS) substance, reinforcing its safety for various applications. The versatility of ZONPs stems from their unique optical, magnetic, morphological, electrical, catalytic, mechanical, and photochemical properties. Importantly, these properties can be finely tuned by modifying particle size, doping with additional compounds, or adjusting synthesis conditions. Notably, as particle size decreases, the desirable characteristics of ZONPs tend to improve, further enhancing their efficacy in dental applications.

Applications also include anti-inflammatory activity in response to pathogens, anti-demineralizing and remineralizing effects on the hard tissues of the tooth, potential as an anti-cancer agent, and many others. The safety of nanoparticles, even those considered biologically safe like ZONPs, absolutely requires thorough examination, especially in the context of long-term use in oral care products. Research into regulatory aspects and potential effects over extended periods is crucial to ensure the overall well-being of users. It's an essential step in the development and deployment of such innovative materials.

The unique properties of ZONPs indeed position them as valuable contenders in cancer treatment. Their high biocompatibility and low toxicity, coupled with the ability to selectively induce apoptosis, make them promising for targeted therapies. The role in diagnosis and drug delivery further enhances their potential in advancing cancer therapeutics. It's exciting to see the ongoing exploration of such nanoparticles for medical applications.

Indeed, incorporating ZONPs into dental resin composites appears to offer a promising approach for enhancing both the mechanical and anti-bacterial properties of restorative materials. The inhibition of *Streptococcus mutans* growth and adhesion, along with minimal impact on mechanical properties, supports the potential of ZONPs in preventing secondary caries and reducing bulk material fractures. This aligns with the ongoing efforts to improve the longevity and performance of dental restorations. The incorporation of ZONPs into dental adhesive systems, as reported, demonstrates a positive impact on antimicrobial properties without compromising bond strength. This dual enhancement is valuable in promoting not only structural integrity but also the prevention of microbial-related issues in dental applications. The observed strong bond at the dentin-resin interface due to zinc particles, along with decreased collagen degeneration, underscores the potential for improved durability in dental applications. Additionally, higher percentages of ZONPs have been associated with increased amalgam alloy hardness, indicating a strengthening effect on the alloy. However, excessive ZONP incorporation can lead to a decline in certain mechanical properties, except for flexural strength, likely due to nanoparticle agglomeration. Furthermore, the relatively short duration of antibacterial activity and limited effectiveness against multi-species biofilms may restrict the overall clinical benefits of ZONP incorporation in composite materials.[4, 64]

The inhibitory effect of ZONPs on *C. albicans* in saliva, starting from a concentration of 0.05 mg/ml, suggests their potential for oral thrush treatment. Additionally, the enhanced antibacterial effect when combined with silver NPs in human and artificial saliva opens up promising possibilities for clinical applications, showcasing the broader impact of these nanoparticles in oral health scenarios. The synthesis of ZONPs from *aloe vera* leaf extract, along with their demonstrated sensitivity to *E. coli, P. aeruginosa*, and methicillin-resistant *S. aureus*, positions them as a promising candidate for nano-antibiotics. Their efficacy against bacterial strains resilient to conventional antibiotics highlights a potential avenue for addressing antibiotic-resistant infections.

The utilization of ZONPs in cancer diagnosis is supported by their ability to detect low-level expression of biomarkers for early detection. In vitro, their inherent preferential cytotoxicity against cancer cells and the induction of reactive oxygen species due to semiconductor properties suggest potential applications in cancer therapy through targeted effects on malignant cells. The ongoing research on utilizing ZONPs for gene silencing and targeted gene delivery presents a potential avenue for combating cancer. Additionally, the enhanced antibacterial activity against *E. coleus* and *S. aureus*, achieved through ZONPs coated with polyethylene glycol, highlights their versatility in addressing various biomedical challenges, from cancer therapy to antimicrobial applications. The efficient role of ZONPs in non-surgical tumor ablation for cancer therapy is notable. The demonstration of synergistic cytotoxic effects on leukemic cells through the combination of ZONPs with the anti-cancer drug daunorubicin and ultraviolet irradiation underscores their potential contribution to advancing therapeutic strategies in cancer treatment.

The wide-ranging applications of ZONPs in various fields of dentistry underscore their potential in diagnosis, treatment, and prevention of oral conditions, as well as enhancing dental materials. Strengthening collaboration between clinicians and materials scientists is crucial in navigating the technology-driven landscape of nano-dentistry. While there are roadblocks, ongoing research and a focus on clinical implications are essential for realizing the complete in vivo potential of ZONPs in dentistry.

## 13.14 CONCLUSION

This chapter provides uses of metal and metal oxide nanoparticles and advancements in the field of diagnosis. They provide more selectivity and sensitivity, and their optical properties have high potential in the expansion of bio-sensing techniques. AgNPs are less stable and compatible than AuNPs; they provide plasmonic biosensors for many pandemic illnesses. Metal oxide nanoparticles have a porous structure with a high surface area and are also applied in bio-sensors. However, there are some limitations, like controlling their morphology and mass production. Also, there are some challenges that need to be faced, including producing a large-scale nanoprobe with high sensitivity and stability. Also, the structure of NPs can affect their toxicity. They offer promising applications in the field of diagnosis due to their unique physical, chemical, and optical properties. They can be engineered to interact with biomolecules, giving sensitive detection and imaging of various diseases. However, there are many challenges, such as stability, toxicity, and scalability, that need to be addressed. Ongoing research in this area holds great promise for advancing diagnostic capabilities and patient outcomes. Every sector needs to contribute and collaborate to bring about this nanotechnology for early diagnosis of illness in the near future.

### Keywords

Bioimaging
Biological applications
Biosensors
Metal/metal oxide nanoparticles
Nanomaterials
Pharmacology
Theranostics

## REFERENCES

1. Jouyandeh, M.; Sajadi, S. M.; Seidi, F.; Habibzadeh, S.; Munir, M. T.; Abida, O.; Saeb, M. R.; Metal nanoparticles-assisted early diagnosis of diseases. *OpenNano.*, 2022, 8, 100104.
2. Yadavalli, T.; Shukla, D.; Role of metal and metal oxide nanoparticles as diagnostic and therapeutic tools for highly prevalent viral infections. *Nanomed J.*, 2017, 13, 219–230.
3. Khorasani, A.; Shahbazi-Gahrouei, D.; Safari, A.; Recent metal nanotheranostics for cancer diagnosis and therapy: a review. *Diagnostics*, 2023, 13, 833.
4. Pushpalatha, C.; Suresh, J.; Gayathri, V. S.; Sowmya, S. V.; Augustine, D.; Alamoudi, A.; Patil, S.; Zinc oxide nanoparticles: a review on its applications in dentistry. *Front. Bioeng. Biotechnol.*, 2022, 10, 917990.
5. Yi-Cheun, Y.; Brian, C.; Rotello, V. M.; Gold nanoparticles: preparation, properties, and applications in bionanotechnology. *Nanoscale*, 2012, 4, 1871.
6. Ralph, A.; Sperling, P.; Rivera, G.; Feng, Z.; Marco, Z.; Parak, W. J.; Biological applications of gold nanoparticles. *Chem. Soc. Rev.*, 2008, 37, 1896.
7. Weibo, C.; Ting, G.; Hao, H.; Jiangtao, S.; Applications of gold nanoparticles in cancer nanotechnology. *Nanotechnol. Sci. Appl.*, 2008, 1, 17.
8. Bursy, D.; Monika, S.; Maciej, M; Paweł, B; Radosław, B.; Nanogold as a component of active drugs and diagnostic agents. *Int. J. App. Pharm.*, 2023, 52.
9. Xiaohua, H; Mostafa, A.; El-Sayed.; Gold nanoparticles: Optical properties and implementations in cancer diagnosis and photothermal therapy. *J. Adv. Res.*, 2010, 1, 13.
10. Maysinger, D.; Nanoparticles and cells: good companions and doomed partnerships. *Org. Biomol. Chem.*, 2007, 5, 2335.
11. Khlebtsov, B. N.; Khanadeev, V. A.; Panfilova, E. V.; Pylaev, T. E., Bibikova, O. A., Staroverov, S. A.; Bogatyrev, V. A.; Dykman, L. A.; Khlebtsov, N. G.; New types of nanomaterials: powders of gold nanospheres, nanorods, nanostars, and gold-silver nanocages. *Nanotechnologies in Russia*, 2013, 8, 209.

12. Jorge, P.; Isabel, P.; Liz-Marzán, L. M.; Paul, M.; Gold nanorods: synthesis, characterization and applications. *Coord. Chem. Rev.*, 2005, 249, 1870.
13. Vincenzo, A.; Roberto, P.; Marco, F.; Onofrio, M.; Maragò., M; Antonia, I.; Surface plasmon resonance in gold nanoparticles: a review. *J. Phys. Condens. Matter.*, 2017, 29, 203002.
14. Yang, Z.; Wang, D.; Zhang, C.; Liu, H.; Hao, M.; Kan, S.; Liu, D.; Liu, W.; The applications of gold nanoparticles in the diagnosis and treatment of gastrointestinal cancer. *Front. Oncol.*, 2022, 11, 819329.
15. Skrabalak, S. E.; Chen, J.; Sun, Y.; Lu, X.; Au, L.; Cobley, C. M.; Xia, Y.; Gold nanocages: synthesis, properties, and applications. *Acc. Chem. Res.*, 2008, 41, 1587.
16. Chen, J.; Saeki, F.; Wiley, B. J.; Cang, H.; Cobb, M. J.; Li, Z. Y.; Au, L.; Zhang, H.; Kimmey, M. B.; Li, X.; Xia, Y.; Gold nanocages: bioconjugation and their potential use as optical imaging contrast agents. *Nano Lett.*, 2005, 5, 473.
17. Skrabalak, S. E., Au, L.; Lu, X.; Li, X.; Xia, Y.; Gold nanocages for cancer detection and treatment. *MedComm – Future Medicine*, 2007, 2, 657.
18. van de Looij, S. M.; Hebels, E. R.; Viola, M.; Hembury, M.; Oliveira, S.; Vermonden, T.; Gold nanoclusters: imaging, therapy, and theranostic roles in biomedical applications. *Bioconjugate Chem.*, 2021, 33, 4.
19. Chattoraj, S.; Amin, M. A.; Mohapatra, S.; Ghosh, S.; Bhattacharyya, K.; Cancer cell imaging using in situ generated gold nanoclusters. *Chem. Phys. Chem.*, 2016, 17, 61.
20. Porret, E.; Le Guével, X.; Coll, J. L.; Gold nanoclusters for biomedical applications: toward in vivo studies. *J. Mat. Chem. B*, 2020, 8, 2216.
21. Zhou, Z.; Zhang, C.; Qian, Q.; Ma, J.; Huang, P.; Zhang, X.; Pan, L.; Gao, G.; Fu, H.; Fu, S.; Song, H.; Folic acid-conjugated silica capped gold nanoclusters for targeted fluorescence/X-ray computed tomography imaging. *J. Nanobiotech.*, 2013, 11, 1.
22. Park, S.; Kim, H.; Lim, S. C.; Lim, K.; Lee, E. S.; Oh, K. T.; Choi, H. G.; Youn, Y. S.; Gold nanocluster-loaded hybrid albumin nanoparticles with fluorescence-based optical visualization and photothermal conversion for tumor detection/ablation. *J. Control. Release.*, 2019, 304, 7.
23. Fan, L.; Wang, W.; Wang, Z.; Zhao, M.; Gold nanoparticles enhance antibody effect through direct cancer cell cytotoxicity by differential regulation of phagocytosis. *Nat. Commun.*, 2021, 12, 6371.
24. Wu, S. Y., Wu, F. G.; Chen, X.; Antibody-incorporated nanomedicines for cancer therapy. *Adv. Mater.*, 2022, 34, 2109210.
25. Davidi, E. S.; Dreifuss, T.; Motiei, M.; Shai, E.; Bragilovski, D.; Lubimov, L.; Kindler, M. J. J.; Popovtzer, A.; Don, J.; Popovtzer, R.; Cisplatin-conjugated gold nanoparticles as a theranostic agent for head and neck cancer. *Head & Neck*, 2018, 40, 70.
26. Pal, K.; Al-Suraih, F.; Gonzalez-Rodriguez, R.; Dutta, S. K.; Wang, E.; Kwak, H. S.; Caulfield, T. R.; Coffer, J. L.; Bhattacharya, S.; Multifaceted peptide assisted one-pot synthesis of gold nanoparticles for plectin-1 targeted gemcitabine delivery in pancreatic cancer. *Nanoscale*, 2017, 9, 15622.
27. Xi, D.; Dong, S.; Meng, X.; Lu, Q.; Meng, L.; Ye, J.; Gold nanoparticles as computerized tomography (CT) contrast agents. *Rsc Adv.*, 2012, 2, 12515.
28. Dreifuss, T.; Barnoy, E.; Motiei, M.; Popovtzer, R.; Design and applications of nanoparticles in biomedical imaging, *J. Nu. Med.*, 2017, 403.
29. Xie, J.; Zhang, F.; Aronova, M.; Zhu, L.; Lin, X.; Quan, Q.; Liu, G.; Zhang, G.; Choi, K. Y.; Kim, K.; Sun, X.; Manipulating the power of an additional phase: a flower-like Au– Fe3O4 optical nanosensor for imaging protease expressions in vivo. *Acs. Nano.*, 2011, 5, 3043
30. Hu, J.; Rivero, F.; Torres, R. A.; Loro Ramirez, H.; Rodríguez, E. M.; Alfonso, F.; Garcia Sole, J.; Jaque, D.; Dynamic single gold nanoparticle visualization by clinical intracoronary optical coherence tomography. *J. Biophotonics.*, 2017, 10, 674.
31. Chemla, Y.; Betzer, O.; Markus, A.; Farah, N.; Motiei, M.; Popovtzer, R.; Mandel, Y.; Gold nanoparticles for multimodal high-resolution imaging of transplanted cells for retinal replacement therapy. *Nanomedicine*, 2019, 14, 1857.
32. Gharatape, A.; Davaran, S.; Salehi, R.; Hamishehkar, H.; Engineered gold nanoparticles for photothermal cancer therapy and bacteria killing. *Rsc. Adv.*, 2016, 6, 111482.
33. Wang, Y.; Yang, Y.; Zheng, X.; Shi, J.; Zhong, L.; Duan, X.; Zhu, Y.; Application of iron oxide nanoparticles in the diagnosis and treatment of leukemia. *Front. Pharm.*, 2023, 14, 1177068.
34. Kim, H. S.; Lee, D. Y.; Near-infrared-responsive cancer photothermal and photodynamic therapy using gold nanoparticles. *Polymers*, 2018, 10, 961.
35. Palaniappan, K.; Clinical progress in gold nanoparticle (GNP)-mediated photothermal cancer therapy. *Curr. Cancer Ther. Rev.*, 2023, 19, 13.
36. Sun, M.; Liu, F.; Zhu, Y.; Wang, W.; Hu, J.; Liu, J.; Gao, W.; Salt-induced aggregation of gold nanoparticles for photoacoustic imaging and photothermal therapy of cancer. *Nanoscale*, 2016, 8, 4452.
37. Park, B.; Park, S.; Kim, J.; Kim, C.; Listening to drug delivery and responses via photoacoustic imaging. *Adv. Drug Deliv. Rev.*, 2022, 184, 114235.
38. Wang, W.; Wang, J.; Ding, Y.; Gold nanoparticle-conjugated nanomedicine: design, construction, and structure–efficacy relationship studies. *J. Mater. Chem. B*, 2020, 8, 4813.
39. Huy, T. Q.; Huyen, P.; Le, A. T.; Tonezzer, M.; Recent advances of silver nanoparticles in cancer diagnosis and treatment. *Anticancer Agents Med. Chem.*, 2020, 20, 1276.
40. Zhang, X. F.; Liu, Z. G.; Shen, W.; Gurunathan, S.; Silver nanoparticles: synthesis, characterization, properties, applications, and therapeutic approaches. *Int. J. Mol. Sci.*, 2016, 17, 1534.
41. Lee, S. H.; Jun, B. H.; Silver nanoparticles: synthesis and application for nanomedicine. *Int. J. Mol. Sci.*, 2019, 20, 865.
42. Natsuki, J.; Natsuki, T.; Hashimoto, Y.; A review of silver nanoparticles: synthesis methods, properties and applications. *Int. J. Mater. Sci. Appl.*, 2015, 4, 325.
43. Yaqoob, A. A.; Umar, K.; Ibrahim, M. N. M.; Silver nanoparticles: various methods of synthesis, size affecting factors and their potential applications–a review. *Appl. Nanosci.*, 2020, 10, 1369.
44. Rafique, M.; Sadaf, I.; Rafique, M. S.; Tahir, M. B.; A review on green synthesis of silver nanoparticles and their applications. *Artif. Cells Nanomed. Biotechnol.*, 2017, 45, 1272.
45. Evanoff Jr, D. D.; Chumanov, G.; Synthesis and optical properties of silver nanoparticles and arrays. *Chem. Phys. Chem.*, 2005, 6, 1221.
46. Baker, C.; Pradhan, A.; Pakstis, L.; Pochan, D. J.; Shah, S. I.; Synthesis and antibacterial properties of silver nanoparticles. *J. Nanosci. Nanotechnol.*, 2005, 5, 244.

47. Liao, C.; Li, Y.; Tjong, S. C.; Bactericidal and cytotoxic properties of silver nanoparticles. *Int. J. Mol. Sci.*, 2019, 20, 449.
48. Kumar, S. S. D.; Rajendran, N. K.; Houreld, N. N.; Abrahamse, H.; Recent advances on silver nanoparticle and biopolymer-based biomaterials for wound healing applications. *Int. J. Biol. Macromol.*, 2018, 115, 165.
49. Salleh, A.; Mustafa, N.; Teow, Y. H.; Fatimah, M. N.; Khairudin, F. A.; Ahmad, I.; Fauzi, M. B.; Dual-layered approach of ovine collagen-gelatin/cellulose hybrid biomatrix containing graphene oxide-silver nanoparticles for cutaneous wound healing: fabrication, physicochemical, cytotoxicity and antibacterial characterisation. *Biomedicines*, 2022, 10, 816.
50. Poon, T. K. C.; Iyengar, K. P.; Jain, V. K.; Silver nanoparticle (AgNP) technology applications in trauma and orthopaedics. *J. Clin. Orthop. Trauma.*, 2021, 21, 101536.
51. Brennan, S. A.; Fhoghlú, C. N.; Devitt, B. M.; O'mahony, F. J.; Brabazon, D.; Walsh, A., Silver nanoparticles and their orthopaedic applications. *Bone Jt. J.*, 2015, 97, 582.
52. Allizond, V.; Comini, S.; Cuffini, A. M.; Banche, G.; Current knowledge on biomaterials for orthopedic applications modified to reduce bacterial adhesive ability. *Antibiotics*, 2022, 11, 529.
53. Yin, I. X.; Zhang, J.; Zhao, I. S.; Mei, M. L.; Li, Q.; Chu, C. H.; The antibacterial mechanism of silver nanoparticles and its application in dentistry. *Int. J. Nanomedicine.*, 2020, 2555.
54. Flores-López, L. Z.; Espinoza-Gómez, H.; Somanathan, R.; Silver nanoparticles: electron transfer, reactive oxygen species, oxidative stress, beneficial and toxicological effects. Mini review. *J. Appl. Toxicol.*, 2019, 39, 16.
55. Sakthi Devi, R.; Girigoswami, A.; Siddharth, M.; Girigoswami, K.; Applications of gold and silver nanoparticles in theranostics. *Appl. Biochem. Biotechnol.*, 2022, 194, 4187.
56. Dadfar, S. M.; Roemhild, K.; Drude, N. I.; von Stillfried, S.; Knüchel, R.; Kiessling, F.; Lammers, T.; Iron oxide nanoparticles: diagnostic, therapeutic and theranostic applications. *Adv. Drug Deliv. Rev.*, 2019, 138, 302.
57. Martinkova, P.; Brtnicky, M.; Kynicky, J.; Pohanka, M.; Iron oxide nanoparticles: innovative tool in cancer diagnosis and therapy. *Adv. Healthc. Mater.*, 2018, 7, 1700932.
58. Montiel Schneider, M. G.; Martín, M. J.; Otarola, J.; Vakarelska, E.; Simeonov, V.; Lassalle, V.; Nedyalkova, M.; Biomedical applications of iron oxide nanoparticles: current insights progress and perspectives. *Pharmaceutics*, 2022, 14, 204.
59. Laurent, S.; Mahmoudi, M.; Superparamagnetic iron oxide nanoparticles: promises for diagnosis and treatment of cancer. *Int. J. Mol. Epidemiol. Genet.*, 2011, 2, 367.
60. Ali, A.; Zafar, H.; Zia, M.; ul Haq, I.; Phull, A. R.; Ali, J. S.; Hussain, A.; Synthesis, characterization, applications, and challenges of iron oxide nanoparticles. *Nanotechnol. Sci. Appl.*, 2016, 49.
61. Laurent, S.; Forge, D.; Port, M.; Roch, A.; Robic, C.; Vander Elst, L.; Muller, R. N.; Magnetic iron oxide nanoparticles: synthesis, stabilization, vectorization, physicochemical characterizations, and biological applications. *Chem. Rev.*, 2008, 108, 2064.
62. Hashem, F. A.; Nasr, M. O.; Ahmed, Y. O.; Preparation and evaluation of iron oxide nanoparticles for treatment of iron deficiency anaemia. *Int. J. Pharm. Sci.*, 2018, 10, 142.
63. Singh, N.; Jenkins, G. J.; Asadi, R.; Doak, S. H.; Potential toxicity of superparamagnetic iron oxide nanoparticles (SPION). *Nano Rev.*, 2010, 1, 5358.
64. Smijs, T. G.; Pavel, S.; Titanium dioxide and zinc oxide nanoparticles in sunscreens: focus on their safety and effectiveness. *Nanotechnol. Sci. Appl.*, 2011, 95.
65. Ruszkiewicz, J. A.; Pinkas, A.; Ferrer, B.; Peres, T. V.; Tsatsakis, A.; Aschner, M.; Neurotoxic effect of active ingredients in sunscreen products, a contemporary review. *Toxicol. Rep.*, 2017, 4, 245.


## LIST OF ABBREVIATIONS

| | |
|---|---|
| **MRI** | Magnetic Resonance Imaging |
| **CT** | Computed Tomography |
| **SERS** | Surface-Enhanced Raman Scattering |
| **PET** | Positron Emission Tomography |
| **GNPs** | Gold Nanoparticles |
| **PDA** | Pancreatic Ductal Adenocarcinoma |
| **OSCC** | Oral Squamous Cell Carcinoma |
| **HCC** | Hepatocellular Carcinoma |
| **Tmab** | Targeting Drug Trastuzumab |
| **HER2** | Human Epidermal Growth Factor Receptor-2 |
| **EGCG** | Epigallocatechin Gallate |
| **HNSC** | Head and Neck Squamous Cancer |
| **PDAC** | Pancreatic Ductal Adenocarcinoma |
| **AUNP** | Gold Nanoparticle |
| **GEM** | Gemcitabine |
| **PTT** | Photothermal Therapy |
| **PDT** | Photodynamic Therapy |
| **OCT** | Optical Coherence Tomography |
| **ER** | Endoplasmic Reticulum |
| **ROS** | Reactive Oxygen Species |
| **PEG** | Polyethylene Glycol |
| **PAI** | Photoacoustic Imaging |
| **SPR** | Surface Plasmon Resonance |
| **MAP** | Molecularly Activated Plasmonic |
| **EGFR** | Epidermal Growth Factor Receptor |
| **PTX** | Paclitaxel |
| **PFH** | Perfluorohexane |
| **FA-BSA** | Folic Acid-Bovine Serum Protein |
| **AGNPs** | Silver Nanoparticles |
| **MRSA** | Including Methicillin Resistant Staphylococcus Aureus |
| **UHMWPE** | Ultra-High Molecular Weight Polyethylene |
| **IONPs** | Iron Oxide Nanoparticles |
| **CKD** | Chronic Kidney Disease |
| **ZONP** | Zinc Oxide Nanoparticles |
| **DCP** | Des-Carboxy-Prothrombin |
| **IDE** | Interdigitated Electrode |
| **GRAS** | Generally Recognized As Safe |

# 14 Microbe-mediated Nanoparticles Synthesis and Applications

*Shweta Kashid and Kalpana Joshi*
Sinhgad College of Engineering, Vadgaon Budruk, Pune, Maharashtra, India

## 14.1 INTRODUCTION

Nanotechnology is a cutting-edge area that has the potential to revolutionize several industries, including electronics, healthcare, and environmental science. Materials with special qualities and functions at the nanoscale, which spans 0.1 to 100 nm, are called nanoparticles. Nanomaterials have high surface areas that improve mechanical, optical, electrical, and thermal stability; these qualities are dependent on their size.[1] These materials have opened up new horizons for innovation, providing promising solutions to complex challenges in areas such as targeted drug delivery, efficient energy conversion, and pollution remediation.

Traditionally, the synthesis of nanomaterials has relied heavily on chemical and physical methods, often involving harsh reagents, high temperatures, and energy-intensive processes. While these approaches have been successful in producing a wide range of metal-based nanomaterials,[2, 3] they are not without their drawbacks, including environmental concerns, high cost, and limited control over material properties.

Microorganisms, by their very nature, are masters of nanoengineering. They manipulate matter at the nanoscale daily as they carry out essential life processes. Harnessing the inherent abilities of microorganisms allows researchers to mimic and amplify these natural processes, resulting in the creation of nanomaterials with tailored properties and applications. From the controlled reduction of metal ions to the precision engineering of nanoscale structures, microorganisms offer an eco-friendly and sustainable path to nanotechnology's future.

This chapter embarks on a journey through the fascinating world of microorganisms and their pivotal role in nanomaterial synthesis. We will explore the unique capabilities of different microorganism types, delve into the mechanisms behind their nanomaterial synthesis prowess, and examine applications across various domains. Moreover, we will discuss the sustainability aspect of microorganism-based synthesis, contrasting it with traditional chemical methods, and highlight the potential of this approach to revolutionize nanotechnology while minimizing its environmental footprint. In both approaches, morphology plays an important role and can be synthesized in various shapes such as rods, cubes, plates, prisms, and tetrapods.[4] Another important aspect of NPs is their composition, as they contain metal ions with valuable properties. For example, one review[5] covered applications of silver nanoparticles in various areas such as catalysts, food storage, drug therapeutics, and optical detection systems. However, it was indicated that during biochemical synthesis of NPs, its difficult to control the precise size and form of metal NPs. Biological synthesis of NPs also includes engineered bacteria which produce metallic nanoparticles, and fungi act as potent reducers and stabilizers. All together, these microorganisms hold immense potential for the sustainable and precise fabrication of metal NPs.

## 14.2 BIOLOGICAL SYNTHESIS OF METAL NANOPARTICLES USING BACTERIA

Biological synthesis of NPs is environmentally friendly and a rapid process. The size and toxicity of nanoparticles is to be controlled precisely, as it affects the survival of microorganisms during the process. Many researchers have explored nanoparticle synthesis using bacteria, fungi, and yeast,[6–13] which are listed in detail in Tables 14.1 and 14.2. Microbial cells have the inbuilt quality to double and maintain growth in varied environments. Some fungi and bacteria have the capacity to produce unique nanomaterials.

### 14.2.1 Synthesis of Gold Nanoparticles

Gold nanoparticles possess exclusive characteristics, making them useful in a variety of applications. They are used in the food industry for labeling and sensing, in medicine for targeted drug delivery and diagnostics, in water purification for removing contaminants, and in various biological applications, such as imaging and cancer treatment. Their versatility makes them valuable in many fields.[34]

Gold nanoparticles have an ability to kill bacteria and fungi. While they are not as famous for this as silver nanoparticles, they still can fight against various harmful microorganisms. Green-synthesized gold nanoparticles, which are made in an eco-friendly way, have also been found to act as antioxidants and fight against fungi, which are tougher to combat than bacteria in multiple ways.[35, 36]

DOI: 10.1201/9781032660639-14

**TABLE 14.1**
**Metallic Nanoparticles Produced Using Bacteria**

| Sr. No. | Microorganism | Type of NP | Size | Shape | Reference |
|---|---|---|---|---|---|
| 1 | *Enterococcus* | Au | 7 nm | Spherical | 14 |
| 2 | *Vibrio alginolyticus* | Au | 50–100 nm | Irregular | 15 |
| 3 | *Cladosporium* sp. | Au | 5–10 nm | Spherical | 16 |
| 4 | *Fusarium oxysporum* | Au | 66 and 10 kDa | Oval/kidney shaped | 17 |
| 5 | *Sclerotium rolfsii* | Au | 28.7 and 19.5 | Spherical | 18 |
| 6 | *Rhodopseudomonas capsulate* | Au | 10–20 nm | Spherical | 19 |
| 7 | *Escherichia coli* | Au | 20–30 nm | Extracellular | 20 |
| 8 | *Pyrobaculum islandicum* | Au | - | Extracellular | 21 |
| 9 | *Pyrococcus furiosus* | Au | 0.8–1.5 um | Extracellular | 9 |
| 10 | *Shewanella algae* | Pt | 1–3 um | Rod shaped | 22 |
| 11 | *Enterobacter cloacae* | AgNPs | - | Rod shaped | 23 |
| 12 | *Pseudomonas aerogenosa JP2* | AgNPs | 66 kDa | Rod shaped | 24 |
| 13 | *Pseudomonas aerogenosa JP1* | AgNPs | 65 kDa | - | 24 |
| 14 | *Acinetobacter* | Ag | - | - | 25 |
| 15 | *Streptomyces laurentii* | Ag | 7–15 nm | Spherical | 26 |
| 16 | *Bacillus licheniformis* | Ag | 10–30 nm | Spherical | 27 |
| 17 | *E. coli* | AgNPs | - | Rod shaped | 28 |
| 18 | *Bacillus subtilis* | AgNPs | 66–116 kDa | Rod shaped | 29 |
| 19 | *Shewanella oneidensis* | AgNPs | - | Rod shaped | 30 |
| 20 | *S. maltophilia* | AgNPs | 0.7–1.8/0.4–0.7 um | Rod shaped | 31 |
| 21 | *Tetrathiobacter* | SeNPs | 90-kDa protein | - | 32 |
| 22 | *D. desulfuriacans ATCC 29577* | PdNPs | 2.5–10 um | Rod shaped | 33 |

**TABLE 14.2**
**List of Fungi-Mediated Metal and Metal Oxide NPs**

| Microorganisms | Type of NP | Size (nm) | Shape | Location | Reference |
|---|---|---|---|---|---|
| *Mycena citricolor* | ZnO-NPs | < 100 | Decahedral | Extracellular | 71 |
| *Colletotrichum* sp. | ZnO-NPs | < 100 | icosahedral | Extracellular | 71 |
| *Phanerochaete chrysosporium* | Nanodiamonds | 3.4 ± 1.1 | Flat | Extracellular | 25 |
| *Beauveria bassiana* | Silver | 36.88–60.93 | Spherical | Intracellular | 72 |
| *Metarizhium Anisopliae* | Silver | 40.14–289.13 | Irregular | Intracellular | 72 |
| *Isaria fumosorosea* | Silver | 51.31–111.02 | Spherical | Intracellular | 72 |
| *Trichoderma Harzianum* | Silver | 10–20 | Irregular | Extracellular | 72 |
| *Aspergillus fumigatus* | Ag | 5–25 | Spherical, triangular | Extracellular | 73 |
| *Agaricus bisporus* | Ag | 5–50 | Spherical | extracellular | 74 |
| *Aspergillus oryzae* | Fe | 10–24.6 | Quasispherical | Intracellular | 75 |
| *Calocybe indica* | Ag | 5–50 | Spherical | Extracellular | 70 |
| *Lentinus edodes* | Au | 5–50 | Spherical | Extracellular | 76 |
| *Penicillium fellutanum* | Ag | 5–25 | Spherical | Extracellular | 77 |
| *Trichoderma viride* | Ag | 5–40 | Spherical, rodlike | Extracellular | 68 |
| *F. oxysporum* | Au, Ag, Si, Ti, Zr, magnetite | 8–15 | Spherical Quasispherical | Extracellular | 68 |
| *F. oxysporum* | Cdse, $BaTi_3$, $Bi_2O_3$, Pt | 10–50 | Spherical, quasispherical | Extracellular | 72 |
| *Trichothecium* sp. | Au | 5–200 | Quasispherical | Extracellular | 17 |
| *Trichoderma* sp. | Ag | 8–60 | Spherical | Extracellular | 74 |
| *Penicillium* sp. | Ag | 25 | Spherical | - | 77 |
| *Pleurotus florida* | Ag | 5–50 | Spherical | Extracellular | 74 |

Gold nanoparticles made from plants are often produced by using a substance called chloroauric acid mixed in an appropriate proportion with plant extracts. This process yields gold nanoparticles of various sizes and forms, including triangles and spheres ranging from 10 to 300 nanometers. These plant-derived gold nanoparticles are less harmful to living cells compared to chemically produced ones. They have a wide range of applications, like coating glass surfaces for bacterial and mold protection, and in food packaging for their antioxidant and antibacterial properties. They can also be used to coat fabrics like cotton, silk, and leather to make them antibacterial. Recent studies have shown that these plant-mediated gold nanoparticles are even better at fighting bacteria due to the natural compounds from plant extracts.[37]

The process of making gold nanoparticles using bacteria involves some interesting steps. It starts with the transfer of electrons from a molecule called NADH through a special carrier. Then, these electrons are used to change gold ions ($Au^{3+}$) into solid gold nanoparticles. Different types of bacteria use specific enzymes like NADPH-dependent reductase to undergo reduction reaction. They also use molecules to hold the nanoparticles together. In one of the case studies, researchers found evidence of proteins and phosphate molecules playing a role in the process. So, bacteria can make gold nanoparticles and use certain molecules to stabilize them in this unique way.[35]

For the very first time, the bacteriogenic mechanism of gold nanoparticle (AuNPs) synthesis was investigated using *Rhodopseudomonas capsulate* culture. In this process, electrons from NADH are moved to gold ions and then reduced to elemental gold ($Au^{3+}$). The bacteria *Stenotrophomonas maltophilia* has also been suggested for AuNP synthesis. FTIR spectrum analysis revealed that the phosphate bond was engaged and that peptides or proteins were involved in the reduction of gold ions.[36]

In one of the studies, the synthesis of gold nanoparticles was carried out using the *Actinobacter* species, in presence of $HAuCl_4$ and an inducer called bovine serum albumin (BSA). The experiment was run in both anaerobic and aerobic settings at two distinct temperatures (25°C and 37°C). The role of varying activity of enzyme proteases in gold nanoparticles synthesis was examined, and the decrease of $HAuCl_4$ was tracked over time using UV-Vis spectra.[38]

The algae extract was mixed with a solution of gold chloride, resulting in the formation of gold nanoparticles. The size and shape of these nanoparticles were measured using a scanning electron microscope, and qualitative and quantitative analysis were performed.

Microbial-derived gold nanoparticles (β-AuNPs) have potential applications in biomedical domains, including cancer therapy. They were fabricated using *Streptomyces clavuligerus* culture supernatant and characterized using various techniques. The nanoparticles were tested for anticancer activity in various cell lines and showed biocompatibility and non-toxic nature. They induced cell cycle arrest and inhibited microtubule assembly in DU145 cells, supporting their potential for future therapeutic applications.[36]

In another study, AuNPs with 20-nm diameter were synthesized using the modified Turkevich method. The process involved heating $HAuCl_4$ solution to 80°C, reducing $Au^{3+}$ solution with sodium citrate, and maintaining the temperature until solution became ruby red.[39, 40]

The biosynthesis of AuNPs involved the addition of leaf extract to $HAuCl_4$ and constant stirring to create colloids in the range of 1–5 gm. The reduction rate increased with an increase in leaf extract quantity. Colloidal solution with 5 gm extract showed fast reduction and completed within 4 hr, resulting in a stable gold colloid.[39, 40]

Different quantities of hydrogen tetrachloroaurate ($HAuCl_4$) were reduced in order to create gold nanoparticles. A microwave oven was used to heat the solution while the bioreduction of metal ions was being observed. The effect of temperature on nanoparticle synthesis was observed. For later usage, the nanoparticles were rinsed twice, centrifuged, concentrated, and then kept at 4°C.[41]

*C. pepo* and *M. crispa* fresh leaves were washed, boiled, and filtered. $HAuCl_4$ solutions were combined with the extracts, and a control configuration was upheld. The setup was maintained under static conditions at room temperature, with a final reaction volume of 50 ml.[42] In another study researchers utilized Mulberry leaf extract to create gold nanoparticles at room temperature. This aqueous synthesis approach is simple, inexpensive, non-toxic, environmentally friendly, and stable over an extended period of time. Due to these qualities, AuNPs are becoming more and more popular. Using different methods for characterization of metal nanoparticles from DLS, TEM, AFM, XRD, FT-IR, and UV-Vis spectroscopy, researchers looked into the morphology and production of AuNPs. Chemically synthesized AuNPs and mulberry leaf extract demonstrated a reasonable zone of inhibition against *Vibrio cholera* and *Staphylococcus aureus*, whereas biosynthesized AuNPs demonstrated a robust zone of inhibition.[39]

#### 14.2.1.1 Antibacterial Activity of Gold Nanoparticles

The anti-bacterial activity assay was carried out between aqueous leaf extract, chemically synthesized AuNPs, and biosynthesized AuNPs against human pathogens, *Vibrio cholera* and *Staphylococcus aureus*. When it came to *Vibrio cholera*, AuNPs made from *Morus alba* leaf extract had the strongest anti-bacterial activity, but the chemically manufactured AuNPs only shown mediocre activity against *Staphylococcus aureus* and *Vibrio cholera*. Aqueous leaf extract was found to have reduced antibacterial action against *Vibrio cholera* and *Staphylococcus aureus*. The higher inhibitory action of biosynthesized AuNPs depends on their size and the capping agent (proteins), which play a crucial role in NPs and bacterial interactions.[43, 44]

The study investigated the antibacterial efficacy of gold nanoparticles against *Klebsiella planticola*, *Bacillus subtilis*, and *Streptococcus* sp. using the well diffusion method. The results showed how well the nanoparticles inhibited the growth of these bacteria.[13]

#### 14.2.1.2 Characterization Techniques of Biosynthesized Gold Nanoparticles

Transmission electron microscopy analysis involves preparing a sample by drop-coating onto a carbon-coated copper grid. An additional analysis of selected area electron diffraction (SAED) was performed. JEOL model 1200EX equipment running at 120 kV accelerating voltage was used to accomplish the tests. The visualization of nanoparticles were carried out at the nanoscale using TEM.

A JASCO dual-beam spectrophotometer (model V-570) was used to measure UV-Vis-NIR spectroscopy at a resolution of 1 nm. This method is employed to evaluate the nanoparticles' optical characteristics and track the bioreduction of $HAuCl_4$ over time.

X-ray diffraction measurements were performed on solution-cast films of the synthesized biogenic gold nanoparticles on glass substrates. A Phillips PW 1830 device running at 40 kV of voltage and 30 mA of current with Cu $K\alpha$ radiation was used for the analysis. X-ray diffraction is an effective method for investigating a material composition and crystal structure. The aforementioned procedures offer significant insights into the optical characteristics, size, shape, and structure of the synthesized gold nanoparticles.[38] *Turbinaria conoides*, a sea brown algae, is used to study green synthesis of gold nanoparticles with potential applications in biomedicine. The potential for environmentally friendly nanoparticle synthesis was demonstrated by testing the antibacterial efficacy of NPs against *Streptococcus* sp., *Bacillus subtilis*, and *Klebsiella pneumoniae*. The nanoparticles were detected using SEM and EDS.

### 14.2.2 Synthesis of Silver Nanoparticle

Silver nanoparticles (AgNPs) are often used because they can kill bacteria, making them useful in various products, but there are concerns about how they affect the environment and bacteria. Researchers found that AgNPs can harm and stress out bacteria like *E. coli* and *Staphylococcus aureus*. Surprisingly, when these bacteria were exposed to AgNPs before, they became better at resisting antibiotics, making the antibiotics less effective. AgNPs also caused changes in bacterial DNA, which can lead to mutations. This means that AgNPs might have unexpected effects on bacteria and the environment, so it's important to use and dispose of them carefully.[45]

A sustainable method uses soil bacteria for creating tiny silver particles. These particles are tested for their size and shape and found to have the ability to fight against MRSA, a drug-resistant bacteria.[46] This study emphasized an environmentally friendly method of producing these silver particles and their potential use in controlling infections acquired in hospitals, as well as in managing diseases. This study appears to be a significant contribution to the realm of healthcare and nanotechnology.[34]

Silver nanoparticles are very popular because they have special properties like being good at fighting germs, not harming our cells much, and having many uses. They're used in things like cosmetics, clothes, cleaning, and even in medicine for things like finding diseases, treating wounds, and delivering medicine. Silver can help in fighting with parasites and viruses and also has potential for treating against cancer and other health issues.[47]

Silver nanoparticles are tiny bits of silver that can help in different ways. They are better at fighting harmful molecules and germs in some cases than a common antioxidant like vitamin C. These silver nanoparticles are used in medical devices, like plastic catheters, to prevent harmful bacteria from growing on them. They can also be used on things like refrigerators or building materials to stop mold from growing. Paints with silver nanoparticles can even keep the walls in our homes clean. These special silver particles are often made from plants and are safer for the environment and people.[48]

In simpler terms, silver nanoparticles are extremely tiny pieces of silver that are good at fighting germs and harmful substances. They are used in medical equipment and can also be applied on things like fridges and walls to keep them clean. These silver nanoparticles are made from plants and are safe for the environment and its application.[46]

Silver nanoparticles can be used to make the insides of buildings safer by applying them on things like walls and decorations. These silver particles last for a long time and remain intact even in hot conditions. When we use silver nanoparticles made from plants, they have extra benefits because the plant molecules can help fight germs and attach to the surface of the nanoparticles, making them even better at keeping things clean. So, using plant-made silver nanoparticles can be a great way to make things germ free.[49]

Silver nanoparticles are incredibly useful because they can kill bacteria and have many applications. People have found different ways to make them using physical, chemical, and biological methods. In this case, they're focusing on using bacteria to make AgNPs. Bacteria can produce these particles inside or outside their cells, and it's easier to get them when they're produced outside. Certain genes, enzymes, and substances released by bacteria are responsible for making AgNPs and keeping them stable. These tiny silver particles made by bacteria have benefits like fighting cancer, acting as antioxidants, and being very effective at killing harmful microorganisms, especially the ones that are resistant to antibiotics. Their size and shape determine how well they work against different germs, making them exciting as new types of germ-fighting agents.[50]

#### 14.2.2.1 Synthesis and Physicochemical Characterization of AgNPs

In one study, silver nanoparticles were produced using an eco-friendly method with an aqueous extract from *Artemisia tournefortiana*. This plant extract served as the nanoparticles' capping and stabilizing agent. The specific combination of chemicals in the plant extract, including enzymes, flavonoids, polyphenols, and other phytochemicals, played a crucial role in reducing silver ions to AgNPs. The color of the solution changing to reddish-brown indicated the successful formation of AgNPs, and this process was quite fast, occurring within 10 minutes.

UV-visible spectroscopy was used to establish the presence of AgNPs, and it revealed a distinctive absorption peak at 420 nm that was in line with earlier research. The presence of biomolecules like hydroxyl and carbonyl groups in the plant extract was shown by Fourier transform infrared spectroscopy, which likely helped to reduce and stabilize AgNPs.

The AgNPs were found to be crystalline by X-ray diffraction examination, which also showed that different diffraction peaks matched the (111), (200), (220), and (311) orientations. Images from high-resolution transmission electron microscopy (HRTEM) and scanning electron microscopy (SEM) revealed that the AgNPs were single-crystalline, spherical, and had an average size of 10–30 nm. Energy-dispersive X-ray (EDAX) examination verified the AgNP elemental makeup and purity.

In simpler terms, this study used a plant extract to make tiny silver particles in a quick and eco-friendly way. The plant chemicals helped to change the color of the solution showing that AgNPs were formed. Scientific tests confirmed the presence of AgNPs, and they were found to be small, spherical, and pure in nature, with a size of around 10–30 nm.[51]

#### 14.2.2.2 Antibacterial Capacity of Fabricated AgNPs

Silver nanoparticles can be used instead of regular antibiotics to kill harmful bacteria, especially when they become resistant to common antibiotics. AgNPs have been tested against different bacteria. The results showed that they worked best against *Streptococcus pyogenes* and *Bacillus subtilis*, with low concentrations needed to inhibit their growth. However, they were less effective against *E. coli* and *Pseudomonas aeruginosa*.

Compared to standard antibiotics like vancomycin and gentamicin, AgNPs were not as strong in killing bacteria, but they still showed promising antimicrobial activity. They were better at killing Gram-positive bacteria like *Streptococcus pyogenes* than Gram-negative ones like *E. coli*, possibly because of differences in their cell wall structures. AgNPs made from *A. tournefortiana* plants seem to have these antibacterial properties because of their unique surface properties.

Si and Mandal's[36] work explained the reductive properties of tryptophan moieties, which can transform tryptophan into a tryptophol radical and can then transfer electrons to metal ions and produce metal nanoparticles. UV-Vis and fluorescence spectroscopy can be used to study this process. AgNPs containing short yeast peptides can be produced under ambient light by aspartic acid and glutamic acid. A study described the environmentally friendly synthesis of AgNPs using nitrate reductase and *Pseudomonas aeruginosa* JP1 culture, with NADH serving as a coenzyme, as shown in Figure 14.1. Enzymes are essential for the formation of AgNPs and play an important role in the reduction of metal salts. The synthesis of AgNPs using *Fusarium oxysporum* occurred utilizing superoxide dismutase. The process of creating metal nanoparticles involves nitrate reductases and NADH as a cofactor. Monodispersity and homogeneity can be achieved in AgNP synthesis by optimizing two important factors, maximal activity and production.

### 14.2.3 Synthesis of Palladium

Another valuable metal that is frequently employed in electrochemical catalysts, the refining industry, and car catalytic converters to reduce the emission of harmful pollutants is palladium. Palladium has been consumed at a rate that has outpaced its supply in recent years. While some chemical techniques are being used to recover palladium in bulk from waste streams, more effective and sustainable methods needs to be developed. The reduction of soluble Pd (II) to Pd (0) using microorganisms has been demonstrated to be an effective and environmentally friendly method of recycling precious metals and producing catalytic nanoparticles. According to Ahmed et al. (2018),[52] numerous studies

**FIGURE 14.1** Synthesis of AgNPs using extracellular enzymatic nitrate reductase in bacterial culture.

have documented the range of applications, and palladium is reduced to palladium nanoparticles (PdNPs) by bacteria.

One sustainable way to recover precious metal catalysts is by using *Geobacter sulfurreducens* to reduce palladium extracellularly to PdNPs. In another recent study, selective electrocatalysis of biofuel molecular oxidation was carried out to synthesize palladium nanoparticles utilizing *Shewanella oneidensis* MR-1.[53]

### 14.2.4 Synthesis of Platinum

The hydrogenase enzyme from sulphate-reducing bacteria (SRB) was used by microorganisms to create the first platinum nanoparticles. Since then, a variety of microorganisms have functioned as PtNP nanofactories: *Shewanella* strains employing a reductive decomposition channel, *Acinetobacter calcoaceticus* utilizing enzymatic reduction, and *Desulfovibrio vulgaris* using bioreduction. When hydrogenase as an enzyme is utilized from SRB, it helped to reduce Pt (IV), and became the first microbial-mediated platinum nanoparticle.[54]

Since then, a wide variety of microorganisms have functioned as platinum nanoparticle nanofactories. While the majority of research reported synthesis based on fungi, *Desulfovibrio vulgaris* resulted in formation of spherical PtNPs in periplasmic gaps using a bioreduction method. Nevertheless, the intracellular synthesis of monodispersed cuboidal PtNPs (5 nm) by *Acinetobacter calcoaceticus* was also observed to be reduced enzymatically. Similar to this, Gaidhani et al.[55] reported intracellular synthesis of PtNPs utilizing a *Shewanella* strain situated within the periplasm of the membrane.

### 14.2.5 Synthesis of Copper

The three most widely utilized metals are copper (Cu), silver (Ag), and gold (Au). Cu is a comparatively inexpensive metal that is more economical than Au and Ag. CuNPs can also be utilized repeatedly and are effective catalysts with high yields and simple product separation. Tyrosinase, cytochrome oxidase, and superoxide dismutase, are made up of these trace elements which are present in human body along with necessary enzymes. The presence of these trace elements has been significantly improved biomedical field in applications like biosensing, imaging, and diagnosis.

A combination of copper and copper oxide nanoparticles synthesized by bacteria is described. *Stibara* sp., a beetle species located in the northwestern Ghats of India, was found to have a midgut containing a Gram-negative bacterium from the genus *Serratia*. A number of different strains isolated from the same insect and common Indian mosquitoes did not result in the creation of nanoparticles, indicating that this particular bacterium is quite specialized for the manufacture of copper oxide nanoparticles. Through methodical reaction monitoring, researchers were able to identify that the production of nanoparticles takes place inside cells. Bacterial cells are killed during the process, nevertheless. The cell wall then breaks down, allowing the nanoparticles to escape. The resulting nanoparticles were extensively studied by TEM, UV-Vis, XRD, and XPS.[56]

### 14.2.6 Metal Oxide Nanoparticles

#### 14.2.6.1 Titanium Dioxide

In recent developments, metal oxide nanoparticles have found a wide range of uses in various industrial, medicinal, and environmental domains in the realm of nanotechnology. $TiO_2$-NPs, metal oxide nanoparticles, have drawn increased interest because of their potential applications across multiple fields, including nanomedicine and biomedicine. Several studies have shown the major benefits of employing natural resources to manufacture NPs through the biosynthetic approach, which is preferred over producing NPs through conventional methods. These benefits include the ability to produce NPs with precise sizes and shapes, a reduction in hazardous pollutants, a reduction in the requirement for further complex chemical synthesis, environmental friendliness, affordability, and stability. Consequently, a great deal of work is being done to put these advantages of biological synthesis processes into practice. There are numerous biological, chemical, and physical techniques for producing $TiO_2$-NPs. Physicochemical techniques are expensive, require a lot of energy, and release large amounts of hazardous substances into the atmosphere. Conversely, the synthesis of $TiO_2$-NPs can be accomplished simply, affordably, consistently, conveniently, and in an environmentally friendly manner using the biological technique. This section includes the biomediated production of TiO2-NPs.[57]

Nanostructures made of titanium dioxide show promising characteristics for antibacterial and optical applications. This study used simple microwave-assisted hydrothermal methods to create $TiO_2$ nanoparticles. The optical characteristics and morphology of the synthesized $TiO_2$ NPs were examined using a variety of methods, such as UV measurements, thermogravimetric analysis (TGA), transmission electron microscopy, Fourier transform infrared, scanning electron microscopy, X-ray diffraction, and energy and dispersive X-ray spectroscopy (EDXS). $TiO_2$ nanoparticles' antibacterial efficacy was evaluated using the pathogenic strain *E. coli* ATCC 8739TM. According to the published results, $TiO_2$ nanoparticles exhibited greater antibacterial activity when tested against pathogenic bacteria, suggesting possible uses for these particles in the biomedical and medical domains.

##### 14.2.6.1.1 *Synthesis of $TiO_2$ Nanorods*

$TiO_2$ nanorods were created with the aid of a microwave during a wet chemical process. $TiO_2$ nanorod synthesis was regulated by two chemical processes. The first involved evaluating microwave irradiation and the hydrothermal technique using a wet chemical route. The second method involved synthesizing the suggested $TiO_2$ nanorods through a straightforward reaction with water. Fifty ml of water and

25 ml of titanium (IV) tetrachloride ($TiCl_4$, 99.9%, bought from Aldrich) were used in the synthesis process. The following microwave hydrothermal technique was employed: Distilled water was combined with titanium tetrachloride. The mixture was kept in an autoclave lined with Teflon. After being sealed and heated to 220°C for 90 minutes using a 700 W microwave furnace, the autoclave was allowed to cool to room temperature.[58]

### 14.2.6.2 Iron Oxide

Hydrothermal cells were used to create low-dimensional iron oxide nanoparticles by introducing urea and ferric chloride. After stirring and adjusting the concentration to 0.5 M, the liquid was heated for 6 hr in a Teflon line autoclave. The optical and structural characteristics of the as-grown iron oxide products were described. The crystal growth behaviors and chemical reactions of iron oxide account for the formation of $Fe_2O_3$ nanoparticles. FeOOH is created when $FeCl_3$ and $NH_4OH$ combine, leading to the production of $Fe^{2+}$ and $OH^-$ ions. The end products' building components, $Fe_2O_3$ nuclei, are created when the FeOOH dissociates. Low-dimensional particles are well-created to produce $Fe_2O_3$ NPs.[59]

In the work, iron sulfate was heated with n-decanoic acid or n-decylamine to create surface-modified iron oxide nanoparticles. It was discovered that the nanoparticles had an organic layer on their surface and were in the $\alpha$-$Fe_2O_3$ or $Fe_3O_4$ phase. At higher temperatures, the chemicals became more soluble, which facilitated the interaction between the organic reagents and the nanoparticles.[60]

Iron oxide nanoparticles are gaining attention due to their unique properties like superparamagnetism and easy separation. They have various applications in bioengineering, commercial, and industrial fields. These NPs can be coated with organic or inorganic molecules and can be directed to organs, tissues, or tumors for hyperthermic treatment.[61]

As iron oxide, a polymorphous mineral, performs well in the thermal breakdown of ammonium perchlorate, it is employed in solid propulsion. It works well for gas adsorption because of its smaller size of particle, active sites, and large surface area. Although there are several ways to synthesize these nanoparticulate oxides, less information is available about their shape, structures, and methods of characterization.

### 14.2.6.3 Zirconium Dioxide

Zirconium oxide nanoparticles ($ZrO_2$ NPs) are created by utilizing extracts from three distinct plants: *Lycopersicon esculentum, Allium cepa*, and *Capsicum annum*. It was discovered that the NPs were nanosized, had good qualities, and all of their crystal forms were Baddeleyite. The average diameters of the NPs generated by *C. annum* were 100.25, 86.66, and 1.17 nm for roughness and 1.98 nm for root mean square. The average sizes of cepa were 105.14, 83, and 0.238 nm for roughness; 1.27 nm for root mean square; and 21.97 nm for crystal size.[61]

The synthesis of $ZrO_2$-NPs involved adding *W. tinctoria* leaf extract to a $ZrOCl_2$-$8H_2O$ solution, centrifuging, washing, drying, and calculating in a muffle furnace to obtain a powder, which was then subjected to further characterization. The study examined the crystalline size and phase of biosynthesized $ZrO_2$-NPs using X-ray diffractometers, surface morphologies using SEM, average particle size and distribution through DLS, and surface charges through zeta potential. Functional groups were evaluated through FT-IR analysis, optical spectra, and fluorescence spectrometer.[62]

Monoclinic zirconia nanoparticles were produced using green synthesis analysis from sunflower seeds. The nanoparticles showed a sharp peak in the UV-Vis spectrum and a 40.59 nm zirconia crystallite. The stability of the nanoparticles was confirmed through zeta potential and particle size distribution. These nanoparticles have superior antimicrobial activity for biomedical applications.[63]

The preparation of zirconium oxide nanoparticles involved dissolving 0.05 M zirconium oxychloride octahydrate in deionized water, adding a metal salt solution to a methanol extract of seeds, centrifuging, washing, drying, and annealing the black powder at 600°C for 4 hr to prepare nanoparticles. The process involved removing impurities, oil content, and drying the residue.[63]

Another synthesis mechanism involves the reaction of nanoparticles from zirconyl chloride reduction and plant phytochemical compounds. The -OH moiety reduces ions in zirconium salt, resulting in nanoparticles of zirconium oxide. The compounds also support the termination process, preventing agglomeration and forming nanoparticles as stabilizers.[64]

The transitional metal element zirconium oxide, also referred to as "ceramic steel", has an electrical configuration of 4d25s2. It is only present in significant amounts when mixed and is not found in its pure form. Zirconium in pure powder form may find utility in biological applications, according to research. Three distinct forms of crystallographic symmetry can be seen in zirconium: cubic (c), tetragonal (t), and monoclinic (m). It is possible to regulate the pressure and temperature that are employed to change the crystalline patterns of pure zirconia. Based on the preparation techniques, zirconia possesses a wide bandgap p-type semiconductor with a bandgap of 3.25 to 5.1 eV. It is sometimes referred to as zircosol, zirconia, and zirconic anhydride. $ZrO_2$ nanoparticles' unique electrical, heat, catalytic, sensing, optical, mechanical, and compatible biological properties have made them interesting. Drug administration, tissue engineering, biomarkers, water treatment, bioconjugation, photocatalysts, refractory cells, gas sensors, solar cells, implant materials, and agriculture are just a few of the many uses for them that are commonly found.[65]

Zirconia can be synthesized using various methods, but these require high temperatures, pressures, and specialized tools, which can have negative environmental impacts. Instead, using biological techniques like enzymatic processes can be a more environmentally friendly method for synthesizing nanoparticles like $ZrO_2$.[66]

Nanoscience and technology are gaining a lot of attention for recent years due to their variety of applications. A study revealed that hydrothermal process led to monoclinic-shaped zirconium dioxide nanoparticles. The band gap of material was observed to be reduced with increasing calcination temperature, probably due to the increase in surface-to-volume ratio. A wide band around 500 $cm^1$ was displayed by the synthesized material, confirming the formation of NPs. The efficient synthesis of size-dependent zirconia nanoparticles resulted in a reduction of both production time and energy consumption.[48]

The hydrothermal method was used to synthesize zirconium oxide nanoparticles, using zirconium oxychloride octahydrate ($ZrOCl_2$-$8H_2O$) as a raw material, deionized water as a solvent, and aqueous NaOH as a precipitation agent. The mixture was stirred, filtered, washed, and dried before being calcined at 400 and 600°C to study the nanopowder's structure, morphology, and resistivity.[67]

#### 14.2.6.4 Copper Oxide

By using unique bacterial strains for efficient nanoparticle manufacturing, the drawbacks of current methods are removed and the detrimental effects on the environment are reduced. *Stenotrophomonas* sp. BS95 cell-free extract was used in one study. Extraction with the cell-free method was applied to yield CuONPs. UV-Vis, Fourier transform infrared, X-ray diffraction, dynamic light scattering, zeta potential, field emission scanning electron microscopy, atomic force microscopy, and transmission electron microscopy were used to analyze the crude and calcined CuONPs. The antibacterial, antioxidant, and cytotoxic properties of biogenic CuONPs were then assessed using the broth microdilution method, the 2,2-diphenyl-1-picrylhydrazyl (DPPH) assay, and the alamar blue assay, respectively. Ultimately, real-time PCR was used to determine the molecular pathways behind CuONPs' anticancer activities.

Surface plasmon resonance peaks were visible in the UV-Vis absorbance spectra of the calcined and crude CuONPs, respectively, at 286 and 420 nm. Bands observed in FTIR spectra, which verified the organic functional groups of bacterial proteins, substantiated the capping and functionalization of CuONPs. Using XRD, it was revealed that the average crystallite diameters of the calcined and crude CuONPs were 21.3 and 18.24 nm, respectively. Crude and calcined CuONPs had average zeta potentials of $-28.57 \pm 5.13$ and $-29.47 \pm 4.78$ mV, respectively, demonstrating their great stability. The average size of the calcined and crude CuONPs was $43.68 \pm 2.31$ and $35.24 \pm 4.64$ nm, respectively, according to electron microscopy. They were generally spherical in shape. Biogenic CuONPs showed antibacterial efficacy against Gram-positive and Gram-negative bacteria, with low inhibiting concentrations ranging from 62.5 to 1,000 μg/ml.

Potential antibacterial, antioxidant, and cancer agents, biosynthesized CuONPs show lethal effects on human colon and gastric adenocarcinoma cells together with significant alterations in apoptosis-related gene expression and antioxidant activity ($83\% \pm 2.64\%$). The study synthesized CuONPs using psychrotolerant *Stenotrophomonas* species, reducing microbial contamination risk. The biosynthesis was validated by monitoring four flasks, revealing instant precipitation of green aggregates over 24 hr, indicating CuONP formation.

The study synthesized copper oxide nanoparticles from *Spirodela polyrrhiza*, a plant known as greater duckweed. The nanoparticles were characterized using ultraviolet spectroscopy and EDX and tested for antibacterial, antifungal, and nematicidal properties. Results showed significant antifungal and nematicidal activity, suggesting they could be a safer alternative for plant disease management.[68]

### 14.2.7 Toxicity Aspects of Bacteria

The recent development of nanoparticles has led to serious concerns about human as well as environmental health. NPs are extremely reactive and autocatalytic, so their applications include drug delivery, electronics, optics, and therapeutics. However, if they are released into the environment, they may enter the food chain through presence of microorganisms. NPs are mainly targeted at bacteria, which can lead to oxidative DNA damage, reactive oxygen species production, and membrane disruption. Nevertheless, toxicity study is essential to assess nanotoxicity.

Microorganisms adapt to their environment and exploit energy resources. Bacteria, found in air, water, and soil, are crucial for the formation and functioning of ecosystems. They serve as decomposers and are employed for a variety of biotechnological applications, such as fermentation, decontamination of food, wastewater treatments, antibiotic production, bioremediation, pollution control, and ore mining. Engineered nanomaterials, generated as by-products or due to accidental release, can indirectly impact ecosystem health, soil quality, and sustainability. The limited understanding of the environmental impact of nanoparticle exposure and interactions with bacterial cells increases the complexity of synthesis methodologies. The bactericidal toxicity of nanoparticles is evaluated, and it varies depending on their form or type of NP.

Engineered metallic core/core-shell nanoparticles are widely used in biology and medicine due to their potential bactericidal activity. However, their microbial toxicities vary due to factors like the source of materials, form, size, shape, surface charge, stabilizing agents, chemical and physical properties, particle behavior, toxicity evaluation procedures, reaction medium, and particle characterization

Uncontrolled accumulation of metal-based nanoparticles in land-based ecosystems, mainly in agriculture, is endangering the surrounding ecosystem and beneficial biofactories, such as soil bacteria and fungi. The reason behind this ecosystem imbalance is poor wastewater treatment and application in agricultural fields. Nanoparticle deposition on lands, low biodegradability, and persistence have adverse impacts on soil-beneficial bacteria and fungi populations.[69]

Silver nanoparticles' toxicity mechanism against microorganisms is a topic of discussion. A study examining the effects of 11-mercaptounde-canoic acid-and citrate-capped silver nanoparticles on various microorganisms revealed that silver nitrate NPs are more efficient against bacteria compared to equimolar mixture of nanoparticles. The choice of capping agent is essential in terms of their toxicity. Per the reported study on medical devices, it was observed silver nanoparticles could prevent microbial colonization in medical devices and control the surrounding environment of the medical device.[49]

## 14.3 FUNGI-BASED SYNTHESIS

It is possible to create nanoparticles with precise dimensions using fungi. Research has suggested that fungi, rather than bacteria, may serve as a source for the large-scale manufacturing of nanoparticles. The fungi-mediated synthesis of various NPs is listed in Table 14.2. Since fungi are known to secrete a far greater amount of chemicals and proteins than other species, the productivity of nanoparticles could be greatly boosted in a biosynthetic process. Reductases are among the specialized enzymes that fungi release, and they raise the exciting possibility of developing a sensible biosynthesis process for nanomaterials with diverse chemical compositions. Green chemistry mediated by actinomycetes and fungi can be used to easily scale up the production of nanoparticles, and with the right mycelia development, it can potentially cover large surface areas very quickly. To produce a natural "nanofactory", fungi were used instead of bacteria. This offered the benefit of significantly simplifying the downstream processing and handling of biomass.[70]

Two methods are utilized to produce fungi-mediated nanoparticles: intracellular and extracellular processes.[70]

### 14.3.1 Fungal-Mediated Synthesis of AuNPs

The green synthesis of NPs using microorganisms has established a distinctive place among other production methods because of its simple culture, quick growth, and capacity to grow in optimized pH, temperature, and pressure conditions with respect to its surroundings. The employment of microorganisms in the production of nanoparticles is currently a matter of ongoing research due to its ease of usage, affordability, and environmental friendliness. This method works well for producing NPs on a higher scale as well. Because they have a variety of reductase enzymes, microorganisms are necessary nanofactories which can store, decrease, and cleanse heavy metals. Microorganisms play a key role as nanofactories in the creation of nanoparticles, including titanium, gold, copper, zinc, nickel, palladium, and silver.[71]

#### 14.3.1.1 Intracellular Biosynthesis of NPs

Cell walls carry negatively or positively charged metal ions which get deposited on them by electrostatic interactions. NPs are produced using reducing ions of nitrate reductase enzyme. Microorganisms that are used in synthesis of intracellular nanoparticles are grown in a suitable medium, and the biomass is collected by centrifugation after specific period of incubation and washed by sterile distilled water. Furthermore, after centrifugation biomass is collected, and mixed with salt solution containing metal ions which produces NPs.[71]

Another study suggested an altered procedure, cultivate fungus in Erlenmeyer flasks under shaking conditions at 200 rpm, 27°C for 96 hr. Following fermentation, the isolated mycelia is washed and resuspended in a 0.001 M aqueous $HAuCl_4$ solution. The mixture is then incubated on a shaker for 72 hr at 27°C. The progress of biotransformation is visually observed, and the samples are periodically taken for UV-Vis spectral analysis. The intracellular production of BioAuNPs requires further steps of extraction and purification, which makes synthesis processes more complex.[78] For further analysis, fungal films were prepared before and after exposure to $AuCl_4$ ions, and UV-Vis spectroscopy and XRD investigations were carried out. These films were created by casting the washed fungal cells on Si (111) and quartz wafers and followed by a thorough drying under N2 flow. UV-Vis spectroscopy measurements on a dual-beam spectrophotometer and Fourier transform infrared spectroscopy studies were performed using a Shimadzu FTIR instrument. XRD analysis was carried out on Si (111) substrates for the Au nano-fungus biofilm and Au nano solution using a Philips PW 1830 instrument.

Gold nanoparticles, despite increasing importance, are less commonly synthesized by fungi than silver nanoparticles. Their small size makes them ideal for catalysts and electronics applications. A study using *Verticillium* sp. synthesis showed intracellular AuNPs on mycelia surfaces, which is depicted in Figure 14.2.[78]

In another study, 10 g of *F. oxysporum* biomass was added to 100 ml of distilled water in a conical flask for a standard reaction. In order to achieve an overall Ag-ion concentration of 0.001 M in the aqueous solution, an amount of $AgNO_3$ was added to the conical flask, and the reaction was carried out in the dark. Two-ml aliquots of the reaction solution were taken out periodically and subjected

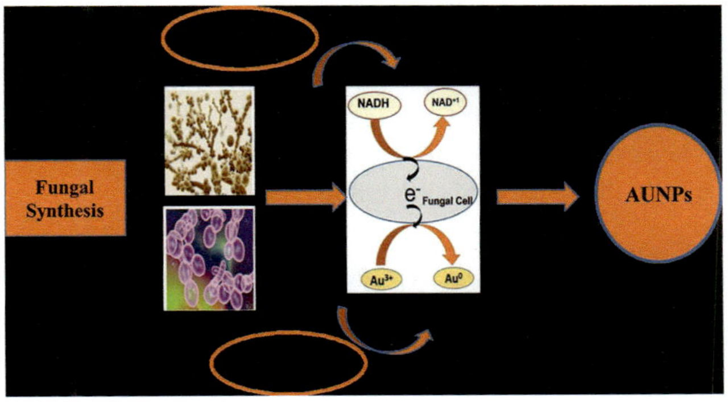


FIGURE 14.2 Fungi-mediated synthesis of gold nanoparticles.

to spectroscopic measurements using UV-Vis and fluorescence. The fluorescence was estimated using a Perkin/Elmer LS 50B luminescence spectrophotometer. The Shimadzu dual-beam spectrophotometer (model UV-1601 PC) was utilized for UV-Vis spectroscopy measurements, with a resolution of 1 nm. Tyrosine residues and tryptophan optical transitions were maximized by selecting 260 nm as the excitation wavelength.[78]

#### 14.3.1.2 Extracellular Synthesis of AU NPs

Nanoparticle synthesis involves the use of cell-free culture supernatant, which contains enzymes and other extracellular components of microorganisms. NADH-dependent nitrate is the primary reductase enzyme which controls the synthesis of NPs. Exopolysaccharides (EPSs) secreted by bacteria, fungi, and yeast act as stabilizing and reducing constituents. Biosurfactants such as emulsan and rhamnolipid offer a surface for NPs. Extracellular proteins help in metallic NP formation via carboxylate and/or amino groups. Metal salts are reduced and converted to NPs with the help of reducing molecules like phenolic compounds and organic chlorocholinchloride (CCC).[37]

Gold nanoparticle synthesis using fungi involves a biotransformation process where fungal mycelia interact with gold ions, leading to the reduction and subsequent formation of AuNPs. Fungi such as *Trichothecium* sp. have shown potential, indicating their ability to generate AuNPs from aqueous $HAuCl_4$ solutions. This biogenic approach offers an environmentally friendly and economical alternative to conventional synthesis using chemicals for AuNP synthesis. An alkalotolerant fungus, *Trichothecium* sp. (AA-UDG-1), with optimum growth at pH 7 and 27°C was maintained on potato-dextrose-agar (PDA) slants by subculturing at monthly intervals. Inoculum was prepared with 4 days incubation, and this alkalotolerant fungus was transferred to 500-ml Erlenmeyer flasks containing 100 ml MGYP medium, which is composed of yeast extract (0.3%), glucose (1%), malt extract (0.3%), and peptone (0.5%), and kept in a stationary phase at 27°C for 10 days. After 10 days of fermentation, a mycelial mat in stationary phase was obtained and separated from the culture broth by centrifugation at 5000 rpm, 20° C for 20 min. The mycelia was aseptically washed 2–3 times using sterile distilled water. The harvested mycelial mat at stationary phase, resuspended in 100 ml of 10–3 M aqueous $HAuCl_4$ solution in 500 ml of Erlenmeyer flasks and incubated at 27°C for 48 hr. The biotransformations that occurred in the flask were routinely monitored by periodic sampling of 2-ml aliquots of the aqueous component and estimated optical density of the solution under the UV-Vis spectra.[71]

### 14.3.2 Synthesis of AgNPs

Nanomaterial synthesis is crucial, as it requires an environmentally sustainable method with nontoxic chemicals. The relationship between inorganic nanoparticles and biological structures is a fascinating field of study, and the size and form of noble metal nanocrystals greatly influences their properties. Numerous techniques, such as hard-template synthesis, bioreduction, and solution phase synthesis, have been employed to create noble metal nanomaterials. It has been possible to use bacteria to make nanoparticles, produce cermets, convert metal ions into metal particles, synthesize magnetite nanoparticles, and enrich ions. Novel biotechnological approaches in nanotechnology are indicated by recent findings in the biosynthesis of metal nanoparticles. Nanoparticles can observe biological processes without interfering by acting as tiny probes. Understanding biological activities at the nano level is an essential component in the progress of nanobiotechnology.

In one of the studies of extracellular biosynthesis of AgNPs, the fungus *Cladosporium cladosporioides* was found to be the most suitable for synthesis of AgNPs in the local environment.[79] The fungus *C. cladosporioides* was cultivated on an MM1 medium containing 5 g/L $KNO_3$, 2.5 g/L $K_2HPO_4$, 1g/L $MgSO_4{\cdot}7H_2O$, 0.001 g/L $MnSO_4{\cdot}H_2O$, 0.003 g/L $CuSO_4{\cdot}5H_2O$, 0.1 g/L $ZnSO_4{\cdot}7H_2O$; 0.0015 g/L $Na_2MoO_3{\cdot}2H_2O$, 0.02 g/L $FeCl_3$, 1g/L$Na_2SO_4$, and 40 g/L glucose. Each of the ingredients was separately dissolved in distilled water to prepare the mineral salts medium. The components were mixed thoroughly in a 2-L beaker; then 4 M NaOH or HCl was added to adjust the pH down to 6.5–7.0. Further distilled water was added to make up the volume. This mixture was heated to a certain extent to dissolve the remaining undissolved solids and was filtered. Culture flasks were incubated for a week at 27°C, then filtered and washed with a distilled water. The filtrate was further combined with a metal ion solution to reduce the metal ions. An Erlenmeyer flask containing 100 mL of double-distilled water and 0.01 mL of Ag+ metal ion solution along with 10 mL supernatant solution was kept on a shaker for 78 hr at 27°C. The reaction solution's aliquots were collected periodically, and UV-Vis spectroscopy was carried out for surface plasmon measurements. Dropwise coat of the colloidal nanoparticles was utilized to create AgNP colloidal solutions on Si (1:1:1)-based substrates. Additionally, the size of the produced crystals was analyzed using XRD pattern. The resulting nanoparticles' size and form were investigated using transmission electron microscopy.[79]

### 14.3.3 Synthesis of Non-Metallic NPs

The antibacterial and therapeutic properties of nanomaterials and nanoparticles can be determined by their size and shape. ZnO NPs in the shapes of hexagons and rods were biosynthesized from fungal extracts and deposited onto cotton textiles. When these NPs were tested against both Gram-positive and Gram-negative bacteria, their antibacterial properties varied. For the purpose of storing NPs, their stability in colloidal media is essential. Because nucleic acid and protein leakage occurs during the generation of reactive oxygen species, zinc oxide nanoparticles have bactericidal effects on both Gram-positive and Gram-negative bacteria in addition to their antioxidant and anticancer properties.[80]

#### 14.3.3.1 Synthesis of Copper NPs

*Penicillium olsonii* was inoculated for 7 days at 28°C on potato dextrose (PD) broth media. After incubation, the biomass was harvested by filtration using filter paper. The pellet obtained was washed twice with distilled water to remove any medium component. From the obtained biomass, 3 gm of fungal biomass was brought in contact with 100 mL of sterile distilled water and 0.02M concentration of blue-colored copper sulfate solution was added. This mixture in the flask was incubated at 25°C for 3 days with control (without Cu).

After 24 hr, 1 ml of sample supernatant was withdrawn, and absorbance under UV-visible spectroscope was measured between 300 and 800 nm, confirming the presence of CuNPs. The size of the CuNPs was characterized using scanning electron microscopy (SEM). Furthermore, the poisoned food technique was used to test the antifungal activity of CuNPs. This method involved preparing a potato dextrose agar medium with different concentrations of 50, 100, and 200 ppm CuNPs. Each petri dish was filled with 25 ml of growth media, which was then left to harden. A 5-mm disc filled with this solution was positioned in the middle of a plate holding a test fungus culture. The plate was then incubated for 7 days at 28 ± 2°C, and the fungal colony's growth was gauged in millimeters. The control was PDA medium without CuNPs.[81]

### 14.3.4 Toxicity Aspects of Fungi

Heavy metal contamination is a global concern since many industrial effluents include harmful metal ions. A metal's degree of use, bioavailability, species of origin, ubiquity, and extent of absorption and distribution all contribute to its toxicity.[82] Due to the extreme persistence of these metal pollutants, they accumulate more in soil and water bodies.[83] Bioremediation is difficult because of the strong persistence of these metal contaminants.[82] The World Health Organization (WHO) states that hazardous metals such as cadmium (Cd), lead (Pb), nickel (Ni), cobalt (Co), mercury (Hg), chromium (Cr), and copper (Cu) have a significant impact on aquatic animals and microbes. In order to determine appropriate criteria for these heavy metals in drinkable water, the WHO has produced certain guidelines. The following are the approximate threshold values for specific heavy metals: Mercury (Hg) 0.006 mg $L^{-1}$, Ni 0.07 mg $L^{-1}$, Cd 0.003 mg $L^{-1}$, As 0.01 mg $L^{-1}$, Pb 0.01 mg $L^{-1}$, Cu 2.0 mg $L^{-1}$, and Cr 0.05 mg $L^{-1}$. Metal occurrence and accumulation in the environment can be attributed to both natural and human-caused processes. Metals are released into the atmosphere by a variety of sectors' wastewater, including the mineral, automotive, tanning, fossil fuel, agricultural, petroleum, and fertilizer industries. These industrial effluents primarily contain residues of manganese (Mn), copper (Cu), mercury (Hg), zinc (Zn), and lead (Pb), along with a few additional components, according to statistics reported by the World Health Organization in 2021.[84]

The physical and chemical characteristics of the growing environment have a major impact on the toxicity of Cd NPs to fungus and other microbial species. Comprehensive discussions of this issue will not be addressed here, aside from a brief overview of evident cases, since in-depth reports on environmental impacts on heavy metal toxicity can be found elsewhere.[85] The level of toxicity interactions could rise or fall.[86] Babich and Stotzky (1977a) found that cadmium toxicity to several fungi was enhanced at alkaline pH values of 8 to 9,[87] while cadmium toxicity to four yeast species was enhanced at low pH values. It may not be evident that external pH and Cd toxicity are generally correlated, and there are a number of possible explanations for apparent contradictions.[88] These include variations in the hydroxylated forms of metal ions' and their chemical behaviors' toxicities, variations in the effects of pH on the physiology of experimental organisms, and the impact of pH on the binding or complexation of metals to other organic elements in the growth medium or environment.[89]

Low external pH levels decreased $Cd^{2+}$ surface binding and intracellular uptake.[90] Any decrease in toxicity at low pH levels may be simpler to understand in cases where such a pH effect is evident. Toxicology can be impacted by additional anions and cations. Soft water has higher levels of cadmium toxicity than hard water for aquatic hyphomycetes and other fungi, with $Ca^{2+}$ being the primary factor influencing this impact.[91] $Mg^{2+}$ has little effect on Cd toxicity, while zinc could decrease it as well.[92] *Saccharomyces cerevisiae* showed reduced toxicity with low $Zn^{2+}$ and $Cd^{2+}$ concentrations, but high $Zn^{2+}$ concentrations increased toxic effects, possibly due to the extra component of $Zn^{2+}$ toxicity.[93]. When high amounts of tilde $Mg^{2+}$ were present, calcium was found to be effective in lowering Cd toxicity with $Mg^{2+}$ and $Zn^{2+}$ towards *Aspergillus niger*. However, substantial quantities of $Mg^{2+}$ were required to offset the toxicity of $Zn^{2+}$.[94] Anions, which have the ability to create insoluble precipitates, like phosphate, sulfide, and carbonate, can also lessen toxicity.[90] Cadmium toxicity towards fungi can also be decreased using increasing chloride concentrations, which showed reduced toxicity of cadmium chloride complexes compared to $Cd^{2+}$.[88]

Because clay minerals bind $ICd^{2+}$, they lessen its toxicity to fungus; the degree of this reduction depends on the clay's ability to interchange cations. As montmorillonite has a far larger cation exchange capacity than kaolinite, it is more successful at decreasing toxicity.[95] Thus, if a soil's cation exchange capacity is high or rises, it is attributed to the types and amounts of clay minerals, organic colloids, and hydrated metal oxides.

## 14.4 APPLICATION OF NANOPARTICLES

### 14.4.1 Antimicrobial Agents

One of the most significant medical discoveries in human history is the discovery of antibiotics. The creation of super-resistant bacteria in the body as a result of the extended misuse of antibiotics in recent years has significantly raised infection-related mortality rates and poses high level risk to public health. Various applications of biosynthesized metal and nonmetal NPS are depicted in Figure 14.3. Bacteria are easily resistant to conventional drugs because they only have one target. In the 1990s, the World Health Organization became aware that antibiotic-resistant bacteria was a global problem. Finding a safe and efficient antibacterial agent has been a research interest due to the ongoing escalation of multi-drug resistance infections, which makes it vital to create new strategies to address this worldwide therapeutic dilemma. New approaches for treating microbial infections are required to address the most critical issues of the post-antibiotic age. Due to its unique chemical and physical characteristics, including its ultrasmall size, high surface area to mass ratio, and increased chemical reaction activity, nano-antibacterial material presents a wide range of application opportunities in antibacterial therapy. Antibacterial nanoparticles can be categorized as having two functional capabilities simultaneously, carrying antibacterial compounds or intrinsically possessing antibacterial properties. Due to their exceptional optical, electrical, and catalytic qualities, precious metal nanoparticles have drawn a lot of interest.[96]

Metal and metal oxide exhibit a variety of antibacterial properties. Apart from changing the permeability of cell membranes, they can also interfere with the functions of proteins and materials that have phosphorus and sulfur in them, including DNA, making it harder for bacteria to become resistant to them. The most popular metal nano-antibacterial agents are zinc oxide, silver, copper, gold, titanium dioxide, and other metal nanoparticles. Metal oxide nanoparticles, such as $Ag_2O$, FeO, CuO, ZnO, MgO, $TiO_2$-CaO, and others, are also thought to have potential antibacterial activity. Nanomaterials can considerably improve the optical properties, mechanical strength, antioxidant activity, and enzyme-like activity of volume equivalents. The substantial catalytic activity of nanomaterials has made their enzyme-like activity a prominent topic for research in recent years.[97]

**FIGURE 14.3** Various applications of biosynthesised metal and nonmetal NPs.

Gold nanoparticles are gaining more attention due to their unique characteristics in various fields. They are thought to be promising options for photothermal converters, colorimetric detection, catalysis, and imaging.[89] Magnesium oxide (MgO-NP) in aqueous solution has shown low toxicity and potent *in vitro* and *in vitro* antibacterial activity. ZnO, CuO, and AgNP have a wide range of possible uses in biosensors, ceramics, fungicides, pharmaceuticals, catalysts, and health products. Recent years have seen a remarkable advancement in AuNP application research. Research indicates that the nanocomposite's physical and chemical properties remain unaffected by the addition of nanogold, and its antibacterial activity can be enhanced as a result of its characteristics. The antibacterial activity and biocompatibility of the material may be affected if starch is added. Antimicrobial peptides (AMPs) exhibit both immunomodulatory and anti-infective properties. Research revealed that the combination of gold nanoparticles with AMPs has a significant impact on anti-drug-resistant bacteria and can help boost peptide serum stability. According to Wang et al.'s research, AuNPs are created when they are modified with non-antibacterial molecules (thiol and amine, for example) but lack any antibacterial activity themselves.[98]

Silver nanoparticles have the same antibacterial properties as colloidal silver. Studies show that the particles can be used to treat skin infections brought on by MRSA and other bacteria because of their potent antibacterial properties. Furthermore, the particles neither develop toxicity nor drug resistance. For this reason, Au-AgNPs are safe *in vivo* as well as *in vitro*. However, it is revealed from research that when nanosilver particles are exposed to bacteria, they form clusters, losing their active surface area and antibacterial efficacy. Scientists have produced graphene and AgNP nanocomposites. The resulting composites exhibit better light stability, colloidal stability, and antibacterial activity against *Escherichia coli* and *Bacillus subtilis* when compared to AgNPs. According to research by Kamradgi et al., drug-resistant *Enterococcus faecalis* and *Pseudomonas aeruginosa* are particularly susceptible to the antibacterial qualities of biosynthetic silver nanoparticles, which exhibit strong antibacterial activity against both Gram-positive and Gram-negative bacteria. When this material's anticancer activity was evaluated against the human hepatocellular carcinoma (HepG2) cell line, AgNPs showed a substantial inhibitory effect.[99]

Studies have demonstrated a direct relationship between their antibacterial, antifungal, and antiviral properties of CuNPs and the microorganisms they kill. Applications of copper nanoparticles are numerous, including, including blood vessel transplantation, burn therapy, dental materials, and the prevention and repair of bacterial and fungal colonization. Studies on antibacterial activity indicate that the primary mechanism of CuNPs' antibacterial action is DNA damage caused by reactive oxygen species produced by bacteria. Copper's strong antibacterial qualities have led to a wide range of applications in areas such as water filter treatment, compound food antibacterial packaging, and medication delivery. According to research, the optimal size range for nanocopper antimicrobials is 1–10 nm. CuNPs' pathogen interaction, potent chemical and biological activity, large specific surface area, and other qualities make them useful in a variety of medical and dental applications. Copper oxide nanoparticles are a type of multifunctional material with a wide range of potential uses in several domains, including photoconductivity, electronic devices, and gas sensors. They can be used, among other things, to make various types of bio-nanocomposite films that are antibacterial and food-active packaging.

### 14.4.2 Drug Delivery and Diagnosis

#### 14.4.2.1 Cancer

Unchecked cell development is one of the symptoms of cancer, which is a lethal disease. There are hazards associated with the traditional methods of treating cancer, which include radiation, chemotherapy, and surgery. Specific medicine delivery to the affected organ in the early stage has some side effects. Therefore, it is imperative to create alternative diagnostic and therapeutic approaches to this illness. According to reports, tailored medication delivery using nanomedicines can be used to diagnose tumors and treat them. For instance, in a 2020 study by Borse et al.,[100] the *in vitro* anticancer activity of biofabricated PtNPs was assessed against *Saccharomyces boulardii* cell lines, MCF-7 and A431. AgNPs biosynthesized from *Cryptococcus laurentii* have demonstrated strong antitumor and anticancer activity, according to Ortega et al.'s 2015[101] investigation into the efficacy of AgNPs against cancer and breast cancer cell lines. Furthermore, biosynthesized SeNPs showed anticancer activity. Hep-G2 and MCF-7 cancer cells are killed by strain EssamA-1 biofabricated Se nanorods (SeNrs), according to Ahmad et al.'s (2015)[52] study of these nanorods against human cancer lines. Ramya et al. (2018)[102] observed that extracellularly generated SeNPs from *Streptomyces minutiscleroticus* M10A62 showed strong antiproliferative action against HepG2 and HeLa cell lines. El-Batal et al. (2015)[103] investigated the efficacy of AuNPs against cell lines that represent human liver (HEPG-2) and breast cancer (MCF7). They found that AuNPs stimulated DNA damage, nuclei disintegration, cytokinesis detention, and mitochondrial apoptosis in cancer cell lines.[98] As a result, NPs have been used in numerous studies due to their ability to cause cancer under controlled circumstances. Before they are commercialized, problems with toxicity, dosages, and the host immune response during treatment still need to be resolved. Using nanomaterials to successfully achieve targeted drug delivery can help increase the bioactivity, stability, and bioavailability of medications. Major liposomes, nanospheres, polymeric micelles, and water are examples of drug delivery nanomaterials where natural antibodies are coated with nanoemulsion, soluble polymers, and NP.

Using nanoparticles rather than more conventional drug delivery systems has several benefits. Targeted drug administration can lower the chance of adverse drug reactions in patients. The potential of zinc oxide nanoparticles produced from *Rhodococcus pyridinivorans* as anthraquinone nanovehicles was investigated by Kundu et al. (2014).[104] ZnONPs loaded with anthraquinone were shown to be cytotoxic at different concentrations, and they were also found to be capable of killing HT-29 colon cancer cells. Based on these findings, it was determined that ZnONPs may be used for targeted drug delivery.[105]

By employing hydrophilic nanoparticles, drug absorption can be enhanced, and drug diffusion can be enhanced to boost cytotoxicity. When doxorubicin and biofabricated AuNPs were tested against HEK293 cancer cell lines, Kumar et al. (2008)[106] found that the drug diffused into the cell lines more quickly when mixed with AuNPs. When AuNPs were introduced, Syed et al. (2013)[107] reported similar outcomes where the effects of doxorubicin on the cells found responsible to cure liver cancer. In 2014, Khan et al.[108] reported on a combination of bio-produced gadolinium oxide nanoparticles and the anticancer drug taxol. Cancer could be successfully treated with this delivery method.[97]

The traditional approaches to disease diagnosis are time consuming, expensive, and labor intensive. NPs can provide affordable, quick, precise, and accurate pathogen and chronic illness detection services. Magnetic nanoparticles and fluorescent nanoparticles (like metallic NPs, quantum dots, and dye-loaded NPs) can be effectively used for identification, tracking of infections, illness development, and imaging. LED nanoparticles can be used to image chronic diseases in their early stages, such as cancer. Magnetic NPs can be used in cutting-edge detection methods like MRI. As an illustration, a biofabricated probe for liver cancer cells has been developed using AuNPs from *Candida albicans*. AuNPs particularly adhered to cancer cells and made it easier to distinguish them from normal cells.[109]

As a result, microbe-based NPs play a key role in disease diagnosis; however, comprehensive research is needed before applying them to the detection and diagnosis of diseases, which is still a new subject.

#### 14.4.2.2 Metal Nanoparticles as Innovative Antiviral Treatment

Viruses are submicroscopic organisms composed of a protein or lipid coating enclosing an RNA or DNA genetic material core. Viral infections such as the common cold, influenza, respiratory syncytial virus (RSV), Ebola, human immunodeficiency virus (HIV), chickenpox, and SARS-CoV-2 are hazardous because they infect normal cells by interfering with their replication machinery. Various viruses impact distinct cell types inside the body, such as those found in the reproductive, gastrointestinal, liver, blood, respiratory, brain, and skin systems. Transmission pathways include, but are not limited to, airborne dissemination, saliva, direct touch, sexual contact, contaminated food or water, and insect vectors.

Because of their inherent antiviral qualities and capacity to attach to cell surface receptors on viral protein coatings, metal nanoparticles are increasingly essential tools in the diagnosis, management, and treatment of many viral infections. The nanoscale dimensions of metal NPs facilitate their traversal through cell membranes, exerting inhibitory effects on viral replication. Importantly, the multifaceted targeting of different viral regions contributes to reduced resistance upon repeated application.[110]

Silver NPs, known for their antiviral prowess, have demonstrated efficacy against a spectrum of viruses, including Sunthemp rosette virus, Dengue virus, feline coronavirus, HIV, influenza A virus, Chikungunya virus, potato virus Y, yellow mosaic virus, tomato mosaic virus, and respiratory syncytial virus. The antiviral mechanism often involves binding to viral glycoproteins, such as gp120, impeding virus penetration into host cells. For instance, AgNPs made from *Cinnamomum cassia* suppressed influenza A virus in Vero cells, whereas those made from *Moringa oleifera* seed extract showed strong antiviral activity against dengue virus.[111]

Gold NPs, when functionalized or conjugated with various agents, have shown antiviral efficacy against HIV, HSV-1, hepatitis, adenovirus, West Nile virus, influenza, and measles virus. Functionalized AuNPs, such as those conjugated with HIV inhibitors or coated with substances like hyaluronic acid, have demonstrated potent antiviral effects by inhibiting viral replication or altering cell susceptibility.[112]

Copper NPs, particularly copper oxide NPs, have exhibited antiviral activity against hepatitis C virus, influenza, norovirus, and coronaviridae. The viral genome and capsid structures are disrupted by the controlled release of copper ions, which causes oxidative stress. Several metal nanoparticles have shown antiviral properties against different viruses, such as iron oxide/ferromagnetic NPs, silica NPs, zinc oxide NPs, cerium dioxide NPs, and titanium oxide NPs. Functionalized NPs also offer potential diagnostic and drug delivery applications.[113]

### 14.4.3 Applications in Food

The extracellular manufacture of metal oxide or zinc nanoparticles from *Streptomyces* species aids in the development of coating materials with bacterial resistance. These materials can extend shelf life and maintaining food quality. Silver nanoparticles are used in safe food processing and packaging. Polymers (biodegradable) could incorporate nanoparticles. Packaging materials (both biodegradable and not) are antibacterial in nature. This type of packaged food improves the safety, shelf life, and taste of food products.

Food packaging and food processing are two significant applications for NPs in the food sector. For instance, Espitiaet et al. (2012) synthesized packaging material by combining zinc oxide nanoparticles with polymeric material. Newly developed tissues showed better antibacterial properties when used in packing. Likewise, other studies have claimed that zinc oxide nanoparticles may be used to create containers and food packaging materials with nutritional covers. Packaging material with nanomaterials keeps food items fresh and free of contamination. AgNPs have the ability to specifically rupture the bacterial cells' biofilm, giving them resilience to stress. AgNPs may therefore be used in the detoxification and food package cleaning procedure.[114]

### 14.4.4 Applications in Agriculture

Actinomycetes have many uses in agriculture, including promoting plant development by increasing the production of IAA (plant growth hormone), antifungals, and antibiotics, biopesticides and biocontrollers. In an attempt to reduce the use of chemical fertilizers and insecticides and to increase the agricultural production of antimycins, well-known antifungals which work by blocking the electron flow in the mitochondrial respiratory chain. Many metals and their oxides are currently used directly in soils as nanoparticles which can diffuse and uptaken by plants. The integument of athe attine ants isolated from dirt include streptomyces. Actinobacteria are a type of ated extracellular component of AgNPs synthesized by Streptomyces.

As nanopesticides, nanoinsecticides, and nanofertilizers, NPs are used in agriculture. Some significant uses of NPs in the agriculture industry are discussed as follows.

#### 14.4.4.1 Nanoparticles as Fertilizers

A significant problem in the agriculture sector is the steadily declining soil fertility brought on by the excessive use of chemical fertilizers. NPs can be employed to create substitute nanofertilizers which are effective, eco-friendly, and non-toxic. In their 2019 study, Bisinoti et al.[115] hypothesized that soil fertilizers based on carbon-based nanoparticles could lessen the demand for chemical fertilizers In 2013, Subbarao et al.[116] showed that potash fertilizer can be coated with nanomaterial which releases fertilizer gradually, reducing the loss of fertilizer and excessive utilization.

#### 14.4.4.2 Nanoparticles as Fungicides

Excellent antifungal capabilities can be found in metal NPs and metal oxides. The most harmful plant pathogens that cause various plant diseases are fungi. Several recent reports state *Streptomyces* sp. were used to biofabricate copper and copper oxide nanoparticles, which have demonstrated effective antifungal properties against a variety of pathogenic fungus, including *Pythium ultimum, Aspergillus niger, Alternaria alternata and Fusarium oxysporum.*[117] In one of the studies, Kaur et al.[118] examined the antifungal activity of synthetically produced biosynthesized AgNPs using mixed cultures of *Pseudomonas* and *Achromobacter* species against infection with *Fusarium oxysporum* in garbanzo bean. Biosynthesized AgNPs were used in this work both in vivo and in vitro. AgNPs were found to have extremely potent antifungal effects in these trials against *Fusarium oxysporum.* Thus, NPs can be used successfully as fungicides to treat fungi that can cause plant diseases. The toxicity and dosage of NPs, however, continue to be significant obstacles to the commercialization of NPs as fungicides.

#### 14.4.4.3 Nanoparticles as Pesticides

Additionally, NPs can be utilized to make nanopesticides. NPs can be found in nanopesticides as micelles, particles, inorganic metal oxides, and organic nanopolymers. Different arrangements were developed in the last few years using metallic nanoparticles, which were found to be effective insecticides against a variety of insects and other pests reducing agricultural yields globally. For example, Wang et al. (2007)[119] showed that the beta-PP (pesticide beta-pypermethrin) carrier might be a unique oil-water (O/W)-based nanoemulsion.

The results of the investigation showed that beta-PP coated on O/W nanoemulsion diffused more effectively than regular beta-PP and decreased the total amount of pesticide applied. In another work, Gawande et al.[120] demonstrated that a number of NPs, including zinc oxide, silver, aluminum oxide, and titanium oxide, may inhibit the baculovirus Bombyx mori nuclear polyhedrosis virus and associated diseases Sitophilus oryzae in silkworms.[121] NPs are therefore an effective pesticide. Nevertheless, NPs' toxicity to plants and dose remain major barriers to their commercialization as pesticides.[122]

#### 14.4.4.4 Bioremediation

Water contaminants that are extremely persistent and xenobiotic, like azo, cationic, and acid dyes, must be removed in order to treat wastewater and its possible applications. These contaminants worsen water contamination and have detrimental effects on aquatic life. NPs can behave in two ways because of their lower size and increased surface area: as a catalyst or to absorb contaminants over a wider portion of their surface. Numerous reports on NPs' catalytic qualities combined with biological elements have been assessed for lowering harmful pollutants.[123] In 2015, Sharma et al. noted that AgNPs can reduce catalytic activity of organic dyes, making them lighter. NPs might be utilized as a catalyst in industries for highly efficient breakdown of organic dyes.[124]

Both Ag and AuNPs are said to have acceptable catalytic activity when it comes to the elimination of organic dyes. These NPs effectively shorten the time needed for color removal and amplify the reaction rate. In 2016, Bhargava et al. showed that AuNPs can be utilized as an organic color adsorbent as well. The effective adsorption of proteins derived from the fungus *Cladosporium oxysporum was found to be similar as* organic dye rhodamine-β when surface-mounted AuNPs were used. Koul and Taak (2018) recently described the functions of nanoclays, nanotubes, bimetallic NPs, magnetic NPs, $TiO_2$NPs, FeNPs, and bioremediation of soil using nanosponges. The made a point of mentioning that the green synthesis of NPs may provide a successful method of addressing soil and water contamination.[125]

### 14.4.5 Bioimaging

The advancement of medical imaging technologies, initially pioneered by X-ray technology, has led to the development and effective functionalization of several noninvasive methods across diverse realms of medical research. These methodologies, including drug discovery, drug delivery, clinical analysis, and cellular biology investigations, contribute significantly to the progress of medicine. Clinical imaging studies span various research domains, offering valuable insights.

Prominent scientists have illustrated how many modified molecular bioimaging technologies – like optical, magnetic resonance, ultrasound, and positron emission tomography – are adapted for application in vivo and in vitro biological treatment. The research on metallic nanoparticles, an enhanced bioimaging technique combining optical and magnetic resonance imaging, has been introduced.

Conde et al. (2012)[126] state that noble metal nanoparticles like gold, silver, and platinum are essential for instantaneous actuation and tracking because biological tissues absorb light in the near-infrared region. Common noble metallic nanoparticles are very active contrasting agents and have demonstrated potential as in vivo imaging agents due to their considerable near-infrared absorption.

Noble metallic nanomaterials exhibit the capability to amalgamate multiple imaging modalities, presenting an avenue to garner substantial information and synergistic benefits surpassing those of individual bioimaging techniques. The emerging trend of three-dimensional imaging can be experienced through tomography in which a computer-aided image gives a clear idea of a three-dimensional structure. Noteworthy examples include the utilization of nanoparticle composites, such as a FeO/Si core and Au shell composite, serving as dual contrast materials for MRIs and computed tomography. These composites demonstrated commendable computed tomography attenuation and enhanced magnetic resonance outcomes, particularly in hepatoma studies. Chauhan et al.[127] demonstrated the applicability of Gd-Au nanocomposites as contrast agents for dynamic multimodal imaging.

In a study by Zhang et al.[128], the application of Si nanoparticles in dye-loaded fluorescence imaging was explored under both in vivo and in vitro conditions. Si nanoparticles displayed excellent hydrophilicity, low toxicity, remarkable biocompatibility, and possible optical transparency. Consequently, Si nanoparticles can serve as substrates for the production of fluorescent probes, a critical element in cellular bioimaging.

The use of lanthanide-doped up-converting nanoparticles in bioimaging has been discussed by some authors. It is believed that these nanomaterials can undergo multiphoton processes to transition from low-energy to higher-energy states. The morphology chemistry of lanthanide-doped up-converting nanoparticles in a variety of biomedical applications – drug administration, imaging guiding, imaging reagents, and phototherapeutic interventions has been clarified by the authors.

Even with the tremendous advancements in bioimaging, some problems still exist. These include the necessity for flexible advancement at the commercial level, the low up-conversion productivity of lanthanide-based nanomaterials, and difficulties in fabricating subnano-sized particles (~10 nm). Addressing these challenges is imperative to ensure the feasibility and reliability of future bioimaging research endeavors.

Reported studies have shown the use of molecularly modified bioimaging equipment for in vivo and in vitro treatments, such as MRI, optical imaging, and position emission tomography. A more advanced bioimaging system that combines optical and magnetic resonance imaging was presented in recent investigations of metallic nanoparticles.[129]

One of the studies[130] conducted using lanthanide-doped nanoparticles in bioimaging suggested that these nanomaterials can undergo multiphoton processes which enable NPs to transform from low-energy to higher-energy states. The shape and chemistry of lanthanide-doped nanoparticles plays important role in biological applications – drug transport, imaging reagents, phototherapeutic interventions, and imaging.

Despite the recent developments in bioimaging, researchers are facing certain challenges which include the fabrication of subnano-sized particles (~10 nm), low conversion and productivity of lanthanide-based nanomaterials and the production at the commercial level. It is crucial to address these issues in order to guarantee the viability and dependability of upcoming bioimaging research.[131]

### 14.4.6 Therapeutic Applications

Metallic nanoparticles or clusters exhibit several noteworthy features that render them valuable for therapeutic applications, as shown in Figure 14.4. Research in this domain has specifically designated metal nanoparticles as potential preventive-therapeutic agents against human immunodeficiency virus. Notably, studies have demonstrated the direct impact of silver nanoparticles on viral infections by interacting with glycoprotein, as elucidated.[64] This form of interaction disrupts various virial bindings, effectively reducing the incidence of HIV infection. As a result, metallic nanoparticles are acknowledged as potent antiviral agents that effectively combat influenza, herpes virus, and respiratory syncytial virus.[132]

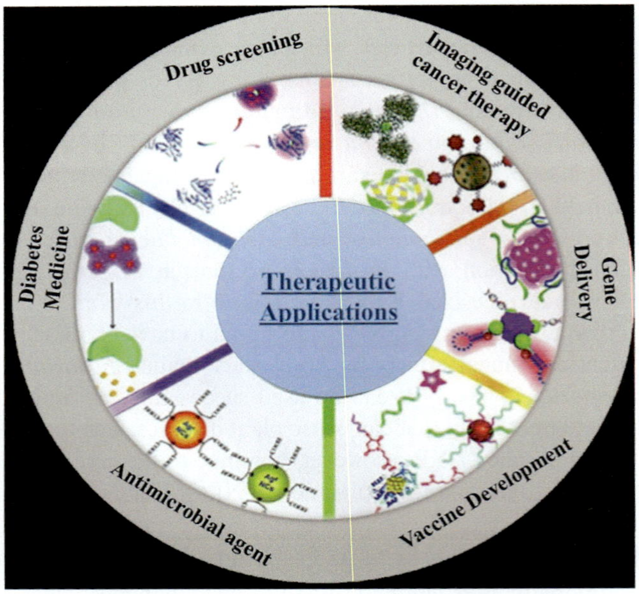


FIGURE 14.4 Therapeutic applications of biosynthesized NPs.

Angiogenesis, the formation of new blood vessels, occurs both normally and under pathological conditions. It is important in the development of disorders such as cancer and arthritis. Under stable conditions, angiogenesis is controlled by various proangiogenic factors, to some extent antiangiogenic factors, such as platelet factor and TSP–1. One of the literature presented, impact of gold nanoparticles on angiogenesis using a mouse ear model introducing adrenoviral vector in a vascular endothelial growth factor (VEGF).[133] Chronic lymphocytic leukemia (CLL), bearing apoptosis resistance, was reported using CLL-based cells with anti-VEGF antibodies. Gold nanoparticles were employed in leukemia therapies to increase the effectiveness of contracting agents, surface area functionalization, and compatibility. The study evaluated the role of VEGF antibodies and Au nanoparticles in targeting CLL cells.

Rheumatoid arthritis, an autoimmune disease affecting joints, results from an inefficient immune system attacking the connecting joints. Researchers from Wollongong University, Australia, have developed a novel method for treating arthritic diseases. They introduced an antidrug that binds with Au nanoparticles, though some limitations exist. Additional research has indicated that Au nanoparticles can target macrophages and inhibit inflammation. The size of Au nanoparticles proves crucial in mitigating toxicity effects; for instance, particles with a size of 50 nm or less demonstrated enhanced cell immunity without side effects.[134]

## 14.5 CONCLUSION

Biological synthesis of NPs using bacteria and fungi has shown remarkable potential in the field of nanotechnology in developing green processes of some promising metallic NPs. Microbial variety possesses some intrinsic characteristics which act as biofactories for the synthesis of reliable NPs. Although studies have revealed various biosynthetic processes and their wide applications, there is a need to understand how physical characteristics of NPs have an influence on choice of microorganisms and optimal growth conditions. In the future, efforts can be made to understand molecular-level interactions during biochemical reactions to improve the synthesis and product texture of metallic NPs.

The optimization of various parameters such as growth, cellular mechanisms, temperature, and pH conditions while performing the growth of microorganisms plays an important role in biosynthesis. The large-scale biosynthesis of NPs, choosing specific biofactories under optimized reaction conditions, can result in stable and reliable green products with the required composition, structure, and sizes of metallic NPs. Biosynthetic routes can be considered an alternatives to hazardous and expensive chemical synthesis.

## REFERENCES


1. Khoshnevisan K, Poorakbar E, Baharifar H, Barkhi M. Recent advances of cellulase immobilization onto magnetic nanoparticles: An update review. *Magnetochemistry*. 2019;5(2). https://doi.org/10.3390/magnetochemistry 5020036
2. Yaqoob AA, Ahmad H, Parveen T, et al. Recent advances in metal decorated nanomaterials and their various biological applications: A review. *Frontiers in Chemistry*. 2020;8. https://doi.org/10.3389/fchem.2020.00341
3. Garba U, Abubakar AS, Saidu M. Nanoparticles: A delivery system for bioactive food components. *International Journal of Current Research*. 2014;6:8032–8035.
4. Yacaman M, Ascencio J, Liu H, Gardea-Torresdey J. Structure shape and stability of nanometric sized particles. *Journal of Vacuum Science & Technology B: Microelectronics and Nanometer Structures*. 2001;19. https://doi.org/10.1116/1.1387089
5. Sweet M, Singleton I. Silver nanoparticles. A microbial perspective. *Advances in Applied Microbiology*. 2011;77:115–133.https://doi.org/10.1016/B978-0-12-387044-5.00005-4
6. Salah Salem S, Fouda A. Green synthesis of metallic nanoparticles and their prospective biotechnological applications: An overview. *Biological Trace Element Research*. 2021;199. https://doi.org/10.1007/s12011-020-02138-3
7. Fariq A, Khan T, Yasmin A. Microbial synthesis of nanoparticles and their potential applications in biomedicine. *Journal of Applied Biomedicine*. 2017;15(4):241–248. https://doi.org/10.1016/j.jab.2017.03.004
8. Castro L, Blázquez ML, Muñoz J, González F, Ballester A. Mechanism and applications of metal nanoparticles prepared by bio-mediated process. *Reviews in Advanced Sciences and Engineering*. 2014;3. https://doi.org/10.1166/rase.2014.1064
9. Jang HD, Kim S, Chang H, et al. Three-dimensional crumpled graphene-based platinum–gold alloy nanoparticle composites as superior electrocatalysts for direct methanol fuel cells. *Carbon*. 2015;93:869–877. https://doi.org/10.1016/j.carbon.2015.06.009
10. Singh P, Kim YJ, Zhang D, Yang DC. Biological synthesis of nanoparticles from plants and microorganisms. *Trends in Biotechnology*. 2016;34(7):588–599. https://doi.org/10.1016/j.tibtech.2016.02.006
11. Saravanan M, Barik SK, MubarakAli D, Prakash P, Pugazhendhi A. Synthesis of silver nanoparticles from Bacillus brevis (NCIM 2533) and their antibacterial activity against pathogenic bacteria. *Microbial Pathogenesis*. 2018;116:221–226. https://doi.org/10.1016/j.micpath.2018.01.038
12. Buszewski B, Railean V, Pomastowski P, et al. Antimicrobial activity of biosilver nanoparticles produced by novel Streptacidiphilus durhamensis strain. *Journal of Microbiology, Immunology, and Infection = Wei mian yu gan ran za zhi*. 2016;51. https://doi.org/10.1016/j.jmii.2016.03.002
13. Sanaeimehr Z, Javadi I, Namvar F. Antiangiogenic and antiapoptotic effects of green-synthesized zinc oxide nanoparticles using Sargassum muticum algae extraction. *Cancer Nanotechnology*. 2018;9. https://doi.org/10.1186/s12645-018-0037-5
14. Nandhini J, Devaraj E, Shanmugam R. An ecofriendly synthesized gold nanoparticles induces cytotoxicity via apoptosis in HepG2 cells. *Environmental Toxicology*. 2020;36. https://doi.org/10.1002/tox.23007
15. Shunmugam R, Renukadevi Balusamy S, Kumar V, Menon S, Lakshmi T, Perumalsamy H. Biosynthesis of gold nanoparticles using marine microbe (Vibrio alginolyticus) and its anticancer and antioxidant analysis. *Journal of King Saud University – Science*. 2021;33(1):101260. https://doi.org/10.1016/j.jksus.2020.101260
16. Munawer U, Raghavendra VB, Ningaraju S, et al. Biofabrication of gold nanoparticles mediated by the endophytic Cladosporium species: Photodegradation, in vitro anticancer activity and in vivo antitumor studies. *International Journal of Pharmaceutics*. 2020;588:119729. https://doi.org/10.1016/j.ijpharm.2020.119729
17. Mukherjee P, Senapati S, Mandal D, Ahmad A, Khan MI, Kumar R, Sastry M. Extracellular synthesis of gold nanoparticles by the fungus Fusarium oxysporum. *ChemBioChem*. 2002;3(5):461–463.
18. Narayanan B, Sakthivel N. Green synthesis of biogenic metal nanoparticles by terrestrial and aquatic phototrophic and heterotrophic eukaryotes and biocompatible agents. *Advances in Colloid and Interface Science*. 2011;169:59–79. https://doi.org/10.1016/j.cis.2011.08.004
19. He S, Guo Z, Zhang Y, Zhang S, Ning G. Biosynthesis of gold nanoparticles using the bacteria Rhodopseudomonas capsulata. *Materials Letters*. 2007;61:3984–3987. https://doi.org/10.1016/j.matlet.2007.01.018
20. Du L, Jiang H, Liu X, Wang E. Biosynthesis of gold nanoparticles assisted by Escherichia coli DH5?? and its application on direct electrochemistry of hemoglobin. *Electrochemistry Communications*. 2007;9:1165–1170. https://doi.org/10.1016/j.elecom.2007.01.007
21. Cao B, Ahmed B, Beyenal H. Immobilization of uranium in groundwater using biofilms. In: Shah V (ed.). *Emerging Environmental Technologies*. Vol. II, Springer. https://doi.org/10.1007/978-90-481-3352-9_1

22. Konishi Y, Ohno K, Saitoh N, et al. Bioreductive deposition of platinum nanoparticles on the bacterium Shewanella algae. *Journal of biotechnology.* 2007;128:648–653. https://doi.org/10.1016/j.jbiotec.2006.11.014
23. Shahverdi AR, Minaeian S, Shahverdi HR, Jamalifar H, Nohi AA. Rapid synthesis of silver nanoparticles using culture supernatants of Enterobacteria: A novel biological approach. *Process Biochemistry.* 2007;42(5):919–923. https://doi.org/10.1016/j.procbio.2007.02.005
24. Srivastava SK, Constanti M. Room temperature biogenic synthesis of multiple nanoparticles (Ag, Pd, Fe, Rh, Ni, Ru, Pt, Co, and Li) by Pseudomonas aeruginosa SM1. *Journal of Nanoparticle Research.* 2012;14:831. https://doi.org/10.1007/s11051-012-0831-7
25. Nadhe S, Tawre M, Agrawal S, Chopade PB, Sarkar D, Pardesi K. Anticancer potential of AgNPs synthesized using Acinetobacter sp. and Curcuma aromatica against HeLa cell lines: A comparative study. *Journal of Trace Elements in Medicine and Biology.* 2020;62:126630. https://doi.org/10.1016/j.jtemb.2020.126630
26. Eid A, Fouda A, Niedbała G, et al. Endophytic Streptomyces laurentii mediated green synthesis of Ag-NPs with antibacterial and anticancer properties for developing functional textile fabric properties. *Antibiotics.* 2020;9:641. https://doi.org/10.3390/antibiotics9100641
27. Sriram M, Kalishwaralal K, Gurunathan S. Biosynthesis of silver and gold nanoparticles using bacillus licheniformis. *Methods in Molecular Biology (Clifton, NJ).* 2012;906:33–43. https://doi.org/10.1007/978-1-61779-953-2_3
28. Cruz L, Wang D, Liu J. Biosynthesis of selenium nanoparticles, characterization and X-ray induced radiotherapy for the treatment of lung cancer with interstitial lung disease. *Journal of Photochemistry and Photobiology B: Biology.* 2018;191. https://doi.org/10.1016/j.jphotobiol.2018.12.008
29. Natarajan KSM, Seetharaman B. A comparative study of morphology, reactivity and stability of synthesized silver nanoparticles using Bacillus subtilis and Catharanthus roseus (L.) G. Don. *Colloids and Surfaces B, Biointerfaces.* 2011;86:378–383. https://doi.org/10.1016/j.colsurfb.2011.04.024
30. Li X, Xu H, Chen ZS, Chen G. Biosynthesis of nanoparticles by microorganisms and their applications. Liang XJ (ed.). *Journal of Nanomaterials.* 2011;2011:270974. https://doi.org/10.1155/2011/270974
31. Oves M., Khan M, Zaidi A, et al. Antibacterial and cytotoxic efficacy of extracellular silver nanoparticles biofabricated from chromium reducing novel OS4 strain of stenotrophomonas maltophilia. *PLoS One.* 2013;8:e59140. https://doi.org/10.1371/journal.pone.0059140
32. Zhou Y, Bastida F, Liu Y, et al. Selenium fertigation with nanobubbles influences soil selenium residual and plant performance by modulation of bacterial community. *Journal of Hazardous Materials.* 2022;423:127114. https://doi.org/10.1016/j.jhazmat.2021.127114
33. Lloyd J, Yong P, Macaskie L. Enzymatic recovery of elemental palladium by using sulfate-reducing bacteria. *Applied and Environmental Microbiology.* 1998;64:4607–4609. https://doi.org/10.1128/AEM.64.11.4607-4609.1998
34. KüünalS, RauwelP, RauwelE. Chapter 14–Plant extract mediated synthesis of nanoparticles. In: Barhoum A, Makhlouf ASH (eds.). *Emerging Applications of Nanoparticles and Architecture Nanostructures.* Elsevier; 2018:411–446. https://doi.org/10.1016/B978-0-323-51254-1.00014-2
35. Bankar A, Joshi B, Ravi Kumar A, Zinjarde S. Banana peel extract mediated synthesis of gold nanoparticles. *Colloids and Surfaces B: Biointerfaces.* 2010;80(1):45–50. https://doi.org/10.1016/j.colsurfb.2010.05.029
36. Ali J, Ali N, Wang L, Waseem H, Pan G. Revisiting the mechanistic pathways for bacterial mediated synthesis of noble metal nanoparticles. *Journal of Microbiological Methods.* 2019;159:18–25. https://doi.org/10.1016/j.mimet.2019.02.010
37. Muvva DP, Batabyal SK, Mamidipudi GK. Highly reactive crystalline (001) facet of BiI3 hexagonal nanoplates for visible light photodetectors. *ACS Appl Electron Mater.* 2023;5(7):3708–3714. https://doi.org/10.1021/acsaelm.3c00433
38. Bharde A, Kulkarni A, Rao M, Prabhune A, Sastry M. Bacterial enzyme mediated biosynthesis of gold nanoparticles. *Journal of Nanoscience and Nanotechnology.* 2007;7.
39. Adavallan K, Krishnakumar N. Mulberry leaf extract mediated synthesis of gold nanoparticles and its anti-bacterial activity against human pathogens. *Advances in Natural Sciences: Nanoscience and Nanotechnology.* 2014;5. https://api.semanticscholar.org/CorpusID:4568232
40. Meghana S, Kabra P, Chakraborty S, Padmavathy N. Understanding the pathway of antibacterial activity of copper oxide nanoparticles. *RSC Adv.* 2015;5(16):12293.
41. Ganesh Kumar C, Poornachandra Y, Chandrasekhar C. Green synthesis of bacterial mediated anti-proliferative gold nanoparticles: inducing mitotic arrest (G2/M phase) and apoptosis (intrinsic pathway). *Nanoscale.* 2015;7(44):18738–18750. https://doi.org/10.1039/C5NR04577K
42. Chandran K, Song S, Yun SI. Effect of size and shape controlled biogenic synthesis of gold nanoparticles and their mode of interactions against food borne bacterial pathogens. *Arabian Journal of Chemistry.* 2019;12(8):1994–2006. https://doi.org/10.1016/j.arabjc.2014.11.041
43. Ramamurthy CH, Padma M, Mareeswaran R, et al. The extra cellular synthesis of gold and silver nanoparticles and their free radical scavenging and antibacterial properties. *Colloids and Surfaces B: Biointerfaces.* 2013;102:808–815. https://doi.org/10.1016/j.colsurfb.2012.09.025
44. Kaviya S, Santhanalakshmi J, Viswanathan B, Muthumary J, Srinivasan K. Biosynthesis of silver nanoparticles using citrus sinensis peel extract and its antibacterial activity. *Spectrochimica Acta Part A: Molecular and Biomolecular Spectroscopy.* 2011;79(3):594–598. https://doi.org/10.1016/j.saa.2011.03.040
45. Kaweeteerawat C, Na Ubol P, Sangmuang S, Aueviriyavit S, Maniratanachote R. Mechanisms of antibiotic resistance in bacteria mediated by silver nanoparticles. *Journal of Toxicology and Environmental Health, Part A.* 2017;80(23–24):1276–1289. https://doi.org/10.1080/15287394.2017.1376727
46. Saleem S, Iqbal A, Hasnain S. Bacterial mediated silver nanoparticles and their efficacy against MRSA. *Tropical Biomedicine*, 2020;37(2):482–488.
47. Li J, Chen Z, Wang XX, Proserpio DM A novel two-dimensional mercury antimony telluride: Low temperature synthesis and characterization of RbHgSbTe3. *Journal of Alloys and Compounds.* 1997;262:28–33.
48. Ravindra S, Murali Mohan Y, Narayana Reddy N, Mohana Raju K. Fabrication of antibacterial cotton fibres loaded with silver nanoparticles via "green approach". *Colloids and Surfaces A: Physicochemical and Engineering Aspects.* 2010;367(1):31–40. https://doi.org/10.1016/j.colsurfa.2010.06.013

49. Parashar UK, Kumar V, Bera T, et al. Study of mechanism of enhanced antibacterial activity by green synthesis of silver nanoparticles. *Nanotechnology*. 2011;22(41):415104. https://doi.org/10.1088/0957–4484/22/41/415104
50. Singh R, Lekhak U, Wadhwani S, Chopade PB. Bacteriagenic silver nanoparticles: Synthesis, mechanism, and applications. *Applied Microbiology and Biotechnology*. 2015;99:4579–4593. https://doi.org/10.1007/s00253-015-6622-1
51. Khan AU, Yuan Q, Wei Y, et al. Photocatalytic and antibacterial response of biosynthesized gold nanoparticles. *Journal of Photochemistry and Photobiology B: Biology*. 2016;162:273–277. https://doi.org/10.1016/j.jphotobiol.2016.06.055
52. Ahmad R, Vaseem M, Tripathy N, Hahn YB. Wide linear-range detecting nonenzymatic glucose biosensor based on CuO nanoparticles inkjet-printed on electrodes. *Anal Chem*. 2013;85(21):10448.
53. Yates MD, Cusick RD, Logan BE. Extracellular palladium nanoparticle production using geobacter sulfurreducens. *ACS Sustainable Chem Eng*. 2013;1(9):1165–1171. https://doi.org/10.1021/sc4000785
54. Martins M, Mourato C, Sanches S, Noronha JP, Crespo MTB, Pereira IAC. Biogenic platinum and palladium nanoparticles as new catalysts for the removal of pharmaceutical compounds. *Water Research*. 2017;108:160–168. https://doi.org/10.1016/j.watres.2016.10.071
55. Gaidhani SV, Yeshvekar RK, Shedbalkar UU, Bellare JH, Chopade BA. Bio-reduction of hexachloroplatinic acid to platinum nanoparticles employing Acinetobacter calcoaceticus. *Process Biochemistry*. 2014;49(12):2313–2319. doi: https://doi.org/10.1016/j.procbio.2014.10.002
56. Saif Hasan S, Singh S, Parikh RY, Dharne MS, Patole MS, Prasad BLV, Shouche YS. Bacterial synthesis of copper/copper oxide nanoparticles. *Journal of Nanoscience and Nanotechnology*. 2008;6:3191–3196.
57. Sagadevan S, Imteyaz S, Murugan B, et al. A comprehensive review on green synthesis of titanium dioxide nanoparticles and their diverse biomedical applications. 2022;11(1):44–63. https://doi.org/10.1515/gps-2022-0005
58. Rahman MM, Khan SB, Jamal A, Faisal M, Aisiri AM. Chapter 3 – Iron oxide nanoparticles. In: Rahman MM (ed.). *Nanomaterials*. IntechOpen; 2011. https://doi.org/10.5772/27698
59. Kumar KY, Muralidhara H, Nayaka Y, Hanumanthappa H, Veena M, Kumar SK. Hydrothermal synthesis of hierarchical copper oxide nanoparticles and its potential application as adsorbent for Pb (II) with high removal capacity. *Sep Sci Technol*. 2014;49(15):2389.
60. Campos EA, Pinto DV, Oliveira JI, Mattos ED, Dutra RD. Synthesis, characterization and applications of iron oxide nanoparticles-a short review. *Journal of Aerospace Technology and Management*. 2015;7(3):267–276.
61. Jalill RD, Jawad MM, Abd AN. Plants extracts as green synthesis of zirconium oxide nanoparticles. *J. Genet. Environ. Res. Conserv*. 2017;5(1):6–23.
62. Al-Zaqri N, Muthuvel A, Jothibas M, Alsalme A, Alharthi FA, Mohana V. Biosynthesis of zirconium oxide nanoparticles using Wrightia tinctoria leaf extract: Characterization, photocatalytic degradation and antibacterial activities. *Inorganic Chemistry Communications*. 2021;127:108507. https://doi.org/10.1016/j.inoche.2021.108507
63. Goyal P, Bhardwaj A, Mehta BK, Mehta D. Research article green synthesis of zirconium oxide nanoparticles (ZrO2NPs) using Helianthus annuus seed and their antimicrobial effects. *Journal of the Indian Chemical Society*. 2021;98(8):100089. https://doi.org/10.1016/j.jics.2021.100089
64. Jain S, Mehata M. Medicinal plant leaf extract and pure flavonoid mediated green synthesis of silver nanoparticles and their enhanced antibacterial property. *Scientific Reports*. 2017;7. https://doi.org/10.1038/s41598-017-15724-8
65. Alsharari SS, Alenezi MA, Al Tami MS, Soliman M. Recent advances in the biosynthesis of zirconium oxide nanoparticles and their biological applications. *Baghdad Science Journal*. 2023;20(1):24.
66. Sana SS, Dogiparthi LK. Green synthesis of silver nanoparticles using Givotia moluccana leaf extract and evaluation of their antimicrobial activity. *Materials Letters*. 2018;226:47–51. https://doi.org/10.1016/j.matlet.2018.05.009
67. Reddy Yadav LS, Ramakrishnappa T, Pereira JR, Venkatesh R, Nagaraju G. Electrical property of zirconium oxide nanoparticle synthesized by hydrothermal method. *Materials Today: Proceedings*. 2022;49:686–689. https://doi.org/10.1016/j.matpr.2021.05.172
68. Khan M, Khan AA, Parveen A, et al. Mitigating the growth of plant pathogenic bacterium, fungi, and nematode by using plant-mediated synthesis of copper oxide nanoparticles (CuO NPs). *Green Chemistry Letters and Reviews*. 2023;16(1):2177520. https://doi.org/10.1080/17518253.2023.2177520
69. Dorobantu LS, Fallone C, Noble AJ, et al. Toxicity of silver nanoparticles against bacteria, yeast, and algae. *Journal of Nanoparticle Research*. 2015;17(4):172. https://doi.org/10.1007/s11051-015-2984-7
70. Michael A, Singh A, Roy A, Islam MdR. Fungal- and Algal-derived synthesis of various nanoparticles and their applications. In: Pandian S (ed.). *Bioinorganic Chemistry and Applications*. 2022;2022:3142674. https://doi.org/10.1155/2022/3142674
71. Senapati S, Ahmad A, Khan MI, Sastry M, Kumar R. Extracellular biosynthesis of bimetallic Au–Ag alloy nanoparticles. *Small*. 2005;1(5):517–520.
72. Santos TS, Silva TM, Cardoso JC, et al. Biosynthesis of silver nanoparticles mediated by entomopathogenic fungi: Antimicrobial resistance, nanopesticides, and toxicity. *Antibiotics*. 2021;10. https://api.semanticscholar.org/CorpusID:236208350
73. Bhainsa KC, D'souza SF. Extracellular biosynthesis of silver nanoparticles using the fungus Aspergillus fumigatus. *Colloids and surfaces B: Biointerfaces*. 2006;47(2):160–164.
74. Bakshi M, Mahanty S, Chaudhuri P. Fungi mediated biosynthesis of nanoparticles and application in metal sequestration. In: *Handbook of Metal-Microbe Interactions and Bioremediation*; 2017:423–434. https://doi.org/10.1201/9781315153353
75. Raliya R, Saran R, Choudhary K, Tarafdar JC. Biosynthesis and characterization of nanoparticles. Current trends in advancement of scientific research and opinion in applied research and opinion in applied microbiology and biotechnology. *Journal of advancement in medical and life sciences*, 2014, Vol 1 (1) ISSN: 2348-294X,1-4; 2013;6.
76. Molnár Z, Bódai V, Szakacs G, et al. Green synthesis of gold nanoparticles by thermophilic filamentous fungi. *Scientific Reports*. 2018;8. https://doi.org/10.1038/s41598-018-22112-3
77. Honary S, Barabadi H, Gharaei-Fathabad E, Naghibi F. Green synthesis of copper oxide nanoparticles using

Penicillium aurantiogriseum, Penicillium citrinum and Penicillium waksmanii. *Dig J Nanomater Bios*. 2012;7:999.
78. Pantidos N, Horsfall L. Biological synthesis of metallic nanoparticles by bacteria, fungi and plants. *Journal of Nanomedicine & Nanotechnology*. 2014;5. https://doi.org/10.4172/2157–7439.1000233
79. Balaji DS, Basavaraja S, Deshpande R, Mahesh DB, Prabhakar BK, Venkataraman A. Extracellular biosynthesis of functionalized silver nanoparticles by strains of Cladosporium cladosporioides fungus. *Colloids and Surfaces B: Biointerfaces*. 2009;68(1):88–92. https://doi.org/10.1016/j.colsurfb.2008.09.022
80. Alavi M, Nokhodchi A. Synthesis and modification of bio-derived antibacterial Ag and ZnO nanoparticles by plants, fungi, and bacteria. *Drug Discovery Today*. 2021;26(8):1953–1962. https://doi.org/10.1016/j.drudis.2021.03.030
81. Abboud Y, Saffaj T, Chagraoui A, et al. Biosynthesis, characterization and antimicrobial activity of copper oxide nanoparticles (CONPs) produced using brown alga extract (Bifurcaria bifurcata). *Appl Nanosci*. 2014;4(5):571.
82. Zaidi J, Pal A. Review on heavy metal pollution in major lakes of India: Remediation through plants. *African Journal of Environmental Science and Technology*. 2017;11:255–265. https://doi.org/10.5897/AJEST2017.2299
83. Altamony H, Tarhini A, Zahran D, et al. The relationship between change management strategy and successful enterprise resource planning (ERP) implementations: A theoretical perspective. *International Journal of Business Management and Economic Research*. 2016;7:690–703.
84. Priyadarshini E, Priyadarshini SS, Cousins BG, Pradhan N. Metal-fungus interaction: Review on cellular processes underlying heavy metal detoxification and synthesis of metal nanoparticles. *Chemosphere*. 2021;274:129976. https://doi.org/10.1016/j.chemosphere.2021.129976
85. Li Q, Liu F, Li M, Chen C, Gadd GM. Nanoparticle and nanomineral production by fungi. *Fungal Biology Reviews*. 2022;41:31–44. https://doi.org/10.1016/j.fbr.2021.07.003
86. Skowronski DM, Astell C, Brunham RC, et al. Severe acute respiratory syndrome (SARS): A year in review. *Annu Rev Med*. 2005;56:357.
87. Płaza G, Łukasik W, Ulfig K. Effect of cadmium on growth of potentially pathogenic soil fungi. *Mycopathologia*. 1998;141:93–100. https://doi.org/10.1023/A:1006991306756
88. Babich H, Stotzky G. Developing standards for environmental toxicants: The need to consider abiotic environmental factors and microbe-mediated ecologic processes. *Environmental Health Perspectives*. 1983;49:247–260. https://doi.org/https://doi.org/10.2307/3429603
89. Bååth E. Effects of heavy metals in soil on microbial processes and populations (a review). *Water, Air, and Soil Pollution*. 1989;47(3):335–379. https://doi.org/10.1007/BF00279331
90. Gadd GM, Stewart A, White C, Mowll JL. Copper uptake by whole cells and protoplasts of a wild-type and copper-resistant strain of Saccharomyces cerevisiae. *FEMS Microbiology Letters*. 1984;24(2):231–234.
91. Naidu CK, Reddy TKR. Protection of cadmium toxicity to B. cereus, *E. coli* and A. niger by Na+, Mg++ and Ca++ ions. *Zentralblatt für Mikrobiologie*. 1988;143(5):383–387. https://doi.org/10.1016/S0232–4393(88)80032-5
92. Krauss G, Baerlocher F, Krauss GJ. Effects of pollution on aquatic hyphomycetes. *Fungal Diversity Research Series*. 2003;10:211–230.
93. Milsom WK, Gilmour KM, Perry S, et al. Control of breathing in ectothermic vertebrates. In: *Comprehensive Physiology*. John Wiley & Sons, Ltd; 2022:3869–3988. https://doi.org/10.1002/cphy.c210041
94. Babich H, Stotzky G. Nickel toxicity to fungi: Influence of environmental factors. *Ecotoxicology and Environmental Safety*. 1982;6(6):577–589. https://doi.org/10.1016/0147-6513(82)90039-2
95. Stotzky G, Broder MW, Doyle JD, Jones RA. Selected methods for the detection and assessment of ecological effects resulting from the release of genetically engineered microorganisms to the terrestrial environment. In: Neidleman S, Laskin AI (eds.). *Advances in Applied Microbiology*. Vol. 38. Academic Press; 1993:1–98. https://doi.org/10.1016/S0065-2164(08)70214-4
96. Zhang YC, Davison S, Brusasco R, Qian YT, Dwight K, Wold A. Preparation and characterization of tetragonal ZrO2. *Journal of the Less Common Metals*. 1986;116(1):301–306.
97. Koul B, Poonia AK, Yadav D, Jin JO. Microbe-mediated biosynthesis of nanoparticles: Applications and future prospects. *Biomolecules*. 2021;11(6). https://doi.org/10.3390/biom11060886
98. Salouti M, Ahangari A. Nanoparticle based drug delivery systems for treatment of infectious diseases. In: Sezer AD (ed.). *Application of Nanotechnology in Drug Delivery*. IntechOpen; 2014. https://doi.org/10.5772/58423
99. Sharma P, Pant S, Dave V, Tak K, Sadhu V, Reddy KR. Green synthesis and characterization of copper nanoparticles by Tinospora cardifolia to produce nature-friendly copper nano-coated fabric and their antimicrobial evaluation. *Journal of Microbiological Methods*. 2019 May 1;160:107–116.
100. Borse V, Konwar AN, Buragohain P. Oral cancer diagnosis and perspectives in India. *Sensors International*. 2020;1:100046. https://doi.org/10.1016/j.sintl.2020.100046
101. Ortega R, Perez-Rodrigo C, López-Sobaler A. Dietary assessment methods: Dietary records. *Nutricion hospitalaria*. 2015;31:38–45. https://doi.org/10.3305/nh.2015.31.sup3.8749
102. Rajendran R, Muthukrishnan M, Nagarajan T. Rabies prophylactic approaches. *Israel Journal of Veterinary Medicine*. 2018;73:3–14.
103. El-Batal A, Elkenawy N, Yassin A, Amin M. Laccase production by Pleurotus ostreatus and its application in synthesis of gold nanoparticles. *Biotechnology Reports*. 2014;5. https://doi.org/10.1016/j.btre.2014.11.001
104. Kundu D, Hazra C, Chatterjee A, Chaudhari A, Mishra S. Extracellular biosynthesis of zinc oxide nanoparticles using Rhodococcus pyridinivorans NT2: Multifunctional textile finishing, biosafety evaluation and in vitro drug delivery in colon carcinoma. *Journal of Photochemistry and Photobiology B: Biology*. 2014;140:194–204. https://doi.org/10.1016/j.jphotobiol.2014.08.001
105. Chauhan A, Zubair S, Tufail S, et al. Fungus-mediated biological synthesis of gold nanoparticles: Potential in detection of liver cancer. *International Journal of Nanomedicine*. 2011:2305–2319.
106. Kumar KY, Muralidhara H, Nayaka Y, Hanumanthappa H, Veena M, Kumar SK. Hydrothermal synthesis of hierarchical copper oxide nanoparticles and its potential application as adsorbent for Pb (II) with high removal capacity. *Sep Sci Technol*. 2014;49(15):2389.
107. Malekjahani A, Sindhwani S, Syed AM, Chan WCW. Engineering steps for mobile point-of-care diagnostic devices. *Acc Chem Res*. 2019;52(9):2406.
108. Khan AU, Yuan Q, Wei Y, et al. Photocatalytic and antibacterial response of biosynthesized gold

nanoparticles. *Journal of Photochemistry and Photobiology B: Biology*. 2016;162:273–277. https://doi.org/10.1016/j.jphotobiol.2016.06.055
109. Garrido C, Simpson CA, Dahl NP, Bresee J, Whitehead DC, Lindsey EA, Harris TL, Smith CA, Carter CJ, Feldheim DL, Melander C. Gold nanoparticles to improve HIV drug delivery. *Future Medicinal Chemistry*. 2015 Jul 1;7(9):1097–1107.
110. Lara HH, Ayala-Nuñez NV, Ixtepan-Turrent L, Rodriguez-Padilla C. Mode of antiviral action of silver nanoparticles against HIV–1. *Journal of Nanobiotechnology*. 2010;8(1):1–10.
111. Udugama B, Kadhiresan P, Kozlowski HN, et al. Diagnosing COVID-19: The disease and tools for detection. *ACS Nano*. 2020;14(4):3822–3835. https://doi.org/10.1021/acsnano.0c02624
112. Alishah H, Pourseyedi S, Ebrahimipour SY, Mahani SE, Rafiei N. Green synthesis of starch-mediated CuO nanoparticles: Preparation, characterization, antimicrobial activities and in vitro MTT assay against MCF-7 cell line. *Rendiconti Lincei*. 2017;28(1):65.
113. Rajivgandhi G, Mythili Gnanamangai B, Heela Prabha T, et al. Biosynthesized zinc oxide nanoparticles (ZnO NPs) using actinomycetes enhance the anti-bacterial efficacy against K. Pneumoniae. *Journal of King Saud University – Science*. 2022;34(1):101731. https://doi.org/10.1016/j.jksus.2021.101731
114. Kumari S, Tehri N, Gahlaut A, Hooda V. Actinomycetes mediated synthesis, characterization, and applications of metallic nanoparticles. *Inorganic and Nano-Metal Chemistry*. 2020;51. https://doi.org/10.1080/24701556.2020.1835978
115. Ragavan K, Rastogi NK. Graphene–copper oxide nanocomposite with intrinsic peroxidase activity for enhancement of chemiluminescence signals and its application for detection of Bisphenol-A. *Sens Actuators, B*. 2016;229:570.
116. Hassan S, Fouda A, Radwan A, et al. Endophytic actinomycetes Streptomyces spp mediated biosynthesis of copper oxide nanoparticles as a promising tool for biotechnological applications. *JBIC Journal of Biological Inorganic Chemistry*. 2019;24. https://doi.org/10.1007/s00775-019-01654-5
117. Kaur P, Thakur R, Duhan J, Chaudhury A. Management of wilt disease of chickpea in vivo by silver nanoparticles; biosynthesized by rhizospheric microflora of chickpea (Cicer arietinum L.). *Journal of Chemical Technology & Biotechnology*. 2018;93. https://doi.org/10.1002/jctb.5680
118. Hassan S, Salah Salem S, Fouda A, Ahmed Awad M, Gamal S, Abdo A. New approach for antimicrobial activity and bio-control of various pathogens by biosynthesized copper nanoparticles using endophytic actinomycetes. *Journal of Radiation Research and Applied Sciences*. 2018;11. https://doi.org/10.1016/j.jrras.2018.05.003
119. Rai M, Ingle A. Role of nanotechnology in agriculture with special reference to management of insect pests. *Applied Microbiology and Biotechnology*. 2012;94(2):287–293. https://doi.org/10.1007/s00253-012-3969-4
120. Gawande MB, Goswami A, Felpin FX, et al. Cu and Cu-based nanoparticles: Synthesis and applications in catalysis. *Chem Rev*. 2016;116(6):3722.
121. Goswami A, Roy I, Sengupta S, Debnath N. Novel applications of solid and liquid formulations of nanoparticles against insect pests and pathogens. *Thin Solid Films*. 2010;519(3):1252–1257. https://doi.org/10.1016/j.tsf.2010.08.079
122. Rawtani D, Khatri N, Tyagi S, Pandey G. Nanotechnology-based recent approaches for sensing and remediation of pesticides. *Journal of Environmental Management*. 2018;206:749–762. https://doi.org/10.1016/j.jenvman.2017.11.037
123. Zhao J, Wu T, Wu K, Oikawa K, Hidaka H, Serpone N. Photoassisted degradation of dye pollutants. 3. Degradation of the cationic dye rhodamine B in aqueous anionic surfactant/TiO2 dispersions under visible light irradiation: Evidence for the need of substrate adsorption on TiO2 particles. *Environ Sci Technol*. 1998;32(16):2394–2400. https://doi.org/10.1021/es9707926
124. Suvith VS, Philip D. Catalytic degradation of methylene blue using biosynthesized gold and silver nanoparticles. *Spectrochimica Acta Part A: Molecular and Biomolecular Spectroscopy*. 2014;118:526–532. https://doi.org/10.1016/j.saa.2013.09.016
125. Panáček A, Prucek R, Hrbáč J, et al. Polyacrylate-assisted size control of silver nanoparticles and their catalytic activity. *Chem Mater*. 2014;26(3):1332–1339. https://doi.org/10.1021/cm400635z
126. Yuan R, Li H, Yin X, Zhang L, Lu J. Stable controlled growth of 3D CuO/Cu nanoflowers by surfactant-free method for non-enzymatic hydrogen peroxide detection. *J Mater Sci Technol*. 2018;34:1692.
127. Chauhan A, Zubair S, Tufail S, et al. Fungus-mediated biological synthesis of gold nanoparticles: Potential in detection of liver cancer. *International Journal of Nanomedicine*. 2011:2305–2319.
128. Zhang W, Du RH, Li B, et al. Molecular and serological investigation of 2019-nCoV infected patients: Implication of multiple shedding routes. *Emerging Microbes Infect*. 2020;9(1):386.
129. Chen G, Wang Y, Xie R, Gong S. A review on core–shell structured unimolecular nanoparticles for biomedical applications. *Advanced Drug Delivery Reviews*. 2018;130:58–72. https://doi.org/10.1016/j.addr.2018.07.008
130. Parashar UK, Kumar V, Bera T, et al. Study of mechanism of enhanced antibacterial activity by green synthesis of silver nanoparticles. *Nanotechnology*. 2011;22(41):415104. https://doi.org/10.1088/0957-4484/22/41/415104
131. Rani R, Sethi K, Singh G. Nanomaterials and their applications in bioimaging. 2019:429–450. https://doi.org/10.1007/978-3-030-16379-2_15
132. Sujitha V, Murugan K, Paulpandi M, et al. Green-synthesized silver nanoparticles as a novel control tool against dengue virus (DEN-2) and its primary vector Aedes aegypti. *Parasitology Research*. 2015;114(9):3315.
133. Friedman L, Friedman H. The new media technologies: Overview and research framework. *SSRN Electronic Journal*. 2008 Apr 1. https://doi.org/10.2139/ssrn.1116771
134. Elechiguerra J, Burt J, Morones-Ramirez J, et al. Interaction of silver nanoparticles with HIV-1. *Journal of Nanobiotechnology*. 2005;3:6. https://doi.org/10.1186/1477-3155-3-6

## LIST OF ABBREVIATIONS

| | |
|---|---|
| **μm** | Micrometer |
| **AFM** | Atomic force microscopy |
| **AgNP** | Silver nanoparticles |
| **AMPs** | Antimicrobial peptides |
| **AuNPs** | Gold nanoparticles |
| **BSA** | Bovine-serum albumin |
| **Beta-PP** | Pesticide beta-pypermethrin |

| | |
|---|---|
| **CCC** | Chlorocholinchloride |
| **CLL** | Chronic lymphocytic leukemia |
| **CRT** | Cathode ray tube |
| **DLS** | Dynamic light scattering |
| **DPPH** | Diphenyl-1-picrylhydrazyl |
| **EPS** | Exopolysaccharides |
| **FESEM** | Field emission scanning electron microscopy |
| **FTIR** | Fourier transform infrared |
| **g/L** | Grams per liter |
| **$HAuCl_4$** | Hydrogen tetrachloroaurate |
| **HepG2** | Hepatocellular carcinoma |
| **INPs** | Iron nanoparticles |
| **kDa** | Kilodalton |
| **KHz** | Kilohertz |
| **Kv** | Kilovolt |
| **MBA** | Mulberry anthocyanins |
| **Mg** | Milligram |
| **MGYP** | Malt extract, glucose, peptone and yeast extract |
| **MION** | Mono-crystalline iron oxide nanoparticles |
| **Ml** | Milliliter |
| **mM** | Millimoles |
| **MMT** | Montmorillonite clay |
| **MRI** | Magnetic resonance imaging |
| **MRSA** | Methicillin-resistant *Staphylococcus aureus* |
| **NADH** | Nicotinamide adenine dinucleotide phosphate |
| **NADPH** | Nicotinamide adenine dinucleotide phosphate hydrogen |
| **Nm** | Nanometer |
| **NMR** | Nuclear magnetic resonance |
| **NPs** | Nanoparticles |
| **PDA** | Potato-dextrose-agar |
| **PdNPs** | Palladium nanoparticles |
| **Pt** | Platinum |
| **ROS** | Reactive oxygen species |
| **Rpm** | Revolutions per minute |
| **RSV** | Respiratory syncytial virus |
| **SEM** | Scanning electron microscopy |
| **SeNPs** | Selenium nanoparticles |
| **SERS** | Surface-enhanced Raman scattering |
| **Si RNA** | Small interfering ribose nucleic acid |
| **TEM** | Transmission electron microscopy |
| **$TiO_2$NPs** | Titanium dioxide nanoparticles |
| **US FDA** | US Food and Drug Administration |
| **UV/V** | Ultraviolet/visible |
| **UV-Vis-NIR** | Ultraviolet visible/near infrared spectroscopy |
| **VEGF** | Vascular endothelial growth factor |
| **XRD** | X-ray diffraction |
| **ZnNPs** | Zinc nanoparticles |

# Index